World Meteorological Organization
Mannava V. K. Sivakumar
Raymond P. Motha
Haripada P. Das

Natural Disasters and Extreme Events in Agriculture

Impacts and Mitigation

Mannava V. K. Sivakumar
Raymond P. Motha
Haripada P. Das
(Editors)

Natural Disasters and Extreme Events in Agriculture

Impacts and Mitigation

With 93 Figures and 23 Tables

Springer

Dr. Mannava V. K. Sivakumar
Agricultural Meteorology Division
World Meteorological Organization
7 bis, Avenue de la Paix
1211 Geneva 2
Switzerland

Dr. Raymond P. Motha
USDA/OCE/WAOB
1400 Independence Ave. SW
Room 4441, South Building
Washington DC 20250
USA

Dr. Haripada P. Das
Meteorological Office
Division of Agricultural Meteorology
Shivajinagar
Pune 411005
India

Library of Congress Control Number: 2005921509

ISBN-10 3-540-22490-4 Springer Berlin Heidelberg New York
ISBN-13 978-3-540-22490-7 Springer Berlin Heidelberg New York

Springer is a part of Springer Science+Business Media
springeronline.com

Printed in The Netherlands

Cover Design: E. Kirchner, Heidelberg
Production: Almas Schimmel
Typesetting: LE-TEX Jelonek, Schmidt & Vöckler GbR, Leipzig, Germany
Printing: Krips bv, Meppel
Binding: Stürtz AG, Würzburg

Printed on acid free paper 30/3141/as 5 4 3 2 1 0

Foreword

Natural disasters cause heavy loss of life and property, forcing humankind to "learn to live" with these calamities. During the period 1992–2001, floods, droughts, tropical cyclones, hurricanes, typhoons, storm surges, landslides and wild fires and other weather- and climate-related calamities have killed over 622,000 people and affected over 2 billion people. For the same period, losses from natural disasters of hydrometeorological origin were estimated at US$ 446 billion, accounting for about 65% of damages due to all natural disasters.

Natural disasters also affect socio-economic activities. Of these, agricultural production is highly dependent on weather, climate and water availability and is adversely affected by the weather- and climate-related disasters. Failure of rains and occurrences of natural disasters such as floods and tropical cyclones could lead to crop failures, flood insecurity, famine, loss of property and life, mass migration and decline in national economy. The growing concern with the possible impact of natural disasters and extreme events on agriculture and forestry has created new demands for information from, and assessment by agrometeorologists. The need for reorienting and recasting meteorological information, fine-tuning of climatic analysis and presentation in forms suitable for agricultural decision-making and helping marginal farmers cope with the adverse impact of natural disasters and extreme events has become more pressing.

Due to recent advances in science and technology, it is now possible to forecast the occurrence of extreme events and the nature of devastation that they may cause with a greater degree of accuracy and with longer lead time. Availability of such crucial information in advance greatly helps in taking effective measures for prevention and mitigation of loss of life and property and avoid human suffering.

Awareness of the need to give greater attention to disaster mitigation, preparedness and management has been growing among decision-makers. Pre-disaster preparedness now forms an integral part of national development planning in many countries. The Commission for Agricultural Meteorology (CAgM) of WMO at its thirteenth session formed an Expert Team on "Reduction of the Impact of Natural Disasters and Mitigation of Extreme Events in Agriculture, Forestry and Fisheries". The Team was invited to provide guidance on the strategies to reduce the destructive effects of the extreme events by stimulating data acquisition for forecasting and early warning systems, and by making improvements in disaster preparedness.

The China Meteorological Administration (CMA) hosted a meeting of the CAgM Expert Team in Beijing, China from 16 to 20 February 2004. The meeting reviewed and discussed the papers prepared by the members of the expert team and developed appropriate recommendations to reduce the impact of natural disasters and mitigate extreme events in agriculture, forestry and fisheries. I hope that the proceedings of this meeting will serve as a useful source of information to all institutions and agencies interested in this subject.

(M. Jarraud)
Secretary-General
World Meteorological Organization

Preface

Over the past few decades, there is an increasing intensity and frequency of natural disasters around the world with severe socio-economic impacts, especially in the developing world. Agriculture is one of the most important sectors heavily impacted by the natural disasters and the challenge in front of the agrometeorologists around the world is that more than ever before, there is a great need to more effectively integrate and deploy the skills to use climate information and products successfully in natural disaster preparedness strategies. There is also a need to develop locally agrometeorological adaptation strategies to reduce the effect of natural disasters especially in vulnerable regions where food and fibre production is most sensitive and vulnerable to climate fluctuations.

The Commission for Agricultural Meteorology (CAgM) of WMO at its thirteenth session held in Ljubljana, Slovenia established an Expert Team (ET) on Reducing the Impact of Natural Disasters and Mitigation of Extreme Events in Agriculture, Rangelands, Forestry and Fisheries. A meeting of this ET was held in Beijing, China from 16 to 20 February 2004 at the kind invitation of the Chinese Meteorological Administration (CMA). The decision to organize this meeting at this point of time is largely due to the increasing concerns with the impact of natural disasters in recent times as highlighted by Mr M. Jarraud in the Foreword. Fifty-four participants from eight countries, including forty-five from China, attended the meeting. In addition to the experts from CAgM, three experts nominated by the Joint Commission on Marine Meteorology (JCOMM) also participated in the meeting. The specific objectives of the meeting were:

- To review the current status of application of climate and weather information in reducing the impacts of natural disasters and mitigation of extreme events in agriculture, forestry and fisheries.
- To assess the potential for improved disaster reduction strategies and relevant agrometeorological applications in different countries for sustainable agricultural development.
- To identify the shortcomings and limitations of current disaster management and mitigation strategies in reducing the risks associated with natural disasters.
- To discuss the resources and strategies, including education and training, required for promotion of sustained efforts in disaster reduction and mit-

igation of extreme events and research activities to better understand the potential risks to agriculture from natural disasters and extreme events.

Altogether there were 10 sessions (including opening and closing session) in the meeting during which 21 invited papers were presented dealing with various aspects on reducing the impact of natural disasters and mitigation of extreme events in agriculture, rangelands, forestry and fisheries. All the participants in the meeting were engaged in discussions on these papers and developed several useful recommendations for all organizations involved in disaster reduction and mitigation of extreme events, in particular the National Meteorological and Hydrological Services. It should be noted that the recommendations listed reflect the considered opinions of the participants at the meeting and we are aware that these recommendations do not address the totality of the needs. We do hope that these will encourage others to suggest further ways in which the science and applications of agricultural meteorology could contribute to reducing the impacts of natural disasters and extreme events in agriculture.

As Editors of this volume, we would like to thank all the authors for their efforts and for their cooperation in bringing out this volume in time. We are most grateful to the Chinese Meteorological Administration (CMA) for hosting this meeting and to the Secretary-General of WMO for his continuous support and encouragement.

M.V.K. Sivakumar
Agricultural Meteorology Division
World Meteorological Organization
7bis, Avenue de la Paix
CP 2300
1211 Geneva 2
Switzerland

R.P. Motha
USDA/OCE/WAOB
1400 Independence Ave. SW
Room 4441, South Building
Washington DC 20250
USA

H.P. Das
Meteorological Office
Division of Agricultural Meteorology
Shivajinagar
Pune 411005
India

Contents

List of Contributors

Bi Baogui

Weather Forecast Office, National Meteorological Center
China Meteorological Administration
46 ZhongGuancun South Avenue, 100081, Beijing, P.R. China
E-mail: bibg@cma.gov.cn

Wang Chunyi

Chinese Meteorological Society
No. 46, Zhongguancun Nandajie, 100081, Beijing, China
E-mail: wcy@cms1924.org

Haripada P. Das

Meteorological Office, Division of Agricultural Meteorology
Shavajinagar, Pune, 411005 India
E-mail: hpd_ag@rediffmail.com

Douglas G. Fox

Center for Investigations and Research of the Atmosphere (CIRA)
Colorado State University
Fort Collins, Colorado 80526 USA
E-mail: dfox@cira.colostate.edu

Bernard Gomez

Department of Water Resources
Banjul, Gambia
E-mail: be63gomez@hotmail.com

R.P.R. Guerreiro

Division of Hydrometeorology and Agricultural Meteorology
Department of Climate and Atmospheric Environment
Institute of Meteorology
Lisbon, Portugal
E-mail: rita.guerreiro@meteo.pt

Guo Jianping

Institute of Eco-environment and Agrometeorology
Chinese Academy of Meteorological Sciences
No. 46, Zhongguancun Nandajie, 100081, Beijing, China
E-mail: guojp@cams.cma.gov.cn

Penehuro Lefale

National Institute of Water and Atmospheric Research
P.O. Box 109-695
Newmarket, Auckland, New Zealand

Sanmei Li

National Satellite Meteorological Center of China (NSMC)
46 Zhongguancun Nandajie, Beijing 100081, China
E-mail: sanmeili@nsmc.cma.gov.cn

Jiao Meiyan

National Meteorological Center, China Meteorological Administration
46 ZhongGuancun South Avenue, 100081, Beijing, P.R. China
E-mail: Jiaomy@cma.gov.cn

Raymond P. Motha

U.S. Department of Agriculture, World Agricultural Outlook Board
1400 Independence Avenue, S.W.
Room 4441, South Building, Washington, D.C. 20250
E-mail: rmotha@oce.usda.gov

Liliana Nuñez

Servicio Meteorológico Nacional
C.P. 1002
Buenos Aires, Argentina
E-mail: lnunez@meteofa.mil.ar

John Jianhe Qu

School of Computational Science (SCS)
GeorgeMasonUniversity (GMU)
4400 University Drive, MS 5C3
Fairfax, VA. 220303 USA
E-mail: jqu@gmu.edu

Allen.R. Riebau

USDA Forest Service Research and Development
Wildlife, Fisheries, Watershed and Air Research (WFWAR)
14th and Independence, S.W., RPC-4
Washington, D.C. 20250-1113 USA
E-mail: ariebau@fs.fed.us

M. James Salinger

National Institute of Water and Atmospheric Research
P.O. Box 109-695
Newmarket, Auckland, New Zealand
E-mail: j.salinger@niwa.co.nz

Wang Shili

Institute of Eco-environment and Agrometeorology
Chinese Academy of Meteorological Sciences
No. 46, Zhongguancun Nandajie, 100081, Beijing, China
E-mail: wangsl@cams.cma.gov.cn

Mannava V. K. Sivakumar

World Meteorological Organization
7 bis, Avenue de la Paix
1211 Geneva 2, Switzerland
E-mail: msivakumar@wmo.int

Chang Tao

Weather Forecast Office, National Meteorological Center
China Meteorological Administration
46 ZhongGuancun South Avenue, 100081, Beijing, P.R. China
E-mail: zhangtao@cma.gov.cn

Donald A. Wilhite

National Drought Mitigation Center
and International Drought Information Center
University of Nebraska
Lincoln, Nebraska 68583-0749 U.S.A.
E-mail: dwilhite2@unl.edu

William J. Wright

National Climate Centre, Australian Bureau of Meteorology
GPO Box 1289K, Melbourne, 3001 Australia
E-mail: w.wright@bom.gov.au

Ming Xu

Shanghai Typhoon Institute
166 Puxi Road, Shanghai, China, 200030
E-mail: xum@mail.typhoon.gov.cn

Qiuzhen Yang

Shanghai Typhoon Institute
166 Puxi Road, Shanghai, China, 200030
E-mail: yangqz@mail.typhoon.gov.cn

Du Yaodong

Climatic and Agrometeorological Center
Meteorological Bureau of Guangdong Province
People's Republic of China
E-mail: yddu@grmc.gov.cn

Ming Ying

Shanghai Typhoon Institute
166 Puxi Road, Shanghai, China, 200030
E-mail: yingm@mail.typhoon.gov.cn

Xu Yinglong

National Meteorological Center, CMA
Beijing, China, 100081
E-mail: xuyl@cma.gov.cn /
xuyinong_china@163.com

Yeping Zhang

National Satellite Meteorological Center of China (NSMC)
46 Zhongguancun Nandajie, Beijing 100081, China
E-mail: yepingzhang@nsmc.cma.gov.cn

Yanxia Zhao

Chinese Academy of Meteorological Sciences(CAMS)
46 Zhongguancun Nandajie, Beijing 100081, China
E-mail: zyx@cams.cma.gov.cn

Huo Zhiguo

Institute of Eco-environment and Agrometeorology
Chinese Academy of Meteorological Sciences
No. 46, Zhongguancun Nandajie, 100081, Beijing, China
E-mail: huozhigg@cams.cma.gov.cn

CHAPTER 1

Impacts of Natural Disasters in Agriculture, Rangeland and Forestry: an Overview

Mannava V. K. Sivakumar

Abstract Natural disasters play a major role in agricultural development and the economic cost associated with all natural disasters has increased 14 fold since the 1950s. Natural disasters are classified into hydro-meteorological and geophysical disasters. Definitions of various types of hydrometeorological disasters such as floods, droughts, cyclones, forest fires, heatwaves were presented. Evidence available from different parts of the world showed that there is a rising trend in the occurrence of natural disasters from 1993 to 2002. Impacts of droughts, cyclones, floods, forest and bush fires on agriculture, rangeland and forestry were described with suitable examples. While the predominant impacts from these disasters are negative, there are some positive impacts as well. Environmental degradation is one of the major factors contributing to the vulnerability of agriculture, forestry and rangelands to natural disasters because it directly magnifies the risk of natural disasters. Some methodological issues concerning the characterization of the impacts of natural disasters in agriculture, rangeland and forestry were described. There is an urgent need to mitigate the effects hydro-meteorological disasters through improved use of climate and weather information and forecasts, early warning systems, and appropriate methods of management of land and natural resources.

1.1 Introduction

Throughout human history, natural disasters have played a major role in the economic development and survival of humanity. Historians now believe that an unusually long and severe drought was a primary cause of the disappearance of the Maya civilization (Hodell et al. 1995).

During the past four decades, natural hazards such as droughts, floods, storms and tropical cyclones and wildland fires have caused major loss of human lives and livelihoods, the destruction of economic and social infrastructure, as well as environmental damages. Deaths since the 1950s increased 50 percent each decade, whereas the corresponding population growth rate was only 20 percent (Kreimer and Munasinghe 1991).

The economic cost associated with all natural disasters has increased 14 fold since the 1950s (World Disasters Report 2001). World wide, annual economic costs related to natural disasters have been estimated at about $50 to 100 billion. According to China's Ministry of Civil affairs, natural disaster costs have averaged US$12 billion or so annually during the past 10 years with an estimated 200 million or so people affected annually. By the year 2050 it is predicted that globally 100,000 lives will be lost each year to natural disasters and the global cost could top $300 billion annually (SEI, IUCN, IISD 2001).

The world land use data (FAO 1999) show that 70% of the global land use is for agriculture, rangeland and forestry with 12% of the land used for arable and permanent crops, 31% for forest and woodlands and 27% for permanent pasture. Agriculture is also the essential source of income in most developing countries. For example, agriculture accounts for 70 percent of full-time employment in Africa, 33 percent of total GDP, and 40 percent of total export earnings. Agricultural production is highly dependent on weather, climate and water availability, and is adversely affected by weather- and climate-related disasters.

In order to ensure sustainable agricultural production and assure the livelihood of millions of people, especially in the developing countries, a better understanding of the natural disasters that impact agriculture, forestry and rangelands is essential. This paper provides an overview of the major issues involved.

1.2 Natural Disasters – Definitions and Types

In simple terms, a natural disaster is a natural event with catastrophic consequences for living things in the vicinity. But, different definitions of natural disasters are often used and some of them are based primarily on loss of life.

The emergencies database (EM-DAT) operated by the Centre for Research on the Epidemiology of Disasters (CRED) classifies an event as a disaster if at least "10 people are killed and/or 100 or more are affected and/or an appeal for international assistance is made or a state of emergency declared" (CRED 2000). Clearly, for agricultural purposes only the last part of this definition is applicable.

According to a 1992 disaster training programme, United Nations (UN) defines a disaster as "a serious disruption of the functioning of society, causing widespread human, material or environmental losses which exceed the capacity of the affected society to cope using only its own resources". With suitable interpretation of some parts, this definition could be used by agriculture.

Anderson (1990) defines natural disasters as temporary events triggered by natural hazards that overwhelm local response capacity and seriously affect the social and economic development of a region.

Susman et al. (1983) describe disasters as the interface between an extreme physical environment and a vulnerable human population. Such definitions emphasize the fact that the socio-economic and political factors are of paramount importance in understanding why populations are vulnerable to the environment and experience disasters.

According to World Disaster Report (2003), natural disasters include hydro-meteorological disasters and geophysical disasters. The hydro-meteorological disasters include landslides/avalanches; droughts/famines; extreme temperatures and heat waves; floods; hurricanes; forest/scrub fires; windstorms; and others (insect infestation and waves/surges). The geophysical disasters include earthquakes and volcanic eruptions. In this paper, only the hydro-meteorological disasters impacting agriculture, rangeland and forestry are dealt with. Definitions of each of these disasters primarily from are given below.

A landslide is a geological phenomenon which includes a wide range of ground movement, such as rock falls, deep failure of slopes, and shallow debris flow. Although gravity acting on an over steepened slope is the primary reason for a landslide, there are other contributing factors. An avalanche is caused when a build up of snow is released down a slope, and is one of the major dangers faced in the mountains in winter. An avalanche is a type of gravity current.

Drought is the consequence of a natural reduction in the amount of precipitation over an extended period of time, usually a season or more in length, often associated with other climatic factors (such as high temperatures, high winds and low relative humidity) that can aggravate the severity of the event. Drought is not a purely physical phenomenon, but instead is an interplay between natural water availability and human demands for water supply. The precise definition of drought is made complex due to political considerations, but there are generally three types of conditions that are referred to as drought.

- Meteorological drought is brought about when there is a prolonged period with below average precipitation.
- Agricultural drought is brought about when there is insufficient moisture for average crop or range production. This condition can arise, even in times of average precipitation, due to soil conditions or agricultural techniques.
- Hydrologic drought is brought about when the water reserves available in sources such as aquifers, lakes, and reservoirs falls below the statistical average. This condition can arise, even in times of average (or above average) precipitation, when increased usage of water diminishes the reserves.

A heat wave is a prolonged period of excessively hot weather, which may be accompanied by excessive humidity. The term is relative to the usual weather in the area, so temperatures that people from a hotter climate find normal can be a heat wave if they are outside the normal pattern for a cooler area. The term is applied both to "ordinary" weather variations and to extraordinary spells of heat which may only occur once a century.

Flood is defined as the condition that occurs when water overflows the natural or artificial confines of a stream of other body of water, or accumulates by drainage over low-lying areas. A flood is a temporary inundation of normally dry land with water, suspended matter and/or rubble caused by overflowing of rivers, precipitation, storm surge, tsunami, waves, mudflow, lahar, failure of water retaining structures, groundwater seepage and water backup in sewer systems.

Forest fire (or bushfire in Australasia) is an uncontrolled fires occurring in vegetation more than 6 feet (1.8 m) in height. These fires often reach the proportions of a major conflagration and are sometimes begun by combustion and heat from surface and ground fires.

Tropical cyclones, hurricanes and typhoons are regional names for what is essentially the same phenomenon. Depressions in the tropics which develop into storms are called tropical cyclones in the south-west Indian Ocean, the Bay of Bengal, and the Arabian Sea, parts of the south Pacific and along the northern coasts of Australia. These storms are called typhoons in the north-west Pacific and are known as hurricanes in the Caribbean, south-east United States and Central America.

Tsunami (in Japanese, big wave in port), often incorrectly called a tidal wave, is a series of massive waves that occur after an earthquake, a seaquake, volcanic activity, slumps or meteorite impacts in or near the sea. Since the constant energy of the tsunami is defined by height and speed, its height increases once its speed is reduced where the wave approaches land. The waves travel at high speed, more or less unnoticed where crossing deep water, but raising to a height of 30 m and more. Tsunamis can cause severe destruction on coasts and islands.

1.3 Natural Disasters – the Rising Trend

Information on natural disasters and trends is basically available from global databases that provide essential information on the occurrence, recurrence and location of disasters and disaster trends over time (World Disasters Report 2003). The emergencies database (EM-DAT), referred to earlier, serves the global community. Other databases such as Sigma and Natcat are operated by insurance companies Swiss Re and Munich Re respectively, but are not always accessible to the public.

There is evidence available from different parts of the world that there is a rising trend of natural disasters from 1993 to 2002 (Fig. 1.1a). Of a grand total of 2,654 disasters during this period, floods and windstorms account for about 70% of the disasters while the remaining 30% of the disasters are accounted for by droughts, landslides, forest fires, heat waves and others (Fig. 1.1b).

At the regional level, in South East Asia and Bangladesh, over the last century, 700 disasters have occurred of which 158 (23%) occurred between 1900 and 1979, and 542 (77%) between 1972 and 1996. For the Latin American and Caribbean region, Charveriat (2000) showed a noticeable trend of increase in the frequency of disasters.

At the national level, Roy et al. (2002) showed that the Orissa state of India has been disaster affected for 90 years; floods have occurred for 49 years, droughts for 30 and cyclones have hit the state for 11 years.

These data together with that of deaths and affected people appear to show that the natural disasters are becoming more frequent and are also causing heavier and heavier consequences.

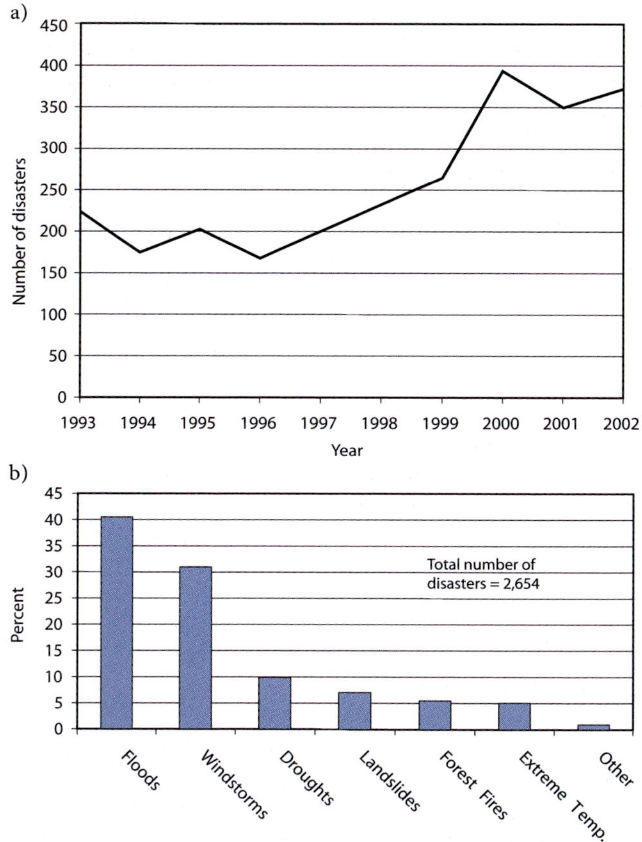

Fig. 1.1. (**a**) Annual variations in the occurrence of hydro-meteorological disasters during 1993–2002 and (**b**) the percentage of different hydro-meteorological disasters as a percent of total number of disasters during 1993–2002

1.4 Impacts of Natural Disasters in Agriculture, Rangeland and Forestry – General Discussion

Impacts from natural disasters on agriculture, rangeland and forestry can be positive or negative. While the impacts are predominantly negative and do affect human society significantly (Joy 1991), there are some positive impacts or benefits that need to be pointed out as well in any discussion on impacts of natural disasters.

As Das (2003a) explained, the impact of natural disasters on agriculture, rangeland and forestry can be direct or indirect in their effect. Direct impacts arise from the direct physical damage on crops, animals and trees caused by the extreme hydro-meteorological event. The impacts may be considered in terms of short-term temporary damage at a particular crop stage to complete crop loss. Within hours of their occurrence, natural disasters produce direct damage to agriculture in terms of total or partial destruction of farm buildings, installations, machinery, equipment, means of transport, storage as well as damage to crop land, irrigation works, dams and destruction of crops ready for harvesting.

Disasters also cause indirect damage which refers to loss of potential production due to disturbed flow of goods and services, lost production capacities, and increased costs of production. Such indirect impacts appear progressively as a result of low incomes, decreases in production, environmental degradation and other factors related to the disaster (Das 2003a).

Anaman (2003) pointed out that the impacts of natural disasters can also be classified as tangible or intangible. Tangible impacts are those that can be easily measured in monetary terms. Intangible impacts are often difficult to measure in monetary terms since they are not purchased or sold in well defined markets and hence direct market values do not exist eg., anxiety or fear of future natural disasters (Oliver 1989), inconvenience and disruption to farm work and stress-induced ill health and human fatalities.

1.4.1 Negative Impacts

Many famines in pre-20th-century Africa, Asia and Europe were triggered by natural disasters – drought, extreme cold, pests and diseases – that devastated crops and livestock (Devereux 2000). According to the EM/DAT data quoted in the World Disasters Report 2003, on average, 246 million people were affected by hydro-meteorological disasters globally each year, between 1993 and 2002. During the same period, these disasters claimed 46,000 lives per year.

Annual variations in the number of people affected by hydro-meteorological disasters during this period (Fig. 1.2a) showed a significant rise in 2002. Data on the percentage of people affected by different hydro-meteorological disasters (Fig. 1.2b) showed that about 56.9% of the 2.46 billion people were affected by floods, 29.8% by droughts, 12.7% by windstorms and 0.6% by the rest of the disasters. On a regional basis (Fig. 1.3), 91% of the people affected were in Asia, due to its huge population. Africa accounted for 6% of the affected people, followed by Americas (2%), Europe ($<1\%$) and Oceania ($<1\%$).

During 1993/2002, hydro-meteorological disasters caused an estimated damage of US$ 41.3 billion per year on average. Estimated damage on annual basis during this period (Fig. 1.4a) varied from a high of US$ 67.7 billion in 1995 to a low of US$ 18.1 billion in 2001. Ranking of the different hydro-meteorological disasters according to the percentage of damage caused by them (Fig. 1.4b) is as follows: floods, windstorms, droughts, forest fires, extreme temperatures, landslides and others. On a regional basis (Fig. 1.5), maximum damage occurred in Asia (49%), followed by Americas (29%), Europe (19%), Oceania (2%) and Africa (1%).

Loss of perennial crops such as banana trees or forests has long-term consequences on the ability to generate income. In the case of agricultural income generating assets, the loss might be temporary or permanent (Charveriat 2000).

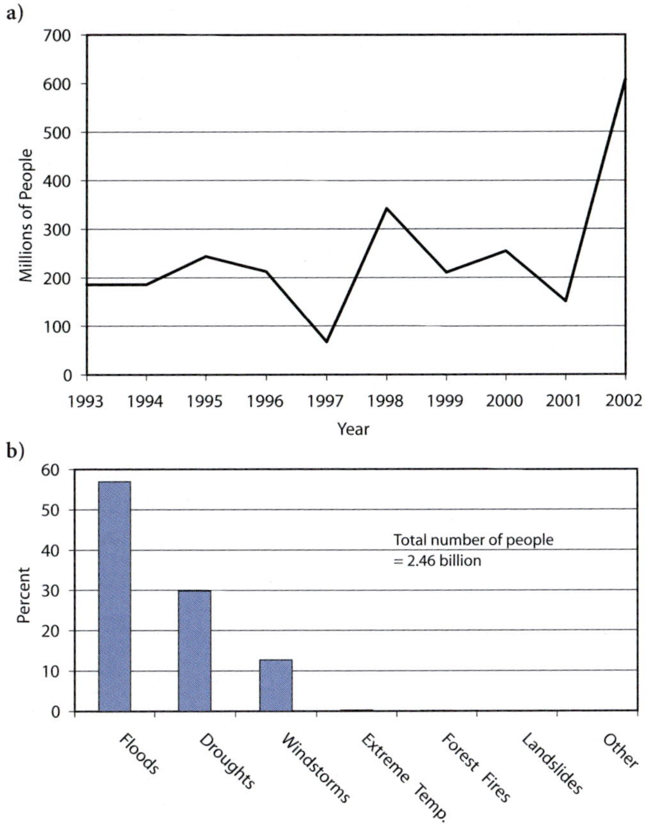

Fig. 1.2. (**a**) Annual variations in the number of people affected by hydro-meteorological disasters during 1993–2002 and (**b**) the percentage of people affected by different hydro-meteorological disasters during 1993–2002

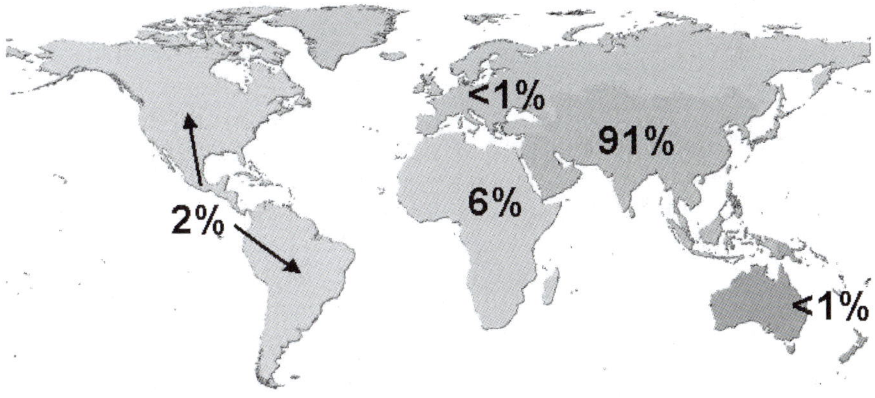

Fig. 1.3. Percentage of total number of people reported affected by hydro-meteorological disasters by region

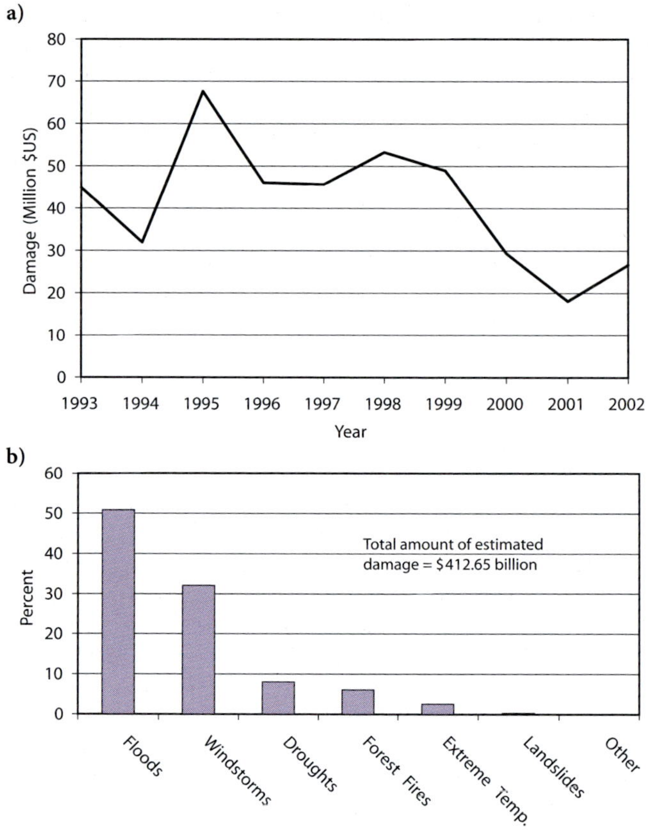

Fig. 1.4. (**a**) Annual variations in the estimated damage due to hydro-meteorological disasters during 1993–2002 and (**b**) the percentage of damage caused by different hydro-meteorological disasters during 1993–2002

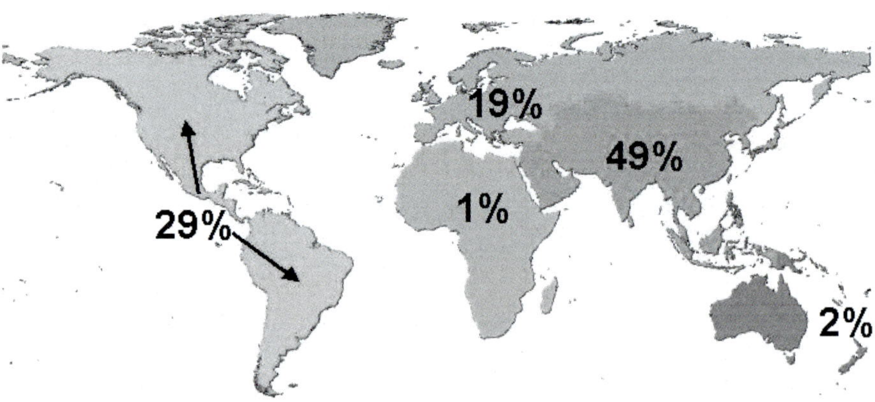

Fig. 1.5. Percentage of total amount of disaster estimated damage by region (2002 prices)

Floods make land unsuitable for agricultural production until waters recede, while hurricanes might wash out arable land or permanently increase its salinity through storm surges and flash floods. Indirect impacts include the evacuation of people in the event of cyclone landfall, disruption to households, stress induced sickness and apprehension (Handmer and Smith 1992; Anaman 1996).

The duration and geographical size of the disaster is an important factor. Localized disasters tend to produce limited aggregate impacts, unlike countrywide natural events such as Hurricane Mitch (Charveriat 2000). Sudden hazards such as storms usually have fewer long/lasting effects than droughts, which are often described as creeping in nature because of the slow rate at which they develop.

Recurrent disasters in the same geographical area might lead to reduced investment due to the perceived risk of asset loss or emigration from stricken areas. Regions repeatedly hit by natural disasters, such as Northeast Brazil, or the coasts of Peru and Ecuador, are usually poorer on average than less hazard-prone areas in a given country.

Poor nations suffer the most from the natural disasters. As Devereux (2000) explained, poor people are more exposed because they tend to live in marginal areas and depend on high-risk, low return livelihood systems such as rainfed agriculture and face many sources of economic vulnerability including little physical infrastructure. Vos et al. (1999) estimated that the poverty incidence in affected municipalities in the coastal province of Ecuador, which already reached 73% before El Niño, rose by 10 percentage points in 1998 due to loss of harvests of poor farmers and rising unemployment among agricultural workers. The UNDP reports that 24 out of 49 least developed nations face a high risk of natural disasters. At least 6 of them have been hit by between 2 to 8 major disasters per year in the last 15 years, with long-term consequences for human development (UNDP 2001).

While damages related with natural disasters are greater in absolute value in developed countries, loss/GDP rates are 20% higher in the developing countries (Funaro 1982). United States experienced more disasters between 1970 and 1999 than any other region, but the impact on national development was not as severe as in some of the developing countries. For example, Hurricane Andrew in 1992 caused a total damage of $ 26.5 billions in the United States, but it was a mere 0.4% of GDP.

Beyond the direct or indirect losses, the economic consequences are of major importance given the repercussions they have on the economic development of the countries (GDP, public finances, foreign trade, price indices). Because of the important role it plays considering the creation of national wealth and the population needs, the agricultural sector appears as a highly vulnerable one. For example, 30.9% of the GNP in Bangladesh was attributed to agricultural activities in Bangladesh while in Cambodia and Laos, it was 44.6 and 54.3% respectively. During the last El Niño in Ecuador, Vos et al. (1999) estimated that around 12,000 workers on banana and sugar cane plantations in the lowlands temporarily lost their jobs. In Honduras, the press reported that the rate of unemployment in the immediate aftermath of Hurricane Mitch had reached an estimated 32%, according to the firm, Asesorias Economicos.

The economic consequences also concern the activities related to international trade, which have become indispensable because of national debt. Export agriculture, tourism, crafts and industrial activities are assumed to bring in foreign currency that is indispensable for the equilibrium of the balance of payments.

The agricultural products hold an even more significant place in exportations. Free zones can be affected by cyclones and floods, with greater probability as they are situated in the coastal plains and on the principal deltas. In Bangladesh, the Chittagong free zone was very seriously affected by the 1991 cyclone (Normand 1991).

1.4.2 Positive Impacts

The positive impacts of natural disasters include increased rainfall to inland areas from tropical cyclones along coastal areas (Ryan 1993), the fixing of atmospheric nitrogen by thunderstorms, the germination of many native plant species as a result of bushfires and the maintenance of fertility of flood-plain soils due to flooding (Blong 1992). The influx of funds into disaster-relief activities after the occurrence of natural disasters can sometimes be positive to local communities, as was shown for the city of Mobile, Alabama after Hurricane Federic (Chang 1984).

1.5 Impacts of Specific Natural Disasters in Agriculture, Rangeland and Forestry

According to Johnson (2003), in a survey of the impacts of extreme weather and climate events on agriculture, the events which were reported by most of the 57 countries around the world which responded included drought (91 per cent), local severe storms (83 per cent), floods (79 per cent), frost (74 per cent) and high winds (72 per cent).

1.5.1 Droughts

Seasonal droughts occur in climates that have well defined annual rainy and dry seasons. Numerous studies have been conducted on the impacts of droughts on crop growth and development at different levels including soil moisture uptake, root growth, shoot growth, various plant processes such as photosynthesis, respiration, plant water uptake and final yield and literature is replete with several good examples. But it is to be understood that the effects of droughts are seriously worsened by human factors such as population growth that forces people into drier and drier regions and inappropriate cropping and herding practices. The impacts of drought are likely to become ever more severe as a result of development processes and population increases (Squires

2001). Droughts often stimulate sequences of actions and reactions leading to long-term land degradation. Droughts may also trigger local food shortages, speculation, hoarding, forced liquidation of livestock at depressed prices, social conflicts and many other disasters associated with famines that may catastrophically affect numerous groups and strata of local populations.

The most prolonged and widespread droughts occurred in 1973 and 1984, when almost all the African countries were affected, and in 1992 when all southern African countries experienced extreme food shortages. In 1973 alone, drought killed 100,000 people in the Sahel (Gommes and Petrassi 1996). It was estimated that hundreds of thousands of people died and nearly half of the entire livestock herds and two million heads of wild animals were killed due to the severe droughts and land desertification at the southern edge of the Sahara Desert. More than six million ecological refugees were forced to emigrate from their homeland to other regions. The whole world was shocked by these disastrous events, and as consequence, the United National Conference on Desertification (UNCD) was sponsored in Nairobi in 1977.

The change in weather patterns during an ENSO event alters regions of high and low pressures around the globe. Descending air of atmospheric circulation cells creates high-pressure centers at the surface. The high surface pressures prevent areas of precipitation from moving into its region. When these abnormal high-pressure patterns persist they lead to drought conditions, depriving the area and its ecosystem of rainfall. Droughts generally occur in the western Pacific during ENSO events, an area normally rich in rainfall. However, droughts in many other regions of the world, including southeastern Africa, India, China and the northeastern region of the South American continent, have been linked to El Niño.

ENSO results in drier conditions in Northeast Brazil during the Northern Hemisphere winter. During the Southern Hemisphere winter, the climatic impact of El Niño is drier conditions in Central America, Colombia and Venezuela. During the 1997–98 El Nino phenomenon, severe droughts and forest fires occurred in northeast Brazil (WMO 1999).

The Committee of Agricultural Organizations in the Union (COPA) and the General Committee for Agricultural Co-operation in the European Union (COGECA) estimate that drought and fires during 2003 have cost European Union farmers €13.5 billion – with Austria, France, Germany, Italy, Portugal and Spain identified as countries most affected.

Currently, South Africa may be heading for a prolonged drought which researchers warn could be among the most severe in decades. It is understood the parts of the country (including KwaZulu-Natal, Limpopo and Mpumalanga provinces) are currently experiencing drought conditions with farmers being forced to sell livestock (recent wildfires have also had a negative impact insofar as they have destroyed large swathes of grazing in KwaZulu-Natal and Mpumalanga provinces). As many as 15 million South Africans could be affected by one of the worst droughts in 10 years. According to Toffee Mokgethi, head of the country's National Disaster Management Centre, "Of those ... at least four million are in need of immediate assistance." Among them are three million communal or subsistence farmers and their dependants. It is claimed that reduced agricultural production in South Africa might impact neighbouring countries since the World Food Programme procures a significant volume of food aid from South Africa for distribution throughout the region.

Benson and Clay (1998) found that the aggregate impacts of droughts could be quite significant in terms of growth. A 50% fall in agricultural GDP would translate into a 10% decrease in GDP for an economy in which agriculture accounted for 20% of total activity in the pre-drought year. Gomes and Vergolino (1995) showed that during 1970, 1983 and 1993, the years of severe drought in Northeast Brazil, when the agricultural GDP decreased between 17.5 and 29.7%, the fluctuations in the region's GDP were explained almost entirely by the occurrence of droughts. For example, the estimated GDP per capita in the Northeast was $ 1,494 in 1993, compared with $ 3,010 in the rest of the country.

One of the major impacts of droughts is regressive distributional effects across communities and across households within communities. Scott and Litchfield (1994) provided evidence of these impacts in the rural regions of Coquimbo in Chile where inequality and poverty at the community level increased significantly in Las Tazas due to the cumulative effects of very low rainfall over the years.

Although drought impacts are primarily negative, it must be mentioned that some times droughts do carry positive impacts. Moderate droughts in the post flowering maturity stage of sugar cane, for instance, helps to increase the sucrose content. Other beneficial impacts include mosquito reduction, reduced cost for snow removal in snowfall regions and other related activities, emergency water conservation leading to permanent and efficient establishment of water use saving patterns etc. (Das 2003a).

1.5.2 Cyclones

Tropical cyclones are among the most destructive of all natural hazards, causing considerable human suffering in about 70 countries around the world. As Das (2003b) described, tropical cyclones are the off-spring of ocean-atmosphere interactions, powered by heat from the sea, driven by the easterly trades and temperate westerlies, the high planetary winds and their fierce energy. An average of 80 tropical cyclones form annually over the tropical oceans, of which, 30 occur in the typhoon region of the western North Pacific (Obasi 1997). The impact of tropical cyclones is greatest over coastal areas that bear the brunt of the strong winds and flooding from rainfall. For example, while the average for the Bay of Bengal and the Arabian Sea is only five, some of the most destructive of these storms have occurred in that region, such as the severe tropical cyclone in Bangladesh in 1970, which claimed 300,000 lives.

El Nino is generally associated with worldwide anomalies in the patterns of precipitation and temperature, as well as with patterns of tropical storms and hurricane activity, the behaviour of subtropical jetstreams, and many other general circulation features over various parts of the world.

The magnitude of hurricanes is assessed with the Saffir-Simpson scale, which takes into account maximum sustained winds and minimum storm pressure. Among the most devastating hurricanes of all times were Hurricane Georges (September 1998) and Hurricane Mitch (October 1998). A category 5 hurricane, Hurricane Mitch was one of the most powerful Atlantic Hurricanes, with 290 km/hr winds and a minimum storm pressure of 906 mb, and had quite a long life span (14.5 days) which explains why it turned out to be the deadliest of the century. It caused loss of life, destruction of property, damage to food production, food reserves and transportation system as well as increased health risks.

The loss to agriculture, rangelands and forests from tropical cyclones can be due to direct destruction of vegetation, crops, orchards and livestock, damage to infrastructure such as canals, wells and tanks and long term loss of soil fertility from saline deposits over land flooded by sea water. Typhoons have been known to inflict severe damage on agriculture: for example, in southern Hainan on 2 October 1999, some 25 million timber and rubber trees were blown down (WMO 1994). A typhoon that struck Thailand on 4 November 1989 wiped out some 150,000 ha of rubber, coconut and oil plantations and other crops (WMO 1997).

Traditionally, small-scale fisheries are also hit by cyclones. In monetary terms, the losses incurred by livestock raising, forestry and fisheries mostly remain below those suffered by crops. But in small islands such as Antigua and Barbuda where fisheries constitute the backbone of the economy, the impact could be quite significant. After Hurricane Hugo in 1989, 47 percent of the losses occurred in fisheries, but crop losses still represented almost 40 percent of the total damage (OSRO 1989).

The loss from a single cyclone may run into millions of dollars. For example, in 1998, El Niño-related weather phenomena caused $ 6.6 billion of damages in Argentina, Peru and Ecuador, while Hurricane Georges caused $ 2.1 billion in damages in the Dominican Republic and Hurricane Mitch produced damages of $ 2.4 billion in Honduras and Nicaragua (Charveriat 2000).

In May 2002, Cyclone Kesiny hit Madagascar affecting more than half a million people, making them homeless or in need of emergency food, shelter and drinking water. Up to 75% of the crops were destroyed, 20 people died and 1,200 were injured (CIDI 2002).

The Orissa cyclones on 17–18 October and again on 29–30 October 1999 in Orissa, India caused devastating damage. The cyclones on 29–30 October with wind speeds of 270–300 km/h for 36 hours were accompanied by torrential rain ranging from 400 to 867 mm over a period of 3 days. The two cyclones together severely affected around 19 million people in 12 districts (Roy et al. 2002). Sea waves reaching 7 metres rushed 15 kms inland. 2.5 million livestock perished and a total of 2.1 million ha of agricultural land was affected.

Not all the impacts of cyclones are negative, some reports cite beneficial effects of tropical cyclones. Ryan (1993) mentioned some important benefits of tropical cyclones in Australia. Increased water availability in water-critical regions makes agricultural production less susceptible to the dry period. Sugg (1968) estimates that nine major hurricanes in the United States since 1932 terminated dry conditions over an area of about 622,000 km^2.

1.5.3 Floods

Floods are among the greatest natural disasters known to mankind. According to historical record, 1092 flood disaster events occurred since 206 BC in China during a period of 2155 years, averaging once in every two years. In China, about 8% of the land area is located in the mid and down-stream parts of the seven major rivers in the eastern and southern parts of the country that are prone to floods. About 50% of the total population of the country lives in these areas (Heng 2004).

The number of people affected by floods during the world during 1991 to 2000 was reported to be around 1.5 billion. According to Roy et al. (2002), the frequency of floods had increased in the Orissa state of India. Between 1834 and 1926, the state experienced floods once in four years, which rose to once in two years after 1926. The state experienced nine bouts of floods within a span of just 15 days in 2001, an all time high, damaging 2.12 million hectares of standing crop.

There are a wide variety of floods from localized flash flood to extensive floods over large areas. There are both positive and negative aspects of floods, which should be factored into land use planning. It is well known that around the world, flood plains are the areas of highest productivity. Some of the most flourishing ancient civilizations were in the flood plains as proximity to rivers enabled them to enhance agricultural productivity. In many areas, annual floods are welcomed by the populations settled in such areas since the increased availability of water resources could be harnessed for greater agricultural output. But such activities are not without risk since they could further increase the risk of flooding hazards.

Floods are a function of the climate (variability in rainfall pattern, occurrence of storms) as well as hydrology (shape of river beds, intensity of drainage, and debit flow of rivers) and soil characteristics (moisture absorption capacity). Flooding occurs primarily when water due to rain from various types of weather phenomena or snowmelt accumulates faster than soils can absorb it or rivers can carry it away. .

More than three quarters of natural disasters in the Americas are high wind and floods. The countries most affected by floods are Brazil (15%), USA (12%), Peru (11%). According to a press release from the European Union, for the period 1980–2002, the greatest number of floods occurred in France (22%), Italy (17%) and the UK (12%) ... the highest number of fatalities occurred in Italy (38%), followed by Spain (20%) and France (17%) ... the greatest economic losses occurred in Germany and Italy (both € 11 billion), followed by Spain and the UK (both around € 6 billion). Floods and high winds represent more than three quarters of the natural disasters in Oceania.

A study of the impact of the 1998 Bangladesh floods revealed that 62% of all micro finance clients had lost their homes, nearly half had lost their every day possessions and over 75% had their ability to generate income at least suspended (Hassan and Hussein 1998). In the same year, many parts of East Africa experienced record rainfall (up to 10 times the usual amount) and disastrous flooding. In Uganda alone, 10,000 people were affected, 40% of the main roads were destroyed and the country became heavily dependent on food imports and aid (NEMA 1999). According to UNEP, the threat of natural disasters is one of the five key environmental issues facing the flood-prone nation that is home to 140 million people.

Johnson (2003) categorized the direct and indirect effects of flooding into those that occur during the non-growing season or fallow period and those that occur during the growing season. Important impacts during the non-growing season include loss of top soil; loss of soil nutrients; soil compaction; soil erosion; permanent damage to perennial crops, trees, livestock, buildings, and machinery; and permanent cessation of farming in floodplains. Impacts during the growing season include waterlogging of crops; lodging of standing crops; loss of soil nutrients; loss of pasture use; soil erosion; greater susceptibility to diseases and insects; interruptions to farm operations; permanent damage to perennial crops, trees, livestock, buildings and machinery etc.,

Johnson (2003) also gave examples of the impacts of flooding on agriculture in the USA. In one example, flooding damage to Californian agriculture from winter rains in 1997 included loss of 24,000 ha of crop with an additional 38,500 ha damaged by rain and flooding with the total losses estimated at $ 245 million.

1.5.4 Forest and Bush Fires

During 1982/83 and 1994/95 El Nino events, South-East Asia experienced severe smoke and haze episodes associated the forest and bush fires due to reduced rainfall and drought conditions. The resulting transboundary pollution compromised the health of downwind populations. Pollution and decreasing air quality caused by the fires contribute to changes in the composition and chemistry of the global atmosphere.

Fires (of which 23% are forest/grassland fire) occur with fairly high frequency especially in Germany (18%), France (15%) and UK (11%). The single largest group in manmade disasters in Oceania are fire (forest/grassland fire as well as building fire) which represent nearly 60 percent of the total in this category. Nearly all of them occur in Australia, the most important in the recent years being the one in January 1994 causing enormous economic losses. Previously fires in Australia have caused damage worth 20 million US$ in 1967, 400 million US$ in 1983 and 215 million US$ in 1985. A mid-July, (Australian) National Association of Forest Industries (NAFI) press release claimed that early-2003 wild fires (bushfires) were "the worst environmental disaster in the history of Australia ... responsible for over three million hectares of destruction, the greatest extent of environmental ruin on record in the nation's history, the bushfire disaster will have long term effects on the environment and the forest estate of this country". According to National Interagency Fire Center (NIFC) published material covering the period January 2003 to early-November 2003 inclusive, around 56,000 United States' wild fires had affected 3.8 million or so acres.

1.6
Environmental Degradation and Impact of Natural Disasters in Agriculture, Forestry and Rangelands

Environmental degradation is one of the major factors contributing to the vulnerability of agriculture, forestry and rangelands to natural disasters because it directly magnifies the risk of natural disasters, or by destroying natural barriers, leaves agriculture, forestry and rangelands more vulnerable to their effects.

Deforestation, land clearing, weed invasion and loss of wetlands could lead to ecosystem alteration, including changes to vegetation cover and composition and the introduction of plant and animal pests. Water erosion, wind erosion, siltation and sedimentation and coastal erosion could result in transport of soil and deposition elsewhere. Soil salinity, degradation of soil structure, soil fertility decline, soil acidification, water logging and soil pollution could lead to soil degradation, involving the alteration of soil properties *in situ*. Clearing of vegetation, rapid abandonment of exhausted cropland, expansion of cropping into more and marginal land set up a vicious cycle that is hard to break.

Poverty and environmental degradation are closely linked, often in a self-perpetuating negative spiral in which poverty accelerates environmental degradation and degradation results in or exacerbates poverty. While poverty is not the only cause of environmental degradation, it does pose the most serious environmental threat in many low-income countries (Pimentel et al. 1994). This problem is compounded because many millions of people, who live near the subsistence minimum, have to exploit natural resources with inappropriate technologies in order to survive. For example, forest fires in Indonesia in 1997–98 were deliberately set, went completely out of control and turned into wildfire surpassing any nation's fire fighting capabilities. They caused hundreds of deaths, devastated some 9 million hectares of forest and cost US$ 10 billion in short term ecological and economic losses. The degree and extent of damage caused by Hurricane Mitch is attributed to the drastic alteration over the years of natural systems that would have provided a buffer effect.

A new report entitled "Outlook for Fish to 2020: Meeting Global Demand" issued by the International Food Policy Research Institute (IFPRI) in October 2003 projects a significant increase in fish consumption during the next 20 years or so – with aquaculture expected to provide more than 40% of all fish consumed by 2020. Expanding aquaculture could also increase pollution and the use of scarce water and land resources, threatening the environment and the poor in developing countries. Substantial increases in fish farming could actually damage already vulnerable wild fisheries, and growth plans for this sector must also consider these potential effects.

1.7 Natural Disasters in Agriculture, Rangeland and Forestry – Some Methodological Issues

One of the major problems in dealing with natural disasters in agriculture, rangelands and forestry is the lack of systematic and standardized data collection from disasters. There is no recognized and acceptable international system for disaster-data gathering, verification and storage.

One good example of the methodological issue is the definition of natural disaster itself. The emergencies database (EM-DAT) of CRED classifies an event as a disaster if at least "10 people are killed and/or 100 or more are affected and/or an appeal for international assistance is made or a state of emergency declared". The NatCat database of Munich Re classifies an event as a natural disaster if there is any property damage and any person is sincerely affected (injured, dead). The Sigma database of Swiss Re regards an event as a natural disaster if there are > 20 deaths, and/or > 50 injured and/or > 2,000 homeless and/or insured losses exceed a given amount.

The above approach to definition of natural disasters was based on the need to respond to development and humanitarian agenda (Guha/Sapir and Below 2002).

Another problem is that different disasters can be classified as different types by different databases. For example, a flood which was a consequence of severe wind storm, may be recorded as one or the other. The reporting of high wind disasters pose a problem for a database indexed over countries since meteorological organisations report an event as a whole covering all the countries in its path.

Given the complexity of damage assessment, estimating the total impact of disasters in monetary value is quite difficult. It is also complicated by the fact that governments and international agencies use different methodologies. Numerous studies have been made on the assessment and display of specific hazards, but there is no standard method for assessing natural hazards in resource evaluations for development planning. Different methods are a response to specific concerns about individual hazardous phenomena.

Current practices of assessing the economic Impact of natural disasters are inadequate. Many existing estimates of the impacts of disasters are based on damage assessments undertaken in the immediate aftermath of individual disasters. Such assessments are typically undertaken without the use of comprehensive guidelines and by untrained assessors. Moreover, they provide an incomplete account of the impact of a disaster, focusing primarily on areas of particular concern to the assessors, and are completed too rapidly to allow a full assessment of the economic consequences.

Poor assessment practices have important impacts on the effectiveness of disaster management more generally. In particular, they can inhibit the design of timely and appropriate relief and rehabilitation programmes; and result in insufficient attention being paid to disaster prevention and mitigation measures, for example by over- or underestimating the true economic costs of disasters.

Post-disaster damage assessments of the impacts of disasters are generally concerned with the assessment of the economic impacts of sudden-onset rather than slow-onset natural disasters, such as drought or pest outbreaks.

Standard methodologies for estimating economic damage from disasters remains an area where much research is still needed. Governments, insurance companies, or other agencies involved in relief generally estimate economic damage according to varying criteria and standards. This presents serious difficulties, not only in comparisons across countries but also over time. Despite the multiple sources that have generated over 9000 events for the CRED database, only 24% have information on economic losses.

1.8 Mitigating the Impacts of Natural Disasters

The Plan of Implementation of the World Summit on Sustainable Development (WSSD) held in Johannesberg in 2002 highlighted the need to mitigate the effects of droughts and floods through such measures as improved use of climate and weather information and forecasts, early warning systems, land and natural resource management, agricultural practices and ecosystem conservation in order to reverse the current trends and minimize degradation of land and water resources. WSSD noted the need to promote the access and transfer of technology related to early warning systems and to mitigation programmes to developing countries affected by natural disasters.

Over the recent years, new technologies have brought about an accelerated increase in our knowledge of the climate system. Satellites for monitoring aspects of the oceans and sparsely populated parts of the globe, ocean buoys and expendable bathythermographs for monitoring the physical and chemical properties of the oceans, hundreds of specially equipped commercial aircraft, and manned and automatic weather stations on land, are all expanding the volume of data and contributing to knowledge base.

Communities that are most exposed to risk from climate extremes and natural disasters and potentially at risk from climate change, are those with limited access to technological resources and with limited development of infrastructure. Countries, especially the geographically smaller ones, cannot be expected to cope alone because each one needs to have information on the full extent and magnitude of natural disasters. Socio-economic losses cannot be entirely eliminated, but timely and appropriate mitigation measures can certainly reduce the impacts. Planning, early warning and well-prepared response strategies are the major tools for mitigating the losses. The longer in advance a warning can be given about potentially damaging conditions, the easier it will be to mitigate and reduce its impact.

Major advances in technology and notable progress in scientific understanding, the accuracy and timeliness of weather and flood warnings have significantly improved over the last few decades. Today, the accuracy of forecasts of large-scale weather patterns for seven days in advance is the same as those for two days in advance only 25 years ago (Obasi 1998). Now forecasts up to 10 days are showing remarkable accuracy, and there is now capability to provide some skillful information on expected weather patterns several seasons in advance.

For example, early information on El Nino episodes is now allowing advanced national planning, with considerable advantage in many sectors of the economy, such as in water resources management, tourism, and fisheries and agricultural production (Obasi 1996). In the case of the 1997–98 El Nino event, advances in El Nino related science and in monitoring the sea-surface temperatures in the Pacific Ocean, enabled scientists in the NMHSs to predict its formation longer in advance than all the previous events. With recent developments in communication technology, nincluding the use of Internet, information on the El Nino was disseminated in a rapid and timely manner throughout the world. These enabled many governments to take appropriate measures, and stimulated international cooperation and integrated efforts to address the associated impacts.

The accuracy of tropical cyclone track forecasts and the timeliness of warnings have been steadily improving in the past few years. Global efforts, especially within the context of Tropical Cyclone Programme of WMO, have resulted in a noticeable improvement in the warning systems in many parts of the world and resulted in saving a lot of lives and limiting property damage. For example, the decrease in the death toll in Bangladesh, from about 130,000 to 500 caused by similar tropical cyclones in 1991 and 1994 respectively, was attributed in large part, by government sources, to improvements in early warning and evacuation systems (Obasi 1997).

The evolving Internet has proven to be an invaluable tool in facilitating the exchange of global and regional climate monitoring and prediction information. However many users require assistance in the selection, interpretation and application of appropriate information.

Effective early warning systems coupled with community education for protective action has reduced the potential human loss from these events. Floods as a disaster also lend themselves well for preparedness measures both structural and legislative (land use laws, zoning plans and urbanisation). Preparedness of life-saving techniques and evacuation plans should be promoted actively in these high risk zones.

1.9
Conclusions

According to the International Federation of Red Cross and Red Crescent Societies (IFRC), natural disasters are on the rise and they continue to target the world's poorest and least-developed and there must be greater investment in disaster reduction rather than high-profile response efforts. Improved data on past disasters would help inform investment and policy decisions and thus help secure more appropriate levels and forms of disaster prevention, mitigation and preparedness. Historical studies would also help inform the development of appropriate methodologies for the assessment of future disasters.

Despite a long history of disasters affecting agriculture, rangelands and forestry, comprehensive documentation of these disasters at the national, regional and international levels has been weak and it is important to develop mechanisms for more efficient assessment and documentation of natural disaster impacts in agriculture. A comprehensive assessment of impacts of natural disasters on agriculture requires a multi-sectoral and integral approach involving key organisations.

Priority should be given to supporting research with practical applications since research is needed to understand the physical and biological factors that contribute to disasters. Since major impact of the natural disasters is on poor farmers with limited means in developing countries, community-wide awareness and education programs on natural disasters should be a priority. Programs for improving prediction methods and dissemination of warnings should be expanded and intensified. Efforts are also needed to determine the impact of disasters on natural resources.

References

Anaman KA (1996). An introductory discussion of cost-benefit analysis applied to climate change issues. Paper presented at the Users of Climate Change Predictions Experts Workshop, 31 May 1995, Macquarie University, Sydney, Australia. Published by the Graduate School of the Environment, Macquarie University, Sydney, Australia, ISBN 1 86408 247 X

Anaman KA (2003) Assessing the economic and social impacts of extreme events in agriculture and use of meteorological information to reduce adverse impacts. In: Agrometeorology related to Extreme Events. WMO No. 943, World Meteorological Organization, Geneva, pp

Anderson M (1990) Analyzing the costs and benefits of natural disaster responses in the context of development. Environment Working Paper 29, World Bank, Washington DC, USA

Benson C, Clay, EJ (1998) The impact of drought on Sub-Saharan economies: a preliminary examination. World Bank Technical Paper No. 401. Washington DC, USA: World Bank

Blong R (1992) Impact of climate change on severe weather hazards – Australia. Department of the Arts, Sports, the Environment and the Territories, Australian Government Publishing House, Canberra, Australia

Chang S (1984) Do disaster areas benefit from disasters ? Growth and Change (October): 24–31.

Charveriat C (2000) Natural disasters in Latin America and the Caribbean: an overview of risk. Research Department Working Paper No. 434, Inter-American Development Bank, Washington DC, USA

CIDI (2002) International Disaster Situation Reports, Center for International Disaster Information. Available at http://www.cidi.org/disaster

CRED (2000) EM-DAT: The OFDA/CRED International Natural Disaster Database 1900–1999. www.md.ucl.ac.be/cred. Centre on the Epidemiology of Disasters (CRED), Louvain, Belgium: Universite Catholique de Louvain

Das HP (2003a) Introduction. In Agrometeorology related to Extreme Events. WMO No. 943, World Meteorological Organization, Geneva, pp

Das HP (2003b) Incidence, prediction, monitoring and mitigation measures of tropical cyclones and storm surges. In Agrometeorology related to Extreme Events. WMO No. 943, World Meteorological Organization, Geneva, pp

Devereux S (2000) Famine in the Twentieth Century, IDS Working Paper 105, Institute of Development Studies, Brighton

FAO (1999) New concepts and approaches to to land management in the tropics with emphasis on steeplands. FAO Soils Bulletin No. 75, Food and Agriculture Organization, Rome

Funaro C (1982) Natural disasters and the development process: a discussion of issues. United States Agency for International Development, Washington DC

Gomes GM, Vergolino JR (1995) A microeconomia do desenvolvimento Nordestino:1960–1994. IPEA Paper No. 372, IPEA, Rio de Janeiro

Gommes R, Petrassi F (1996) Rainfall variability and drought in Sub-Saharan Africa since 1960. FAO Agrometeorology Series Working Paper No. 9, Food and Agriculture Organization, Rome

Guha-Sapir D, Below R (2002) The quality and accuracy of disaster data – a comparative analyses of three global data sets. Provention Consortium, The Disaster Management Facility, World Bank, Washington DC

Handmer J, Smith, DI (1992) Cost effectiveness of flood warnings. Report prepared for the Australian Bureau of Meteorology by the Centre for Resource and Environment Studies. Australian National University, Canberra

Hasan ME, Hussain, F (1998) Effects and implications of high impact emergencies on microfinance: experiences from the 1998 floods in Bangladesh. Dhaka: South Asian Network of Microfinance Initiatives

Heng L (2004) Flood management in China. Extended Summary. WMO/GWP Associated Programme on Flood Management.

Hodell DA, Curtis JH, Brenner M (1995) Possible role of climate in the collapse of classic maya civilization. Nature 375:391–393

Johnson G (2003) Assessing the impact of extreme weather and climate events on agriculture, with particular reference to flooding and rainfall. In Agrometeorology related to Extreme Events. WMO No. 943, World Meteorological Organization, Geneva

Joy CS (1991) The cost of natural disasters in Australia. Paper presented at the Climate Change Impacts and Adaptation Workshop, Climate Impacts Centre, Macquarie University, Sydney

Kreimer A, Munasinghe M (eds.) (1991) Managing natural disasters and the environment. Environmental Policy and Research Division, World Bank, Washington DC

NEMA (1999) State of the Environment Report for Uganda 1998, Kampala, Uganda: NEMA

Normand J-M (1991) Le Bangladesh à la dérive. Le Monde, Dossiers et documents, 22 juin 1991

Obasi GOP (1996) Climate, climate change, variability and predictability. RGICS Paper No. 36. Rajiv Gandhi Institute for Contemporary Studies. New Delhi. 35 pp

Obasi GOP (1997) Address at the opening of the Second Joint Session of the WMO/ESCAP Panel on Tropical Cyclones and the ESCAP/WMO Typhoon Committee, 20 February 1997, Phuket

Obasi GOP (1998) Address at the opening of the International IDNDR-Conference on Early Warning Systems for the Reduction of Natural Disasters, 7 September 1998, Potsdam

Oliver J (1989) A survey of public interpretation and opinions in Queensland on the present tropical cyclone warning system, Part 1: Case study of tropical cyclone Winifred and Part 2: Case study of tropical cyclone Charlie. Disaster Management Studies Centre, Cumberland College of Health Sciences, East Lidcombe, Sydney, New South Wales

OSRO (1989) Evaluation of the agriculture situation in eastern Caribbean countries affected by hurricane Hugo. OSRO 03-89-E, FAO, Rome

Pimentel D, Harman R, Pacenza M, Pecarsky J and Pomentel M (1994) Natural resources and an optimum human population. Population and Environment: a Journal of Interdisciplinary Studies 15(5): 347–369

Roy BC, Mruthyunjaya, Selvarajan S (2002) Vulnerability of climate induced natural disasters with special emphasis on coping strategies of the rural poor in Coastal Orissa, India. Paper presented at the UNFCC COP 8 Conference organized by the Government of India, UNEP and FICCI, 23 October to 1 November 2002, New Delhi, India

Ryan CJ (1993) Costs and benefits of tropical cyclones, severe thunderstorms and bushfires in Australia. Climatic Change 25:353–367.

Sugg AL (1968) Beneficial aspects of tropical cyclone. J. Appl. Meteorol. 7:39–43

Scott CD, Litchfield JA (1994) Inequality, mobility and the determinants of income among the rural poor in Chile, 1968–1986. London School of Economics, London

SEI, IUCN, IISD (2001) Coping with climate change: environmental strategies for increasing human security. (Source: MunichRe and UNEP)

Susman P, O'Keefe P, Wisner B (1983) Global disasters, a radical interpretation. In: Hewitt K (ed) Interpretations of calamity from the view point of human ecology. Allen and Unwin, London

Hasan ME, Hussain F (1998) Effects and implications of high impact emergencies on microfinance: experiences from the 1998 floods in Bangladesh. South Asian Network of Microfinance Initiatives, Dhaka

UNDP (2001). Disaster profile of the least developed nations

Vos R, Velasco M, De Labastida E (1999) Economic and social effects of El Niño in Ecuador. Inter-American Development Bank, Washington DC

WMO (1994) Climate variability, agriculture and forestry. Technical Note 196, World Meteorological Organization, Geneva

WMO (1997) Extreme agrometeorological events. CAgM Report No. 73, TD No. 836, World Meteorological Organization, Geneva

World Disasters Report (2001) International Federation of Red Cross and Red Crescent Societies, Geneva

World Disasters Report (2003) International Federation of Red Cross and Red Crescent Societies, Geneva

CHAPTER 2

The Role of Disaster Preparedness in National Planning with Specific Reference to Droughts

Donald A. Wilhite

Abstract Drought is a slow-onset, insidious natural hazard. Vulnerability to drought is increasing in all drought-prone nations and the traditional crisis management approach to address the impacts of these events has proven to be ineffective, untimely, and poorly coordinated. A more risk-based management approach has been gaining acceptance worldwide because it addresses the underlying cause of impacts – vulnerability. This approach includes development of national policies and mitigation plans directed at identifying the causes of drought impacts and improved early warning systems that provide information to decision makers in a timely manner. This paper discusses the conceptual framework for improved drought preparedness planning with an emphasis on reducing societal vulnerability to future episodes of drought.

2.1 Introduction

Drought is considered by many to be the most complex but least understood of all natural hazards, affecting more people than any other hazard (Hagman 1984). However, there remains much confusion within the scientific and policy communities about its characteristics. It is precisely this confusion that explains, to some extent, the lack of progress in drought preparedness in most parts of the world.

Drought is a slow-onset, creeping natural hazard that is a normal part of climate for virtually all regions of the world; it results in serious economic, social, and environmental impacts. Drought onset and end are often difficult to determine, as is its severity. The impacts of drought are largely nonstructural and spread over a larger geographical area than are damages from other natural hazards. The nonstructural characteristic of drought impacts has certainly hindered the development of accurate, reliable, and timely estimates of severity and, ultimately, the formulation of drought preparedness plans by most governments. The impacts of drought, like those of other hazards, can be reduced through mitigation and preparedness.

Drought preparedness planning should be considered an essential component of integrated water resources management. Increasing society's capacity to cope more effectively with the extremes of climate and water resources variability (i.e., floods and droughts) is a critical aspect of integrated water resources management. Drought preparedness planning will also provide substantial benefit in preparing for potential changes in climate. Historically, more emphasis has been given to flood management than drought management. With increasing pressure on water and other natural resources because of increasing and shifting populations (i.e., regional and rural to urban), it is imperative for all nations to improve their capacity to manage water supplies during water-short years.

Drought risk is a product of a region's exposure to the natural hazard and its vulnerability to extended periods of water shortage (Wilhite 2000). If nations and regions are to make progress in reducing the serious consequences of drought, they must improve their understanding of the hazard and the factors that influence vulnerability. It is critical for drought-prone regions to better understand their drought climatology (i.e., the probability of drought at different levels of intensity and duration) and establish comprehensive and integrated drought EWSs that incorporate climate, soil, and water supply factors such as precipitation, temperature, soil moisture, snow pack, reservoir and lake levels, ground water levels, and stream flow.

Vulnerability to drought is dynamic and influenced by a multitude of factors, including increasing and regional shifts in population, urbanization, technology, government policies, land use and other natural resource management practices, desertification processes, water use trends, and increasing environmental awareness. Therefore, the magnitude of drought impacts may increase in the future as a result of an increased frequency of occurrence of the natural event (i.e., meteorological drought), changes in the factors that affect vulnerability, or a combination of these elements. All drought-prone nations should develop national drought policies and preparedness plans that place emphasis on risk management rather than following the traditional approach of crisis management, where the emphasis is on reactive, emergency response measures. Crisis management decreases self-reliance and increases dependence on government and donors.

In the past decade or so, drought policy and preparedness has received increasing attention from governments, international and regional organizations, and nongovernmental organizations. Simply stated, a national drought policy should establish a clear set of principles or operating guidelines to govern the management of drought and its impacts. The policy should be consistent and equitable for all regions, population groups, and economic sectors and consistent with the goals of sustainable development. The overriding principle of drought policy should be an emphasis on risk management through the application of preparedness and mitigation measures. Preparedness refers to pre-disaster activities designed to increase the level of readiness or improve operational and institutional capabilities for responding to a drought episode. Mitigation is short- and long-term actions, programs, or policies implemented during and in advance of drought that reduce the degree of risk to human life, property, and productive capacity. These actions are most effective if done before the event. Emergency response will always be a part of drought management because it is unlikely that government and others can anticipate, avoid, or reduce all potential impacts through mitigation programs. A future drought event may also exceed the "drought of record" and the capacity of a region to respond. However, emergency response should be used sparingly and only if it is consistent with longer-term drought policy goals and objectives.

A national drought policy should be directed toward reducing risk by developing better awareness and understanding of the drought hazard and the underlying causes of societal vulnerability. The principles of risk management can be promoted by encouraging the improvement and application of seasonal and shorter-term forecasts, developing integrated monitoring and drought EWSs and associated information delivery systems, developing preparedness plans at various levels of government, adopting mitigation actions and programs, and creating a safety net of emergency response programs that ensure timely and targeted relief.

The traditional approach to drought management has been reactive, relying largely on crisis management. This approach has been ineffective because response is untimely, poorly coordinated, and poorly targeted to drought stricken groups or areas. In addition, drought response is post-impact and relief tends to reinforce existing resource management methods. It is precisely these existing resource management practices that have often increased societal vulnerability to drought (i.e., exacerbated drought impacts). The provision of drought relief only serves to reinforce the status quo in terms of resource management. Many governments and others now understand the fallacy of crisis management and are striving to learn how to employ proper risk management techniques to reduce societal vulnerability to drought and, therefore, lessen the impacts associated with future drought events.

Two important trends in drought management are: (1) improved drought monitoring tools and early warning systems and (2) an increased emphasis on drought preparedness and mitigation. Recent trends in each of these areas are discussed below.

2.1.1
Monitoring Drought: Unique Challenges and Recommendations

Effective drought EWSs are an integral part of efforts worldwide to improve drought preparedness. Timely and reliable data and information must be the cornerstone of effective drought policies and plans. Monitoring drought presents some unique challenges because of drought's distinctive characteristics, as previously discussed. In addition, several types of drought exist, and the factors or parameters that define it will differ from one type to another. For example, meteorological drought is principally defined by a deficiency of precipitation from expected or "normal" over an extended period of time, while agricultural drought is best characterized by deficiencies in soil moisture. This parameter is a critical factor in defining crop production potential. Hydrological drought, on the other hand, is best defined by deficiencies in surface and subsurface water supplies (i.e., reservoir and ground water levels, streamflow, and snowpack) and impacts generally lag the occurrence of meteorological and agricultural drought. These types of drought may coexist or may occur separately.

An expert group meeting on early warning systems for drought preparedness sponsored by the World Meteorological Organization (WMO) and others recently documented the status of drought EWSs in several countries, the shortcomings and needs of drought EWSs, and recommendations on how these systems can help in achieving a greater level of drought preparedness (Wilhite et al. 2000). This meeting was organized as part of WMO's contribution to the UNCCD meeting in Bonn, Germany, in December 2000. This report documented recent efforts in drought EWS in countries such as Brazil, China, Hungary, India, Nigeria, South Africa, and the United States, but also noted the activities of regional drought monitoring centers in eastern and southern Africa and efforts in West Asia and North Africa. The shortcomings of current drought EWS were noted in the following areas:

- *Data networks*: inadequate density and data quality of meteorological and hydrological networks and the lack of data networks on all major climate and water supply parameters;
- *Data sharing*: inadequate data sharing between government agencies and the high cost of data limits the application of data in drought preparedness, mitigation, and response;
- *Early warning system products*: data and information products are often not user friendly and users are often not trained in the application of this information to decision making;
- *Drought forecasts*: unreliable seasonal forecasts and the lack of specificity of information provided by forecasts limit the use of this information by farmers and others;
- *Drought monitoring tools*: inadequate indices for detecting the early onset and end of drought, although the Standardized Precipitation Index (SPI) was cited as an important new monitoring tool;
- *Integrated drought/climate monitoring*: drought monitoring systems should be integrated and based on multiple indicators to fully understand drought magnitude, spatial extent, and impacts;
- *Impact assessment methodology*: lack of impact assessment methodology hinders impact estimates and the activation of mitigation and response programs;
- *Delivery systems*: data and information on emerging drought conditions, seasonal forecasts, and other products are often not delivered to users in a timely manner;
- *Global early warning system*: no historical drought data base exists and lack of a global drought assessment product based on one or two key indicators could be helpful to international organizations, NGOs, and others.

Participants of the expert group meeting on drought EWS made several recommendations. Participants recommended that early warning systems be an integral part of a drought preparedness and mitigation plan and that priority should be given to improving existing observation networks and establishing new meteorological, agricultural, and hydrological networks.

2.1.2 Drought Policy and Preparedness

There are four key components of an effective drought risk reduction strategy (O'Meagher et al. 2000). These include the availability of timely and reliable information on which to base decisions; policies and institutional arrangements that encourage assessment, communication, and application of that information; a suite of appropriate risk management measures for decision makers; and actions by decision makers that are effective and consistent. In 1992, Australia adopted a National Drought Policy that applied these components through three objectives. These objectives were: (1) to encourage primary producers and other sections of rural Australia to adopt self-reliant approaches to managing for climatic variability; (2) to maintain and protect Australia's agricultural and environmental resource base during periods of extreme climate stress; and (3) to ensure early recovery of agricultural and rural industries, consistent with long-term sustainable goals (O'Meagher et al. 2000). Australia's national drought policy is widely known and its philosophy often replicated in other settings.

In the United States, there has been significant progress as well in addressing the impacts of drought through the development of preparedness plans. The most noticeable progress has been at the state level, where the number of states with drought plans has increased dramatically during the past two decades. In 1982, only three states had drought plans in place. In 2004, thirty-seven states have developed plans and four states are at various stages of plan development. The basic goal of state drought plans should be to improve the effectiveness of preparedness and response efforts by enhancing monitoring and early warning, risk and impact assessment, and mitigation and response. Plans should also contain provisions (i.e., an organizational structure) to improve coordination within agencies of state government and between local and federal government. Initially, state drought plans largely focused on response efforts aimed at improved coordination and shortening response time; today the trend is for states to place greater emphasis on mitigation as the fundamental element of a drought plan. Thus, many plans are more pro-active, adopting a more risk management approach to drought management.

The growth in the number of states with drought plans suggests an increased concern at that level about the potential impacts and conflicts associated with extended water shortages and an attempt to address those concerns through planning. Initially, states were slow to develop drought plans because the planning process was unfamiliar. With the development of drought planning models (Wilhite 1991; Wilhite et al. 2000) and the availability of a greater number of drought plans for comparison, drought planning has become a less mysterious process for states. As states initiate the planning process, one of their first actions is to study the drought plans of other states to compare methodology and organizational structure.

The rapid adoption of drought plans by states is also a clear indication of their benefits. Drought plans provide the framework for improved coordination within and between levels of government. Early warning and monitoring systems are more comprehensive and integrated and the delivery of this information to decision makers at all levels is enhanced. Many states are now making full use of the Internet to disseminate information to a diverse set of users and decision makers. Through drought plans, the risks associated with drought can be better defined and addressed with proactive mitigation and response programs. The drought planning process also provides the opportunity to involve numerous stakeholders early and often in plan development, thus increasing the probability that conflicts between water users will be reduced during times of shortage. All of these actions can help to improve public awareness of the importance of water management and the value of protecting our limited water resources.

2.1.3 Drought Mitigation Planning: Examples and Perspectives

Drought mitigation plans have three essential components regardless of whether they are developed at the provincial, national, or regional scale. First, a comprehensive early warning system provides the basis for many of the decisions that must be made by a wide range of decision makers as drought conditions evolve and become more severe. Equally important, early warning systems must be coupled to an effective delivery system that disseminates timely and reliable information. As drought plans incorporate more mitigation actions, it is imperative that these actions be linked to thresholds (e.g., reservoir levels, climate index values) that can serve as triggers for mitigation and emergency response actions. Second, a critical step in the development of a mitigation plan is conduct of a risk assessment (Knutson et al. 1998). The purpose of the risk assessment is to determine who and what is at risk and why. This is successfully accomplished through an analysis of historical and recent impacts associated with drought events. This risk assessment task is accomplished as part of a broader planning process as defined by Wilhite et al. (2000). Third, after impacts have been identified and prioritized, the next step is to identify appropriate mitigation actions that can help to reduce the risk of each impact for future drought events. Mitigation actions currently being employed by some states are referenced later in this paper. In many cases, appropriate response actions are also identified through this process but these actions should not conflict with the basic goal of the drought mitigation plan – to reduce vulnerability to drought events. Some response actions may increase reliance on government and encourage the continuation of inappropriate resource management practices.

One of the current needs is quantifying the advantages of a risk-based drought mitigation planning effort over the current crisis management approach. For those that have been actively involved in drought management, the former holds numerous obvious advantages over crisis management. Still, questions remain in the minds of policy makers that must commit resources up-front to develop and implement mitigation actions versus dealing with impacts during the crisis and in the post-drought period. This field of research is important both in terms of methodology development and conclusions that would emanate from this research. In most cases the costs associated with mitigation actions are minimal when compared with the costs of drought, which often are in the billions of dollars. Clearly, we must improve our estimates of drought impacts, both direct and indirect, in all sectors. It must also be understood that failure to develop mitigation actions aimed at reducing drought impacts will result in a continuation of the spiraling costs of the impacts associated with this natural hazard.

The National Drought Mitigation Center (NDMC) has been promoting the concept of mitigation in its work with state, local, and tribal governments in the United States, as well as to foreign governments and international organizations working in this field. The NDMC has served as a conduit for gathering and disseminating these actions to a broad audience of policy and other decision makers in the United States and elsewhere from the local to national level (Wilhite 1997). These mitigation actions illustrated in Table 2.1 have helped states visualize possible actions to consider in association with the drought plan development process.

Assessment programs adopted by states cover a broad range of actions. Examples include development of improved criteria or triggers for the initiation of specific mitigation actions in response to drought, to the establishment of new data collection networks. Automated weather networks have become commonplace in many states and these networks have significantly improved the capability to monitor climate and water supply conditions in near real time.

Legislative actions have included the passage of measures to protect instream or environmental flows and guarantee low-interest loans to farmers. Low-interest loans, a common federal response to drought in the U.S., are not generally state financed. Many U.S. states have been reexamining aspects of water rights doctrine in response to growing water use and associated conflicts. Some states have formalized state drought plans through legislative actions (e.g., South Carolina and Nebraska). Water banks have been used in some states (e.g., California) as a means of temporarily modifying water allocation procedures (i.e., water rights) during water shortages. MacDonnell et al. (1994) present a review of water banking in the west United States.

Augmentation of water supplies during recent droughts included rehabilitating reservoirs to operate at design capacity and reviewing reservoir operation plans. New reservoirs are being considered in some locations and other cities have expanded current water supply capacity. Cities also worked with self-supplied industrial users on programs to reallocate some water for emergency public water supplies.

One of the key responsibilities of state government during periods of drought is to keep the public aware of the severity of the situation through timely reports. These reports must provide a clear rationale for mitigation actions imposed on either a voluntary or mandatory basis. During recent droughts, states organized informational meetings for the media and the public, implemented water conservation awareness programs, prepared and distributed informational materials, and organized workshops on drought-related topics. Sample ordinances on water conservation were also prepared and distributed to municipalities and rural suppliers.

Table 2.1. Drought-related mitigative actions of state government in response to recent episodes of drought

Category	Specific Action
Assessment programs	Developed criteria or triggers for drought-related actions Developed early warning system, monitoring program Conducted inventories of data availability Established new data collection networks Monitored vulnerable public water suppliers
Legislation/public policy	Prepared position papers for legislature on public policy issues Examined statutes governing water rights for possible modification during water shortages Passed legislation to protect instream flows Passed legislation providing guaranteed low-interest loans to farmers Imposed limits on urban development
Water supply augmentation/ development of new supplies	Issued emergency permits for water use Provided pumps and pipes for distribution Proposed and implemented program to rehabilitate reservoirs to operate at design capacity Undertook water supply vulnerability assessments Inventoried self-supplied industrial water users for possible use of their supplies for emergency public water supplies Inventoried and reviewed reservoir operation plans
Public awareness/ education programs	Organized drought information meetings for the public and the media Implemented water conservation awareness programs Published and distributed pamphlets to individuals, businesses, and municipalities on water conservation techniques and agricultural drought management strategies Organized workshops on special drought-related topics Prepared sample ordinances on water conservation for municipalities and domestic rural supplies Established drought information center as a focal point for activities, information, and assistance
Technical assistance on water conservation and other water-related activities	Provided advice on potential new sources of water Evaluated water quantity and quality from new sources Advised water suppliers on assessing vulnerability of existing supply system Recommended that suppliers adopt water conservation measures
Demand reduction/ water conservation programs	Established stronger economic incentives for private investment in water conservation Encouraged voluntary water conservation Improved water use and conveyance efficiencies Implemented water metering and leak detection programs

Table 2.1. (continued)

Category	Specific Action
Emergency response programs	Established alert procedures for water quality problems Stockpiled supplies of pumps, pipes, water filters, and other equipment Established water hauling programs for livestock from reservoirs and other sources Compiled list of locations for livestock watering Established hay hotline Provided funds for improving water systems, developing new systems, and digging wells Provided funds for recovery programs for drought and other natural disasters Lowered well intakes on reservoirs for rural water supplies Extended boat ramps and docks in recreational areas Issued emergency surface water irrigation permits from state waters Created low-interest loan and aid programs for agricultural sector Created a drought property tax credit program for farmers Established a tuition assistance program to enable farmers to enroll in farm management programs
Water use conflict resolution	Acted to resolve emerging water use conflicts Negotiated with irrigators to gain voluntary restrictions on irrigation in areas where domestic wells were likely to be affected Established a water banking program Clarified state law regarding sale of water Clarified state law on changes in water rights Suspended water use permits in watersheds with low water levels Investigated complaints of irrigation wells interfering with domestic wells
Drought contingency plans	Established statewide contingency plans Recommended to water suppliers the development of drought plans Evaluated worst-case drought scenarios for possible further actions Established natural hazard mitigation council

Most U.S. states lack the financial resources necessary to provide drought relief to individual citizens during times of emergency. However, it is often within the mission and capability of state agencies to provide technical assistance to municipalities and others. During recent droughts, states assisted by providing advice on potential new sources of water and evaluating the quality and quantity of those supplies. Agencies also assisted municipalities in assessing the vulnerability of water supply systems. States encouraged the adoption of voluntary water conservation measures and established stronger economic incentives for water conservation within the private sector. Water metering and leak detection programs were implemented.

Emergency response programs are not considered by some to be mitigation actions. However, if these measures are implemented to reduce immediate impacts or the risk of future impacts as part of a long-term mitigation program, they represent a proactive approach to drought management. State responses included a wide range of measures such as lowering of well intakes on reservoirs for rural water supplies, establishing water hauling programs for livestock, extending boat ramps in recreational areas, and creating a tuition assistance program to enable farmers to participate in farm management classes.

Conflicts between water users increase during water-short periods. Timely intervention to resolve these conflicts will become increasingly necessary as demands on limited water supplies continue to expand in number and complexity. The best approach is to anticipate these conflicts well in advance of drought and initiate appropriate actions to avoid conflict. Many of the actions taken focused on the growing conflicts between municipal and agricultural water use.

The State of Georgia in the U.S. is an excellent example of how mitigation actions can be identified to reflect state-specific needs. Georgia recently developed a drought planning framework and identified a broad range of pre-drought strategies that could be used to lessen the state's vulnerability to future drought events. These strategies are divided by sector into municipal and industrial, agriculture, and water quality. Selected examples of these actions are provided in Table 2.2. These examples are quite illustrative of the types of actions identified by states that have recently completed the drought mitigation planning process.

With the tremendous advances in drought planning at the state level in the United States in recent years, it should come as no surprise that states have been extremely frustrated and dissatisfied with the lack of progress at the federal level. Early into the 1995–1996 drought that affected a large portion of the south-western and south-central portions of the country, the lack of leadership and coordination at the federal level quickly became obvious and continued with subsequent drought episodes. Recent initiatives toward development of a national drought policy are the direct result of those frustrations (Wilhite 2001).

The U.S. Congress and the president are currently considering actions that could be taken in response to recommendations issued in May 2000 by the National Drought Policy Commission (NDPC). These recommendations were in response to the National Drought Policy Act, passed by the U.S. Congress in July 1998 and directed at developing a national drought policy that would emphasize preparedness and mitigation in future drought management efforts. One of the NDPC's recommendations strongly endorses drought planning at all levels of government. An interim National Drought Council has been formed, composed of federal and nonfederal members, and legislation was introduced in the U.S. Congress in 2003 that will lead to a more permanent

Table 2.2. Summary of selected pre-drought strategies included in the Georgia Drought Management Plan (Georgia Department of Natural Resources, 2003)

MUNICIPAL AND INDUSTRIAL	AGRICULTURE	WATER QUALITY, FLORA, AND FAUNA
State Actions	*Farmer Irrigation Education*	*State Actions*
Formalize the Drought Response Committee as a means of expediting communications among state, local, and Federal agencies and non-governmental entities.	Recommend that farmers attend classes in best management practices (BMP) and conservation irrigation, prior to (i) receiving a permit, (ii) using a new irrigation system, or (iii) irrigating for a coming announced drought season.	Encourage all responsible agencies to promote voluntary water conservation through a wide range of activities.
Establish a drought communications system between the state and local governments and water systems.	Provide continuing education opportunities for farmers.	Monitor streamflow and precipitation at selected locations on critical streams.
Provide guidance to the local governments and water supply providers on long-term water supply, conservation and drought contingency planning.	Encourage the use of BMPs, conservation irrigation, efficient use of irrigation systems, and the Cooperative Extension Service's water conservation guidelines.	Monitor water quality parameters, such as temperature and dissolved oxygen at selected critical streams.
Review the local governments and water supply providers' conservation and drought contingency plans.	Develop electronic database for communicating with water use permit holders.	Provide the streamflow and water-quality data in real time for use by drought managers and work with drought managers to optimize information delivery and use.
Work with the golf course and turf industry to establish criteria for drought-tolerant golf courses.	Encourage development and distribution of information on water efficient irrigation techniques.	Evaluate the impact of water withdrawals on flow patterns, and the impact of wastewater discharges on water quality during drought.
Encourage water re-use.	*Field/Crop Type Management* Encourage the use of more drought resistant crops.	Investigate indicators and develop tools to analyze drought impacts for waterways such as coastal ecosystems, thermal refuges such as the Flint River, and trout streams.

Table 2.2. (continued)

MUNICIPAL AND INDUSTRIAL	AGRICULTURE	WATER QUALITY, FLORA, AND FAUNA
Provide water efficiency education for industry and business.	Encourage the use of innovative cultivation techniques to reduce crop water use.	Improve the agencies' capabilities and resources to monitor land-disturbing activities that might result in erosion and sedimentation violations.
Conduct voluntary water audits for businesses that use water for production of a product or service.	Conduct crop irrigation efficiency studies.	Identify funding mechanisms and develop rescue and reintroduction protocols for threatened and endangered species during extreme events.
Identify vulnerable water dependent industries, fund research to help determine impacts and improve predictive capabilities.	Provide farmers with normal year, real time irrigation, irrigation scheduling, and crop evaporation/transpiration information.	Develop and execute an effort to identify pollutant load reduction opportunities by wastewater discharge permit holders.
Develop criteria for a voluntary certification program for landscape professionals.	Monitor soil moisture and provide real time data to farmers.	Develop and execute an effort to identify opportunities for industry to decrease water use during drought periods.
Develop and implement a statewide water conservation program to encourage local and regional conservation measures.	***Irrigation Equipment Management*** Encourage the installation of water efficient irrigation technology.	Evaluate the impact of water withdrawals on flow regimes and the impact of wastewater discharges on water quality during drought.
Develop and implement an incentive program to encourage more efficient use of existing water supplies.	Retrofit older irrigation systems with newer and better irrigation technology. Update any system over 10 years old.	Develop and promote implementation of sustainable lawn care programs based on selected BMPs and/or integrated pest management practices.

Table 2.2. (continued)

MUNICIPAL AND INDUSTRIAL	AGRICULTURE	WATER QUALITY, FLORA, AND FAUNA
Local/Regional Actions Develop and implement a drought management and conservation plan.	Encourage farmers to take advantage of available financial incentives for retrofitting and updating older or less efficient systems.	Encourage protection and restoration of vegetated stream buffers, including incentives for property owners to maintain buffers wider than the minimum required by state law.
Assess and classify drought vulnerability of individual water systems.	Recommend irrigation system efficiency audits every 5 to 7 years.	Provide for protection of recharge areas through measures including land purchase or acquisition of easements.
Define pre-determined drought responses, with outdoor watering restrictions being at least as restrictive as the state's minimum requirements.	***Government Programs*** Improve irrigation permit data to create a high degree of confidence in the information on ownership, location, system type, water source, pump capacity, and acres irrigated for all irrigation systems to determine which watersheds and aquifers will be strongly affected by agricultural water use, especially in droughts.	Encourage and explore wildland fire mitigation measures.
Establish a drought communications system from local governments and water supply systems to the public.	Improve on the agriculture irrigation water measurement and accounting statewide.	Enhance programs to assist landowners and farmers with outdoor burning.
	Improve communications and cooperation among farmers and relevant state and Federal agencies regarding available assistance during drought conditions.	
	Support legislation and efforts to enhance the ability of farmers to secure adequate water supplies during drought conditions.	

national drought council and a national drought policy. Key components of this bill include an emphasis on risk management, preparedness planning, and the improvement of the nation's drought monitoring system and forecasting capabilities. A recently completed project, the National Integrated Drought Information System (NIDIS), is intended to provide the foundation for the development of an improved drought monitoring system when the bill is passed by the U.S. Congress.

2.2 Summary and Future Challenges

Drought is an insidious natural hazard that is a normal part of the climate of virtually all regions. It should not be viewed as merely a physical phenomenon. Rather, drought is the result of interplay between a natural event and the demand placed on water supply by human-use systems.

There are many challenges before us if we are to improve our management of droughts. First, drought must be accepted as a natural hazard within the natural hazard community of scientists and policy makers. Because of its slow-onset characteristics and lack of structural impacts, it is often disregarded. A second challenge is to build awareness of drought as a normal part of climate. It is often considered to be a rare and random event, a departure from normal climate; thus the lack of emphasis on preparedness and mitigation. Improved understanding of the different types of drought and the need for multiple definitions and climatic/water supply indicators that are appropriate to various sectors, applications, and regions is a critical part of this awareness-building process. A third challenge is to erase misunderstandings about drought and society's capacity to mitigate its effects. Many people consider drought to be purely a physical phenomenon rather than the result of interplay between the natural event and human activities, i.e., all factors that influence vulnerability. Although it is important to understand the hazard (i.e., the climatology of drought from temporal and spatial perspectives and its causes), it is equally if not more important to understand the factors behind our vulnerabilities and take proper steps to reduce those vulnerabilities. Reducing drought risk is largely about changing behaviors, thus social scientists must play a key role in the drought planning process. A fourth challenge is to convince policy and other decision makers that investments in mitigation are more cost effective than post-impact assistance or relief programs. Evidence from around the world, although sketchy, illustrates that there is an escalating trend of losses associated with drought in both developing and developed countries. Also, the complexity of impacts is increasing. It seems clear that investments in preparedness and mitigation will pay large dividends in reducing the impacts of drought.

A growing number of countries are realizing the potential advantages of drought planning. Governments are formulating policies and plans that address many of the deficiencies noted from previous response efforts – efforts that were largely reactive. Most of the progress made in drought preparedness and mitigation has been accomplished in the past decade or so. Although the road ahead will be difficult and the learning curve steep, the potential rewards are substantial. The crisis management approach of responding to drought has existed for many decades and is ingrained in our culture and reflected in our institutions. Moving from crisis to risk management will certainly require a paradigm shift. The victims of drought have become accustomed to government and donor assistance programs. In many instances, these misguided and misdirected government programs and policies have promoted the nonsustainable use of natural resources. Many governments have now come to realize that drought response in the form of emergency assistance programs only reinforces poor or nonsustainable actions and decreases self-reliance.

Policies that encourage self-reliance and the sustainable use of natural resources will be more effective in the long term and will reduce the need for government and donor intervention. A critical first step is to identify and quantify the sectors and peoples at risk from drought. Once this step is completed, policies, plans, and mitigation programs can be formulated to address these vulnerabilities in a systematic manner. Virtual networks linking the broad range of institutions with interest in drought management at the national, regional, and global scale can greatly accelerate drought preparedness through the sharing of lessons learned.

References

Georgia Department of Natural Resources (2003) Georgia drought management plan. Atlanta, Georgia

Hagman G (1984) Prevention better than cure: report on human and natural disasters in the Third World. Swedish Red Cross, Stockholm, Sweden

Knutson C, Hayes MJ, Phillips T (1998) How to reduce drought risk. National Drought Mitigation Center, University of Nebraska, Lincoln, Nebraska

MacDonnell LJ, Howe CW, Miller KA, Rice TA, Bates SF (1994) Water banks in the West. Natural Resources Law Center, University of Colorado, Boulder, Colorado

O'Meagher B, Stafford Smith M, White DH (2000) Approaches to integrate drought risk management. In: Wilhite DA (ed) Drought: a global assessment, vol 2. Routledge Publishers, London, UK, pp 115–128

Wilhite DA (1991) Drought planning: a process for state government. Water Resour Bull 27(1):29–38

Wilhite DA (1997) State actions to mitigate drought: lessons learned. J Am Water Resour Assoc 33(5):961–968

Wilhite DA (2000) Drought as a natural hazard: concepts and definitions. In: Wilhite DA (ed) Drought: a global assessment, vol 1. Routledge Publishers, London, UK, pp 3–18

Wilhite DA (2001) Moving beyond crisis management. Forum for Appl Res Pub Policy 16(1):20–28

Wilhite DA, Hayes MJ, Knutson C, Smith KH (2000) Planning for drought: moving from crisis to risk management. J Am Water Resour Assoc 36(4):697–710

Wilhite DA, Sivakumar MVK, Wood DA (Eds) (2000) Early warning systems for drought preparedness and drought management. Proceedings of an Expert Group Meeting held in Lisbon, Portugal, 5–7 September 2000. Geneva, Switzerland: World Meteorological Organization.

CHAPTER 3

The Occurrence and Predictability of Extreme Events over the Southwest Pacific with Particular Reference to ENSO

M. James Salinger, Penehuro Lefale

Abstract Extreme events in the Southwest Pacific (SWP) are widespread and largely centre on tropical cyclones, droughts and floods. These events are directly influenced by the state of the El Niño Southern Oscillation (ENSO). The impacts of climate extremes are dramatic, particularly on Small Island Developing States (SIDS) of the SWP region. They cause destruction and widespread damage to human, socio, economic and development well being of local communities, examples being tropical cyclones Ofa, Val and more recently Heta, Olaf and Percy that devastated Samoa, Niue, Tonga and the Cook Islands and the severe impacts of the 1997–1998 El Niño event on Papua New Guinea and Fiji. Although warming in the SWP region is decreasing, extremes of low temperature, the trend of more El Niño episodes in the 20th century, based on observations, has increased both dry periods and extreme rainfalls in significant parts of the region, especially near the Date Line. The phase of ENSO also significantly affects temperature and rainfall anomalies and extreme across the region, with sharp gradients between contrasting areas of above and below average departures. ENSO also changes tropical cyclone track, density and occurrence, with El Niño events increasing the frequency and displacing tropical cyclones more towards the central South Pacific, and La Niña episodes decreasing the frequency and confining tracks to the west of the Date Line.

Seasonal to Inter-annual (SI) prediction has become a major research area and application tool that offers much promise for early warning and disaster preparedness to extreme events. In the SIDS of the SWP region, the Island Climate Update (ICU), a multinational virtual type regional climate prediction tool, with success hit rates (percentage of prediction being correct after comparing it with what actually happened) greater than 70 percent being achieved in some parts of the region since it was launched in September 1999, provides an extremely useful regional mechanism for preparing and predicting seasonal rainfall departures and other extreme events associated with ENSO for SIDS. The ICU could be used as a model in other regions.

3.1 Introduction

The Southwest Pacific (SWP) region is highly susceptible to climate related extreme events such as droughts, landslides and coastal flooding associated with the El Niño Southern Oscillation (ENSO) and tropical cyclones (Krishna et al., 2001). These extreme events are an integral part of the climate of the region and have dominant influences on the socio, economic and developmental well being of local communities. Experiences from the 1997-1998 El Niño event such as water shortages in Papua New Guinea and an unusual increase in outbreak of diseases such as malaria in some parts of region highlight the vulnerability of the region to extreme events (Shea et al., 2001; Hamnett et al., 2002).

On inter-annual timescales, the ENSO causes much climate variations throughout many areas of the South West Pacific and the globe (Philander, 1990, Kidson and Renwick, 2002). The impacts of ENSO varied considerably from one island to another (Hamnett et al., 2002), including wetter than normal or drier than normal seasons and changes in extreme events.

This chapter briefly assesses the occurrence, impacts and predictability of extreme events in the SWP. A brief overview of the vulnerability of the SWP region to extreme events is provided by assessing the economic costs to selected Small Island Developing States (SIDS) of the region associated with recent extreme events. Trends in extreme daily temperature and rainfall using climate indices for selected areas of the region anre examined, and a review is given of the progress to date of climate prediction (seasonal to interannual (SI)) in the region using knowledge of Sea Surface Temperatures (SSTs) anomalies in the tropical Pacific region (Niño 3.4 region) to predict climate outlooks (temperature, rainfall and tropical cyclones) for the tropical SWP region. This new area of climate science is still in its infancy and developmental phase. The chapter concludes with a brief analysis of a pilot project, the *Island Climate Update* (ICU) implemented in the tropical SWP since September 1999 to provide climate outlooks, including outlooks on occurrence of tropical cyclones, for SIDS in the region.

3.2 Impacts of Extreme Events

Of the extreme events in the Southwest Pacific, tropical cyclones are particularly significant (Krishna et al. 2001). Tropical cyclone Ofa caused an estimated US$ 120 million damages in Samoa in 1990 (or about 25% of its GDP) while tropical cyclone Val caused damages of about US$ 200 millions or 45% of its GDP in 1991. These two cyclones alone set back the development of Samoa by at least twenty years. In Fiji, tropical cyclone Kina in 1993 caused an estimated US$ 120 million damages, about 2.4% of Fiji's GDP. The damage caused by tropical cyclone Heta in January 2004, a category 5 tropical cyclone, is only now being assessed. Property, crops, roads and bridges were destroyed or damaged on Samoa with losses equivalent to US$ 226 million on American Samoa with roads washed away, and other damage to property and infrastructure. In the Cook Islands, 6 metre swells affected the west coast of Rarotonga. However, Heta devastated the tiny island nation of Niue (population 1200). One person died, much property was damaged or destroyed, roads were destroyed, infrastructure cut and crops destroyed because of intense rainfall and high winds (last recorded speed of 150 km/h and gusts to 200 km/h before equipment failure, G. Clarke, pers.comm).

El Niño events produce widespread impacts on communities across the Southwest Pacific, as instanced by the 1997–98 event (Shea et al. 2001). Drought severely affected Fiji, Papua-New Guinea, the Solomon Islands, Tonga and the Marquesas Islands of French Polynesia. In Fiji, by October 1998, 54,000 people were receiving food supplies and 400,000, half the population, were receiving water deliveries (Hamnett et al. 2002, Lightfoot 1999). In contrast Kiribati was wetter than normal. Fisheries were impacted with a general shift of the catch to the east, followed by a shift westwards (Lehodey et al. 1997, Lefale et al. 2003). Catches declined for Papua-New Guinea and the Solomon Islands, and increased for Kiribati. The impacts were severe on eastern regions of New Zealand where there were significant losses in agriculture and water (Basher 1998).

3.3 Trends in Extremes

Trends in daily temperature and rainfall indices for New Zealand have been examined for frequencies and values of daily maximum and minimum temperatures above and below specified percentile levels, as well as frequencies of these above and below fixed temperature thresholds (Salinger and Griffiths 2001). There were no significant trends in maximum temperature extremes ('hot days'), but there was a significant increase in the value, and decrease in frequency, of minimum temperature ('cold nights') extremes over the period 1951–1998. Compared with the 1961–1990 reference period, there was a de-

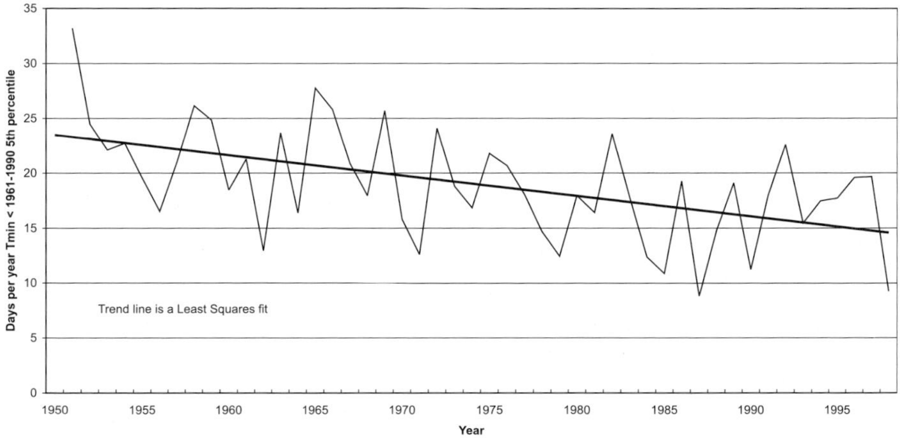

Fig. 3.1. Nationally averaged frequency of days with minimum temperatures below the 1961–1990 mean 5th percentile (*frequency of cold nights*) over the period 1951–1998 for New Zealand (from Salinger and Griffiths, 2001)

cline in frequency of the minimum temperature 5th percentile ('cold nights') of 10 to 20 days a year in many locations (Fig. 3.1). Trends in rainfall indices show a zonal (west-east) pattern of response, with the frequency of 1-day 95th

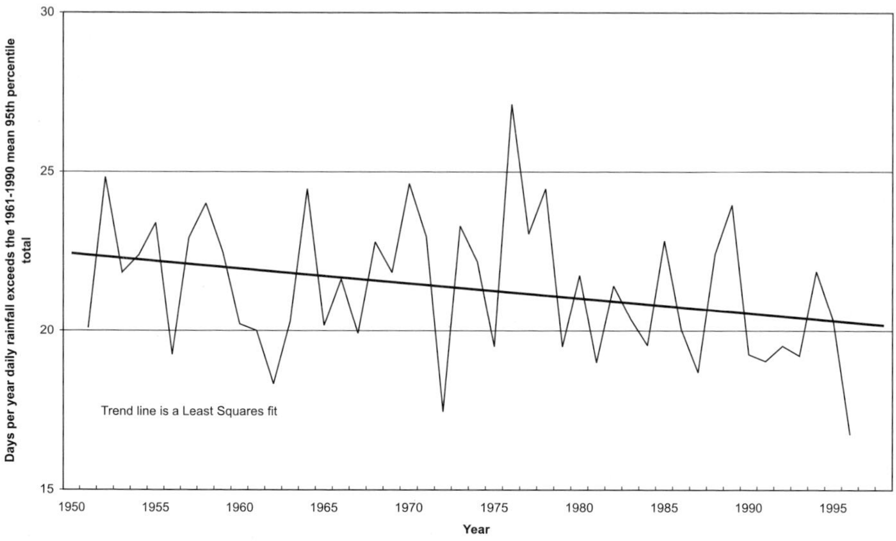

Fig. 3.2. Trend of nationally averaged frequency of daily rainfall exceeding the 1961–1990 mean 95th percentile (*extreme frequency*) for days with rainfall totals of 1 mm or more over the 1951–1996 period for New Zealand (from Salinger and Griffiths, 2001)

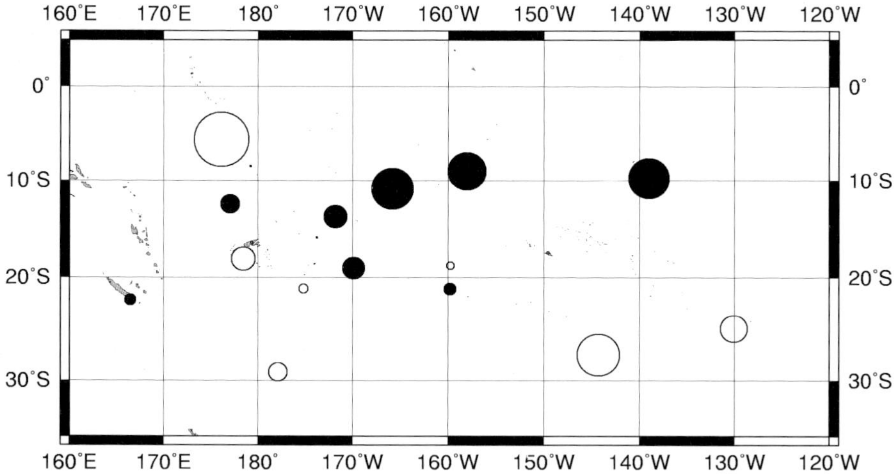

Fig. 3.3. Normalized trends in the extreme intensity of rainfall over 1961–2000. The circle size is proportional to the size of the normalized linear trend per decade. *Black circles* represent positive trends, *white circles* represent negative trends (from Griffiths et al., 2003)

percentile extremes decreasing in the north and east, and increasing in the west over the 1951–1996 period, but decreasing overall when taken nationally (Fig. 3.2). This equates to a higher frequency of ENSO events in the late 20th century. Frost day frequencies decreased between 5 and 15 days a year in many localities, with little change in the west of the South Island, and at higher elevation locations.

For the South Pacific, analyses of daily rainfall records for extremes over the period 1961–2000 have been assessed (Griffiths et al. 2003). Clear spatial patterns emerged in the trends of most indices of extreme rainfall, with a major discontinuity across the South Pacific Convergence Zone (SPCZ). The position of the SPCZ is known to be influenced by ENSO (Folland et al. 2002). Precipitation has become more extreme in the SPCZ zone between 180° and 170°W (Fig. 3.3) over the period 1961–2000, with both dry spell length increasing and daily extreme high rainfalls increasing in magnitude as well as frequency, including extended duration extreme rainfall. The majority show a change in the 1970s or 1980s – coincident with an increase in ENSO episodes.

3.4 ENSO Impacts on Climate in the Southwest Pacific

ENSO is the primary global mode of natural climate variability in the 2–7 year time band defined by sea surface temperature (SST) anomalies in the eastern tropical Pacific. Since the mid-1970s El Niño events have been more frequent than in the previous 30 years, and in each subsequent event global temperature anomalies have been higher (Trenberth and Hoar 1996). Figure 3.4 shows the Southern Oscillation Index (SOI, the Tahiti minus Darwin normalized pressure index) since 1920, which measures whether the climate system is in the El Niño or La Niña state. A significantly negative index indicates the El Niño state, and a positive index the La Niña state.

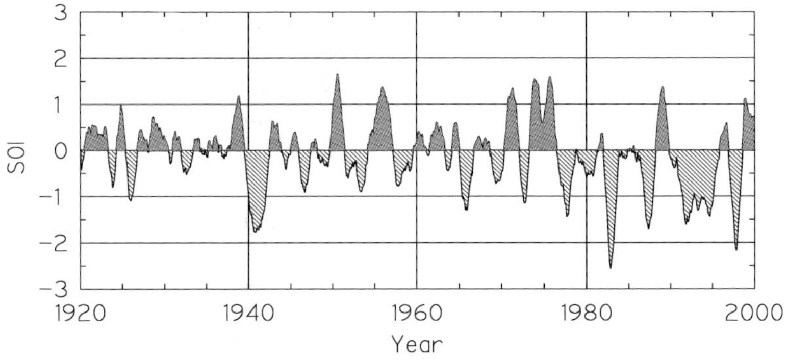

Fig. 3.4. Times series of 12-month running averages of the Southern Oscillation Index, 1920–2000

Relationships between the ENSO state and temperature and rainfall in the Southwest Pacific (Fig. 3.5) using correlations with the SOI (Salinger et al. 1995) are shown. In the El Niño (La Niña) state above average (below average) temperatures occur in the central South Pacific north east of the SPCZ including Samoa, Tokelau, eastern Kiribati, northern Cook Islands and French Polynesia, with below average (above average) temperatures to the south west. Rainfall anomalies are more zonal with all of Kiribati, Tuvalu, Tokelau, northern Cook Islands, the Tuamotu and Marquesas groups of French Polynesia, areas mainly north of 10°S, being wetter (drier) during the El Niño (La Niña) state, and those areas of the Southwest Pacific subtropicals between 15–35°S being drier (wetter).

Tropical cyclones develop in the South Pacific over the wet season, usually from November through to April. Peak cyclone occurrence is usually during January, February and March based on historical tropical cyclone data analysis. Those countries with the highest risk include Vanuatu, New Caledonia, Fiji, Tonga and Niue. Taken over the whole of the South Pacific, on average nine tropical cyclones can occur during the November to April season, but this can range from as few as three in 1994/95, to as many as 17 in 1997/98, during the last very strong El Niño. The mean frequency of tropical cyclones for the 1970–2000 period for El Niño episodes is 11.5, and La Niña events 8.6 per season. The tropical cyclone track densities vary depending on the ENSO state (Fig. 3.6). During El Niño episodes a higher frequency of tropical cyclone tracks occur near Vanuatu and Fiji, and their occurrence spreads further east to 160°W to affect the Cook Islands and most of French Polynesia. In contrast, during La Niña events the maximum occurrence is largely confined to the Coral Sea area of the Southwest Pacific centring on 160°E, 20°S and affecting New Caledonia in particular.

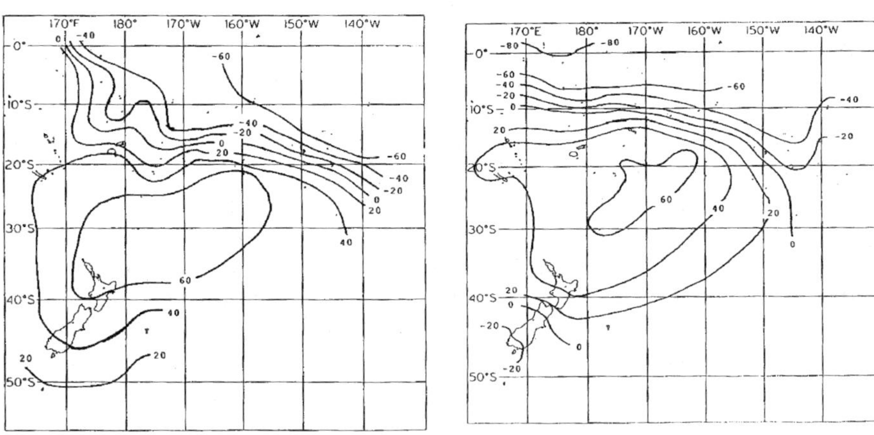

Fig. 3.5. Annual correlation (∗100) between (a) temperature and (b) rainfall and the Southern Oscillation Index. All isopleths < −35 or > 35 are at least significant at the 95% confidence level

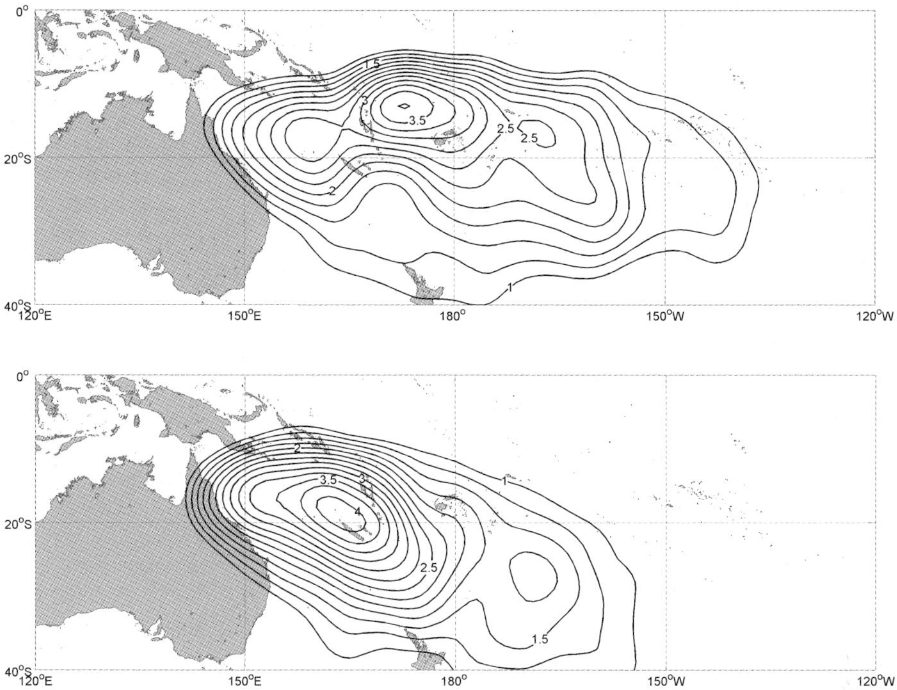

Fig. 3.6. Tropical cyclone track density for (**a**) the 11 strongest El Niño seasons and (**b**) the 11 strongest La Niña seasons (by SOI). Contour interval is 0.25, starting at 1.0

3.5 Predictability of Extreme Events and ENSO

Seasonal to interannual climate forecasting is used to predict ENSO and extreme events related to ENSO (Glantz et al. 1991). Two main groups of models are used (Goddard et al. 2001) – empirical models are developed using historical data and attempt to represent statistical links between one or more predictors, and computer-based dynamical models of the ocean and/or atmosphere, often similar to, or even the same as, the models used for weather or climate change prediction, form the other group. The variety of models targeted on predicting ENSO-related tropical sea surface temperature variations in the Pacific Ocean all have measurable skill out to perhaps nine months to a year, in some case perhaps a little longer. It is these that form the basis of ENSO predictions and the extreme events associated with the ENSO state.

3.6 Predictability of Southwest Pacific Climate

Operational seasonal climate forecasting for this region only commenced in the late 1990s through the Pacific ENSO Applications Center for the North Pacific.

The Australian Bureau of Meteorology produce seasonal outlook reference material for the South Pacific. The National Institute of Water and Atmospheric Research (NIWA) commenced production of the Island Climate Update (ICU) for the Southwest Pacific in September 2000. This is an overview of the climate in the South Pacific, with an outlook for coming months including forecasts of the state of ENSO. It is now a multi-national project with important contributions

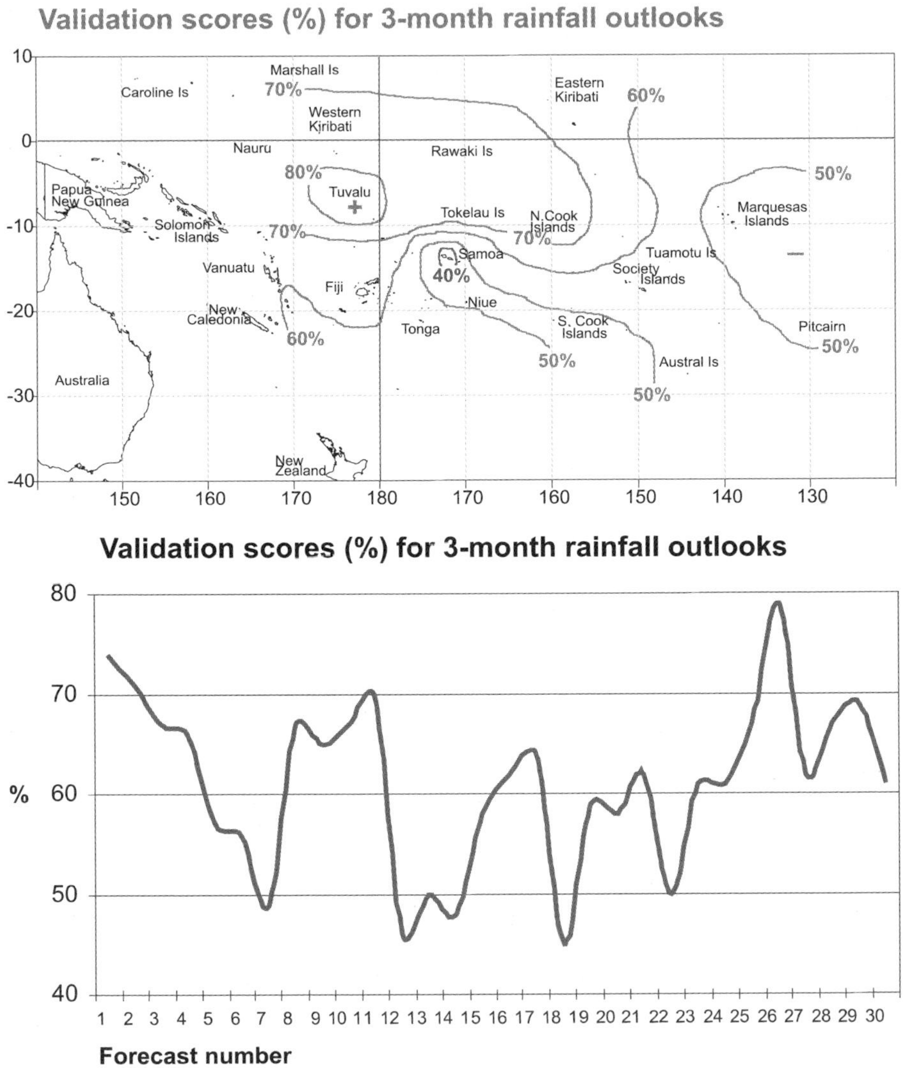

Fig. 3.7. (**a**) Frequency of correct 3-month rainfall outlooks and (**b**) validation score for the 30 (3-month) rainfall outlooks. This shows the percentage correct forecasts. Scores of 60% or more indicate significantly better outcomes than by chance

from the meteorological services of 13 Pacific Island countries, Australia, New Zealand and the United States.

Validation of the first 30 ICU outlooks for the Southwest Pacific (Fig. 3.7a) indicates how often the forecasts have been correct, in a categorical sense. Taking the ICU forecast category (below, average or below, average, average or above, or above average) to be that assigned the highest probability, the contours and numbers show the percent frequency of correct forecasts (or the "hit" rate). The overall hit rate is equivalent to the percentage frequency of correct forecasts. The hit rate has been highest (> 70%) in the 0–10°S latitudes west of about 155°W, in the region from Western Kiribati to the Northern Cook Islands, including Tokelau, and especially over Tuvalu (81%). However, scores have been much lower in areas around and to the south and east of Samoa. For the 30 issue period (Fig. 3.7b), the average hit rate for the whole of the forecast region has ranged between about 45 and 80%. The distribution of forecast success is affected by some extent to the ENSO state and the location of the SPCZ. With much of the period being above 50%, useful skill has been achieved.

3.7 Conclusions

Extreme climatic events produced by tropical cyclones, floods and droughts in the southwest Pacific are particularly devastating and can demolish economies and communities. Those caused by tropical cyclones and ENSO events are particularly significant. There are trends in climate extremes that are detectable throughout the region. Particularly evident is the decrease in cold nights in New Zealand because of general regional warming (Folland et al. 2003). At the same time rainfall extremes in region show a direct response to more ENSO events in the latter part of the 20th century with both dry spells and daily extreme rainfalls increasing near the Date Line in the tropical Pacific, and the latter decreasing in the north east, and increasing in the west of New Zealand.

ENSO, in both the El Niño and La Niña states produces very significant influences on the mean climate, and associated weather and climate extremes of the Southwest Pacific. El Niño events produce the more significant impacts on extreme events. These also affect tropical cyclone tracks and numbers with the El Niño (La Niña) episodes moving the tracks north east (south west) with increases (decreases) in frequency and rainfall distribution.

Seasonal to interannual prediction in the six years during and since the 1997/98 El Niño event, has become a major research and application issue in the Southwest Pacific. The results so far from the ICU work have demonstrated unequivocally that short-range climate prediction is achievable with skill. In the Southwest Pacific with the development of the ICU seasonal climate forecasting project useful skill has been achieved in the seasonal forecasting of seasonal rainfall departures and ENSO events with hit rates greater than 70% being achieved in some parts of the region. This new technology offers much promise for disaster preparedness strategies for all end users in the region.

Acknowledgements

This research was funded by the New Zealand Foundation for Research, Science and Technology (FRST) under contract No. CO1X0202. We would like to thank Dr Jim Renwick and Ms Ashmita Gosai for helpful comments in the preparation of this manuscript.

References

Basher RE (1998) The 1997/98 El Niño event: impacts, responses and outlook for New Zealand : a review prepared for the Ministry of Research, Science and Technology. Wellington, 28 pp.

Folland CK, Renwick JA, Salinger MJ, Mullan AB (2002) Relative influences of the Interdecadal Pacific Oscillation and ENSO on the South Pacific Convergence Zone. Geophysical Research Letters 29(13): 10.1029/2001GL014201, 21-1–21-4.

Folland CK, Salinger, MJ, Jiang N, Rayner N (2003) Trends and variations in South Pacific island and ocean surface temperatures. Journal of Climate 16 (17): 2859–2874.

Glantz MH, Katz RW, Nicholls N (1991) Teleconnections linking worldwide climate anomalies, Cambridge University Press, Cambridge, 535 pp.

Goddard L, Mason SJ, Zebiak SE, Ropelewski CF, Basher R, Cane MA (2001). Current approaches to seasonal to inter-annual climate predictions, International Journal of Climatology 21: 1111–1152.

Griffiths GM, Salinger MJ, Leleu I (2003) Trends in extreme daily rainfall across the South Pacific and relationships to the SPCZ. International Journal of Climatology 23: 847–869.

Hamnett M, Anderson CL, Gosai, A (2002) Pacific Islands Summary. Pages 38–42 in Preparing for El Niño: Advancing regional plans and interregional communication. International Research Institute for Climate Prediction, New York, USA.

Kidson JW, Renwick JA (2002) The Southern Hemisphere evolution of ENSO during 1981–1999. Journal of Climate 15: 847–863.

Krishna R, Lefale PF, Sullivan M, Young E, Pilon JC, Schulz C, Clarke G, Hassett M, Power S, Prasad R, Veitch T, Turner K, Shea E, Taiki H, Brook R (2001) A needs analysis for the strengthening of Pacific Island Meteorological Services, Australian Agency for International Development (AusAID), Published by the South Pacific Regional Environment Programme (SPREP).

Lefale PF, Lehodey P, Salinger MJ (2003) Impacts of ENSO on Tuna stocks and coral reefs of the Pacific Islands. NIWA Report AKL 2003-16.

Lehodey P, Bertignac M, Hampton J,. Lewis, J. (1997) El Niño/Southern Oscillation and tuna in the western Pacific, Nature 389, 715–718.

Lightfoot C. (1999) Regional El Niño social and economic drought impact assessment and mitigation. Disaster Management Unit, South Pacific Geoscience Commission, Suva, Fiji, 39 pp.

Philander SGH (1990) El Niño, La Niña, and the Southern Oscillation. Academic Press, 289 pp.

Salinger MJ, Griffiths G (2001) Trends in annual New Zealand daily temperature and rainfall extremes. International Journal of Climatology 21: 1437–1452.

Salinger MJ, Basher RE, Fitzharris BB, Hay JE,Jones PD , MacVeigh J P, Schmidelly-Leleu I (1995) Climate trends in the South West Pacific. International Journal of Climatology 15: 285–302.

Shea EL, Dolcemascolo G, Anderson CL, Barnston A, Guard CP, Hamnett MP, Kubota ST, Lewis N, Loschnigg J, Meehl G (2001) Preparing for a changing climate: The potential consequences of climate variability and change. A report of the Pacific Islands Regional Assessment Team for the U.S. Global Change Research Program. East-West Center, 102 pp.

Trenberth KE, Hoar TJ (1996) The 1990–1995 El Niño-Southern Oscillation event: longest on record, Geophysical Research Letters 23: 57–60.

CHAPTER 4

Accessibility of Database Information to Facilitate Early Detection of Extreme Events to Help Mitigate Their Impacts on Agriculture, Forestry and Fisheries

R.P.R. Guerreiro

Abstract Extreme events can cause severe damage in several sectors such as agriculture, forests and fisheries. In order to facilitate early detection of these harmful episodes, adequate climate and agrometeorological databases must be ensured. Some observational data and products necessary for early detection are presented in this paper. It briefly reviews the main features of proper databases that provide quality controlled data and products, useful to the end-users, easily accessible and in a timely manner. The data can be accessed through the standardization of database management and electronic accessibility. The main features and importance of data collection, automatic weather stations (AWS), database management and relational database management systems (RDBMS) are described. Examples of agrometeorological databases, database management systems and their applications and accessibility are given. Remote sensing (geostationary satellites, NOAA-Advanced Very High Resolution Radiometer (AVHRR), radar and lighting detectors) offers a valuable source of spatial information and can be complementary or even alternative to ground-based observations. Due to the processing of data from various sources in agrometeorology, and the need to display them in maps, geographical information systems (GIS) are in wider use today. GIS can help to identify the risk, extent and severity of many extreme events. Some examples of applications of remote sensing and GIS are presented. Finally, training, interdisciplinary collaboration and communication between users and developers of products are referred as essential means to achieve these goals.

4.1 Introduction

Extreme events that most affect agriculture, rangelands, forests and fisheries include mainly extreme meteorological events and weather conditions leading to the development of agents such as pests and deseases that affect these systems. Among these events, one can mention droughts, floods, heatwaves, frost, high wind, severe storms, like tropical storms, hailstorms, thunderstorms and duststorms. All of them cause severe damage to agriculture or may even have disastrous effects. Extreme events do less damage when they strike a resilient community; so, it is possible to mitigate their effects with an early detection. To reach this, adequate climate and agrometeorological databases are necessary.

Forest fires, which are considered as indirect effects of extreme meteorological events, constitute an important cause for the loss of forest produce in many countries, next to wind, hail, etc. (Gommes 1997). In Portugal, a heatwave, which took place in August 2003, significantly contributed to the worst forest fires in the last decades. An area of 280,550 ha was destroyed during this month (13th Provisional Report – Forest Fires 2003), similar to the area of Luxemburg. Monitoring of the forest fires is done by using the Canadian Fire Index. The computation and operationality of this Fire Weather Index (FWI) is possible due to the accessibility of databases of daily meteorological observations, as well as an atmospheric mesoscale model. Warnings were issued to the National Fire Brigade and Civil Protection Service, because of the high risk of this situation. However, this did not seems to be adequate. Although this system has proved to give fairly good results, the huge areas were burnt and a few casualties occurred due, among other factors, to some kind of lack of coordination and internal communication between Civil Protection and Fire Brigade, as it was their first year of working in the same organization. Besides this, the access to the information was restricted to official entities and not available to every user. This single example demonstrates the importance of an efficient and automated access to agrometeorological databases, in the prevention and early detection of extreme events, namely forest fires.

One of the main duties of Meteorological and Hydrological Services is to provide public warning of weather conditions conducive to extreme events, which can be dangerous to the population and can cause severe damages in sectors like agriculture, forestry and fisheries. Timely information is essential in the case of extreme events as information is needed at the right time by the users (Anaman 2003). It is indeed very important, since information supplied at the wrong time may be worthless.

Appropriate communication channels for transmitting advice and warnings to users are important. Agrometeorological bulletins transmitted via Internet or web pages dedicated to agromeorological information, with timely warnings of adverse weather conditions are essential to farmers and agricultural authorities. Recent developments in computer technology, database software and telecommunication systems provide the essential tools to facilitate the accessibility to databases and to the development of preparedness measures in the case of extreme events.

Interactions between meteorological conditions, agriculture, forestry and fisheries are complex, as they involve products from different sources. The building of a database of meteorological, phenological, soil and agronomic information is inevitably a major priority (Doraiswamy et al. 2000). The building of these databases depends on the availability of technical personnel and on the necessary hardware and software.

The recent developments in computer technology, communications and database software makes it possible to the users to access large data sets. The World Wide Web (WWW) allows users to access text, images, data or other information that are linked together electronically.

4.2 The Need for Observational Data and Products

The need for observational data necessary for an early detection of extreme events depends on the type of hazard considered. In any case, meteorological observations are always necessary. Some examples are presented below.

4.2.1 Floods

In the case of floods, precipitation is the most important parameter and is used as input for the hydrological models. Real time data collection from an automated raingauges network is an essential condition for flood forecasting improvement (Balint et al. 2001). A suitable communication network is designed in order to exchange the hydrological data in real time. Besides precipitation, other observations such as water level and stream flow are required. The use of radars and remote sensing from satellites are essential for flood forecasting, as they can provide information on areal precipitation and movement of the storms. Hydrological forecasting systems usually contain a database, where data management is necessary to provide standardized formats, quality control, analysis, risk management and dissemination. At the Portuguese Meteorological Institute a model was developed, in connection with precipitation forecasts from the European Centre for Medium-Range Weather Forecasts (ECMWF), which processes automatically an Hydrometeorological Bulletin, with observed and forecasted daily rainfall amounts in the main river basins. This information is available to several entities responsible for water resources, environment and civil protection.

4.2.2 Droughts

As far as agricultural droughts are concerned, different indices are often used to quantify them. Percentage of normal and deciles are simple indices based on precipitation amounts. Palmer Drought Severity Index (PDSI), Crop Moisture Index (CMI) and Standardized Precipitation Index (SPI) also include data from a water balance model. Geographical Information Systems (GIS) are used to map these indices, in order to monitor their behaviour and provide information about areas experiencing drought and drought severity. Figure 4.1 shows the PDSI distribution in Mainland Portugal, in January 2000, where the whole country was affected by an intense period of drought, with some regions in extreme drought. With these maps, it is possible to monitor the spatial and temporal evolution of droughts across Mainland Portugal (Szalai et al. 2002).

In Portugal, PDSI and SPI were implemented and applications were developed to perform the calculations, archive, analyse the results and make graphs. This quick access to the data and products derived is essential for an early detection of droughts, mainly in drought prone areas. In a near future these

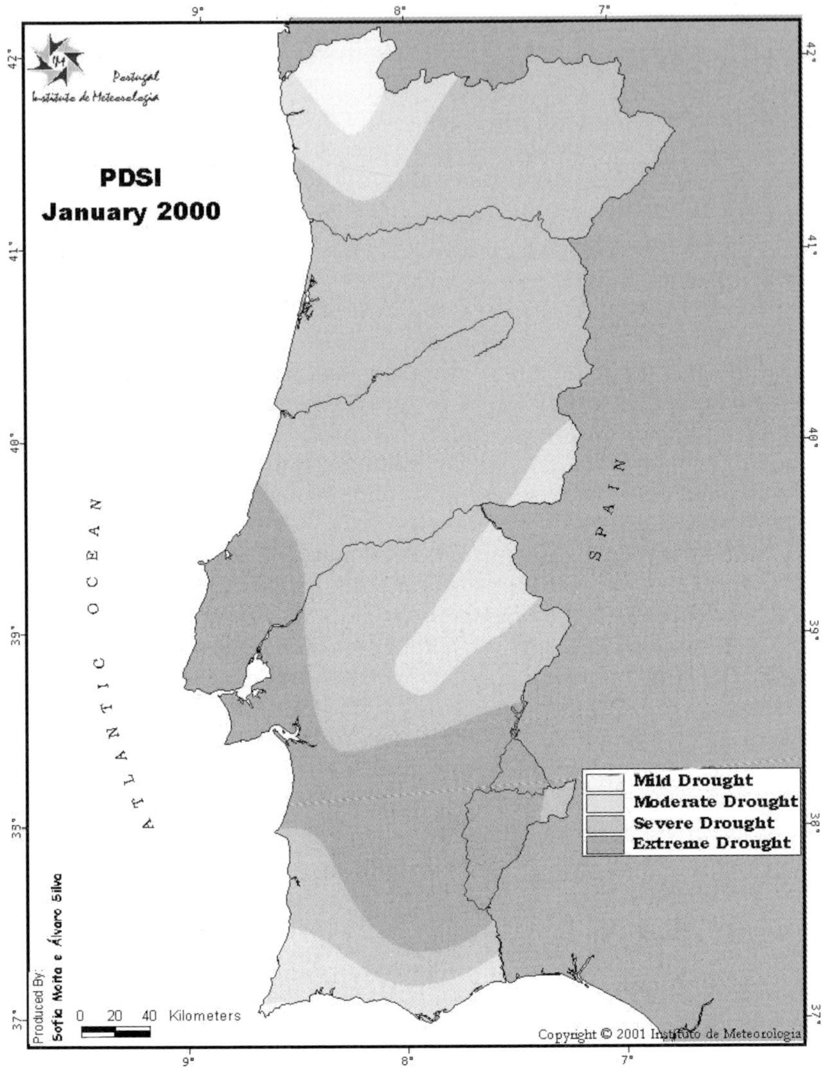

Fig. 4.1. PDSI map for January 2000 in Mainland Portugal (after three months with very short rainfall amounts). *Dark grey* represents extreme drought

products will be available in a web site for direct access by users. It is hoped that, this way, information reaches people in time for them to use it in making decisions (Wilhite and Svoboda 2000).

Besides the indices referred to above, as well as others, one should bear in mind that for agricultural drought assessment, data on soil type, its texture, water holding capacity, slope of the surface, soil bulk density, cultivar characteristics, irrigation and crop management are needed (Das 2003).

4.2.3
Lightning Storms

Lightning storms, due to electric atmospheric discharges in their extreme form, are amongst the most devastating atmospheric phenomena, which can cause loss of lives, hail and heavy rain, causing severe damage on agriculture, forestry and fisheries. Usual methods include meteorological observations, satellite images, data from meteorological radars and results obtained by numeric forecast models. The Portuguese Meteorological Institute has installed four lightning detectors (Fig. 4.2), which measure the variations of the atmospheric electromagnetic field and transmit them to a concentrator and processor. Remote terminals connected to the concentrator make it possible to visualize the information available. These data are automatically introduced in a database, where every detail about the storm is stored. Lightning detection is extremely important to determine the location of the electric discharge and to predict the path of the nebulous system that generates it, highly contributing to early detection of lightning storms.

Fig. 4.2. Lightning detectors installed by the Meteorological Institute in Portugal

4.2.4
Data Needs for Fisheries

As far as fisheries are concerned, the most important observation is water temperature. In the case of inland fisheries and aquaculture, meteorological data that serve for agriculture are also relevant for these fisheries, although the relative importance of parameters and thresholds may be different (Kapetsky 2000). Taylor (2001) refers that besides sea surface temperature, sea surface height, chlorophyll and sediments are of interest to fisheries and aquaculture. These data are available directly from high resolution satellites in real time. If timeliness, resolution and predictive capabilities for agrometeorological data can be improved, considerable benefits will come to fisheries.

4.2.5 Automated Weather Stations

The use of Automated weather stations (AWS) is becoming more and more widespread, as they provide real-time meteorological data from places in agricultural areas with very scarce stations. Data from AWSs have become essential to provide information for the assessment of risk management and for decision making.

During the 90's the Portuguese Meteorological Institute started the installation of a network of AWSs in Mainland and Islands with the following objectives:

- to increase the frequency and to improve the quality of the observations;
- to overcome the lack of non-professional people at the climatological stations;
- to increase data availability during night-time period;
- to allow real-time data availability in digital format at the Institute's central systems.

In total, 93 standard AWSs were installed until 2002. They generate 10 minutes records, for several meteorological and a few agrometeorological parameters, which are collected at the central office by a set of concentrators that automatically call the AWSs to retrieve the data. The transmission of data to the central system is made by using fixed and mobile telecommunications operators . Evaluation of data quality is done in near real-time through the analysis of data graphs and tables and with the application of automatic procedures consisting of algorithms to check and validate the records. The objectives of data processing and validation are to prepare them for long-term archive into a database with compatible format (Nunes et al. 2003). In this system, the degree of automation of the validation procedures still needs to be increased. To achieve this, a new generation system is being prepared, based on more powerful software and better data quality communication system. Another database has been developed and maintained to archive all the features of the AWSs, where detailed metadata have been introduced, including AWSs identification, location, equipment and even the local people contacts. All the changes that occurred at the AWSs are also recorded, including the actions performed by the maintenance company, namely the replacement of sensors and other components. Metadata relating to the operating characteristics and performance of the observational station are an essential part of the data record (Motha 1999).

4.2.6 Agrometeorological Databases

A basic requirement in agrometeorological hazard assessment and mitigation is an adequate agrometeorological database. This should include not only meteorological data, but also agricultural (phenological, crop management, pest and desease cycle), hydrological, land use, soil types, economic and other relevant information. Inclusion of derived data, maps and graphs is a requirement too. With efficient agrometeorological databases, including quality controlled historical data, it is feasible to estimate the risk of extreme events in quantitative terms, which is an important information in risk assessment. According to Weiss et al. (2000), one useful purpose of agrometeorological databases is to transform data into useful information.

Some limitations still exist. For example, detailed phenological information is often sporadic and not very widespread. In Portugal, like in many other countries, such data are generally archived in agronomic research institutes and are not centralized in an agrometeorological database, what makes them not very easily available.

Databases of extreme events, or better, of agricultural disasters resulting from extreme events should also be encouraged. This represents a basic requirement in hazard assessment for extreme events. Such databases should contain the name, type, location, severity, extension and timing of each event.

According to Sinha Ray (2002), the basic components of the databases should have descriptions of the long term effects including recovery, description of extreme factor, impact assessment, production loss, environmental loses, etc. The purpose of a database of extreme agrometeorological events is, of course, to identify patterns of impacts on agriculture with a view to improving impact assessments, including impact forecasting, mitigation, adaptation and emergency operations whenever feasible (Gommes 2003). This kind of database should thus be regarded essentially as an operational tool.

4.3 Database Management

4.3.1 Database Design

The management of climatic and agrometeorological data has become faster and more efficient in the last years. Yet, management of the data is perhaps one of the most critical processes in any design concept (Doraiswamy et al. 2000). A consistent database consists of a complex set of processes involving the collection, transmission, quality control, archiving, processing, analysing and displaying of useful information that can be utilized by users. In fact, large amounts of collected data are essentially useless without a system to manage them and provide analysis (Monnik 2001). When data are processed or transformed into forms useful to people, they become information (Anaman 2003). According to Weiss et al. (2000), information is the result of utilizing a statistical simulation or other type of model, such that the output can be used to make a decision.

Proper database management systems should store, temporally and spatially, data from different instrumentation and sources, including traditional meteorological stations, automated weather stations, radar, satellite imagery, special agricultural and hydrological networks and others (Fig. 4.3).

Database design must provide standardized formats and, preferably, modular structure, since all users do not require or need the same information input (Motha 1999). Moreover, standardized data makes it easier to share and exchange data. Checking the data quality is a necessary step in agrometeorological applications. Databases should store only validated, homogenized long-term datasets and with the smallest possible number of missing data.

Developing homogeneous time series requires complete and accurate station metadata (Murnane et al. 2002). However, known metadata are generally incomplete. The importance of collecting and managing metadata was recognized a few years ago. These data about data must reflect how, where, when and by whom the observational data were collected and what happened between the moment they were observed and the moment they were archived (Plummer et al. 2003). Station metadata should include any kind of changes, not only alterations in the environment around a station, but also instrument upgrades and relocations as well as planned modifications. Metadata relating to the operating characteristics and performance of the observational station are an essential part of the data record (Motha 1999).

Crop growth simulation models are very widely used nowadays to assess the impact of extreme events on crop yields. Mechanistic models simulate crop development from sowing to maturity, based on meteorological conditions, soil type and crop characteristics (Supit and van der Goot 2002). These models are very sensitive to drastic departures from normal conditions (Bedson 1997).

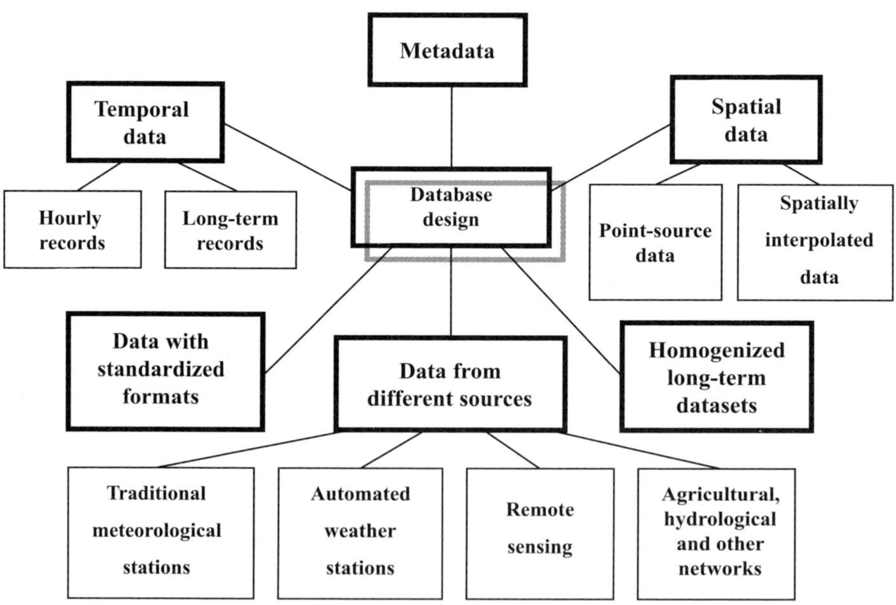

Fig. 4.3. Main components of an agrometeorological database design

Quite often they take advantage of seasonal or inter-annual forecasts in order to estimate crop yields and losses due to extreme events. In Portugal, WOFOST model has been used at the Institute of Meteorology in collaboration with the Faculty of Sciences of Lisbon (Guerreiro and Pires 2004) and DSSAT models at the Faculty of Agronomy in Lisbon (Pinto and Brandão 2002), not only to assess the impacts of extreme events on crops, at a regional scale, but also for climate and climate variability impact assessment and adaptation measures. These actions are included in the CLIVAR project (Santos et al. 2003), which also includes the impact of the use of seasonal forecasts in the process of decision making of crop production. These models usually require daily meteorological data, regional crop parameters and soil data, as inputs. The introduction of meteorological data into the models is still time consuming, due to flaws in the system and also the need to estimate values by means of spatial interpolation of daily data. A database management system (DBMS) , allowing an adequate collecting, analysis, archiving and access to these data can facilitate these tasks.

4.3.2
Real-time Information

It is important to develop the structures and devices for real-time agrometeorological information, in order to face and manage climate hazards, namely extreme events. With the recent advances in the automation technology, information is more and more easily accessible. Internet represents the quickest way to disseminate agrometeorological information to the end-users in a timely manner. Large data volumes and useful information can be transferred quite easily with the File Protocol Transfer (FTP), nowadays. However, the availability of Internet hosts is extremely limited in many developing countries. Information and communication technologies (ICT) offer great scope for innovative dissemination of agrometeorological information to farming communities (Weiss et al. 2000). In these countries, the effectiveness of these tools can be enhanced by linking them to media which are more accessible to farmers, such as rural radio, which allows timely advice to the end-users.

Inclusion of seasonal forecasts in agrometeorological products helps farmers and governments improve their preparedness for extreme events. In fact, an increased demand on climate prediction is greatly due to an increased frequency of extreme events. Meinke et al. (2000) refer that the Australian experience demonstrates that applied seasonal climate forecasts can increase farmer's profits and reduce their production risk. One of the future challenges is to operationalize the forecasts and provide them to the users in a timely and easily understandable manner (Sivakumar 2000).

An example of a good climate information service for rural areas through the Internet was developed in Australia by the Queensland Departments of Natural Resources and Primary Industries. This website offers a range of decision-support information services, tools and training to help clients to better manage climatic risks (Laughlin and Clark 2000). The main themes presented are: seasonal climate outlooks, rainfall and pasture growth, drought monitor and satellite fire monitor (http://www.dnr.qld.gov.au/longpdk).

4.3.3 Relational Database Management Systems

The use of Relational Database Management Systems (RDBMS) by Meteorological Services is increasing and it is quite useful, as the access to reliable and meaningful information becomes much more efficient and user-friendly. It contributes to the development of communication links between users and meteorological and agrometeorological networks, which is specially important in the case of early warning systems of extreme events. The three major RDBMS types are: ORACLE, INFORMIX and EMPRESS (Pérarnaud 2001).

The Meteorological Institute in Portugal has used INFORMIX to create a national relational database, with daily and hourly meteorological and agrometeorological observations and rainfall in short periods. The data were originated from historical archives and from AWSs, since 1920 up to now, and were submitted to quality control. It also includes metadata. Recently, satellite images have been added to this database. Structured Query Language (SQL) was used to create this database, and the access is through a graphical user interface. This database represents a great advance in data access, although it is still disseminated only in Intranet. Before this, the only way to retrieve archives was by means of FORTRAN programs. More derived parameters will be introduced and the transfer of data and other pertinent information to the end-users will be made via Internet in a near future.

An example of an extensive use of data and methods in agrometeorology in Europe is the Crop Growth Monitoring System (CGMS) implemented by the MARS (Monitoring Agriculture with Remote Sensing) project, run by the Agriculture and Regional Information Systems (ARIS), at the Joint Research Centre (JRC), in Ispra, Italy. It is devoted to predict output of the major crops at national level. Every ten days, crop growth conditions, such as occurrences of droughts, excess rain, heat stress, etc., are assessed. The CGMS is based on the deterministic model WOFOST. Input data include daily weather data, topography, regional crop parameters, soil data and historical crop statistics (Supit and van der Goot 2002). Input and output data as well as system definition data are stored in an ORACLE RDBMS, which constitutes the core of the CGMS.

A good example of an up-to-date and efficient database management system is the Unified Climate Access Network (UCAN), derived from US Department of Agriculture. This framework allows climate information users to request information through a unified climate access network that automatically sends requests for specific datasets and climate products to a network of computer systems that maintain the data archives for the requested product (Motha 2000). This system provides users with virtual access via the Internet to several climate datasets. Each participating organization maintains a computer that serves its climate data to the UCAN (Doraiswamy et al. 2000). The various UCAN data servers involved in the processes are coordinated through an efficient catalog of all meteorological stations from which data are archived and disseminated. This system represents a rapid access to quality controlled climate data and products, leading to benefits in several activities. The efficiency of this process is mainly due to the standardization of database management and electronic accessibility.

4.4 Remote Sensing

Remotely sensed data can provide useful information specially in those regions where no ground data are available. Besides, they give a spatial coverage of agrometeorological parameters, contributing to an early detection of droughts, frost, forest fires and other extreme events that affect agriculture, forestry and fisheries.

Geostationary satellites, like METEOSAT and GOES, can be used to monitor severe weather conditions like active frontal systems or tropical depressions and allow real time information about potential storm damage. They collect operational meteorological, hydrological and oceanographic data. Using data from these satellites, several products, such as net radiation, actual evapotranspiration and rainfall estimates are regularly available. They can provide continuous monitoring of the Earth's surface and atmosphere, contributing to an early detection of extreme events.

Advanced Very High Resolution Radiometer (AVHRR) on NOAA polar-orbiting satellites, with a spatial resolution of 1 km, is very useful for estimating vegetation stress, by means of derived indices. They have been used for drought monitoring, detection and impact assessment in agriculture. The Global Vegetation Index (GVI) developed by Kogan (1995) provides information about changes in seasonal land cover by comparing the data with normal dynamics of vegetation. A new numerical method of drought detection and impact assessment from NOAA operational environmental satellites was developed (Kogan 2000). With this new method, drought can be detected several weeks earlier than before, more accurately and in any part of the world. From high resolution satellite data that can track the course of hurricanes, fires, droughts or floods, huge amounts of multidisciplinary observational data can now be fed into computer models and prediction algorithms, or directly into decision processes (Sarewitz and Pielke 2003).

In the MARS project, for the activity on monitoring vegetation conditions and yield indicators NOAA-AVHRR data has been the main source of information. For that purpose, a dedicated software system was implemented (SPACE) to process the data and to derive composite products which were the input for the Advanced Agricultural Information System, namely for the Crop Growth Monitoring System (Perdigão 2000).

Improved access to satellite data is necessary. Several countries have already established databases of information collected by meteorological satellites with high resolution. Monnik (2000) refers that South Africa has been developing an entirely calibrated database of daily NOAA-AVHRR images, including NDVI.

The Portuguese Meteorological Institute is routinely calculating the Normalized Difference Vegetation Index (NDVI) based on NOAA-AVHRR data, in order to quantify vegetation stress and monitor droughts (Fig. 4.4).

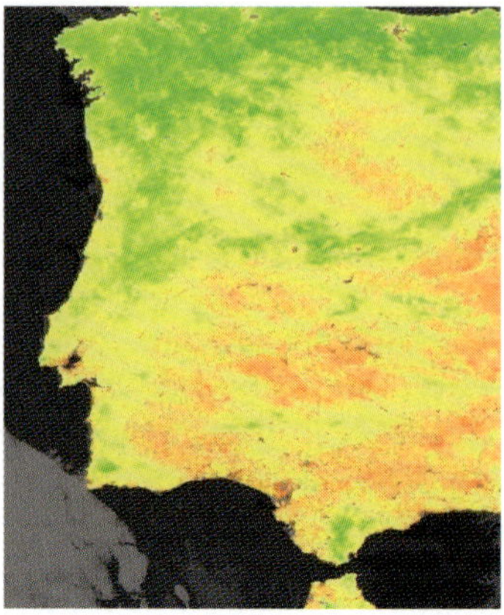

Fig. 4.4. NDVI image for the 18th of June 2000, derived from NOAA-AVHRR sensor, prepared at the Portuguese Meteorological Institute

Satellite Application Facilities (SAFs) are specialized centers within EUMETSAT and are hosted by the European Meteorological Services. The Portuguese Meteorological Institute is responsible for the development of the Land SAF (IM 2003). The main purpose of this SAF is to enhance the benefits of the EUMETSAT new generation of satellite systems, MSG and EPS data related to land,

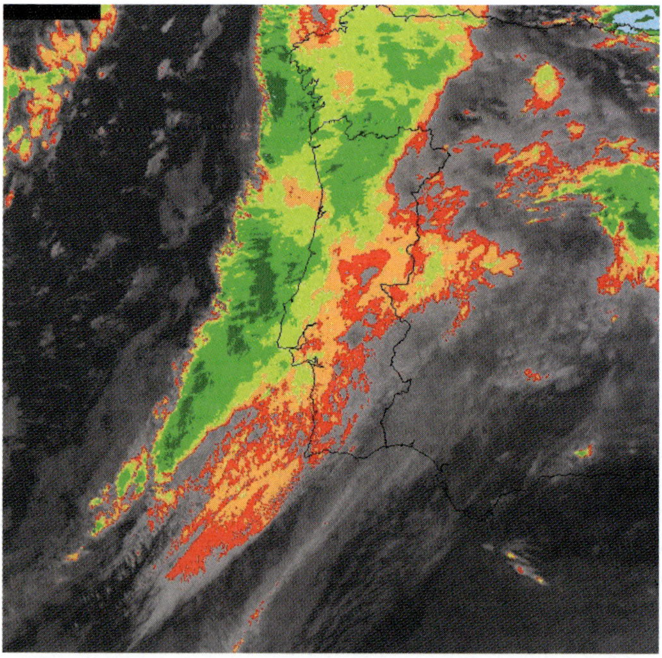

Fig. 4.5. NOAA image showing an active frontal surface crossing Mainland Portugal on 26th December, 2002

land-atmosphere interactions and biophysical applications, by developing the techniques, products and algorithms for more effective use of the satellite data. The Land SAF involves near real-time generation, archiving and distribution of a coherent set of products that may characterize the land surface by surface temperature, scattered radiance and albedo, evapotranspiration, snow/ice cover, soil moisture and vegetation parameters that are specially relevant for vegetation health monitoring and drought management.

Besides meteorological satellites (Fig. 4.5), radar and lightning detectors are remote sensing devices that provide information on storm cells, as mentioned above, contributing significantly to early detection of severe storms. Real-time satellite and radar images are available at the Meteorological Institute web site.

4.5 Geographical Information Systems

In agrometeorology, the processing of data from various sources (meteorology, soil science, topography, etc.) and the need to display them in maps, at regional or national level, led to growing use of geographical information systems. A GIS uses a database management system for the acquision, storage, analysis and display of georeferenced data. It can manipulate information from different sources, like climate and agrometeorogical databases, remote sensing or digital maps, and simulate the behaviour of complex systems. GIS allows datasets to be overlaid over one another, which can help to assess the risk and identify the areas and severity of many extreme events. Besides, this capability enables users to statistically evaluate and quantify relationships, such as the percentage of a corn-producing area that is experiencing severe drought (Shannon and Motha 2002).

The success of any GIS project depends on the data quality, namely accuracy and density of station network, updated time series and ability to manage different types of data and formats. Estimation of values in places where there are no measurement stations is done by interpolation. GISs have several interpolation methods and other statistical applications that combined with other intrinsic capabilities (like data conversion, edition, spatial analysis, modelling and fast corrections or modifications) constitute important tools in climatological and agrometeorological matters (Silva 2002). At the Institute of Meteorology in Portugal, a few interpolation methods are always tested and validated in order to use the model which presents the lowest errors.

Figure 4.6 shows two maps, one for extreme maximum temperatures and one for extreme minimum temperatures for the period 1–14 August 2003, considered of heatwave in Mainland Portugal. Temperature distribution is overlaid with relief, AWSs utilized, districts and main rivers.

The frost risk map and the desertification climatic index are examples of spatial modelling with GIS, produced at the Portuguese Institute of Meteorology. Figure 4.7 represents the overlay of the frost risk and the municipalities in Mainland Portugal and the resulting average risk in each one of them. The frost risk map was elaborated taking in consideration not only the spatial

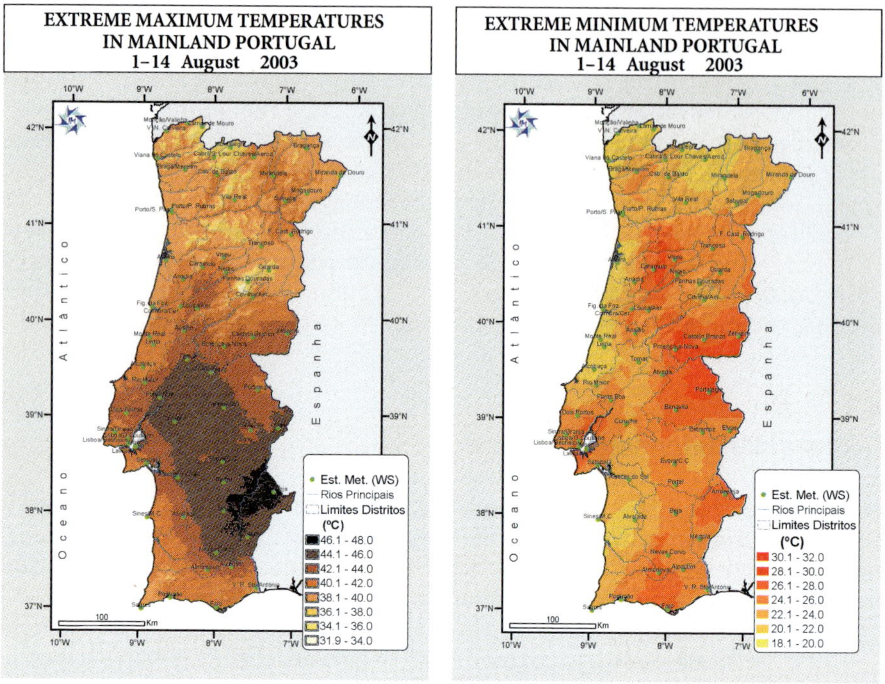

Fig. 4.6. GIS maps with the distribution of extreme maximum and minimum temperatures for the period 1–14 August 2003 in Mainland Portugal

distribution of agroclimatic elements, but also frost physical features that condition frost occurrence (Moita and Silva 2003). To make this map, the following information was necessary:

- Statistical values (1961–1990);
- Digital elevation model GTOPO 30 (USGS);
- Corine Land Cover map.

Remote sensing allows a spatial assessment of crop growth and development and provides important agrometeorological information to GIS. However, in several countries, including European ones, remotely sensed data are not easily accessible yet.

GISs can be extremely useful tools in the analysis of flood-prone areas (Johnson 2003). A GIS in conjunction with digital elevation model (DEM) data can quickly determine slope and aspect of a region and can be used to provide geospatial analysis of multiple spatial layers: elevation, slope, soil characteristics, precipitation, temperature, vegetation and other factors.

A good example of the integration of several information products (soils, crops, meteorology) is the *ISOP* (*Information et Suivi Objectif des Prairies*) project in France, which is operational since January 2000. This project, which

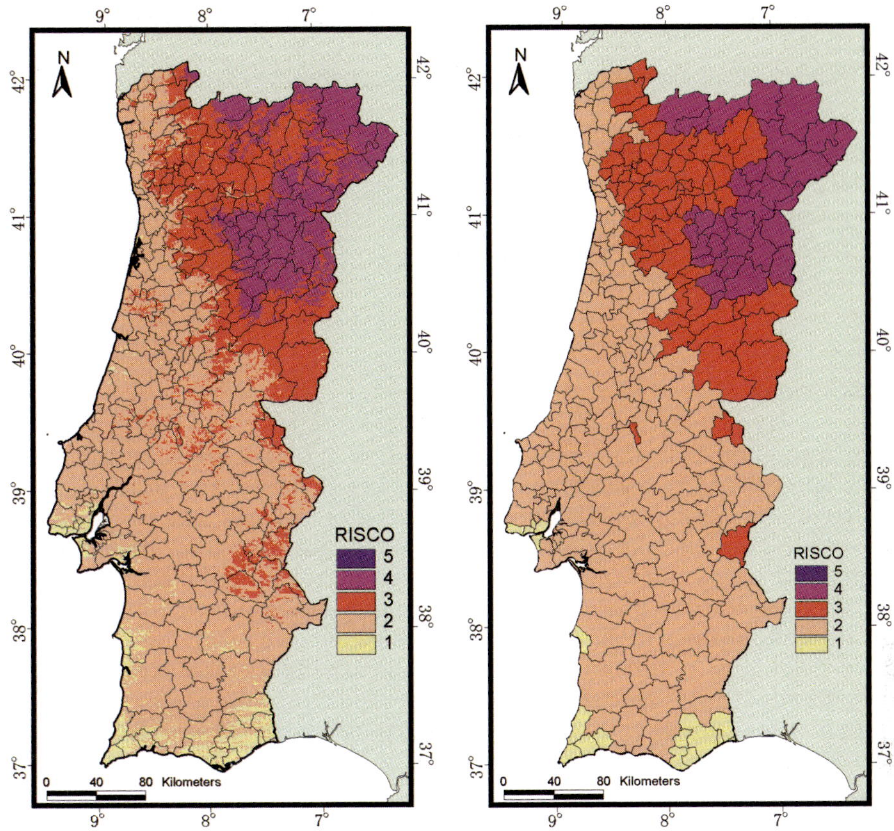

Fig. 4.7. Overlay of the frost risk and the municipalities in Mainland Portugal and the resulting average risk in each one of them (Moita and Silva 2003)

results from a combined effort of *METEO FRANCE*, *INRA* and *Ministère de l'Agriculture*, is an integrated system for real-time assessment of forage production variability over France (Donet et al. 1999). Its objectives are to detect crisis situations, like significantly low productions due to unfavourable soil-climate conditions; moreover, its purpose is also to produce reliable estimates of forage production, in order to objectively estimate real production losses in case of local or global drought (Huard and Pérarnaud 2001). Input data are multiple, including spatialized daily meteorological parameters, percentage of soil types, nitrogen status and amount and frequency of mowing or grazing, estimated from a national survey. The *STICS* crop model, developed by *INRA*, is applied and the results of all GIS operations are stored in a database. This information is synthesized in alert maps and temporal graphs for selected drought-stricken areas. The *ISOP* end-users are several departments of the Ministry of Agriculture and Fishery, namely the bureau for agricultural disasters and the central department for sample survey and statistical studies.

The described examples show the advantages that GIS can provide, namely:

- Fast access and data processing;
- Easy integration of data types and formats;
- Access to error and uncertainty of predictions;
- Spatial queries and analysis;
- Good graphical capabilities;
- Help in fast decision making.

4.6 Conclusions

New technologies in data collection and data management have led to the development of databases where quality controlled data and products from different sources can be displayed in a user-friendly way and electronically accessible. However, the information generated must match (or be adapted to match) the needs and capabilities of the users (Sarewitz and Pielke 2003).

In some countries, phenological, soil, remote-sensed and other types of data are not concentrated in the same database, but scattered in separate ones, quite often belonging to different organizations, which makes their access not very easy. Besides this, these data are not always available to the users.

Improving climate and agrometeorological databases is important to facilitate early detection of extreme events. This initiative includes:

- Developing procedures to identify user needs;
- Taking advantage of the latest technology in database management and standardizing it;
- Creation of databases which include pertinent information besides agrometeorological and climatological;
- Displaying useful products to the end-users, not only for farmers but also government policy makers and other decision makers;
- Dissiminating the products in a user-friendly way and making them readily accessible through the Internet; this will help reach more users;
- Information displayed to the users should include seasonal forecasts in order to help them take the right decisions to reduce or even avoid risks associated with extreme events;
- Strengthening the use of remote sensing and GIS;
- Establishing databases of agricultural disasters, resulting from extreme events, with thorough information about the agrometeorological conditions, including remotely sensed data, and also their economic and social impacts.

To achieve these purposes, the following actions should be undertaken:

- To reinforce training in the new technological applications, such as AWS, DBMS, telecommunication systems, remote sensing and GIS.

Murnane RJ, Crowe M, Eustis A, Howard S, Koepsell J, Leffler R, Livezey R (2002) The weather risk management industry's climate forecast and data needs. A workshop report. American Meteorological Society, August 2002: 1193–1198.

Nunes LF, Neto J, Carvalho R, Carvalho F, Prior V, Henriques D (2003) Automatic and interactive methods for filtering automatic weather stations database. 3th International Conference on Automated Weather Stations, February 2003, Torremolinos, Spain.

Pérarnaud V (2001) Review and summary of operational agrometeorological techniques. In: Report of the CAgM Working Group on Weather and Climate Related to Agricultural Production. CAgM Report No.87, WMO/TD-No.1079, World Meteorological Organization, Geneva, pp 5–30.

Perdigão V (2000) Information Derived from Earth Observation and Ground Surveys. In: Perdigão V (ed) MARS Project, Ten Years of Demand-Driven Technical Support. Proceedings of a Conference, 22–23 April 1999, Brussels. SAI/ARIS, Joint Research Centre, European Commission, pp 27–31.

Pinto PA, Brandão AP (2002) Agriculture. In: Santos FD, Forbes K, Moita R (eds) Climate Change in Portugal. Scenarios, Impacts and Adaptation Measures. SIAM Project. Lisbon, pp 223–239.

Pires VC (2003) Intensity and frequency of extreme meteorological phenomena associated to precipitation. Developing of a monitoring system for drought in Mainland Portugal (in Portuguese). MSc Thesis. University of Lisbon.

Plummer N, Aguilar E, Seng TL, Wong K (2003) Climate data and data management. World Meteorological Organization Bulletin, Volume 52, No.3: 229–236.

13th Provisional Report – Forest Fires 2003, Direcção Geral de Florestas, Ministério da Agricultura. Available on-line: http://www.dgf.min-agricultura.pt/ (accessed in December 2003).

Santos FD, Miranda P, Pires C, Trigo R, Pinto PA, Guerreiro R, Valente A, Brandão A, Esteves M (2004) Project CLIVAR – CLImate VARibility and Change: Patterns and Impacts at the Regional Scale. 4ª Assembleia Luso Espanhola de Geodesia e Geofísica, 3–7 February, Figueira da Foz, Portugal. IGIDL.

Sarewitz D, Pielke Jr R (2003) Extreme Events: A Research and Policy Framework for Disasters in Context. Forthcoming in International Geology Review.

Shannon H, Motha R (2002) Applicattion of GIS technology for agrometeorological bulletins. In: Sivakumar MVK (ed) Proceedings of the Inter-Regional Workshop, 15–19 October 2001, Bridgetown, Barbados. AGM-5, WMO/TD No. 1108, World Meteorological Organization, Geneva, pp 249–260.

Silva A (2002) Mean air temperature estimation in Mainland Portugal. Discussion Papers within the Working Group of the COST 719 WG2 Meeting (restricted), 20–21 September, Oslo.

Sinha Ray KC (2002) A structure for an expert system on extreme meteorological events. In: Report of the CAgM Working Group on the Impacts of Desertification and Drought and of other Extreme Meteorological Events. CAgM, Thirteenth Session. World Meteorological Organization, Geneva, pp 78–84.

Sivakumar MVK (2000) Issues, Challenges and Solutions from the Workshop Papers in the Context of CLIMAG. In: Sivakumar MVK (ed) Climate Prediction and Agriculture. Proceedings of the START/WMO International Workshop, 27–29 September 1999, Geneva. Washington DC, USA: International START Secretariat, pp 187–194.

Supit I, van der Goot E (eds) (2002) Updated system description of the WOFOST crop growth simulation model as implemented in the Crop Growth Monitoring System (CGMS) applied by the European Commission. http:// www.iwan.-supit.cistron.nl/˜iwan-supit/contents/ (accessed in January 2002).

Szalai S, Santo FE, Pires VC (2003) Drought Investigations. Final Report of Project no.4 in the framework of the climatological projects in the application area of ECSN. MeteoSwiss, Zurich.

Taylor P (2001) Impact of the environmental data on fisheries and aquaculture in New Zealand. Report for the Commission for Agricultural Meteorology, World Meteorological Organization, Geneva.

Weiss A, van Crowder L, Bernardi M (2000) Communicating agrometeorological information to farming communities. Agriculture and Forest Meteorology, Vol. 103/1–2: 185–196.

Wilhite DA, Svoboda MD (2000) Drought Early Warning Systems in the Context of Drought Preparedness and Mitigation. In: Wilhite DA, Sivakumar MVK, Wood DA (eds) Proceedings of an Expert Group Meeting, 5–7 September 2000, Lisbon, Portugal. WMO/TD No.1037, World Meteorological Organization, Geneva, pp 1–16.

CHAPTER 5

Tools for Forecasting or Warning as well as Hazard Assessment to Reduce Impact of Natural Disasters on Agriculture, Forestry and Fisheries

Liliana Nuñez

Abstract Disasters induced by natural disasters carry significant consequences for agriculture, forestry and fishery sectors. In this paper, an attempt was made to give a brief account of tools available for forecasting or warning of natural disasters including floods, droughts, tropical cyclones, forest fires, and volcanoes. Use of such tools is important to decrease the impact on natural disasters on agriculture and forestry. Amongst the most well known predictions schemes is the El Niño – Southern Oscillation (ENSO). This and other prediction tools have been discussed with suitable examples.

5.1 Introduction

Disasters induced by natural events may deeply affect the social structure of the area struck and causes severe consequences for men and environment, agriculture, forestry, rangelands and fisheries. Mature society may fall apart, places of employment may be destroyed, and infrastructure build up during several decades may break down. The poorer a region, the deeper are the soil effects of disasters.

The prevailing agriculture in a region is largely determined by the climate. This includes temperature (winter minimum, summer means and maximum, ranges, variability), precipitation (total amount, temporal distributions, extremes, variability), wind, solar radiation, humidity and other factors.

Traditional, ground-based observation networks should be spatially dense enough to capture the horizontal dimension of the extreme event. In regions where available, the use of radar and satellite-derived data fields can be extremely useful. The remote sensed data fields are keys to timely and accurate forecasts and warnings. Statistical analysis can be performed on both point and spatial data sets. An ideal database management system would store temporally and spatially congruent data from a variety of instrumentation and sources, including standard weather networks, radar and satellite imagery, special agricultural and hydrological networks and others. It also would be ideal to have forecasts and warnings stored.

Inappropriate management of agroecosystems, compounded by severe climatic events such as recurrent droughts in West Africa, have tended to make the drylands increasingly vulnerable and prone to rapid degradation and hence desertification. Even in the high rainfall areas, increased probability of extreme events can cause increased nutrient losses due to excessive runoff and water logging.

As every farmer knows, the weather next year will do something unexpected. The hope that it might be predicted in advance of planting is hardly new. The longevity of the Farmer's Almanac bears witness to that. But until the last two decades of the 20th century there was no sound scientific basis for believing that such predictions might be possible.

Some of the year-to-year variations in climate are the result of random sequences of events, just as a series of coin flips will occasionally produce a long run of heads. A region may experience a dry spell because no storms happen to pass that way for a time. Prediction of such stochastic events is not possible. We now see, however, that many climatic variations are part of large scale, slowly evolving patterns. Skillful prediction may then be possible, particularly if the patterns are forced by observable changes in surface conditions such as the sea surface temperature (SST). The most dramatic, most energetic, and best-defined pattern of interannual variability is the global set of climatic anomalies referred to as ENSO (El Niño and the Southern Oscillation). The progress in ENSO prediction is built on advances in understanding, advances in coupled ocean-atmospheric modeling, and the development of an ocean observing system. Prospects for improving forecasts of ENSO and its global consequences will then be considered in this paper. We emphasize that modern long-range forecasts have developed from a new understanding of climate variability and from a vastly enhanced ability to monitor the present state of the climate system. In this paper attempts have been made to give a brief account of tools available for forecasting or warning some of the natural disasters for reducing their impacts on agriculture.

5.2 Floods

Inundations cause the great damage worldwide. Disasters are caused by floods resulting from heavy precipitation or snow melt, by storm surges created through the interaction of strong wind and tidal waves, and by the occurrences of tsunamis, caused by the massive displacement of water bodies after submarine earthquake. Storm surges have threatened the inhabitants of low-lying coastal flatlands for centuries, and a sea level rise as a result of climate change will increase the risk in these areas.

The number of people affected by floods from 1991 to 2000 is reported to be around 1.5 billion. Disasters caused by floods have been increasing in the recent past, mainly as a consequence of the expansion of settlements and the growth of investment in flood plains. If the risk of flooding is assessed, and the use of flood plain is well managed, then losses can be reduced significantly. Many lives can also be saved provided adequate warnings of floods are given to those under threat. Here national Hydrological and Meteorological Services work together to ensure that accurate and timely forecasts are provided.

It should not be forgotten that floodwater is also a resource. We should, therefore, learn to live with floods and use them for the benefit of fisheries, wetlands and irrigation. In addition, the potential of the flood plains for socio-economic development cannot be neglected. In fact they are one of the areas with the highest development potential.

5.2.1 Forecasting/Warning of Flood for Reduction of Risk Potential

The application of active warning, protection, and safety measures requires advance-warning systems that permit the forecasting of flood disasters well ahead of time. Because the meteorological factors used to determine runoff are highly stochastic processes, the possibilities for forecasting are quite restricted. In hydrology and water resources planning, however, there are well tested methods to stimulate processes in river basins using mathematical models based on these deterministic rainfall-runoff models, forecasts of future runoff can be made, according to the chosen approach ranging from short-term via medium-to-long-term forecasts. Other than in the geo-sciences, the term "forecast" in the hydro-sciences is used for estimates of the temporal and spatial size of an event that is ongoing or anticipated. In forecasting the progress of an extreme flood event that is already happening, the crucial factor is the determination of the future development of precipitation, together with the modeling of the system in question. For this purpose, stochastic precipitation forecasts, as well as radar and satellite-derived fields of estimated precipitation for large-scale observations may be applied. A flood-forecast system has to work operationally (i.e., present information on the state of the river basin will be transferred on-line to the runoff model, thus providing the basis for a continuously updated forecast). For some regions such a runoff-forecast system already exists. As a measure for reducing damage potential, these systems should be combined with warning systems, which are aimed at launching response activities and the realization of already developed scenarios for action in case of disaster.

Forecast systems must be supplemented by control strategies for existing flood retention basins. Regulating discharge from these basins permits maximum advantage to be taken of the available storage capacity, taking into account the present and forecasted flow. The specification of flood plains and the introduction of restriction in flood prone areas (building and use prohibition) may prevent further increase of existing flood potential.

In the design of flood protection structures, standardized design values are useful (extreme values of precipitation and discharge).

5.2.2 Quantification of Risk Potential and Hazard Assessment

Worldwide, floods have the greatest damage potential. Even comparatively minor natural events may induce major changes because of concentration of property in flood-prone regions. Knowledge of damage potential is fundamental in planning protection measures. Quantification is based on damage functions that provide information depending on use, region, or land-on maximum damage to be expected during an extreme event. For damage assessment, land-use catalogues are needed that allows classification of localised areas of the regions under investigation. The necessary instruments for the assessment of damage potential and the quantification of flood discharge constitute methods of differing complexity. These methods range from simple regression or black-box methods to spatially and temporally high-resolution simulations and forecast models. Damage analysis is aimed at the determination of widely standardized damage functions by investigating as many damage scenarios as possible.

Based on data from historical events, land-use-dependent damage function may be deduced. The final determination of flood damage, as a function of the assumed hydrologic design loads, can be carried out for different regions and varying scenarios through use of computer models.

5.2.3 Forecasting System as a Part of Disaster Management

The establishment of an efficient disaster management system presupposes an optimal exchange of information and communication between the institutions involving disaster management. Deficits are evident worldwide in both communication and the allocation of competence.

Forecasting systems, if operationally available, are regarded as a part of disaster management. Disaster management is characterized by the necessities to make decisions when time is pressing and knowledge of consequent effect is still (at some level) insufficient. Therefore, it is advisable to develop specific disaster scenarios beforehand and to practice them in advance. This can be done by simulation, using the same models as that in water resources management. Possible instruments to meet the demands for an interdisciplinary approach to the respective disaster management are knowledge-based systems in disaster management, it is possible to analyse the system interaction for flood-prone regions and to test the mechanism of different strategies of action during exercises and during a disaster. These techniques can give the people involved (disaster staff) the opportunity to rehearse and realistically simulate different disaster scenarios, especially interaction with the people affected (population). Additionally, such systems provide valuable experience in selecting the best probable approach during a disaster.

5.2.4 Impacts of Flood on Agriculture

Some of the numerous direct and indirect effects that flooding and heavy rainfall can have on agriculture:
Non-growing season or fallow period:

- Loss of top soil
- Loss of soil nutrients

- Soil compactation
- Soil erosion
- Deposition of undesirable materials
- Permanent damage to perennial crops, tress, livestock, buildings and machine
- Displacement of persons
- Breakage of levees and other retention structures
- Anaerobic processes
- Permanent cessation of farming in flood plains
- Permanent diversion/realignment of rivers, streams, other bodies of water and settlements

Growing season:

- Water logging of crops
- Lodging of standing crops
- Loss of soil nutrients
- Loss of pasture use
- Soil erosion
- Greater susceptibility to diseases and insects
- Interruption to tillage, planting, crop management, harvesting
- Permanent damage to perennial crops, trees, livestock, buildings and machines
- Soil temperature reduction and/or retardation
- Necessity of installation of expensive drainage systems
- Loss of livestock and/or habitat
- Transportation interruption
- Grain spoilage, in field and off-site.

5.3 Droughts

Although floods, as a rule, produce a sudden onset of disaster with a clearly defined beginning, a shortage of water develops disastrous effects only over longer periods. In the beginning, the shortage does not cause significant harm, but it may develop to more and more perceptible and finally serious catastrophes.

A drought may thus be called a "silent disaster", and, in this case, it is difficult to determine at what point in time shortages take on the proportions of disaster. Extremely long drought periods, possibly intensified by human activities, may cause the desertification of a region, which, in contrast to drought, will produce irreversible damage.

Drought is undoubtedly the most far-reaching of all natural disasters, it shows their effects mainly through famines. They are of increasing importance to social groups and regions vulnerable to this phenomenon. The fight against drought must issue a high priority and research on the interactions between climate, the hydrological regime and drought in the context of climate variability and change and water resources scarcity must be promoted.

Drought is a regional manifestation of a general atmospheric circulation pattern, which favours subsidence over the region. In some regions, an important tool to predict seasonal drought is El Niño Southern Oscillation (ENSO). The link with ENSO can be formalised by calculating precipitation probability distribution conditional on the estate of ENSO (Ropelewski 1995).

5.3.1 The Nature and Prediction of ENSO

ENSO has two extremes, often referred to as warm (or El Niño) events and cold events. For both pragmatic and historical reasons related to the nature of the impacts of the warm events, discussion of the ENSO phenomenon often centers on the characteristics of El Niño events. Over a very long time, the frequency of warm and cold events is similar. On the average there is an El Niño event about every four years, but the cycle is highly irregular. Sometimes there are only two years between events, sometimes almost a decade. There are great variations in amplitude. Though each warm episode has its own peculiarities, all follow a general pattern. At an early stage, anomalously warm surface waters are found in the western equatorial Pacific. Associated with the warmer surface temperatures is an increase in convective activity, and at a certain stage, a persistent slackening of the normally westward flowing trade winds. Following this is a dramatic and expansive warming of the tropical Pacific Ocean from the dateline to the South American coast, and a further disruption of the trade winds. Very heavy rains fall in the normally arid regions of Peru and Ecuador, while droughts are experienced in Indonesia, Australia and southern Africa, and anomalous tropical cyclones occur in regions such as French Polynesia and Hawaii. Further away, there are often disruptions of the Indian monsoon. The seasonal rains of northeast Brazil, and regional climates over much of East Asia, North America, and Africa are usually affected. Cold events, sometimes referred to as La Nina events, occur with similar spacing and have characteristics that are approximately the inverse of warm episodes.

The first model to successfully simulate ENSO, that of Zebiak and Cane (1987), was based explicitly on the Bjerknes-Wyrtki hypothesis.

One of the most significant results of the model simulations was the recurrence of ENSO at irregular intervals as a result of strictly internal processes; that is, without any imposed perturbations. Analysis of the model helped in developing a now widely accepted theory that treats ENSO as an internal mode of oscillation of the coupled atmosphere-ocean system, perpetuated by a continuous imbalance between the tightly coupled surface winds and temperatures on the one hand, and the more sluggish subsurface heat reservoir on the other.

This theory has a number of implications for the prediction of El Niño events. First, since the essential interactions take place in the tropical Pacific, data from that region alone may be sufficient for forecasting. Second, the memory of the coupled system resides in the ocean. Anomalies in the atmosphere are dissipated far too quickly to persist from one El Niño event to the next. The surface layers of the ocean are also too transitory. Hence the memory must be in the subsurface ocean thermal structure. The crucial set of information for

El Niño forecasts is the spatial variation of the depth of the thermocline in the tropical Pacific Ocean.

In the 1980s the Lamont model was the only physically based forecasting system with moderate level of skill, though there were (and are) comparably skillful statistical schemes. More recently a number of other models have been developed for ENSO forecasting, many of them based on the comprehensive physical models of the atmosphere and the ocean known as general circulation models (GCMs). The GCMs are generally agreed to offer the greatest promise for accurate prediction, because they simulate the climate system with the greatest verisimilitude. In addition, unlike models of the tropical Pacific alone, they make it possible to predict the global impacts of ENSO. Their complexity, however, makes them less forgiving of errors and imperfections than the more simplified models. It was thus a considerable achievement to bring them to the same general skill level as the Lamont model and the statistically based schemes.

5.3.2
Drought and its Forecasts Based on ENSO

This section provides results from the applications to Mozambique in Africa and Argentina Pampas region of Latin America.

5.3.2.1
Application to Mozambique

Drought occurs in Mozambique about once every seven years (Rojas and Amade 1997). As mentioned above, drought events tend to be associated with El Niño events. In drought years, there is a clear link between rainfall and agricultural production. Specifically, crop production declines. In non-drought years, the link between rainfall and agricultural production becomes more tenuous. Above average rainfall years are not necessarily associated with above average production years. Since current prediction technology relies heavily on the presence or absence of El Niño events, it seems less capable of forecasting particularly favorable climate for agricultural production in Mozambique. The main utility appears to come from providing an estimate of the probability of drought. Hence, we focus on drought in this analysis. Table 5.1 shows the simulations presented. Experiment one stimulates an unanticipated drought. In this experiment, resources in agriculture are fixed at their base (expected) levels. We assume that an agricultural production plan is adopted and cannot be modified based on realized climate outcomes. Experiment two illustrates the benefits of a perfectly foreseen drought. In this scenario, farmers are assumed to receive a perfect forecast and react to it. It is important to note that, in this experiment, the forecast is assumed to have no impact on the efficiency of the marketing sector.

In Mozambique, agricultural production is essentially the process of converting labor into agricultural production. The vast majority of production is

Table 5.1. Simulations

Base	Base 1995 Data
Exp. 1	Drought with capital and agricultural labor fixed (no forecast)
Exp. 2	Drought with non-agricultural capital fixed (perfect foresight)

smallholder based using rudimentary technology. Input use is essentially confined to seed, capital use confined to rudimentary tools, and land is generally abundant. In this environment, drought can be adequately simulated by shocking technology. For practical purposes, this implies that the same amount of labor applied to a particular activity yields reduced production in a drought year. The exact declines in technology (production if input use, including labor and capital, does not change) are listed in Table 5.2.

The figures in Table 5.2 reflect existing time series data on production as analyzed by Bacou (1999), the change in technology parameters estimated by Arndt et al. (1999) for agricultural in 1992 from their backcasting work, interviews with experts on Mozambican agriculture, and the judgement of the authors. From a value of forecast perspective, the crucial element is not so much the average productivity decline in agriculture as the dispersion across activities. The figure in Table 5.2 represents a weighted average productivity decline of approximately 20%. If we simply applied a shock of 0.8 to all of the activities shown in the table, we would have roughly the same average productivity decline but no possibility to reallocate resources from drought intolerant to more drought tolerant activities. With the differential figures in Table 5.2, farmers faced with a drought forecast have clear incentives to shift resources from, for example, production of grains to production of cassava (price changes will also influence cropping choice as discussed below).

Table 5.2. Declines in technology used to simulate drought

Grain	0.67
Cassava	0.85
Raw cashew	0.85
Raw cotton	0.85
Other exports	0.67
Basic food exports	0.75
Livestock	0.85
Forestry	1.00

Table 5.3 illustrates the macroeconomic effects of the shock. As expected the drought results in a decline in GDP and absorption. The consumer price index is held fixed at a constant level in the simulations (it is numeraire); consequently, this absorption figure is an appropriate welfare indicator. Recall that the agricultural share in GDP is about 25%. The decline in agricultural

Table 5.3. GDP and welfare

	100 bn Mt Base Run	Percent deviation from base values	
		Exp. 1	Exp. 2
Real GDP	172.1	−5.7	−6.7
Absorption	223.1	−5.6	−6.3

productivity is 20%. This implies a GDP decline is of 5%. The actual GDP decline is only somewhat greater at 5.7%. This is consistent with the work of Benson and Clay (1998) mentioned earlier. Mozambique is a poor country with relatively weak links between primary agriculture and non-agricultural sectors. As a result, the drought-induced decline in agriculture is relatively isolated and does not spill over substantially into other sectors. The overall GDP decline is only slightly greater than the shrinkage of primary agriculture in isolation.

The ability to reallocate crops has a very minor effect on GDP but a discernible impact on absorption (welfare). Aggregate welfare is 0.4% higher in the presence of the forecast. Generally, the complete accounting structures imposed on CGE models limit the welfare gains (or losses) generated by this class of model. For example, in an often cited analysis of the impact of full implementation of the trade policy reforms agreed to under the Uruguay round of the GATT, Harrison et al. (1997) estimate the welfare gains to be on the order of 0.5% or roughly the gains achieved here.

5.3.2.2
Application to Argentinian Pampus Region

Studies conducted in the Argentinian Pampas region indicate that during El Niño years, precipitation tends to be higher than normal between November and January, but in La Nina years rainfall tends to be lower than normal between October and December (Tanco and Berri 1996; Moschini et al. 1996).

Although all these studies indicate the existence of significant spatial variability in southeastern South America, they also reveal a general pattern with positive rainfall anomalies during October-March in El Niño years, and negative rainfall anomalies during the austral late winter and spring months in La Nina years.

The first detailed studies conducted in south-eastern South America reveal the existence of a near symmetry between impacts of El Niño and La Nina on precipitation as well as on crop productivity. Positive rainfall anomalies prevail in El Niño years, and negative rainfall anomalies prevail in La Nina years, during the austral spring and/or summer months. Some research results also suggest that the impact of La Nina are stronger and/or less variable in both, rainfall and crop yields than the impacts of El Niño. Finally, the work conducted by Diaz et al. (1998) suggests the needs to consider the connection between SST anomalies in the Atlantic Ocean and precipitation anomalies in southeastern South America.

5.3.2.3
Observed Impacts of ENSO Phases on Crop Productivity

Often the annual variability observed in crop and pasture productivity is an excellent indicator of the impacts of large scale phenomena (e.g. ENSO) on regional climatic conditions. Crops and pastures are 'integrators' of the general climatic conditions (rainfall, temperatures) during their growing seasons. In addition, most crops present growth stages in which they are specially sensitive to environmental conditions (e.g. flowering, grain filling, etc.). When ENSO phases affect the climatic conditions during those critical crop growth stages, the impacts are very evident. For example, Myneni et al. (1996) found strong negative anomalies in the NDVIs of the rangelands of Uruguay during the 1988/89 La Nina, and much weaker positive NDVI anomalies in that same region during the 1982/83 El Niño.

Baethgen (1998) found stronger negative maize yield anomalies in Uruguay during La Nina years than positive maize yield anomalies during El Niño years. Similarly, Podesta et al. (1998) reported that in Argentina the depression of maize and sorghum yields during La Nina years were larger and less variable than the yield increases during El Niño years. The typical correlation found between maize and sorghum yields and rainfall during the flowering months (December-February) reinforce the findings of positive rainfall anomalies during El Niño years and (possibly stronger) negative anomalies during La Nina years.

More recent studies conducted in the Argentina Pampas revealed that in general, maize and soybean yields tend to be higher (lower) than normal during El Niño (La Nina) years. The same trend was detected in wheat but only in the southern portion of the region, while sunflower response was opposite and weaker (Magrin et al. 1999). Maize yield reductions in La Nina years were associated with negative precipitation anomalies during November, while yield increments in El Niño were positively related to December increments in precipitation (Travasso et al. 1999).

5.3.3
Risk Management

Activities which provide information aimed at helping decision makers apply climate forecasts to minimize negative consequences of climate variability and to take advantage of favorable conditions. In order to benefit from climate forecasts, decision makers must possess flexibility to change their management practices in response to the improved information (Sonka et al. 1986). For that reason, activities in the region aim to specify alternative management options that are feasible and reasonable from the perspective of stakeholders. Research is focused on crop production decisions that are sensitive to future weather conditions. Simulation models are being used to identify optimal management with and without climate forecasts. The selected optimal options allow estimating the efficacy and value of climate forecasts, and identifying options to present to decision makers. Following this approach, a number of activities were carried out to evaluate the acceptance and value of ENSO-based climate forecasts for agricultural decision making. In a preliminary study, Hansen et al. (1996) found that predicted benefits of tailoring soybean planting dates to forecasted ENSO phases ranged from US$ 2.4–32.4, according to location and soybean prices.

Messina (1999) found "optimal crop combinations" for each ENSO phase in different Pampas' locations which depend on location, risk aversion and initial wealth. On the other hand, Margin et al. (1999) concluded that the best management option in La Nina years for both maize and soybean was to delay the sowing date. Inversely, early sowings were more likely to optimize yields and incomes in El Niño years. These researchers also found that nitrogen rates in maize should be higher during El Niño events to maximize expected maize yield and profit. However, due to important intra-phase climate variability (e.g. rainfall during flowering is not always lower in La Nina years), making crop management recommendations based on mean values is still leading to negative results (Magrin et al. 1999).

5.4 Tropical Cyclones

Damage induced by cyclones is a phenomenon primarily of tropical and subtropical coastal regions. These cyclones have relatively large-scale effects and affect about a third of the world population. Damage is caused not only by immediate effects of the wind, but to a high degree is induced by the effects of storm surges and ocean waves directly generated by tropical cyclones.

The tropical cyclone causes irreparable damage to the agriculture, ranches and forest.

The principal causes of the cyclones are:

- strong winds,
- torrential rain and the associated flooding and
- high storm tides (the combined effect of storm surge and astronomical tides) leading to coastal saline inundation.

The loss to an agriculture system is due to direct and indirect effect. The direct effects are:

- destruction of vegetation, crops, orchards and livestock,
- damage to irrigation facilities such as canals, wells and tanks and
- long-term loss of soil fertility from saline deposits over land flooded by the sea.
- disruption of the transportation system
- loss of a portion of the future harvests due to the destruction of standing crops.

- loss of human life
- damage to property
- damage to fishing
- damage of off-shore and on-shore installations
- loss in productivity due to disruption of the work force and to other activities

In addition of the effects of the wind is the damage caused by airborne sea salt, which occurs within a few hundred metres of the coast. Winds, which blow from the coast seas, spray salt on coastal areas, making it impossible to grow crops sensitive to excessive salt.

Fields inundated by the storm surge suffer a loss of fertility due to salt deposition, even after the seawater has receded. The affected land takes few years to regain its original fertility. A rise of sea level would adversely affect a coastal belt, comprising the coastal ecosystem, increasing coastal salinity and reducing crop productivity. Small-scale fisheries are also hit by cyclones. In agriculture, there can be large losses in primary production on account of delay in the recovery of the land.

5.4.1 Cyclone Prediction and Warning

Tropical cyclone (TC) prediction constitutes three aspects viz. cyclone intensity in respect of wind, land-fall in respect of time and space, and associated storm surge in respect of maximum heights of tidal wave. The prediction of cyclone is usually done in 4 time scales – long range (more than 10 days or from the day of formation of low pressure area in the sea), intermediate (3–10 days), short range (1–3 days) and very short range (a few hours-'Now casting'). It may be realised that essentialities of cyclone forecast are timeliness, accuracy of landfall and overall reliability. Essentially in this end is credibility of observation of current atmospheric conditions that are accurate, quality controlled and of high resolution in both space and time, and good and reliable models of cyclones on the scales observed.

Forecasting the movement of Tropical Cyclones is universally considered to be the most important task. A large number of statistical/empirical methods have been developed in the past and, despite the numerical models which have been commissioned in most of TC forecast offices, they are still being used and improved upon. The first step in making a TC task forecast involves fixing its position using all available data resources. Highly accurate positioning is important for accurate short-term forecast.

Some major and near-catastrophic forecast errors have been made by analysis using incorrect positions. Some of the TCs have the presence of small-scale oscillations in their track, which are however generally smoothed to some extent, and biased by the available ad hoc observations. The best positions of TCs are furnished by research aircrafts, which are very expensive and, therefore, not cost effective. There is an imperative need to keep an operational track in the operational scenario, because TC fixes are derived from variety of observing platforms. In the absence of any accurate fix of a TC, the forecast position provides the first guess for this purpose. This is considered good, especially when using satellite data with poorly-defined systems. In the event of better quality observations becoming available, one needs to redraw the track for at least the last 24 hours, using a degree of smoothing appropriate to the situation. Holland and Elsberry (1993) provide an indication of the impact of fitting curves of various orders to cyclone tracks. Such curves can also provide a short-term forecast method.

Dynamical methods have proved successful because of the increase in the power of computing, the development in numerical prediction techniques, the improved understanding of physical processes, improvements in observing systems, and objective analysis and initialization. Considerable success has been achieved using dynamical models (Dastoor and Krishnamurthy 1991).

Recently, a quasi-Lagrangian Model (QLM) for cyclone track prediction in the Indian seas has been implemented in the India Meteorological Department (IMD). The model is an adapted version of the hurricane prediction model of the National Center for Environmental Predictions (NECP – the erstwhile National Meteorological Center) Washington. It is a multi-level primitive equation, a fine-mesh model cast in the sigma coordinate system ($p = p/ps$; pressure divided by surface pressure). The model has a limited area domain using a Cartesian grid. The horizontal grid spacing is 40 km and the integration domain consists of 111×111 grid points in a $4{,}400 \times 4{,}400\,\text{km}^2$ domain that is centered on the initial position of the cyclone. The QLM uses 16 layers in the vertical. The model incorporates physical processes, which include surface frictional effects, sea-air exchange of sensible and latent heat, convective release of latent heat, divergence damping, horizontal diffusion, and isobaric condensation of water vapor. Radiation and turbulent processes, which only have a marginal impact on the development, are currently excluded to minimize computational time. The numerical integration of the model is carried out by using the socalled quasi-Lagrangian method.

Track forecasts based on QLM model were carried out with respect to cyclonic storms that formed during the three-year period from 1998 to 2000 in the Bay of Bengal and the Arabian Sea (Kalsi 2001). A quantitative assessment of the performance of the forecast model can be made by the computation of track prediction errors. Two types of prediction errors have been attempted. Direct position errors (DPE) have been calculated by taking the geographical distance between the predicted position in each case of forecast and the corresponding observed position. The second type of error is the angular deviation between the observed and predicted track vectors, starting from a given initial position of the storm. While the former gives a measure of the absolute error of prediction, the latter provides an indication of the closeness of the predicted direction of movement and the observed direction.

5.4.1.1
Winds, Heavy Rains and Storm Surges Caused by TC

Among all natural disasters, TC-caused events are always the worst both in terms of the death toll and the economic losses. These disasters are mainly caused by strong winds, torrential rains and storm surges. A mature TC may have a wind speed of 60–70 m/s. This is main driving force for the generation of storm surges, which incidentally depend on the offshore bathemetry. Although a TC usually reduces in intensity after the landfall, an enhanced supply of water vapor and interaction with moderately cold air may re-intensify the TC and release heavy rainfall. A TC does not have to be intense to produce heavy rainfall; it all depends on the speed and size of the TC.

Pattern recognition techniques are used not only for the movement of TCs, but also for forecasting their other attributes such as strong winds and heavy rains. Satellite and radar images too go a long way in meeting these requirements. Eye-walls and spiral bands are areas of intense weather and are the most conspicuous features in the satellite and radar images. Attempts are being made to simulate these features and also the severe weather associated with them, using the high-resolution, limited-area model.

The WMO GTS provides the backbone of the telecommunications for the relay of warnings, forecasts, observational data and related information within the meteorological community and to some major external users and, in some instances, supports early warning for non-weather hazards. However, some of the other communication systems cited above are more appropriate for distributing warnings to the local population and external agencies, particularly when speed is essential. In general, the effective dissemination of warnings to the public and lower level administrators requires communication systems which have a very broad public reach, such as radio and television stations and community warning facilities. The dissemination of warnings through these external agents and facilities is, however, carefully coordinated to ensure timeliness and accuracy and, as noted elsewhere, experience confirms that there must be a single official issuing authority for warnings to minimize confusion.

Although tropical cyclone motion forecasts have improved over the years, the intensity forecasts have still a lot to be achieved. Using the frequently (hourly to half-hourly) available satellite imagery, a lot of experience has been gathered in the analysis of tropical storm intensity. Since the launch of modern satellites, the intensity of cyclones has been well-captured. This has enabled cyclone experts to issue realistic storm surge forecasts. However, our knowledge of structural changes in tropical storms on account of land interactions is still in its infancy. The development of localized zones of strong winds, tornadoes and very heavy rains in some pockets continues to be elusive. There has been a vast improvement in forecasting techniques and cyclone warning services in recent years. This helped in considerable reduction of the loss of human lives and damage to agricultural products caused by cyclones.

5.4.1.2
Cyclone Warning System

One of the short-term mitigation measures against tropical cyclone disaster is an efficient cyclone warning system. The requirements for an efficient cyclone warning system (Das 2003) are:

1. Advanced accurate and detailed forecasts of dangerous conditions;
2. A rapid and dependable distribution system for the forecasts, advisories and warnings to all interested parties; and
3. Prompt and effective utilization of warnings by the government and the public.

To have fairly accurate forecast, it is necessary to have (i) maximum high-quality data; (ii) Forecasters with sufficient ability, training, experience, and time for data preparation; and (iii) foolproof techniques for preparing accurate forecasts of the storm's movement, changes in intensity and storm tides. The US Weather Research Program

for Hurricane Landfall (OFCM 1996, Elsberry and Marks 1999) promises improved forecasts of track, intensity, surface wind and rainfall, as well as research on decision-making and the technology transfer necessary to convert advances in science and technology into products useful to society. The communication system for the distribution of advisories and warnings should be one that can dependably deliver the advisory information to all concerned in the shortest possible time, even in cyclonic conditions of strong wind, heavy rain, floods, etc.

An essential element of a warning service is that there should be certainty that the warnings will reach the intended recipients promptly. The supporting communications system, including back-up facilities, should therefore be planned and implemented in full detail. The meteorological warning besides giving precise information about the tropical cyclone itself and the winds and rainfall to be expected, might also serve as a preliminary warning of a flood and storm surge. Such preliminary warnings should be confirmed or amended in due course by the forecast centre, in conjunction with hydrographers in the case of a storm surge warning.

5.5 Forest Fires

Forest fire is one of the main environmental hazards to face at present in many places, and causes injuries and economical losses, with significant impact on ecosystem degradation, soil erosion and flood occurrence. In some countries the problem occurs during the dry season and mainly affects pine forests, bushlands and sometimes cultivated fields.

5.5.1 Fire Danger Ratings

Fire danger rating systems are used as fire management guides such as staffing for fire control, scheduling prescribed fire, and fire prevention. They are also used to track the fire season and allow the comparison of one season to another. Fire danger rating do not predict how a specific fire will behave but measure the potential of fire occurrence. Several fire danger rating systems are being discussed here including application from Russia, Australia, The United States, and Canada.

5.5.1.1
Nesterov Index (Russia)

The most widely used fire danger rating system in Russia is a relatively simple ignition index called Nesterov Index (NI). It provides the general index of ignition potential (Stocks et al. 1996). The daily weather requirements for this index are:

- Dry bulb Temperature
- Dew point Temperature (calculated from relative humidity and temperature)
- Precipitation

The index is initialized at zero and is determined by taking the difference between the daily air (dry bulb) and dew point temperatures, multiplying this difference by the air temperature and then cumulatively summing up the values over the number of days since 3mm of precipitation has fallen. Once 3 mm or more of daily precipitation has fallen, the index returns to zero (Buchholz and Weidemann 2000).

5.5.1.2
Keetch-Byram Drought Index (U.S.)

Keetch and Byram (1968) developed a drought index specially to assess fire potential. It is a continuous index representing the net effect of evapotranspiration and precipitation in cumulative moisture deficiency in deep duff and upper soil layers, as well as the flammability of organic matter. Duff is the humus layer on the forest floor consisting of decomposing litter (needles, leaves and other dead vegetation) and mineral soil (Pyne et al. 1996). The Keeth-Byram Drought Index (KBDI) estimates the amount of precipitation necessary to return the soil to full field capacity. It is scaled from 0 to 800 units to represent 0 to 8 inches of soil moisture. At zero, the KBDI assumes saturation at 8 inches of water and therefore no moisture deficiency. At 800, the KBDI indicates the maximum drought possible.

The KBDI was incorporated into the United States National Fire-Danger Rating System (NFDRS) in 1988 to modify the amount of dead fuel available for consumption (Burgan 1988). A modified KBDI has been used for a Fire Danger Rating System in East Kalimantan on the island of Boreno in Indonesia for five years on an operational basis (Bucholz and Weidemann 2000) and for fire management in Sabah, Malaysia (Solibun and Lagan 1998). The KBDI is also used in the calculations of the Australian McArthur fire danger meters for grasslands and forests.

5.5.1.3
McArthur Fire Danger Meters (Australia)

In Australia, McArthur (1967) developed a widely used fire danger and behavior index. It is based on over 800 experimental fires in a wide variety of fuel types, including eucalypt. The index is calculated by using fire danger meters for forest and grasslands. CSIRO based the forest fire danger index (FFDI) equations from Noble et al. (1980).

The FFDI is based on a scale from 0–100, where a rating of 1 represents a fire that will not burn, or that will burn so slowly that control is very easy, and 100 represents a fire that will burn so fast and so hot that control is virtually impossible. The meter also can be used to determine the rate of forward spread of fire on level and sloping ground, flame height, and the distance of spotting from flame front. The advantage of these meters is that they can be used by fire managers in the field using real-time or hourly weather data.

5.5.1.4
Canadian Forest Fire Danger Rating System (CFFDRS)

This rating system has been under development since 1968 by the Canadian forest Service and consists of two modules: The Canadian Forest fire Weather Index (FWI) System and the Canadian forest fire Behavior Prediction System (FBP) System (Stocks et al. 1989). The FWI System has been used in Canada since 1970 and consists of six detailed moisture and fire behavior codes that account for the effects of fuel moisture and wind on fire behavior in a standardized fuel type (mature pine stand). The following description of the CFFDRS was based from Stocks et al. (1989), Pyne et al. (1996), and Van Wagner (1990).

The FWI system components need the following daily (noon local standard time) weather parameter inputs:

- Dry-bulb Temperature
- Relative humidity (%)
- 10-meter Wind Speed
- 24 hour cumulative precipitation

5.5.2
Fire Behavior Prediction

According to Pyne et al. (1996), fire behavior can be defined as the dynamics of a specific fire event. Three elements of the fire environment triangle influence fire behavior: weather, fuel, and topography. Therefore these elements need to be incorporated into any fire behavior prediction system or model. Fire behavior prediction can be considered a combination of art and science that attempts to estimate the spread and intensity of a specific fire; be it for an ongoing fire or a what-if scenario for planning purposes (Pyne et al. 1996). The key of fire behavior prediction is linking the analytical calculations from the models to judgements bade o experience (i.e. historical analogs). These applications have utility for use in prescribed burning planning and managing unplanning wildfires. The following fire behavior prediction systems can be mentioned here:

5.5.2.1
Behave (U.S.)

Behave is a widely used system for fire behavior prediction in the United States. It consists of models for fire spread, moisture, intensity, and fire size (Pyne et al. 1996). It is ideally suited for real-time predictions of wildfire behavior or unplanned ignition prescribed fires (Andrews 1986).

5.5.2.2
Canadian Forest Fire Behavior Prediction System (FBP)

The Canadian Forest fire Behavior Prediction System (FBP) is the second major subsystem of the CFFDRS. The FBP is based on field data of readily measured variables on over 400 experimental fires and well-documented prescribed and wildland fires (Stocks et al. 1989).

5.5.2.3
Sirofire (Australia)

In Australia, the CSIRO's Bushfire Behavior and Management Group developed a Bushfire Spread Simulator called Sirofire (CSIRO 1997). Sirofire was constructed around the McArthur fire danger rate system meters and the new Grassland Fire Spread Meter. Sirofire, unlike the meters, estimates the rate of spread for the entire perimeter of interest.

In complex terrain areas where meteorological parameters are extremely variable, meteorological conditions derived from neighbouring stations are usually not representative of fire conditions. Then a technique based on NDVI index derived from MOAA/AVHRR satellite data would be applied. There exist some GIS application (Geographic Information System) application for forest fire prediction, that integrates input data on spatial fuel properties, topographic information and temporal weather data to form a computer-animated interactive scenario for fire prediction.

5.6
Volcano

The location of potential volcanic eruptions is usually rather well known, and long-term regional planning near high risk volcanoes may avoid great disasters. Volcanic eruptions often have far-reaching effects on the environment; they incriminate water courses and induce gases specially SO_2, the source of formation of widespread aerosols in the stratosphere that may influence the global climate.

Ashfalls during volcanic eruption may cause considerable damage to agriculture, cattle and industry even at distances up to tens of kilometres from a vent, but eruptions generally do not directly endanger life, although the collapse of roofs and houses under the ash load are not uncommon. The ash can cover the land, then the crops are lost, and the animals can't find nourishment and died.

5.6.1 Prediction of Volcanic Eruptions

The development of methods to predict volcanic eruptions is extremely important to provide for early evacuation of densely populated regions. Hazard and risk potential of volcanoes can be localised reasonably well, unlike some other types of natural disasters (earthquakes, storms). Reliable predictions, to a minimum degree, however, are only possible for volcanoes that are well studied and sufficiently instrumented. A prediction based on the statistics of previous eruptions is too vague for specific and short-term prediction of an eruption. A forecast is a general announcement that a volcano will probably erupt in the near future (e.g., by qualitative signs of unrest). A prediction is a relatively precise statement that describes the part of a volcano that is likely to erupt, the time of the eruption, and the presumable type of eruption. Such predictions must be made public with utmost caution in order to gain credibility within the concerned population, thus enabling adoption of preparedness measures. Our increasing understanding of processes inside volcanoes and their measurable effects put predictions more and more on a deterministic basis.

The careful analysis of the history of a volcano is the most important method in assessing the long-term probability of the occurrence of a specific eruption type and its eruptive energy. Volcanic eruptions are often announced years, months, days, or hours before (e.g., by harmonic tremors in the deeper conduit system). This microseismic activity commonly increases prior to an eruption and is characterised by relatively constant amplitudes and wavelengths that are possibly caused by the turbulent motion of the magma ascending to the surface from a magma chamber.

The relatively slow ascent of viscous magma to the upper crust generates a surface expansion that can be measured with modern geodetic instruments. Temperature increases within a volcano as a result of ascending magma can be detected by infrared signals via satellite. Heat conductivity and the magnetic field are changing. An increase of SO_2, emissions often has been observed before eruptions. The characteristic behaviour of a volcano can be identified with the help of intensive monitoring by satellite.

5.7 Conclusion

Climate variability affects all economic sectors, but agricultural and forestry sectors are perhaps two of the most vulnerable and sensitive activities to such climate fluctuations. Many sectors are currently not optimally managed with respect to today's natural climate variability because of the nature of policies, practices and technologies currently in vogue. Decreasing the vulnerability of agriculture and forestry to natural climate variability through a more informed choice of policies, practices and technologies will, in many cases, reduce the long-term vulnerability of these systems to climate change. For example, the introduction of seasonal climate forecasts into management decisions can reduce the vulnerability of the agriculture to floods and droughts caused by the El Niño – Southern Oscillation (ENSO) phenomena.

Some of year to year variation in climate are the result of random sequences of events, just as a series of a coin flips will occasionally produce a long run of heads. A region may experience a dry spell because no storms happen to pass that way for a time. Prediction of such stochastic events is not possible. We now see, however, that many climatic variations are apart of large scale, slowly evolving patterns. Skillful prediction may then be possible, particularly if the pattern are forced by observable changes in surface conditions such as the sea surface temperature (SST). ENSO events are now predicted well in advance with useful (albeit less than perfect) skill. The progress in ENSO prediction is built on advances in understanding, advances in coupled ocean-atmospheric modeling, and the development of an ocean observing system.

Beyond the changes in the physical climate system one would like to predict impacts on agriculture, health, economic system and so on, and then develop response strategies based on this knowledge.

References

Andrews PL (1986) BEHAVE: fire behavior prediction and fuel modeling system: BURN subsystem, Part 1. General Technical Report INT-194. Ogden, UTUS Department of Agriculture, Forest Service, Intermountain Research Station, p 130.

Arndt C, Jensen HT, Robinson S, Tarp F (1999) Marketing margins and agricultural technology in Mozambique. International Food Policy Research institute, trade and Macroeconomics Division Discussion Paper No. 43, July.

Bacou M (1999) Rainfall variability and income distribution in Mozambique – A general equilibrium approach. Unpublished M.Sc. thesis, Purdue university.

Baethgen WE (1998) El Niño and La Nina impacts in south-eastern South America. Review on the causes and consequences of cold events: A La Nina Summit. In: Glantz M (ed) Proc NCAR, Boulder, CD.

Benson C, Clay E (1998) The impact of drought on Sub-Saharan African economies – A preliminary examination. World Bank Technical Paper No. 401.

Buchholz G, Weidemann D (2000) The use of simple fire danger rating systems as a tool for early warning in forestry. FAO/ECE/ILO International Forest Fire News No. 23, pp 33–36.

Burgan RE (1988) 1988 Revisions to the 1978 National Fire-Danger Rating System. Res. Pap. SE-273. Asheville NC: U.S. Department of Agriculture, Forest service, Southeastern Forest Experiment Station, p 39.

Commonwealth Scientific and Industrial Research Organization, CSIRO (1997) SiroFire: The CSIRO Bushfire Spread Simulator.

Das HP (2003) Incidence, prediction, monitoring and mitigation measures of tropical cyclones and storm surges. In: Agrometeorology related to Extreme Events. WMO No. 943, Geneva, Switzerland, World Meteorological Organization.

Dastoor A, Krishnamurthy TN (1991) The landfall and structure of a tropical cyclone. The sensitivity of model predictions to soil moisture parameterizations, Boundary layer Meteorology, vol 55. pp 345–380.

Diaz A, Studzinski CD, Mechoso CR (1998) Relationships between precipitation anomalies in Uruguay and southern Brazil and sea surface temperature in the Pacific and Atlantic oceans. J. Climate, vol 11. pp 251–271.

Elsberry RL, Marks Jr FD (1999) The hurricane landfall workshop summary. Bull Amer Meteor Soc, vol 80. pp 683–685.

Hansen JW, Jones JW, Magrin GO, Meira SG, Guevara ER, Travasso MI, Diaz RA, Marino M, Hordij R, Harwell C, Podesta G (1996) ENSO effects on yields and economic returns of wheat, corn and soybean in Argentina. Proc CONGREMET VII-CLIMET VII, pp 89–90.

Harrison GW, Rutherford TF, Tarr DG (1997) Quantifying the Uruguay Round. Economic Journal, 107:1, vol 405(1). p 430.

Holland GJ, Elsberry RL (1993) Tropical cyclones as natural hazards: a challenge for the IDNDR. In: Lighthill J, Emanuel K, Holland GJ, Zhang Z (eds) Tropical cyclone disasters, pp 17–30.

Kalsi SR (2001) Role of IMD and Scientific Methods of Prediction of Tropical Cyclones and Advance warning Systems, paper presented in the National Seminar on Habitat Safety against Earthquakes and Cyclones, organized by HUDCO, BMPTC and MOUD, Govt. of India, from 2–3 May 2001 at Vigyan Bhawan, New Delhi.

Keetch JJ, Byram G M (1968) A drought index for forest fire control. Res. Paper SE-38. Asheville NC: U.S. Department of Agriculture, Forest service, Intermountain Research Station, p 32.

Magrin GO, Grondona MO, Travasso MI, Boullon DR, Rodriguez GR, Messina CD (1999) ENSO impacts on crop production in the Argentina's Pampas Region, In: Proc. 10th Symposium on Global Change Studies, 10–15 January 1999 Dallas TX, by the AMS, Boston MA, pp 65–66.

McArthur AG (1967) Fire Behavior in eucalypt forest. Common wealth of Australia Forestry and Timber Bulletin Leaflet, vol 107. p 32.

Messina CD (1999) El Niño-Southern Oscillation y la productividad de cultivos en la zona Pampeana: Evaluation de estrategias para mitigar el riesgo climatico. Tesis Magister scientiae, Produccion Vegetal. Escuela para Graduados, Facultad de Agronomia, UBA, p 132.

Moschini RC, Grondona MO, Vila DA (1996) Influencia del ENSO sobre la distribucion de las probabilidades condicionales diarias de precipitation en algunas estaciones de la Region Pampeana. En Impacto de las variaciones climaticas en el desarrolo regional un analisis interdisciplinario-VII Congreso Latinoamericano e Iberico de Meteorologia, Paginas 307–308.

Myneni RB, Los SO, Tucker CJ (1996) Satellite-based identification of linked vegetation index and sea surface temperature anomaly areas from 1982–1990 for Africa, Australia and South America. Geophysical Research Letters, vol 23. pp 729–732.

Noble IR, Bary GA, Gill AM (1980) McArthur's fire-danger meters expressed as equations. Australian Journal of Ecology, vol 5. pp 201–203.

OFCM (1996) National plan for tropical cyclone research and reconnaissance. Office of the Federal Coordinator for Meteorology, FCM-P25, p 159.

Podesta GR, Messina CD, Grondona MO, Magrin GO (1998) Associations between grain crop yields in central-eastern Argentina and El Niño-Southern Oscillation. J. climate.

Pyne SJ, Andrews PL, Laven RD (1996) Introduction to Weildland Fire. John Wiley and Sons, p 769.

Rojas O, Amade J (1997) O impacto de El Niño-Oscilacao Sul (ENOS) na agricultura de Mozambique. Nota Technica, Setembro sistema Nacional de aviso Previo para a Seguranca Alimentar. DINA/MAP, Maputo.

Ropelewski CF (1995) Quantification of ENSO precipitation relationships, J. climate, vol 7. pp 1041–1059.

Solibun H, Lagan P (1998) Fire Management Plan: Deramokot Forest reserve. Deramokot Forest Reserve, Sabah, Malaysia.

Sonka ST, Lamb PJ, Hollinger SE, Mjelde JW (1986) Economic use of weather and climate information: Concepts and an agricultural example. J. Climatol, vol 6. pp 447–457.

Stocks BJ, Lawson BD, Alexander ME, Van Wagner CE, McApline RS, Lynham TJ, Dube DE (1989) The Canadian Forest Fire Danger Rating System: An overview, The Forestry Chronicle, vol 65(6). pp 450–457.

Stocks BJ, Cahoon DR, Levine JS, Cofer III WR, Lynham TJ (1996) Major 1992 forest Fires in Central and Eastern Siberia: Satellite and Fire Danger Measurements. In : Goldammer JG, Furyaev VV (eds) Fire in Ecosystems of Boreal Eurasia, vol 48. of Forestry Sciences. Kluwer Academic Publishers, pp 481–487.

Tanco RY, Berri GJ (1996) Acerca del efecto del fenomeno El Niño sobre la precipitacion en la Pampa humeda argentina, Proc. VII Congreso Argentino de Meteorologia and VII Congreso Latinoamericano e Iberico Meteorologia, 2–6 September 1996, pp 319–320.

Travasso MI, Magrin GO, Grondona MO (1999) Relations between climatic variability related to ENSO and maize production in Argentina, Proc 10th AMS Symposium on Global Change Studies, 10–15 January 1999, Dallas TX, Boston MA, pp 67–68.

Van Wagner CE (1990) Six decades of Forest Science in Canada. The Forestry Chronicle, vol 66(2). pp 133–137.

Zebiak SE, Cane MA (1987) A model El Niño/Southern Oscillation Monthly Weather Review, vol 115. pp 2262–2278.

Agrometeorological Impact Assessment of Natural Disasters and Extreme Events and Agricultural Strategies Adopted in Areas with High Weather Risks

Haripada P. Das

Abstract Agricultural production is highly dependent on weather, climate and water availability and is adversely affected by the weather and climate related disasters. Impacts of extreme events on the agricultural sector can be positive or negative. While it is easier to contemplate negative impacts of extreme events such as droughts, tropical cyclones and floods, there are several positive impacts or benefits of extreme events. In order to assess the impact of weather disaster on agriculture, one must link two fundamental aspects, first, the disaster proper i.e. the destructive power of the event and secondly, the characteristics of the agricultural system which has been hit. Agricultural planning and practices need to be worked out with consideration to the overall water requirements within an individual agroclimatic zone. Sustainable strategies must be developed to alleviate the impact of natural disasters on crop productivity. All such strategies are location, time, crop, crop stage and socio-economic condition specific. Remote sensing satellite information helps minimise damages e.g. the death of cattle, humans etc. and the damage of agricultural production in time of natural calamities by early warning system.

6.1 Introduction

Agriculture, being a vital source of livelihood, was essentially dependent on the natural seasons. Disasters were believed to be an inseparable part of nature and a series of undesirable fluctuations in climate could endanger food security. People looked upon disasters as a part of rural-agrarian life. And they tried to cope with undesirable climate by offering sacrifices to gods, and by inventing various small-scale devices to ensure the security and their lives, lands and livelihoods. They made their rustic disaster policy, with which they were able to manage the risk, damage and stress of disasters. A majority of the people of the developing countries still depends on agriculture for its livelihood. Agriculture includes farming and also animal husbandry, pastoral activities, fishing and harvesting the forest.

In spite of recent technological developments that have helped to increase agricultural production in many countries, growth of plants and animals continue to depend to a large extent on the weather conditions. Each plant has its own climatic requirements for growth and development and any large-scale deviation from it exert negative influence.

Growth of plants is most sensitive to temperature just above a threshold value and near the maximum value, where growth normally stops. Therefore, periods of extreme temperature i.e. low temperatures below the threshold value and high temperatures above the maxima are hazardous to plant development and growth. Extreme temperature conditions during cold spells cause stress and frost; high temperatures lead to heat stress, and both affect agricultural production. Snow and ice storms in late spring or early autumn are very hazardous to many temperate crops.

Similarly, extremes of moisture conditions caused by dry dessicating winds, drought episodes and low moisture conditions as well as very humid atmospheric conditions including wet spells adversely affect agriculture. High soil moisture in situations of water logging and flooding associated with heavy rainfall and tropical storms have adverse effect on plant growth and development since they influence the rate of transpiration, leaf area expansion and ultimately plant productivity. Drastic changes in rainfall variability can have very significant impact, particularly in climatically marginal zones such as arid, semi-arid and sub-humid areas where incidence of widespread drought is frequent. Dry, dessicating and strong winds reduce agricultural production as a result of very high evapotranspiration rates. It also causes mechanical damage to such plants as sugarcane and bananas with weak stems caused by lodging.

6.2 Agrometeorological Impact of Extreme Events

In the past twenty years, earthquakes, volcanic eruptions, landslides, floods, tropical storms, droughts and other natural calamities have killed over three million people, inflicted injury, disease, homelessness and misery on one billion others, and caused billions of dollars of material damage. Currently, annual global economic costs related to disaster events average US$ 880 billion per year. These figures provide a grim picture. By the year 2025, 80 per cent of the world's population will reside in developing countries, and it has been estimated that up to 60 percent of the people will be highly vulnerable to floods, severe storms and earthquakes. Nearly 90 percent of the natural disasters and 95 percent of the total disaster-related deaths worldwide, occur in the developing countries. Since the 1960s, economic losses caused by disasters have increased at least fivefold. These losses are growing largely because of the increasing concentration of population and investments in vulnerable locations coupled with inadequate investment in measures to reduce risk.

More than any human event, a disaster traumatically brings into focus all the basic problems in a society. It reduces all issues to their most fundamental level and strips away the ancillary issues that obscure or confuse the fundamental questions that must be faced. Critical decisions, previously unaddressed, can no longer be ignored, and choices must be made.

The very face of the land may be altered by a disaster. Droughts have led to famines. Famines have led to migrations. When the people ceased to cultivate the areas that were behind, creeping deserts swallowed the arable land and made it untenable. But positive changes have also been recorded. Better cropping patterns have often followed droughts. Reforestation efforts and better use of contours for hillside agriculture have often followed rain-induced landslides. Where contour planting has been introduced, economic benefits have often followed. For the society, disasters often bring changes in the structure of community leadership. New organizations may be born out of necessity to deal with the disasters and remain to continue the work of bringing economic changes to the community.

The CAgM Working Group of WMO on "Agrometeorology related to extreme events", recently developed a survey of countries regarding extreme weather events and their agricultural impacts. The survey identified a number of extreme events including drought, desertification, cold/frost, floods and heavy rainfall, high winds and severe storms, tropical storms, forest and range fires, etc. throughout the globe. Figure 6.1 provides a graphical representation of these results. Impacts of extreme events on the agricultural sector can be positive or negative.

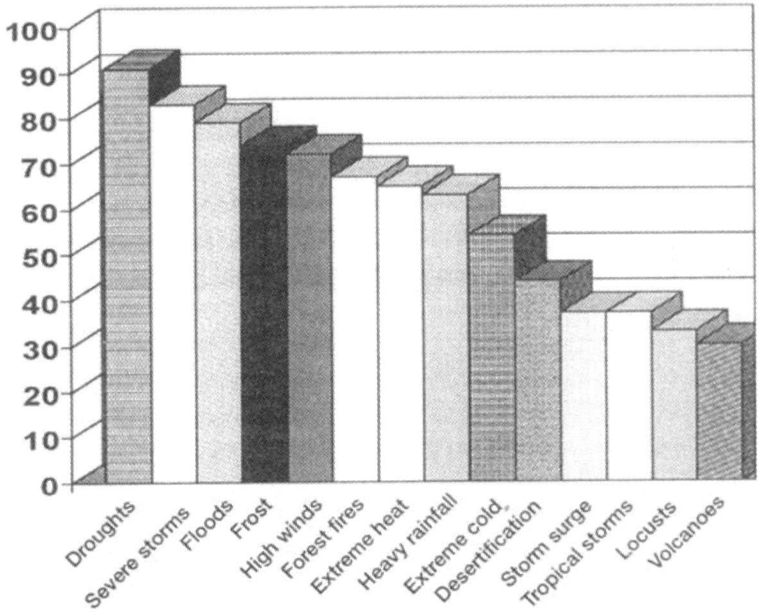

Fig. 6.1. Percentage of countries reporting agricultural impacts from extreme agrometeorological events

6.2.1
Positive Effect on Agriculture of Extreme Events

While it is easier to contemplate negative impacts of extreme events such as droughts, tropical cyclones and floods, there are several positive impacts or benefits of extreme events. Increased rainfall in coastal areas from tropical cyclones, fixing of atmospheric nitrogen by thunderstorms, the germination of many native plant species resulting from bushfires and the maintenance of the fertility of the basin soils due to river flooding are some of the positive impact of extreme meteorological events. There are some advantages of drought at certain times in the development of some crops such as the sugarcane where a brief dry spell is essential during the pre-harvest stage that helps to increase the sucrose content of the cane. Also, there is often a lower incidence of pests and diseases in periods of drought. Transport firms involved in shifting fodder and livestock and tourist organizations do derive benefit from drought.

6.2.2
Negative Effect on Agriculture of Extreme Events

It is, however, the negative or adverse impacts of extreme events referred to as damages or losses which are more pronounced and do affect agricultural operations significantly. Impacts of extreme events can be direct or indirect. Direct impacts arise from direct physical contact of the events with people, their animals and property. Indirect impacts of extreme agrometeorological events are those induced by the events. Indirect impacts often occur away from the scene of the extreme event or after the occurrence of the event. Indirect impacts include evacuation of people in the event of cyclone landfall, disruption to household and leisure activities, stress induced sickness, and apprehension and anxiety of future extreme events like floods or bush fires.

In order to assess the impact of weather disaster on agriculture, one must link two fundamental aspects: the disaster proper (i.e., the destructive power of the event) and, the characteristics of the agricultural system which has been hit. This is illustrated below in the example of Bangladesh (Rahman 1985). Between 1970 and 1984, three major weather disasters struck Bangladesh. In mid-November 1970, one of the worst cyclones in history; in 1978–1979 a severe drought befell the country, followed in 1984, by extended heavy floods lashing from May to September. Each of these disasters resulted in a sharp decrease of the rice production, the staple food of the country (Gommes 1997).

Weather disaster interactions with agriculture are complex and, of course, are likely to involve non-agricultural factors as well. Table 6.1 shows some weather factors that may negatively affect agriculture (Gommes 1997).

As stated earlier, two broad categories of effects on the agricultural sector can be identified: direct and indirect effects. Direct effects to a farmer could be, for example, the loss of his current crop and damage to his irrigation facilities. Indirect effects on agriculture appear progressively, as a result of low income,

Table 6.1. Weather factors that affect agriculture negatively

Weather factor	Negative effects on agriculture of extreme Values (both direct and indirect)
Rainfall	Direct damage to fragile plant organs, like flowers; soil erosion; water logging; drought and floods; land slides; impeded drying of produce; conditions favourable to crop and livestock pest development; negative effect on pollination and pollinators.
Wind	Physical damage to plant organs or whole plants (e.g. defoliation, particularly of shrubs and trees); soil erosion; excessive evaporation. Wind is an aggravating factor in the event of bush or forest fires.
Air moisture	High values create conditions favourable to pest development; low values associated with high evaporation and often one of the most determinant factors in fire outbreaks.
High temperatures	Increased evapotranspiration; induced sterility in certain crops; poor vernalization; survival of pests during winter. High temperatures at night are associated with increased respiration loss. "Heat waves", lengthy spells of abnormally high temperatures are particularly harmful.
Low temperatures	Destruction of cell structure (frost); desiccation; slow growth, particularly during cold waves; cold dews.
High cloudiness	Increased incidence of diseases; poor growth.
Hail	Hail impact is usually rather localized, but the damage to crops-particularly at critical phenological stages- and infrastructure may be significant. Even light hail tends to be followed by pest and disease attacks.
Lightning	Lightning causes damage to buildings and the loss of farm animals. It is also one of the causes of wildfire.
Snow	Heavy snowfall damages woody plants. Unseasonable occurrence particularly affects reproductive organs of plants.
Volcanic eruptions, avalanches and earthquakes	The events listed may disrupt infrastructure and cause the loss of crops and farmland, sometimes permanently. A recent example of carbon dioxide and hydrogen sulphide emissions from a volcanic lake in Cameroon caused significant loss of human life and farm animals.
Air and water pollution	Air pollutants affect life in the immediate surrounding of point sources. Some pollutants, like ozone, are however known to have significant effects on crop yields over wide areas. In combination with fog, some pollutants have a more marked effect on plants and animals. Occurrences of irrigation water pollution have been reported.

decrease in production and other factors related to the weather disaster. The farmer may well have to pay high prices for seeds because of increased demand and disruption of the transportation system. He might also lose a portion of his future harvest because of storm surge-related salinization of soil or the destruction of perennial plantation crops that sometimes take 5 to 10 years to re-establish. Indirect effects are thus difficult to quantify and therefore are often termed "invisible" effects. Conditions conducive to the development of pests and diseases are to be regarded as indirect effects. In fact, the conditions that trigger pest and disease development are rarely directly harmful per se. They are usually a combination of moisture or temperature conditions that do not directly affect crops. Typical examples are desert locust outbreaks or increased disease incidence in sugarcane after hurricanes. Plants weakened by adverse weather are much more susceptible to cryptogamic diseases or pest attacks, such as the explosion of coconut black beetle on wind-damaged coconut trees. A perennial crop needs a number of years before it is restored to its normal situation. For example, it is taking 30 years for certain timber varieties to re establish after hurricane Allen hit Jamaica in August 1980 (FAO 1982). It is frequently from 6 to 12 months for tea and banana plants and 4 to 5 years for coffee and sugarcane. Even for crops that regenerate easily after partial damage, harvesting is usually made difficult by the "abnormal" morphology, thus further reducing yield. In agriculture there can be large losses in primary production on account of delays in the recovery of arable land that has been inundated. For an annual crop, the office of the United Nations Disaster Relief Co-ordinator estimated a global loss (direct and indirect) at 1.5 to 2 times the direct loss (the lost harvest value). For a perennial crop, the global loss was evaluated at 5 to 7 times the value of the lost harvest. The following paragraphs starts with an attempt to list extreme events that can potentially interfere with agricultural productions.

6.2.2.1
Tropical Cyclone and Storm Surge

As regards the impact of cyclones, out of its three accompanying effects of heavy rain, strong wind and storm surge, it is the storm surge of tremendous height and force, which causes most of the damage including loss of human lives and livestock, destruction of property and agricultural and industrial production. Apart from short-term problems, storm surge creates long-term problems because the salt water makes the soil unproductive. The impact of cyclones, with frequent occurrences and substantial magnitude, on the marginal landowners and landless people happens to be tremendous. These affect livelihood assets and seriously challenge the efforts of the country towards self-reliance.

For most developing countries, loss to life, property and the crop represents a bigger part of the damage caused by a cyclone. In monetary terms, the losses incurred by livestock raising, forestry and fisheries mostly remain below those suffered by agriculture. In Madagascar, for instance, according to FAO, a cyclone in 1983–84 caused crop losses of 85% of the total damage to the agricultural sector, whereas the damage to the infrastructure and equipment (drainage, irrigation channels, etc.) barely reach 15%. The traditional small scale fisheries were also hit by the cyclone. The losses incurred by livestock raising, forestry and fisheries mostly remain below those suffered by crops

(OSRO 1984). A rather different impact pattern occurs in small islands like Antigua and Barbuda where fisheries constitute the backbone of the economy. In eastern Caribbean countries affected by Hurricane Hugo in 1989, 47 percent of the losses occurred in fisheries, but crop losses still represented almost 40 percent of the total damage (OSRO 1989).

It is worth mentioning that the losses affecting cash crops, which constitute a major source of export earnings in a number of developing countries is rather high. In Nicaragua, for example, it is reported that direct losses of export crops due to Hurricane "Juana" (Joan) in late 1988 amounted to 21% of the total losses in the agricultural sector (MIDINRA 1988). Coffee and bananas suffered a direct loss of their fruits and mechanical damage to their plants. Nonetheless, food crop losses were estimated to be higher (35%), while the livestock sector was less affected (8%, of which one fifth was poultry).

The effects of strong winds in coastal areas are often stunted and very sculpted trees, providing unmistakable evidence of the direction of the strong winds. In addition to the battering effects of winds, there is the additional damage caused by airborne sea salt that occurs within a few hundred metres of the coast. Winds that blow from coastal seas spray a lot of salt on coastal areas, making it impossible to grow crops sensitive to excessive salt.

Fields inundated by the storm surge suffer a loss of fertility due to salt deposition, even after the sea water has receded. The affected land takes a few years to regain its original fertility. The period of high water can last from 6 hours to several days, if the drainage is poor, and may leave the soil saline and unfit for crops. Saline soils are predominantly observed in the coastal areas of India. A rise in the sea level would adversely affect the 7,000-km coastal belt of India, comprising 20 million ha of coastal ecosystem, increasing coastal salinity and reducing crop productivity.

The effects of cyclonic storms are not entirely bad; benefits focus principally on the precipitation associated with them that may have considerable value to agriculture. The heavy rain associated with cyclonic storms guarantees a longer period of water availability and provides possibilities for off-site extra storage in rivers, lakes and artificial reservoirs (on farms or at the sub-catchment level) giving an improved rural water supply and expanded or more intensive irrigated agriculture and inland fisheries. The extra precipitation due to cyclonic storms on land helps to increase plant growth improving the protection of the land surface and increasing rainfed agricultural production.

6.2.2.2
Drought

Droughts have an immediate effect on the recharge of soil moisture resulting in reductions of streamflow reservoir levels and irrigation potential and even the availability of drinking water from wells.

In the event of a prolonged drought, farmers, either alone or with the entire family, may abandon their land in search of work and food in nearby cities. Fewer and weaker family members remain to till the land, affecting the area under cultivation (McCann 1986). The acreage planted to food crops is also affected by land quality. Due to the uncertainty of rains during the drought, farmers sometimes make several attempts at sowing of seeds leading to a drastic reduction in seed reserves, which in due course are neither sufficient for planting nor for consumption. The farmer is then obliged to borrow offering labour or perhaps a portion of the future harvest as payment for the loan.

Drought not only exposes and accelerates existing land quality problems, it also initiates new ones. The cultivation of lands subject to a high degree of rainfall variability makes them extremely susceptible to wind erosion (and desertification) during prolonged drought episodes, as the bare soil lacks the dense vegetative cover necessary to minimize the effects of aeolian processes. As the fertility of the land and crop yields decline, farmers search for new land to cultivate. Farmers sometimes are forced to cultivate lands considered marginal from the viewpoint of quality, terrain slope and rainfall (Glantz 1994). These newly cultivated lands are high risk areas in the long run for rainfed agriculture.

6.2.2.3
Flooding and Heavy Rainfall

Two of the principal hydrometeorological events that often have deleterious effects on agriculture are heavy rainfall and flooding. It is recognized that in some cases these events can have positive effects. Examples include heavy rainfall from a tropical storm ending a long agricultural drought, or recurring, annual flooding that replenishes topsoil and soil nutrients, as was the historic situation along the lower stretches of both the Nile and Ganges rivers. Soil erosion, disruption to critical agricultural activities, the water-logging of crops, increased moisture leading to increased problems with diseases and insects, soil moisture saturation and runoff, soil temperature reduction, grain and fruit spoilage and transportation interruption are among the more significant agricultural impacts from heavy rainfall.

Direct damage to growing plants from flood is most often caused by depletion of oxygen available to the plant root zones. Flooding creates anaerobic soil conditions that can have significant impacts on vegetation. Root and shoot asphyxia, if prolonged, typically leads to plant death. Chemical reactions in anaerobic soils lead to a reduction in nitrate and the formation of nitrogen gas. The denitrification can be a significant cause of loss of plant vigour and growth following flooding (Foth 1978).

Looking only at temporal characteristics of rain storms and their consequent impact on agriculture, the concept of intensity-duration-frequency (idf) becomes important. If any of these three rain storm characteristics increases, the potential for agricultural impacts increases. Heavy rainfall at a location can thus be defined using the idf convention and is framed by the agroclimatology of a location or region in question. Very intense (extreme) rainfall can result in catastrophic flood damage even though it occurred for a relatively short period of time, and/or at the proper location, at the proper time of year, etc. In general, greater damage to agriculture occurs from rain storms having higher intensities with sufficient duration, compared with low intensity, long duration storms.

6.2.2.4
Sand Storm/Dust Storm

The arid regions are characterised by frequent and strong winds that are due partly to considerable convection during the day. The usually sparse vegetation is not capable of slowing down air movement, so that dust storms and sand storm are frequent concomitant of wind movement. Winds in dry climatic zones also affect growth of the plant mechanically and physiologically. The sand and dust particles carried out by wind damage plant tissues. Emerging seedlings may be completely covered or alternatively, the roots of young plants may be exposed by strong winds. Winds also cause considerable losses by inducing lodging, breaking the stalks and shedding of grains and ultimately decreasing the yield. Many agricultural lands have been lost through wind action by the encroachment of sand, dunes, dust storms and sand storms that carry away the top humus soil, leading to deterioration and degradation of the landscape and desertification.

6.2.2.5
Extreme Cold Weather Including Frost Injury

Long periods of extreme cold weather combined with other meteorological phenomena are detrimental to many sectors of the economy, especially agriculture and can result in the loss of winter crops, fruit crops and vineyards due to frost injury. Frost injury is the most widespread type of cold injury to affect crops. Low soil temperature at the depth of plant roots can cause frost injury. Such reductions in soil temperature occur with strong frosts, in the absence of snow cover and with deep freezing of the soil.

Most frost injury to winter crops takes place in the first half of winter before sufficient snow cover that would afford protection has formed. In the second half of winter frost injury happens in regions with unstable snow cover.

There have been many studies of plant injury caused by extreme cold weather. Under low temperatures, a plant basically dries out and the protoplasm (the living part of cells) dies. This happens because the concentration of cell sap rises, the distance between macromolecules is altered and the process of energy interchange is disrupted. The toxic products of metabolism accumulate within the plant's tissues and it is this that causes the protoplasm to die. The degree of damage depends on the intensity and duration of dangerous frosts as well as the stage of the plant's development. Damage to the part of a plant does not always result in damage or destruction of the whole plant. A determining factor is the degree of frost injury to a tillering node. If it is heavy, the whole plant will perish. The winter crops most frequently destroyed by frost are those grown on upland areas, where snow cover is less and the depth of soil freezing is greater.

The main agrometeorological factor influencing frost damage in winter crops is low soil temperature at the depth of the tillering node. Cooling to the critical temperature of frost injury, even for one day, and especially after a thaw, results in thinning out of crops. Long (three days or more) and intensive cooling causes complete devastation.

6.2.2.6 Forest Fires

Forest fires are considered a potential hazard with physical, biological, ecological and environmental consequences. Forest fires occur frequently in tropical countries particularly in the dry and hot seasons causing serious damage to the forest resources and agricultural production. Since the number of forest fires are increasing every year, continuous monitoring is of great importance, not only to understand present trends but also to devise a model to predict the possibility of fires in future.

6.3 Strategies Adopted in Areas with High Weather Risk

The agrometeorological impacts of natural disasters have a pervasive societal ramification particularly in developing countries. Associated human misery underscores our vulnerability to these natural hazards. A disastrous event does not pose much of the threat and ceases to be a disaster, if suitable and adequate mitigation measures are adopted well in advance. Prevention of the formation of tropical cyclones is not in the realm of possibility, but much of their disastrous potential can be reduced, restricting thereby the loss of human life and loss of property, by adopting appropriate strategies and taking timely precautions on the receipt of weather warnings. Climatological data helps in the advance preparation of long-range policies and programmes for disaster prevention. The following paragraphs list the agricultural strategies to be adopted in the events of some major natural disasters.

6.3.1 Drought Management

In arid, semi-arid, and marginal areas with a probability of drought incidence of say, at least once in ten years, it is important for those responsible for planning of land-use, including agricultural programmes, to seek expert climatological advice regarding rainfall expectations. Drought is the result of the interaction of human pattern of landuse and the rainfall regimes. Thus, there is urgent need for a detailed examination of rainfall records of these regions. In this regards, the development of methods of predicting many weeks or months in advance the occurrence of rainfall deserves high priority.

Since technological inputs quickly reach an optimum level, more emphasis should be placed on drought management policies, especially in dryland farming areas. Agricultural planning and practices need to be worked out with consideration to the overall water requirements within an individual agroclimatic zone. Crops that need a short duration to mature and require relatively little water need to be encouraged in drought prone areas. Irrigation, through canals and groundwater resources, need to be monitored with optimum utilization, avoiding soil salinity and excessive evaporation loss. A food reserve is needed to meet the emergency requirements of up to two consecutive droughts. A variety of policy decisions on farming, human migration, population dynamics, livestock survival, ecology, etc. must be formulated (Das 1999).

Sustainable strategies must be developed to alleviate the impact of drought on crop productivity. In areas of recurring drought, one of the best strategies for alleviating drought is varietal manipulation, through which drought can be avoided or its effects can be minimised by adopting varieties that are drought-resistant at different growth stages.

If drought occurs during the middle of a growing season, corrective measures can be adopted; these vary from reducing plant population to fertilisation or weed management. In high rainfall areas where there are a series of wet and dry spells, rainfall can be harvested in either farm ponds or in village tanks and can be recycled as lifesaving irrigation during a prolonged dry spell. The remaining water can also be used to provide irrigation for a second crop with a lower water requirement, such as chickpea. However, no one strategy can be adopted universally. In fact, all such strategies are location, time, crop, crop stage and (to some extent) socio-economic condition specific. Developing such strategies for each specific factor can help make agriculture sustainable.

6.3.2 Cyclone Preparedness in Agriculture System

Disaster preparedness for impending cyclones, as we know, refers to the plan of action needed to minimize loss to human lives, damage to property and agriculture. The effectiveness of disaster preparedness ultimately depends on the effectiveness of planning and response at a district or local government level (Oakley 1993).

Preparedness for cyclones in the agriculture system can include early harvesting of crops, if matured, safe storage of the harvest, etc. Irrigation canals and embankment of rivers in the risk zone should be repaired to avoid breaching. Beyond this, as the storm approaches the area, nothing can be done, but to secure as much of the property as possible and find safety.

An example of crop diversification may be cited here as a long-term measure to reduce the crop damage during the period of cyclone in coastal districts of Andhra Pradesh of India (Naidu 2001). Paddy is the principal *kharif* crop in coastal districts of Andhra Pradesh. About 50 percent of the paddy-cropped area of the state is situated in the districts of Nellore, Prakasam, Guntur, Krishna, West Godavari and east Godavari. The two traditional varieties of paddy cultivated by nearly 50 percent of the farmers namely 'Akkullu' and 'Massouri' are of five to six month duration. During 1977 and 1996, they were planted in June–July to be harvested in November and December. But unfortunately the cyclone struck at the worst possible time of crop growth, just after the grain had set but before the harvest had commenced. In the cyclone hit areas, the high impact of wind caused majority of the grains to blow away, leaving behind only the stalks. Two reasons could be attributed for the loss of grains. One the traditional varieties are heavy shredders by nature and are susceptible to lodging easily on impact. Secondly, the high winds caused friction among the ear-heads resulting in grain shed again. As a long-term measure, the farmers should be advised to take up the cultivation of those short-duration variety crops which are not easy grain shredders.

Another example of a project "Coastal Environment Preservation" undertaken by Vietnam Red Cross may be illustrated here to show how mitigation programme was useful to the local population living along the seacoast engaged in agricultural and aquaculture. With eight to ten typhoon storms striking the coast of Vietnam annually, tidal flooding often breaches sea dykes causing economic losses to the local population engaged in aquaculture. Under the Environmental Preservation Project, the planting of mangroves were undertaken as mitigation programme to address the suffering of the people living along the sea coast in Thai Thui district of Thai Binh province (Goodyear 2001).

The planting of mangroves serves two important purposes: (a) It acts as a buffer zone in front of the sea dyke system and reduces the water velocity, wave strength and wind energy to protect coastal land, human life and assets invested in development, and (b) it contributes to the production of valuable export products such as shrimps and crabs, high-value marine fish species in cages, mollusk farming and seaweed culture for agar and alginate extraction. By contributing to environmental support to coastal fisheries and aquaculture through the development of 2,000 hectares of mangroves, the Thai Binh Red Cross also helps to provide a livelihood benefit to the vulnerable population through new employment opportunities for their labour.

An evaluation of the project in 1996 indicated that: "By helping to protect the sea dykes, the mangroves were contributing to the economic stability of the communes. All members of the community stand to benefit as their homes, livestock and agricultural land are better protected from the risk of flooding. Poor families with little money to repair or replace material losses from storm damage, are the greatest potential beneficiaries". The project area was struck by the worst typhoon in a decade two month prior to the project evaluation. Lack of any significant damage to the sea dyke and aquaculture pond systems in Thai, provided the best possible indicator of the effectiveness of the mangroves. The perceptions of the local population believed that the mangroves planted would:

- Lessen the frequency of storms, protect flooding of sea dykes and ponds, protect property and the coastal inhabitants;
- Improve aquatic production and the environment, and prevent saline intrusion into agricultural land;
- Expand the land area for the national benefit.

6.3.3 Mitigation of Damage on Agricultural Sector due to Flood and Heavy Rainfall

Preconditioning of an area is very important for determining how significant and damaging a flood will be. Important considerations here are soil, vegetation and water supply factors. Soils that are saturated prior to an extreme weather event will more likely to result in a damaging flood than soils that are relatively dry. Fields that have recently been tilled and are devoid of vegetation are much more susceptible to soil erosion. Vegetation that is able to use much of the water and that can act as a barrier to moving water (horizontally and vertically) can reduce flood severity and impacts. Water storage systems (rivers, lakes, reservoirs, etc.) that are able to capture and hold most of the incoming water will be effective in reducing flood damage. Thus, water supply managers in snow-fed water supply regions of the world typically draw down reservoirs as much as possible prior to the normal beginning of the snowmelt run-off season. In rain-fed agricultural systems, managers typically anticipate rainfall during the growing season sufficient to naturally or artificially irrigate crops. In both situations, however, there is often a balance needed between retaining enough water for agricultural production and environmental health and maintaining enough available storage volume to capture incoming water and prevent floods. Here, analysis of past weather and water data are critical for estimating average conditions and inherent variability. Crops like rice that can function effectively in saturated and even submerged conditions are appropriate for locations that flood regularly and the system becomes dependent upon regular flooding. Many other crops (e.g. corn) would not be adaptable to such conditions and would not be appropriate alternatives to rice.

Geographical Information System (GIS) can be extremely useful tools in the analysis of flood-prone areas. A GIS in conjunction with digital elevation model (DEM) data can quickly determine slope and aspect of region and can be used to provide geospatial analyses of multiple spatial layers (elevation, slope and aspect-all at various scales or resolutions along with soil characteristics, precipitation, temperature, vegetation, and other factors). GIS are being used to develop new floodplain maps (at various frequency and severity levels) and delineate wetlands in many regions and countries, using these types of spatial layers as well as others, including aerial photos. Such information will certainly assist in the best design of agricultural systems, while accounting for reasonable risk.

Floodplain maps with appropriate information about probabilities (return periods) of certain amounts of precipitation and/or depth of flooding water should be developed and used in risk assessments and agricultural planning. Based on these analyses, economic studies can be conducted to determine if certain types of agriculture are appropriate and justified in certain flood regions. For instance, even though analyses indicate a region is subject to significant flooding every five years, it may still be economically advantageous to farm the area, due to its high value of return. Prudent planning would ensure that structures and other items that would be damaged or destroyed by these frequent floods need not be built in the flooded area, but in less vulnerable areas nearby. It is also important that future flood planning looks not only to structural solutions but also to land use planning, zoning and other solutions that encourage agricultural production in less vulnerable areas.

6.3.3.1 Better Land-use and Crop Planning

For development in the floodplains, a careful management of technology so as not to obstruct but to use floods to good purpose through better land-use and crop planning has to be an essential ingredient of any policy. In other words, it is necessary to promote ecologically appropriate policies for human settlements and agriculture in the floodplains.

Suitable short duration strains of paddy and other crops, which can withstand flooding for a few days, have to be developed. Much work has been done in the agricultural universities in India and Bangladesh and it could be utilized to a considerable extent.

People in the eastern floodplains of the Ganga and Brahmaputra in India and those in Bangladesh have a long tradition of living in harmony with floods. The living style, habitations, and crops grown were all evolved taking into consideration the climate and the flood-proneness of the area. Ancient people inhabiting the floodplains took care not to block the natural drains, preserved the natural beds and depressions, and cultivated only those crops that could stand submergence. This tradition is gradually vanishing with changing economic and social conditions. There can be no doubt that the countries of the Ganga-Brahmaputra-Meghna region must adopt ecologically appropriate policies for human settlement, agricultural and industrial development, besides promoting the protection of wetlands for flood moderation (Rangachari 2001).

6.3.3.2 The Application of Nutrients During Flood

Prolonged flooding of soils causes several physical, chemical and biological changes, some of which are not reversible. The application of Nitrogen, Phosphorous and Potassium in appropriate amounts could be especially important to build up soil nutrients for plants weakened by prolonged flooding and especially prone to development of disease problems and provides a perspective on the development of flood mitigation strategies in the future.

6.3.4 Protection of Crop Against Wind

Crop damage by winds may be minimised or prevented by the use of windbreaks (shelterbelts). These are natural (e.g. trees, shrubs, or hedges) or artificial (e.g. walls, fences) barriers to wind flow to shelter animal or crops. Properly oriented and designed shelterbelts are very effective in stabilizing agriculture in regions where strong wind cause mechanical damage and impose severe moisture stress on growing crops. Windbreaks save the loose soil from erosion and increase the supply of moisture to the soil in spring.

6.3.5 Protection of Crops from Dust Storm/Sand Storm

In most countries, field afforestation is the main measure to protect the soil from dust storm. Improving soil resistance to erosion can be achieved by careful selection of cultivation methods, applying mineral and organic fertilizers, sowing grass and spraying various substances that enhance soil structure. It is also important to reduce the areas where a dust can gather, especially in tracts in areas characterized by erosion. One major protection strategy is to establish well-developed plant cover before the dust storms period. This can encourage a reduction in the wind speed in the layer next to the ground by forming an effective buffer.

When assessing the impacts of the dust storm on agricultural crops, it is necessary to take into account the degree of the development of the plants. On well-tilled crops, the deposition of soil moved by airflow is observed more often than soil carried by wind erosion over long distances.

When looking at the conditions in which dust storms developed and data on storm-induced damage, it is evident that measures to reduce the wind speed at the soil surface and to increase the hooking of soil particles are both crucial. Such measures include the establishments of tree belts and wind breaks. Leaving stubble in fields, non-mouldboard ploughing, application of chemical substances promoting the hooking of soil particles, soil-protective crop rotation using perennial grasses and seeding of annual crops are also important.

In regions with intensive wind erosion, especially on wind shock slopes or on light soils, stripe cultivation may be used. On fallow lands bare fallow stripes of 50–100 m can be alternated with stripes of grain crops or perennial grasses; spring crops can be alternated with winter crops.

6.3.6 Protection of Crops from Cold Injury and Frost

A key factor in protection of crops from cold injury is stable air temperatures and snow covers throughout the winter. Thaws, resulting in packing or disappearing snow cover, worsen dormancy conditions and reduce or destroy the protective properties of snow cover. Long thaws can result in the renewal of vegetation, which is accompanied by the consumption of carbohydrates and hence by increase in critical temperature and decrease in winter resistance.

The prevention of crop damage by frost can be controlled by breaking up the inversion that accompanies intense night time radiation. This may be achieved by heating the air by the use of oil burners that are strategically located throughout the agricultural farm. Other methods of frost protection include sprinkling the crops with water, brushing (putting a protective cover of craft paper over plant) and the use of shelterbelts (windbreaks).

6.3.7 Fire Prevention Measures

The use of periodic seasonal data of certain specific forest areas prone to forest fires could be used by the forest managers to enable taking certain fire prevention measures or mounting fire operation measures in advance to mitigate possible damages. On the basis of the weather fire relationship, a system of fire danger rating has been evolved to guide the fire management people in their day to day activity and also to provide a basis for comparing weather and fire behaviour throughout the nation or region. Such system usually include book-keeping schemes for keeping track of the moisture contents from one to three size classes of forest fuels, plus indices of spread rate, fuel quantity consumed and energy output rate of the fire front. Forest fire danger rating system in Canada and universal system of fire danger rating along with fire weather forecasting in USA are providing valuable information to mitigate possible damage due to forest fire.

6.4 Agricultural Strategies for Community Capacity Building

The process of community capacity building for disaster risk reduction starts with community-based risk assessment. Assessment of prevalent hazards and the vulnerabilities and capacities of the community is done with the active involvement of local people. By applying participatory techniques, people are engaged in a process of the analysis of the situation and the risks they face. Imparting awareness to farmers on drought and flood resistant crops is another instance in this line. Some of the agricultural strategies to be adopted are mentioned below (Murshed 2003).

6.4.1 Diversification of Crops According to Different Planting Seasons

Many disaster prone communities depend on only one type of crop, which is harvested only once a year. Due to the occurrence of hazards, this crop is damaged and as a result, they regularly face acute food shortages and people are forced to depend on seasonal out-migration. It is therefore important to identify the potential to harvest other crops in different months of the year. It may involve the assessment of suitability of land for cultivation of other crops, availability of water and orientation and training of farmers to harvest new crops. This can help mitigate the effects of failure of the main crop on farming families due to prevalent hazards; e.g. drought or flood. If the main crop fails, the farmers still can harvest products during the two other seasons. As a result, the number of months people face food shortage could be reduced. Interesting work is being done in this front in Philippines and Vietnam by the agriculture ministries and extensions services.

6.4.2 Propagation of Disaster Resistant Crops

Disasters disrupt farming and damage certain crops, which may result in form of food shortage. In old times in the face of food shortages due to disasters, people depended on wild crops like certain beans, leaves, and root crops. These crops can still be found in many regions although some of them have become scarce and are gradually disappearing. Therefore, it is essential to propagate these, especially among the younger generation, who is hardly aware about the existence of such crops. It is important that the knowledge and skills of old people are tapped regarding these plants for their promotion of mitigation measures. The re-introduction of the already forgotten indigenous crops and the campaign to cultivate them could awaken and remind local residents of the importance of such crops. They are easily grown and do not require a lot of inputs. Their use can be propagated by planting community nurseries/demo-farms and through educational campaigns. Disaster resistant crops and other indigenous crops can serve as a staple food source in times of disasters.

6.4.3 Seed Banks and Nurseries

Establishment of seed banks and nurseries at the community level can ensure a stable supply of seedlings, seeds, cuttings and other plant materials. Seed stocks can be used in times of emergency for rehabilitation of damaged croplands. Most seed banks focus on traditional rice and corn seeds, which are more resistant to pests and less sensitive to changing climatic conditions. These varieties are slowly disappearing because hybrid varieties are promoted in the market. This highlights the importance of community-based seed banks. Nurseries propagate fruit trees, forest trees, forage trees, bamboo, and other plants useful for people in times of crisis. Cultivating utility trees around homes and land, maintaining forest reserves for food, animal feed and cash, improving watersheds, and protecting water sources are long-term interests of vulnerable people. Seed banks and nurseries could strengthen people's existing livelihoods and increase the presence of fallback resources in the community's seed bank and nursery. They should be provided training on seed bank and nursery management, record keeping. They can also formulate policies for the approval of the farmers regarding repayment of seeds, operation and maintenance of the nursery, and the sustainability of the seed bank and the nursery.

6.4.4
Post Harvest Facilities

Marginalised communities cannot always enjoy the benefits of a harvest. After paying the landlord back, they face problems of inaccessible and expensive grain mills, storage problems (attacked by rats and insects, or rotting due to damp conditions), or they have to sell seasonal perishable crops at low price. It is essential to help communities to address post-harvest problems to reduce losses, to reduce expenses for milling, and to maximize the availability of vegetables, fruits and root crops by processing and preserving them for times that they are less abundant. In this way people's coping strategies are strengthened. This can be done by helping construct appropriate storage facilities to avoid rats from entering and eating the rice grains, the main problem in times of drought. Seeds can further be protected from insects and damp conditions by adding charcoal and pinewood splints. Provision of cheap and easily accessible milling facilities could be a strategy. This can improve the availability of food in seasons of disaster.

6.4.5
Encourage Proper Land Use Management and Sustainable Agriculture Practices

The fast deterioration of environment due to outside pressures (logging, mining, multinational plantations, and encroaching settlers) increases the vulnerability of people to various hazards and many times leads towards introduction of new hazards. Therefore, conservation of environment through tree plantation, rehabilitation of watersheds, mitigation of soil erosion, and replenishment of forest reserves is important. It can be done through organizing tree-planting campaigns to raise awareness on environmental issues among a broader public. It is important that people have land use policies which should indicate where remaining forest should not be touched, where 'slash and burn' can still be practised, where permanent farms can be tapped for irrigated farming. In times of crisis, forest products can be used to a limited extent. In this way, local people could try to control and manage their direct surroundings for future generations.

6.4.6
Community Participation for Traditional Rain Water Harvesting

Ever since the beginning of civilization, human beings have adhered to some practices or the other to protect themselves against drought and floods. The traditional rain water harvesting techniques in different parts of the world bear testimony to sincere efforts towards community action and water preservation.

The success story of 'Sukhomajri' (India) is an indicator of the power of traditional wisdom, people's participation and community's deep sensitivity to the environment (Dhameja 2001). A crumbling hillock called Sukhomajri in Haryana state of India is now a farmer's paradise. The ash-ridden saline soil around the area is now yielding fruits, vegetables and even timber. It draws its sustenance from the *catch the rain formula*. The idea behind this is to dig a pit and let the rainwater accumulate. The water can be used to green adjacent tracts. The dried up tank can be used as silted bed for agriculture. Only an inch of deep water penetration is necessary per acre and then the soil can be enriched organically by rotating crops. Sukhomajri today has three earthern dams built across natural gullies, which store rainwater. The water is distributed equitably to each family in return for the commitment to protect the hills (Agarwal et al. 1999). Joint Forest Management, 'Van panchayats' and 'Pani panchayats' are some more examples of effective people's participation and their faith in traditional practices, which, if revived, could strike the right harmony between the human beings and the nature.

6.5 Disaster Risk Reduction Through Livelihood Concerns

Community awareness on disaster preparedness assumes predominance because disaster prevention, preparedness, mitigation and relief are largely dependent on the capability of the communities to withstand disasters. Disaster preparedness needs to be the way of life in such communities if they have to survive the aftermath of different catastrophes. Many studies on droughts, landslides, floods, cyclones, land erosion, etc. conducted in the developing countries found out that disaster-prone communities are living at virtually subsistence levels and very limited livelihood options and opportunities are available to the at-risk communities in these countries.

Most of the disaster-prone population still depends on agriculture as a major source of earning. The continuous onslaught of nature's fury has not only led to the loss of lives but has also led to economic losses. For example, the saltpan workers in Kandla (India) did not have any livelihood option left with them after the Gujarat cyclone. In Kot Murad, a flood prone village in Pakistan where farming is the main livelihood, 87 percent of the total households are landless. Livelihood support measures become more important especially in a disaster scenario. The findings from the case studies in Sri Lanka indicate that the main source of income for 72 percent of the population in Andarawewa and 56 percent of the population in Mahameddawa is from subsistence agriculture during normal years. Off farm income is the major source of income during periods of drought. During such periods people often have to migrate to other places in search of wage labour. Women suffer the most in such situations as they are unable to leave their families in search of employment. Hence, the total income earning opportunity for a family, as a whole is lower in the drought years compared to that of a normal year. Rainfed highland cultivation is the only income source for a majority of the families in villages under study (Bhatti 2003).

Yet, livelihood concerns are practically missing from the disaster policies and plans of many countries. The research argues that besides human lives, livelihood security is the primary concern that comes under severe threat in the case of natural or man-made disasters. This concern becomes graver in the case of communities of developing countries where economic opportunities are generally very few. The limited economic opportunities make disaster-prone communities more vulnerable and they find no way to exit from the burning hell of miseries.

Diversification of livelihood sources can mitigate the effect of disruption of the people's main livelihood. Some livelihoods are extremely vulnerable to particular disaster. To reduce people's vulnerability, support could be provided to these communities with alternative livelihoods that match the people's capacities and that are not sensitive to the existing disaster. This can help them earn an income during the time when their main source of production was down due to the occurrence of disaster (Murshed 2003). One should be conscious of gender roles and responsibilities when selecting an alternative livelihood. Whether the settlement was urban or rural also play an important role in availability and provision of alternative sources of livelihoods.

Some agriculture oriented measures for building more secure livelihoods after the cyclone are mentioned below (Sahni and Dhameja 2003).

6.5.1 Creation of Alternate Livelihood Options

- Measures to improve agricultural production
- More focus on animal husbandry
- Provision of adequate seeds and fertilizers
- Draught animals to be provided to farmers
- Arrangements to be made for providing tractors or power tillers on a hire basis
- Dairy cattle to be produced
- Arrangements for restoring insemination centers for local breeds of cattle have to be made
- Alternative cropping pattern in areas affected by salinity needs to be implemented
- Reclamation and treatment of soils
- Ecological balance needs to be maintained through large-scale plantation
- Wage employment under various Rural Development Schemes

The important sources of livelihood during rehabilitation identified by the community include:

- Supply of seeds
- Supply of fertilizers
- Horticulture plantation
- Livestock rearing
- Food for Work Programme
- Assistance for purchase of tools
- Assistance for repair or construction of work-sheds.

A *Livelihood* and *Employment Restoration Programme* launched by Oxfam in collaboration with the European Union aimed at:

- Restoring the traditional livelihoods of the affected populations through the provision of food/cash for work.
- Facilitating work for community-based rehabilitation and restoration activities, such as reclamation of agricultural land.
- Providing a model for Food-for-Work Programme that is capable of reaching the most vulnerable and marginalized so that other large-scale programmes can integrate vulnerability and equity perspectives.

The support provided by a number of NGOs working in Orissa state of India during rehabilitation and the support extended by the Government of Orissa state has led to the strengthening of the following livelihood options in two major sectors-the farm and non-farm sectors:

Activities in the farm sector:

- Cultivation of paddy, wheat, ragi and groundnut
- Growth of vegetables and potatoes
- Kitchen garden nurseries
- Plantation of mangoes, mangroves, and pipal.

Activities in the non-farm sector for engaging more people in the following occupations: (i) Fishing (ii) Fish drying (iii) Pig rearing (iv) Coir production (v) Rearing of cows, buffaloes, bulls (vi) Bamboo production (vii) Lime making (viii) Poultry (ix) Goat rearing (x) Boat building (xi) Making fishing nets (xii) Shrimp farming (xiii) Basket making (xiv) Mat weaving (xv) Broom making (xvi) Masonry (xvii) Carpentry (xviii) Plumbing (xix) Electrician work (xx) Blacksmith work (xxi) Toy making (xxii) Pottery (xxiii) Applique work (xxiv) Petty trading.

6.5.2
Livelihood Strategies in Disaster Risk Reduction – A Case Study in Bangladesh

Traditionally people have developed different kinds of coping strategies related to their livelihood. Of the indigenous responses to flooding in Bangladesh, a significant number relate to agriculture coping measures including selecting appropriate variety of rice and other crops-depending on the timing-and water level as well as type of soil. People also tend to reduce the magnitude of economic loss through crop loss from flood and erosion by cultivating low cost varieties. Housing techniques are also adapted according to the risks posed by floods and erosion. Houses are built on raised lands or earthen platforms so water cannot reach the plinth during normal floods. Poorer people use thatch, bamboo and corrugated iron sheets as housing material. Plantation of water friendly plats/trees like bamboo, banana, Hoga, Kolmi and others beside the homestead is very common to protect the house from erosion as well as for the use of these plants/trees during floods and after recession of the flood (Khatun 2003). During the dry season, the kitchen is made in the open courtyard, but during floods they make bamboo platforms that can be raised when the water level is increasing and use portable stoves for cooking. Food, household items and crops are stored on raised platforms. Selling animals, grains and other assets like jewelry, sale and lease of land, advance sale of labour and fishing are the common coping mechanism to survive during the crisis period.

6.6 Remote Sensing as a Tool for Disaster Risk Management

Although disasters may be natural phenomena, the damage caused by them is largely due to improper human activity and reckless land exploitation. For instance, the occurrences of floods and landslides can be substantially reduced by proper afforestation of the susceptible region. Thus, accurate mapping of land use and land cover patterns at regular intervals becomes a pre-requisite to effect preventive measure to avoid a disaster and to assess the damages after it has occurred. Remote sensing data at spatial resolutions ranging from 10 m to 100 m available at varied temporal resolutions have been successfully utilised to produce land use land cover maps at a desired accuracy of at least 85% (Arora 2003). Table 6.2 shows the applications of some current satellite sensors for disaster management. The remotely sensed derived land use

Table 6.2. Applications of some current satellite sensors for disaster management

Satellite	Sensor	Spatial resolution	Type of disaster
METEOSAT	VIS	2.5 km, day and night observations	Prediction/monitoring of cyclones, tornadoes and volcanic eruptions, floods, avalanches
NOAA	AVHRR	1.1 km, twice a day	Land cover, detection and monitoring of fires, drought, volcanic eruptions
LANDSAT	MSS TM ETM	30 and 80 m	Land use, flood extent, drought, landslides, fire
SPOT	PAN HRV	10 and 20 m	3-dimensional mapping, flood extent, damage assessment, crop identification, drought, landslide
IRS	PAN LISS WiFs	6, 23 and 188 m	3-dimensional mapping, flood damage assessment, drought, fire, landslide
ERS	SAR	All weather, 25–500 m	3-dimensional mapping, flood extent, damage assessment, night coverage, earthquake, fire, landslide
RADARSAT	SAR	All weather, 10–100 m	3-dimensioanl mapping, flood extent, damage assessment, night coverage, earthquake, fire, landslide
IKONOS	PAN MSS	1 and 4 m	High resolution mapping, infrastructure identification, terrain analysis, property damage assessment

land cover maps are frequently used as input, to a GIS for several disaster related studies, for example landslide hazard zonation and flood risk management.

Now, let's clarify the role of remote sensing and GIS in some commonly occurring hazards.

6.6.1 Cyclones and Tornadoes

The weather satellites (geo-stationary) such as INSAT and GOES provide almost continuous data on day-to-day basis for the prediction of weather and monitoring climate changes. This is beneficial for providing information for issuing early warning and specifying the geographical location of hazards such as severe thunderstorms and tornadoes, thereby assisting the emergency response teams for timely evacuation of the people in the hazard-prone areas.

In an other application, the visual interpretation of SPOT images before and after the cyclone in the Cook Islands sufficiently demonstrated the potential of the remote sensing data sets to identify the large scale changes in coastal and shallow marine areas produced by the cyclones (Loubersac et al. 1991). The large-scale damage due to cyclone visible in SPOT images was helpful in planning emergency aid and reconstruction.

6.6.2 Drought

During drought, physiognomic changes of vegetation may become apparent. Satellite sensors are capable of discerning many such changes through spectral radiance measures and manipulation of such measures into vegetation indices that are sensitive to the rate of plant growth as well as to the amount of growth (Curran 1990). Such indices are also sensitive to the changes in vegetation affected by moisture stress. The visible and near infrared bands on the satellite multi-spectral sensors allow monitoring of the greenness of vegetation. This property is used in the case of monitoring drought, as the stressed vegetation and other bare ground, water, etc. reflect differently.

Reliable drought interpretation requires a geographical information system (GIS) approach, since the topography, soil type, spatial rainfall variability, crop type and variety, irrigation support and management practices are all relevant parameters.

6.6.3 Flood

Remote sensing and GIS tools can be used in flood analysis to determine and delineate the floodplain. Further, these may also assist in mapping and monitoring flood inundated areas, assessing damages due to floods and in flood hazard zoning. The flood plain management and zonation would make for alerting the risk-prone residents and commercial establishments in advance about the areas that are threatened by the floods.

AVHHR images obtained before and after the flood can be classified to generate the flood maps of the region. These maps can be compared to assess the damage due to floods. The data from these maps along with those from other maps (namely soil, geology, land use and contour) can be analysed in GIS to prepare the flood risk map indicating the zones of high to low flood risk (Rahman 1991).

6.6.4 Forest Fires

Forest fires are another natural disaster where remote sensing and GIS can be employed successfully. The detection and monitoring of fires in the quickest possible time helps the emergency management agencies to prevent large-scale damage to life and property. Also, the residents in an area can be notified in advance so that they may have sufficient time to vacate the places.

6.7 Crop Insurance

Insurance is a mechanism for spreading the cost of losses both over time and over a relatively large number of similarly exposed risks. The introduction of disaster linked insurance should be actively pursued and insurance cover should be available not just for life but also for household goods, cattle, structures and crops.

Since insurance premiums are based on the location of a structure within the disaster-prone areas, and are determined primarily by the extent of risk, higher rates may be applied to structures subject to high risks.

In keeping with the impact of natural disasters, disaster insurance could be a critical instrument of development in the field of crop production, providing financial support to the farmers in the event of crop failure. It can encourage farmers to adopt progressive farming practices, better technology in agriculture, besides providing significant benefits not merely to the insured farmers, but to the entire community directly and indirectly through spill over the multiplier effects in terms of maintaining production and employment generation of market fees, taxes etc. and net assertion of economic growth. Crop insurance could also streamline loss assessment procedures and help in building up huge and accurate statistical base for crop production.

One of the difficulties in promoting disaster insurance is that those who are at highest risk have the least capacity to pay the premium. Possibilities of group/community insurance should also be available, particularly for the marginalised communities. The insurance agency may promote a Community Rating System to encourage communities to go beyond the require standards of minimum safety. The incentive can be a reduction in insurance premium for policyholders within communities that take appropriate actions to reduce disaster losses.

6.8 Conclusion

Damage caused by extreme agrometeorological events to the agriculture sector may be significant, sometimes of the order of magnitude of the GNP growth. For many disaster-prone countries, agricultural losses due to exceptional weather events are a real constraint on their global economy. Indirect effects may continue to affect agriculture negatively long after the extreme event took place. The time needed to recover from some extreme agrometeorological events ranges from months to decades.

Environmental planning would be necessary to avoid or mitigate losses from disasters, by using such instruments as land-use planning and disaster management. Natural disaster reduction measures are in place in a significant number of the nations surveyed and ongoing research and development to improve and expand these measures are also a feature of many national strategies to minimize adverse effects of extreme events on agriculture. Steps are being taken to significantly reduce the vulnerability of people and their communities to natural disaster; this can only be done through mitigation.

Acknowledgements

The author expresses his gratefulness to Dr. S.K. Srivastav, Director General of Meteorology, New Delhi for his nomination and participation at this expert meeting.

References

Agarwal A, Narain S, Sen S (1999) "State of India's Environment: The Citizen's Fifth Report (Part I)", Centre for Science and Environment (CSE), New Delhi

Arora MK (2003) Role of Remote Sensing in Disaster Risk Reduction. In: Sahni P, Ariyabandu MM (eds.) Disaster Risk Reduction in South Asia, Prentice-Hall of India Private Limited, New Delhi, pp 84–100

Bhatti A (2003) Disater Risk Reduction through Livelihood Concerns and Disaster Policy in South Asia. In: Sahni P, Ariyabandu MM (eds.) Disaster Risk Reduction in South Asia, Prentice-Hall of India Private Limited, New Delhi, pp 57–66

Curran PJ (1990) Principles of Remote Sensing, Addison Wesley Longman, New York

Das HP (1999) Management and mitigation of adverse effects of drought phenomenon. In: Natural disasters – some issues and concerns. Natural Disasters management Cell, Visva Bharati, Shantiniketan, Calcutta, India, pp 87–97

Dhameja A (2001) Droughts and floods: a Case for 'Dying wisdom'. In: Sahni P, Dhameja A, Medury U (eds.) Disaster Mitigation, Prentice-Hall of India Private Limited, New Delhi, pp 76–91

FAO (1982) Salvage and rehabilitation in the forestry sector of the hurricane affected areas of the eastern region, Jamaica. FAO, Rome, p 16

Foth HD (1978) Fundamentals of soil science. John Wiley and Sons, p 436

Glantz MH (1994) Drought follows the plow. Cambridge University Press

Gommes R, (1997) An overview (extreme agrometeorological events). CAgM Report No. 73, TD No. 836, WMO, Geneva, pp 1-9

Goodyear EJ (2001) Challenges to Raise the Capacity of At-Risk Populations in Coping with Natural, Social and Economic Disasters. In: Sahni P, Dhameja A, Medury U (eds.) Disaster Mitigation, Prentice-Hall of India Private Limited, New Delhi, pp 112–123

Khatun H (2003) Livelihood Strategies in Disaster Risk Reduction in Bangladesh. In: Sahni P, Ariyabandu MM (eds.) Disaster Risk Reduction in South Asia, Prentice-Hall of India Private Limited, New Delhi, pp 264–280

Loubersac L, Dahl AL, Ozouville LD (1991) Comparison of two SPOT images acquired before and after a cyclone. In: Murai S (ed.) Applications of Remote Sensing in Asia and Oceania-Environmental Change Monitoring, Asian Association on Remote Sensing, Japan, pp 269–273

McCann J (1986) The social impact of drought in Ethiopia. In: Glantz MH (ed.) Drought and hunger in Africa: denying famine a future, Chapter 11, Cambridge University Press, Cambridge, England

MIDINRA (1988) Danos en el sector agropecuario provocados por el huracan "Juana". Evaluation preliminar. Managua, November 1988, p 39

Murshed Z (2003) Community Capacity Building for Risk Reduction in South Asia. In: Sahni P, Ariyabandu MM (eds.) Disaster Risk Reduction in South Asia, Prentice-Hall of India Private Limited, New Delhi, pp 145–161

Naidu BR (2001) Disaster Mitigation Measures for Cyclone-Prone Areas of Andra pradesh. In: Sahni P, Dhameja A, Medury U (eds.) Disaster Mitigation, Prentice-Hall of India Private Limited, New Delhi, pp 67–75

Oakley WJ (1993) A national disaster preparedness service. In: Natural disaster, Merriman PA, Browitt CWA (eds.) Thomas Telford, London, pp 270–281

OSRO (1984) Madagascar, evaluation de la situation agricole a la suite des cyclones. OSRO. TCP/MAG/4404, FAO, Rome, p 43

OSRO (1989) Evaluation of the agriculture situation in eastern Caribbean countries affected by hurricane "Hugo", OSRO 03-89-E, FAO, Rome, p 63

Rahman SM (1985) Agroclimatic methods for Bangladesh. Bangladesh Meteorol. Department, Dhaka, p 41

Rahman SM (1991) Flood Risk Mapping in Bangladesh, In: Murai S (ed.) Applications of Remote Sensing in Asia and Oceania-Environmental Change Monitoring, Asia Association on Remote Sensing, Japan, pp 263–268

Rangachari R (2001) Disaster Preparedness and Mitigation in Flood Management: Cooperation and Networking among Countries. In: Sahni P, Dhameja A, Medury U (eds.) Disaster Mitigation, Prentice-Hall of India Private Limited, New Delhi, pp 55–66

Sahni P, Dhameja A (2003) Disaster Risk Reduction through Capacity Building of the Community and Panchayati Raj Institutions. In: Sahni P, Ariyabandu MM (eds.) Disaster Risk Reduction in South Asia, Prentice-Hall of India Private Limited, New Delhi, pp 162–184

CHAPTER 7

Damage Assessment of Agrometeorological Relevance from Natural Disasters: Economic and Social Consequences

Allen R. Riebau, Douglas G. Fox

Abstract Damage from catastrophic events is both social and economic. Perhaps one of the most challenging aspects of modern life is our increasing ability to communicate, especially in the developed countries. This communication has brought an increased sense of connectedness but also made people more aware of the occurrence and consequences of natural disasters. In this paper we explore the potential for changes in frequency and magnitude of weather-related natural disasters and some of their connections to climate change and variability. Although changes in climate may have great consequence, increasing climate variability may become the most immediate challenge for world society. Both observations and simulation models show that as the potential for climate variability increases, natural disasters have become more frequent and of deeper economic impact. In areas of the world where these devastating events appear to occur and then quickly reoccur in a different guise, the sharing of information and resources becomes a pivotal issue for collective security. New technologies can help us understand these events and prepare for them. New social institutions may also be needed to cope with these issues on a global scale.

7.1 Introduction

Natural disasters can be classified into those with a climatic component and those without one. But of course, agrometeorologists or climatologists understand that even if a disaster, such as an erupting volcano, has no climatic instigator, weather and climate often influence the magnitude of consequences of the disaster (Jones, 1999). Many have come to believe that climate change is the underpinning of natural disasters that are the result of severe weather or climate variability. In the 21st century, societies must be vigilant about climate change since it is emerging as a reality to be dealt with. A growing sense that the issue is immediate exists within the natural resources management community, but management responses often are bereft of inclusion of climate variability as a factor, with examples in the USA including rangeland management (Fig. 7.1) and forestry.

During the past fifty years data shows an ever-increasing trend towards higher global surface temperatures (Fig. 7.2). These data in records that range

Fig. 7.1. Green Mountain, Common Allotment (Wyoming, USA) showing 2001 forage conditions. Some sources believe rangelands in the USA have reached a disastrously degraded state (Source: rangenet.org)

well beyond the historic direct measurement time scale show a high correlation with increasing concentrations of lower-atmospheric radiatively active trapping gases, especially carbon dioxide, nitrous oxide and methane. The data are not very controversial. There are however many viewpoints on the consequences of these increases, issues such as to how much the earth may warm, what regional differences are likely, how will precipitation dynamics be affected, and how fast it all might occur. An important point of consideration in climate change, in addition to changes in mean or average conditions, is climate variability and the frequency and magnitude of extremes.

The first aspect that can be explored is economic. Although a warmer more carbon-dioxide rich earth could have beneficial aspects for some areas, increases in variability are likely to cause economic stress world-wide. A simple trend in temperatures might be slowly adjusted to by changing agricultural practices. Forestry and grazing land management practices might also be slowly adjusted, even though the adjustments might mean changes in forest structure or grazing carrying capacity. Climate variability has already shown itself to major economic disruptor in developed areas of the world including the eastern United States, Europe's Elbe basin (Fig. 7.3), and central Russia. Although not often stated in the popular press, extreme variation will ultimately be more economically disruptive than a simple warming or precipitation reduction trend. The results of such variations from normal conditions are so

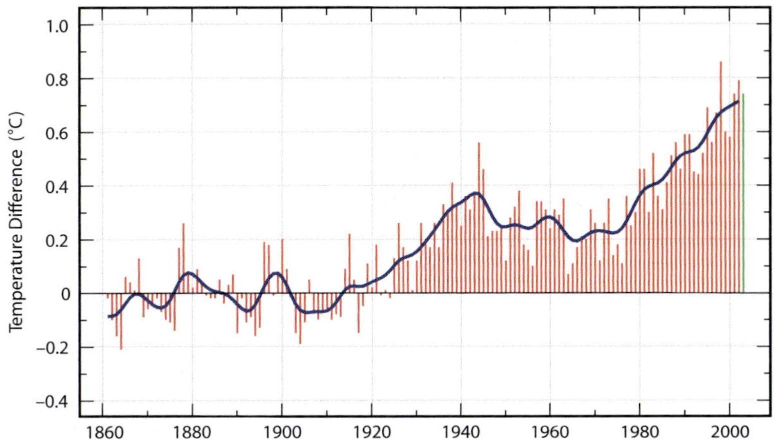

Global near-surface temperatures averaged over the land and ocean. Differences are expressed relative to the end of the 19th century.

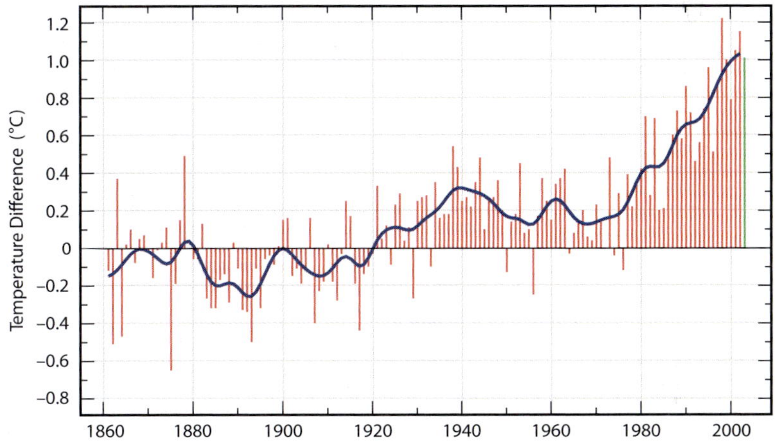

Global near-surface temperatures averaged over the land only. Differences are expressed relative to the end of the 19th century.

Fig. 7.2. Global surface temperatures are rising (Source: Hadley Climate Center, UK)

large scale that they can viewed from space and new remote sensing tools are clearing showing the magnitude of these impacts express in vegetative cover (Fig. 7.4).

The second aspect that can be explored is social. Although in the strictest sense economic impacts are social impacts, climate variability and natural disasters have social impacts that are beyond economic. In extreme natural disasters human life is often lost. At a lesser extreme, issues such as health,

Fig. 7.3. One hundred year flood of the Elbe in 2002 inundates the countryside and small villages near Prague (Source: Mlada fronta dnes (MFD)

longevity, sense of wellbeing, family stability, and quality of life are all factors that are influenced by weather extremes. Ultimately, climate variability may influence the stability of political systems through disruptions in energy supply, fresh water, food, and transport of materials. Societies whose foundations rest on their direct use of natural resources, such as some developing nations on the African continent and their endemic ethnic groups, will find climate variability as the major agent of social change. In fact, variability may reshape the agro-ecological setting of entire continents and thus redraw social relationships world-wide. An example of social disruption from climate variability is a response to the year 2000 drought conditions in the United States. Forests, rangelands, and agricultural lands were greatly impacted by the extent of this situation, which was precipitated by ENSO events as a driving factor in climate variability. Wildfire occurrence in the USA is an example of these

Fig. 7.4. This image derived from ASTER data shows the difference between the amount of vegetation in July 2000 and the average July vegetation for North America. Of particular interest are the dry conditions in the western United States, which resulted in one of worst fire seasons in US history (Source: NASA)

social and economic consequences. In that year 8,422,237 acres were burned in wildfires (about twice the yearly average for the last decade). Fire suppression costs reached almost $1.6 billion dollars. As a result the country has been engaged in deep political debate about the future of forests and their management, which has resulted in the passage of the Healthy Forests Act (Public Law No: 108-148). This law proposes to revolutionize the manner in which publicly owned forests are managed and will have wide reaching social implications, even to such activities as energy generation, tourism and recreation, and wildlife management. Other areas have suffered similar fire situations and are considering new laws or policies to avoid the situations repeating, with the fires of Central Russia in 2002 causing the potential for new zoning regulations a prime example (Fig. 7.5).

Fig. 7.5. Drought and elevated air temperature resulted in massive fire smoke events in Moscow, Russian. The smoke caused as yet uncalculated public health impacts, exposing perhaps 20,000,000 people to high concentrations of particulate matter, and resulted in visibility in the city reduced to 300 meters and below. Although smoke or air quality episodes are not often considered natural disaters, the problem is growing and has tremendous economic implecations (Source: N. Chubarova, Moscow State University)

7.2 Climate Change and Climate Variability

The climate change research community has identified sensitivity, adaptability and vulnerability as dimensions for consideration of climate change impacts, including changes in variability and extremes. Sensitivity means the degree to which a system is affected, either negatively or positively by climate related stimuli. Adaptability identifies the capacity of a system to adjust to a changing climate to moderate damage, take advantage of opportunities or otherwise cope with the consequences. Vulnerability identifies the degree to which a system is susceptible or unable to cope with adverse effects of climate. For the purposes of this paper we define a difference between climate change, which is a long term change to a different climate regime through the accretion of small changes of trend and climate variability, which is the swing from one extreme to another along the progression of weather to a new stable set of climate conditions. This type of variability has always been part of the climate record, but during the past several decades it has increased with great social economic consequence (Aber et al., 2001). Climate variability is perhaps the main factor in adaptability and vulnerability of both human and ecological systems under climate change.

The earth system represents a complex set of interacting feedbacks that we are only beginning to understand. Fig. 7.6, originally developed by Dr. Robert Watson, Chief Scientist of the World Bank, illustrates some of the linkages associated with climate change, agriculture, forestry and other renewable natural resource issues. Although this illustrates feedbacks and linkages, it does not address impacts simply resulting from variability and extremes, especially from variability that is beyond the variability we have come to understand is the norm. That there is this omission is not indicative of any deficiencies in the Dr. Watson's reasoning, but it does illustrate that variability is perhaps difficult to portray succulently within the context of climate change.

We have seen increases in variability result in weather changes that range from local to continental scales. The increase in variability has expressed itself in seasons of drought, in severe flooding, in highly energetic tropical storm systems, and even in changes in the balance of solar radiation reaching the earth (Chubarova et. al., 2003). If this sort of variability continues to escalate, its economic consequences will overshadow any that that might occur from

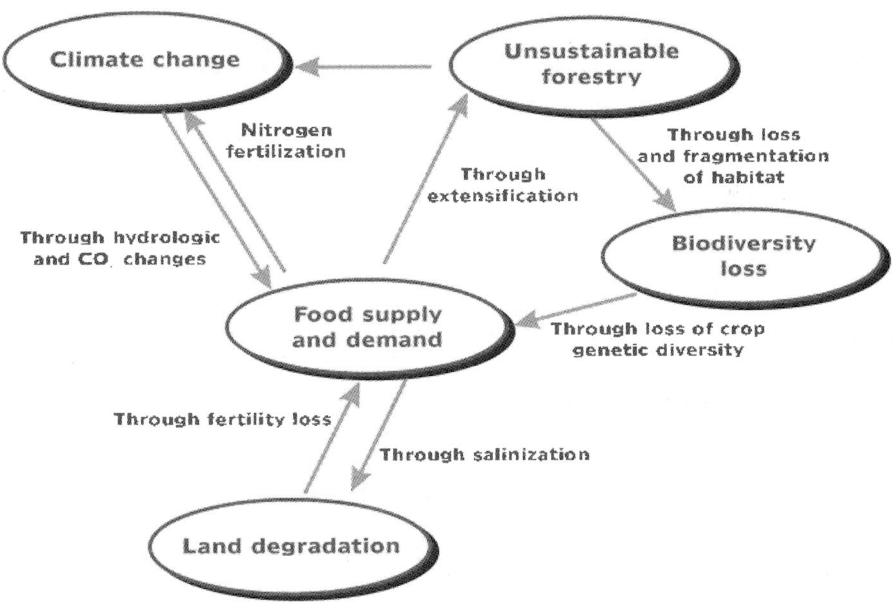

Fig. 7.6. Linkages among food production and environmental issues and climate change (Source: World Bank)

a simple global warming trend (Fig. 7.7). Since nonlinearities of the linked earth system are largely unpredictable, such consequences may be expressed in unanticipated ways.

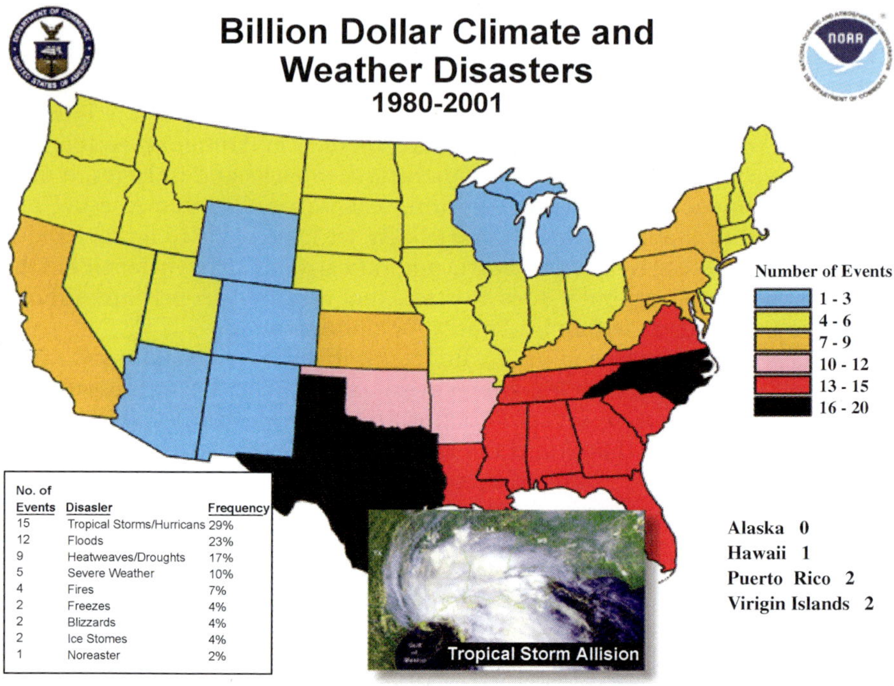

Fig. 7.7. Climate variability resultant weather has caused many natural disaster events in the United States that have resulted in $ 1 billion or more losses per event (Source: NOAA NCDC)

7.3 Variability Shifting Continental Ecosystems

The most severe mid-latitude cyclonic storms have serious impacts on people and property. However, these are fairly rare and small-scale weather events are difficult to detect in historical observations, especially with daily observations that can miss the fastest moving severe storms. In the United Kingdom Hadley Climate Centre scientists have recently studied changes in storm characteristics over the past 50 years or so using three-hourly measurements of surface pressure from the United Kingdom and Iceland. Pressure changes were used instead of winds because the results are less sensitive to observation site moves and instrumentation changes. Observations were available from twenty-one observation sites in the UK and seven in Iceland (Fig. 7.8). The average number of storms (per station) shows a significant increase in the United Kingdom

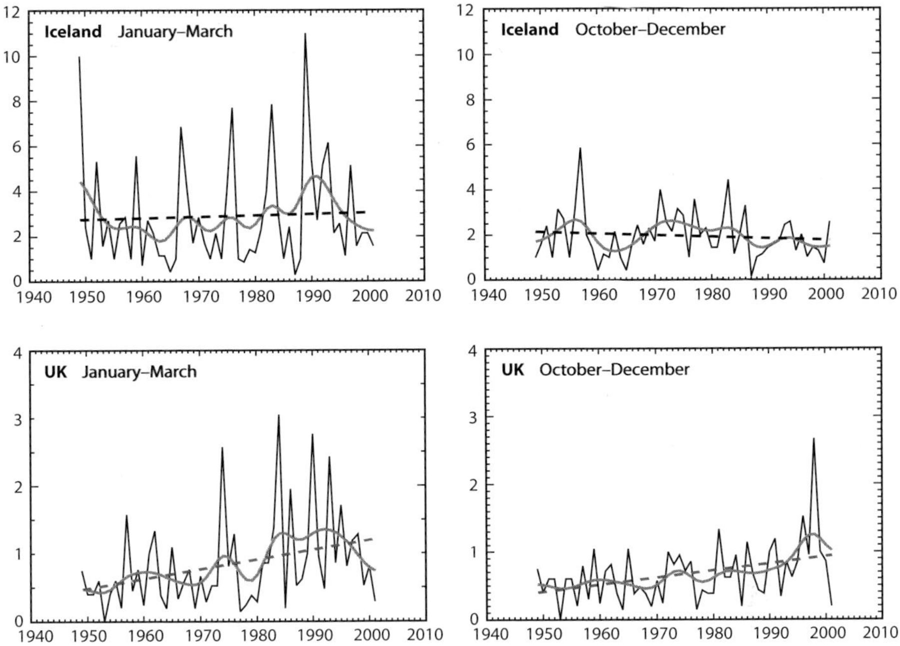

The number of storms per observation station over Iceland and the United Kingdom during (a) January to March and (b) October to December. A red trend line indicates that the trend cannot be explained by natural variability (as estimated from these records) alone.

Fig. 7.8. Storm frequency has been observed to be increasing along with intensity in many regions, with clear evidence seen from data collected in the United Kingdom (Source: Hadley Climate Center)

winter period (October to March). Regional analysis shows that the largest increases occur over the southern UK. Iceland has experienced a slight reduction in the number of storms (between October and November), although this reduction cannot be separated from natural variability with any degree of certainty. A reduction in storm frequency in the north and an increase in the south is consistent with a southerly movement of the North Atlantic storm track.

A more widely used measure of the storm frequency affecting Western Europe is the North Atlantic Oscillation (NAO) index, which we have compared with the new storm rate analysis. Changes in the NAO index correspond to large-scale changes in the north-south pressure difference across the north-eastern Atlantic. Although there is a similar upward trend in the NAO, there is quite a poor correlation between this figure and the storm rate calculated from the 28 pressure measurement sites. This implies that the severe storms over the UK are more related to strong local gradients of pressure than to the large-scale pressure differences over the Atlantic. However, it is likely that the local severe storms are modified by the long-term changes on the large-scale, which are seen in the NAO index. It is also important to place these new results in context. Evidence of storm frequency from daily indices and measurements of wave heights suggest that although it has increased in recent times, the magnitude of storminess at the end of the 20th century was similar to that at the start. This could mean that natural variations in the magnitude of storminess on timescales of several decades or more are responsible for all or part of the trends seen in these new results and that data covering a longer period is needed in order to distinguish a climate change trend from the natural variability. Some of the issues to be considered are:

- It is very likely that precipitation has increased by 0.5 to 1% per decade in the 20th century over most mid- and high latitudes of the Northern Hemisphere continents, and it is likely that rainfall has increased by 0.2 to 0.3% per decade over the tropical (10°N to 10°S) land areas. Increases in the tropics are not evident over the past few decades. It is also likely that rainfall has decreased over much of the Northern Hemisphere sub-tropical (10°N to 30°N) land areas during the 20th century by about 0.3% per decade. In contrast to the Northern Hemisphere, no comparable systematic changes have been detected in broad latitudinal averages over the Southern

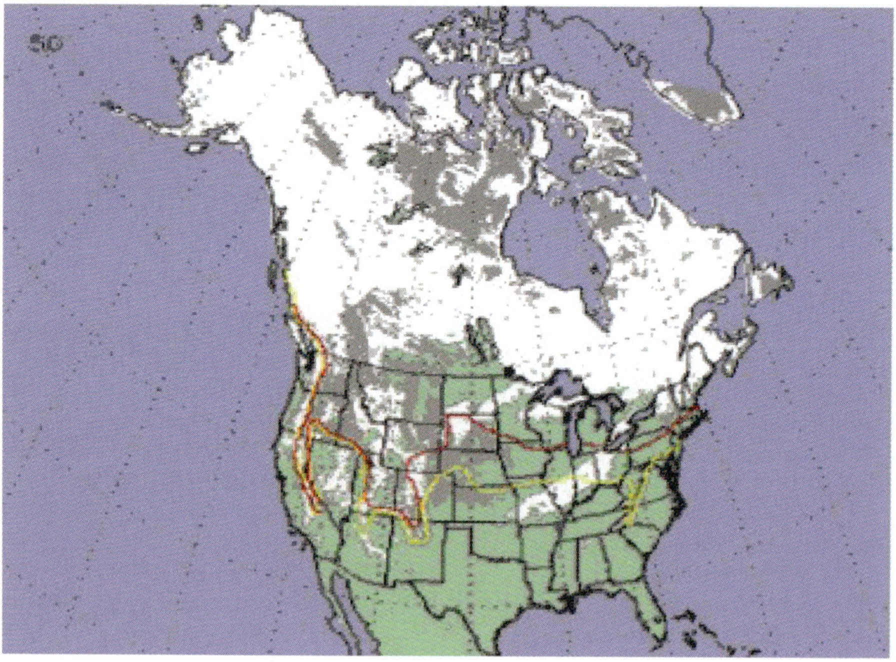

Fig. 7.9. The March 2000 snowline was much farther north than the average March snowline (*red line*) as depicted by the NASA MODIS instrument in a eight day composite image (Source: NASA)

Hemisphere (Holland, 1997). There are insufficient data to establish trends in precipitation over the oceans.

- In the mid- and high latitudes of the Northern Hemisphere over the latter half of the 20th century, it is likely that there has been a 2 to 4% increase in the frequency of heavy precipitation events. This may also be reflected in snowfall and accumulation (Fig. 7.9). Increases in heavy precipitation events can arise from a number of causes, e.g., changes in atmospheric moisture, thunderstorm activity and large-scale storm activity.
- It is likely that there has been a 2% increase in cloud cover over mid- to high latitude land areas during the 20th century. In most areas the trends relate well to the observed decrease in daily temperature range.

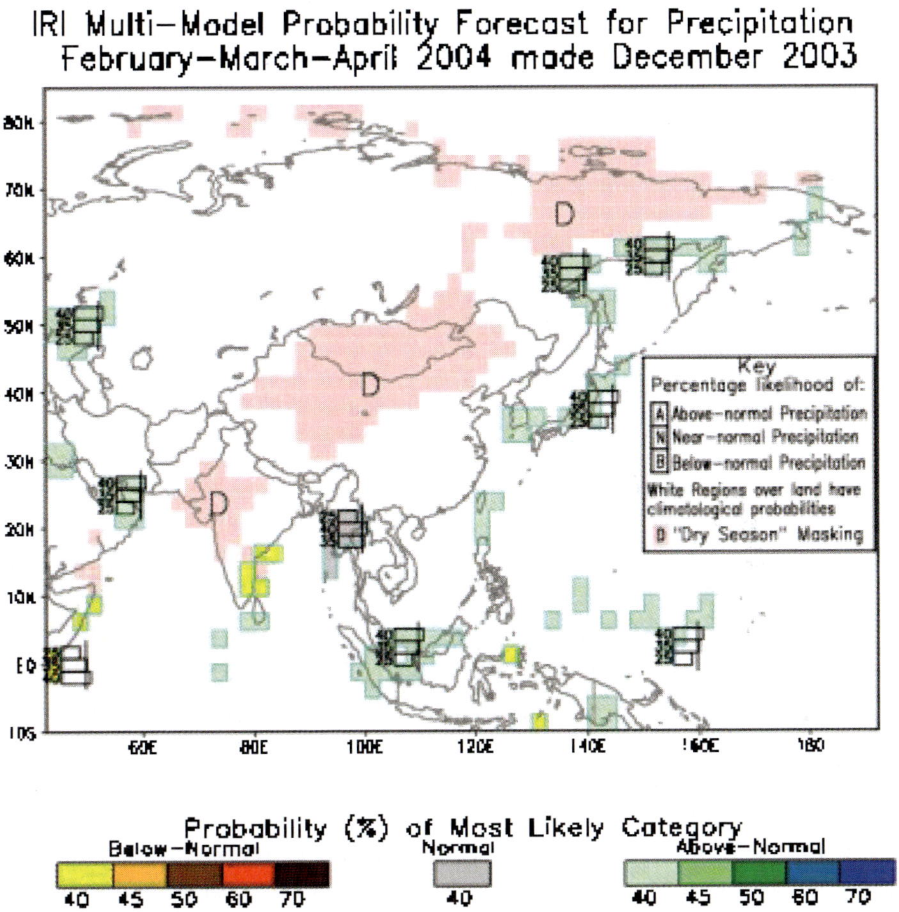

Fig. 7.10. Skill at predictions of seasonal weather has increased, especially utilizing ensemble modeling approaches and incorporating ocean data, as this prediction from Columbia University exemplifies (Source: International Research Institute – IRI)

- Since 1950 it is very likely that there has been a reduction in the frequency of extreme low temperatures, with a smaller increase in the frequency of extreme high temperatures.
- Warm episodes of the El Niño-Southern Oscillation (ENSO) phenomenon (which consistently affects regional variations of precipitation and temperature over much of the tropics, sub-tropics and some mid-latitude areas) have been more frequent, persistent and intense since the mid-1970s, compared with the previous 100 years and skill is increasing in longer-term predictions in many areas (Fig. 7.10).
- Over the 20th century (1900 to 1995), there were relatively small increases in global land areas experiencing severe drought or severe wetness. In many regions, these changes are dominated by inter-decadal and multi-decadal climate variability, such as the shift in ENSO towards more warm events.
- In some regions, such as parts of Asia and Africa, the frequency and intensity of droughts have been observed to increase in recent decades.

7.4 Interdecadal Climate Variability

Changes in climate occur as a result of both internal variability within the climate system and external factors (both natural and anthropogenic). The influence of external factors on climate can be broadly compared using the concept of radiative forcing. A positive radiative forcing, such as that produced by increasing concentrations of greenhouse gases, tends to warm the surface. A negative radiative forcing, which can arise from an increase in some types of aerosols (microscopic airborne particles) tends to cool the surface. Natural factors, such as changes in solar output or explosive volcanic activity, can also cause radiative forcing. Concentrations of atmospheric greenhouse gases and their radiative forcing have continued to increase as a result of human activities. This forcing on the larger scale may result in shifts in ocean temperature regimes, which then may be expressed at climate variability episodes.

Figure 7.11 demonstrates the ability of contemporary climate models to simulate the variability of major patterns such as the El Nino Southern Oscillation (ENSO, amplitude and frequency of temperature swings in the equatorial Pacific.) Climate models show conflicting results, varying from a slight decrease (Tett 1995; Knutson et al., 1997; Noda et al., 1999; Collins, 2000b; Washington et al., 2000; Fig. 7.11b) to a small increase in amplitude (Timmermann et al., 1999; Collins, 2000a; Fig. 7.11a). Kitzberger et. al. 2001 have shown that years characterized by widespread fires in the south-western United States (14 events between 1914–1987) and in northern Patagonia (9 events between 1938–1996) are similarly related to ENSO variation indicators. Fire event years are preceded by approximately 1–2 years of El Niño conditions leading to 1 year of La Niña conditions suggesting that major fire years follow a complete El Niño/La Niña cycle in both regions. Simulations of future wildfire areas in the US, eliminating management considerations, suggest that increasing ENSO variability may result in increasing the potential to change vegetation distribution and carbon budgets (Bachelet et. al., 2001).

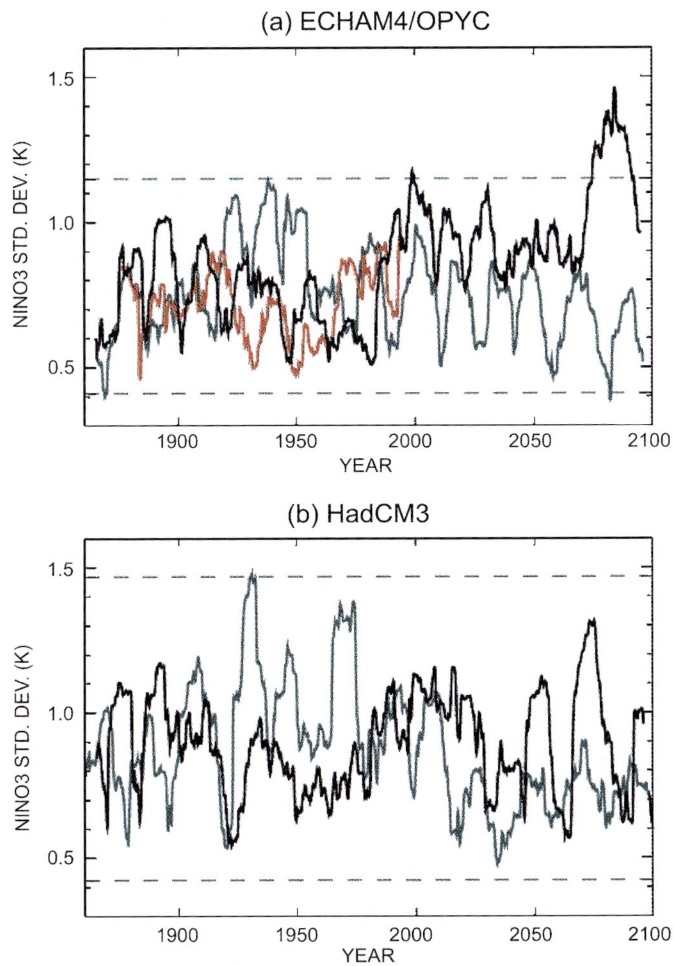

Fig. 7.11. Standard deviations of Niño-3 SST anomalies (Unit: °C) as a function of time during transient greenhouse warming simulations (*black line*) from 1860 to 2100 and for the same period of the control run (*green line*). Minimum and maximum standard deviations derived from the control run are denoted by the *dashed green lines*. A low-pass filter in the form of a sliding window of 10 years width was used to compute the standard deviations. (**a**) ECHAM4/OPYC model. Also shown is the time evolution of the standard deviation of the observed from 1860 to 1990 (*red line*). Both the simulated and observed SST anomalies exhibit trends towards stronger interannual variability, with pronounced inter-decadal variability superimposed, (reproduced from Timmermann et al., 1999), (**b**) HadCM3 (Collins, 2000b). (From IPCC 2001: Figure 9.26)

7.5 Economic and Social Consequences

The costs of extreme weather events have exhibited a rapid upward trend in recent decades. Yearly global economic losses from large events increased from US$3.9 billion year in the 1950s to US$40 billion year in the 1990s (all 1999 US$, uncorrected for purchasing power parity). Approximately one-quarter of the losses occurred in developing countries. The insured portion of these losses rose from a negligible level to US$9.2 billion annually during the same period. Including events of all sizes doubles these loss totals (Fig. 7.12). The costs of weather events have risen rapidly, despite significant and increasing efforts at fortifying infrastructure and enhancing disaster preparedness. These efforts dampen to an unknown degree the observed rise in loss costs, although the literature attempting to separate natural from human driving forces has not quantified this effect. As a measure of increasing insurance industry vulnerability, the ratio of global property/casualty insurance premiums to weather-related losses – an important indicator of adaptive capacity – fell by a factor of three between 1985 and 1999. Part of the observed upward trend in historical disaster losses is linked to socioeconomic factors – such as population growth, increased wealth, and urbanization in vulnerable areas – and part is linked to climatic factors such as observed changes in precipitation, flooding, and drought events. Precise attribution is complex, and there are differences in the balance of these two causes by region and by type of event. Notably, the growth rate in human-induced and non-weather-related losses has been far lower than that of weather-related events.

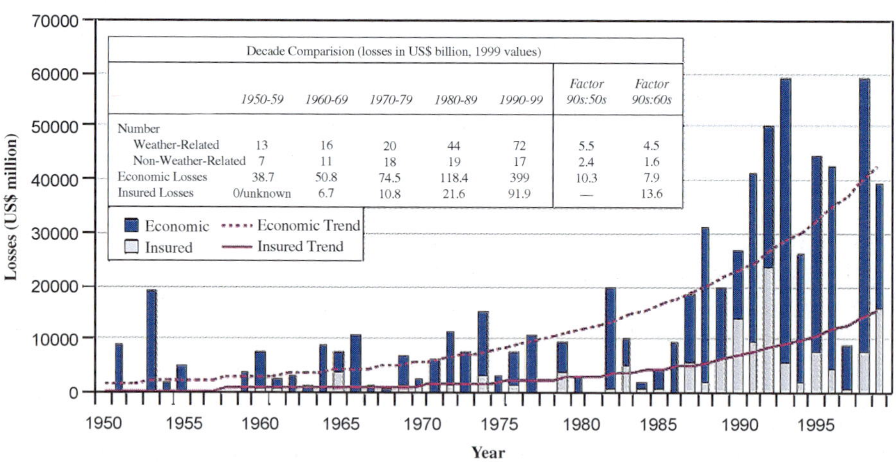

Fig. 7.12. Increases costs of natural disasters are another indication of the complex relationships between climate variability and world society (Source: IPCC)

By 2100 the concentration of carbon dioxide in the atmosphere could be between 90% and 250% higher than in 1750. Increases in temperature over this century without precedent in the last 10,000 years, with the earth's temperature expected to be up to 6 degrees above 1990 levels by 2100. By 2025 up to two thirds of the world's population may experience some form of water stress and also mean ever more extreme rainfall and flooding, and increasingly severe tropical storms.

By 2025 the number of people affected by desertification is expected to double to 1.8 billion, and many will be in Africa. Large parts of Africa and South America could lose their tropical rain forest by 2080. There is likely to be increased flooding, soil erosion, decreased crop yields, increased risk of epidemics. Deserts will grow. Disease will spread. Many species of plant and wildlife are expected to become extinct. Coastal settlements like Senegal, Egypt and Bangladesh are likely to face inundation; in other countries, like Tanzania, rivers will start to run dry. The cost of adapting to these changes is likely to put back development still further.

7.6 Conclusions

When the WWII generation was growing up the world's population was under 3 billion. During the next 60 to 100 years, it is likely to exceed 9 billion. The six warmest years of the twentieth century occurred in the last decade and 25% of the world's land area is affected by soil erosion or other land degradation. Snow and ice cover is estimated to have decreased by 10% since the 1960s according to satellite photography. Since 1980 10% of the forests in the developing world have been lost. 27% of the world's coral reefs have already been lost. This process is accelerating. For some parts of the world, particularly the poorer parts, the effects will be catastrophic. Climate variability, brought on by decal shifts in ocean temperature regimes, often drives the occurrence of weather related natural disasters. As populations increase and agricultural systems are placed under more pressure, events which in stable social systems were historically an annoyance will become increasingly dire.

New techniques, such as remote sensing and sophisticated computer models may give warning of when and where such events are most likely to occur. Information of this sort will not be of greatest value until societies work together on a planet-wide basis to share resources and response to disasters. This will necessitate not only organizations such as the United Nations taking a more active role, but also developed nations viewing world economic stability (with an alleviation of poverty in underdeveloped nations) as the cornerstone of peace and collective security. Although questions remain about climate change and climate variability in the role in the increasing frequency of natural disasters, no question can be allowed to remain as to the every human being's right to clean water, clean air, proper diet, and a sense that world leadership is collectively concerned with the long-term amelioration of the adverse impacts of weather related natural disasters.

References

Aber John, Neilson RP, McNulty S, Lenihan JM, Bachelet D, and Drapek RJ. Forest Processes and Global Change: Predicting the Effects of Individual and Multipule Stressors. Bioscience. Vol. 51 No. 9. 2–001. 735–751.

Bachelet, Dominique, Neilson RP, Leniham JM, and Drapek RJ. Climate Change Effects on Vegetation Distribution and Carbon Budget in the United States. Ecosystems. Vol. 4. 2001. 164–185.

Collins, M., 2000a: The El-Niño Southern Oscillation in the second Hadley Centre coupled model and its response to greenhouse warming. J. Climate, 13, 1299–1312.

Collins, M., 2000b: Understanding uncertainties in the response of ENSO to greenhouse warming. Geophys. Res. Lett., 27, 3509–3512.

Chubarova, Ye N, Prilepsky NG, Riebau AR, Shoettle AW, Musselman R, Uliumdzhieva NN, Yurova, AYu, Potter B, and Rublev AN. A Natural Fire Experiment in Central Russia: Meteorology, Radiative Transfer and Optical Properties of the Atmosphere and Resulting Effects on Boreal Forest Plants. AMS Orlando Florida Specialty Conference on Wildfire. 2003. 5 pgs.

Holland GT, 1997: The maximum potential intensity of tropical cyclones. Journal of Atmospheric Science, 54, 2519–2541.

Intergovernmental Panel on Climate Change (IPCC). Climate Change 2001: Impacts, Adaptation and Vulnerability. 2001. 1005 pages.

Intergovernmental Panel on Climate Change (IPCC), 2001: Climate Change 2001: The Scientific Basis. Contribution of Working Group I to the Third Assessment Report of the Intergovernmental Panel on Climate Change [Houghton, J.T., Y. Ding, D.J. Griggs, M. Noguer, P.J. van der Linden, X. Dai, K. Maskell, and C.A. Johnson (eds.)]. Cambridge University Press, Cambridge, United Kingdom and New York, NY, USA, 881 pgs.

Intergovernmental Panel on Climate Change (IPCC). IPCC Workshop on Changes in Extreme Weather and Climate Events. IPCC Working Group II. Meeting Report. 2002. 107 pgs.

Jones RN, Hennessy KL, Page CM, Walsh KJE, and Whetton PH, 1999: An Analysis of the Effects of Kyoto Protocol on Pacific Island Countries, Part 2: Regional Climate Change Scenarios and Risk Assessment Methods. A research report prepared for the South Pacific Regional Environment Programme, Apia, Samoa and CSIRO Atmospheric Research, Canberra, Australia, 61 pp.

Kitzberger ,T; T.W. Swetnam; T.T. Veblen 2001. Inter-hemispheric synchrony of forest fires and the El Niño-Southern Oscillation. Global Ecology & Biogeography (2001) 10: 315–326

Knutson, T.R., S. Manabe and D. Gu, 1997: Simulated ENSO in a global coupled ocean-atmosphere model: multidecadal amplitude modulation and CO_2-sensitivity. J. Climate, 10, 138–161.

Noda, A., K. Yoshimatsu, S. Yukimoto, K. Yamaguchi and S. Yamaki, 1999b: Relationship between natural variability and CO_2-induced warming pattern: MRI AOGCM Experiment. 10th Symposium on Global Change Studies, 10–15 January 1999, Dallas, Texas. American Meteorological Society, Boston. Mass. pp. 359–362

Tett, S.F.B., 1995: Simulation of El Niño-Southern Oscillation-like variability in a global coupled AOGCM and its response to CO_2-increase. J. Climate, 8, 1473–1502.

Timmermann, A., J. Oberhuber, A. Bacher, M. Esch, M. Latif and E. Roeckner, 1999: Increased El Niño frequency in a climate model forced by future greenhouse warming. Nature, 398, 694–696.

Tonkin H, Landsea C, Holland GJ, and Li S, 1997: Tropical cyclones and climate change: a preliminary assessment. In: Assessing Climate Change Results from the Model Evaluation Consortium for Climate Assessment [Howe W and Henderson-Sellers A (eds.)]. Gordon and Breach, Sydney, Australia. pp. 327–360.

Washington, W.M., J.W. Weatherly, G.A. Meehl, A.J. Semtner Jr., T.W. Bettge, A.P. Craig, W.G. Strand Jr., J.M. Arblaster, V.B. Wayland, R. James and Y. Zhang, 2000: Parallel climate model (PCM) control and transient simulations. Clim. Dyn., 16, 755–774.

Impacts of Tropical Cyclones on Chinese Lowland Agriculture and Coastal Fisheries

Ming Xu, Qiuzhen Yang, Ming Ying

Abstract Tropical cyclones are perhaps the most devastating of all natural disasters and the vulnerability to tropical cyclones is becoming more pronounced because of high population growth rates in the tropical and subtropical coastal regions. Tropical cyclones cause a range of destruction in the coastal areas. Various features of tropical cyclones affecting China and their impacts on lowland agriculture and coastal fisheries were described.

8.1 Introduction

The natural disaster is a crucial issue in human society nowadays, and tropical cyclones are perhaps the most devastating of natural disasters both because of the loss of human life they cause and the large economic losses they induce. On an average, 80–100 tropical cyclones happen over the world and they cause 20,000 deaths and the economic losses worth 7 billion dollars per year. In November 1970, a tropical cyclone hit Bangladesh and more than 300 thousand people died and made 10 million people homeless. Vulnerability to tropical cyclones is becoming more pronounced because of the fastest population growth in tropical and subtropical coastal regions.

A tropical cyclone is the generic term for a non-frontal synoptic-scale low pressure system originating over tropical or subtropical waters with organized convection and definite cyclonic surface wind circulation. Tropical cyclones with maximum sustained surface winds of less than 17 m/s are generally called "tropical depressions." Once a tropical cyclone achieves surface wind strengths of at least 17 m/s, it is typically called a "tropical storm" or "tropical cyclone" and assigned a name. If the surface winds reach 33 m/s, the storm is called a "typhoon" (the northwest Pacific Ocean), a "hurricane" (the North Atlantic Ocean and the northeast Pacific Ocean), or a "severe tropical cyclone" (the southwest Pacific Ocean and southeast Indian Ocean). Tropical cyclones derive energy primarily from evaporation from the ocean and the associated condensation in convective clouds concentrated near their center, as compared to mid-latitude storms that primarily obtain energy from horizontal temperature gradients in the atmosphere. Additionally, tropical cyclones are characterized by a "warm core" (relatively warmer than the environment at the same pressure level) in the troposphere. It is this unique warm-core structure within a tropical cyclone that produces very strong winds near the surface and causes damage to coastal regions and islands through extreme wind, storm surge, and wave action.

As each attack of tropical cyclone causes huge damages, it is therefore of great significance to study the forecasting and mitigation of tropical cyclone disasters. In this paper, we use the information we have collected to analyze the feature of impacts of tropical cyclones on Chinese lowland agriculture and coastal fisheries.

8.2 Destruction Caused by Tropical Storms in Coastal Areas

The tropical cyclone causes irreparable damage to the agriculture ranches and forests on coastal areas. The loss to an agriculture system can be categorised (Das 2003) as follows:

- Destruction of vegetation, crops, orchards and livestock in coastal areas;
- Damage to irrigation facilities such as canals, wells and tanks; and
- Long-term loss of soil fertility from saline deposits over land flooded by sea water.

The loss caused by a single storm may run into millions of dollars (Holland and Elsberry 1993). This is particularly so in the case of developing countries. Coastal areas in developing countries suffer great loss of life. These areas are also densely populated and are centres of brisk marine activity. Most of the population dwell in temporary thatched houses and the farmers have small land holdings. The lack of cyclone shelters and proper escape routes, the slow mode of transportation and the low elevation of the estuarine area all contribute to the regular catastrophes which occur here.

8.3 Features of TCs Affecting China and TC Disasters

Preferred regions of tropical cyclone formation include the western Atlantic, eastern Pacific, western North Pacific, north Indian Ocean, south Indian Ocean, and Australian–southwest Pacific. Almost 1/3 of all the tropical cyclones around the world are formed in western North Pacific, so this area is visited by tropical cyclones most frequently.

Located on the west coast of western North Pacific Ocean with a 18,000 km coast line, China is greatly affected by tropical cyclones. There are two types of tropical cyclones affecting China, one coming from the western North Pacific ocean and the other forming in the South China Sea region.

Tropical cyclones landing in China always carry grave consequences to the country. From 1949 to 2000, there were altogether 478 tropical cyclones attacking China, with south China and east China being most fragile to tropical cyclones' attacks. Figures 8.1 and 8.2 show the monthly distribution and yearly variation of number of tropical cyclones landing in China. A single peak in August is shown in Fig. 8.1, which means China is easily attacked by tropical cyclones in August. In Fig. 8.2, interannual variation of number of tropical cyclones landing in China is obvious, and a slightly decreasing trend and inter-decadal variation can also be detected. Further analysis shows that, China met less tropical cyclones than normal in the year of El-Niño, while it met more tropical cyclones in the year of La-Niña. So the number of tropical cyclones landing in China varies with the atmospheric circulation and sea condition, which means prediction of number of landing tropical cyclones perhaps can be partly made by analyzing atmosphere and sea condition (Henderson-Sellers et al. 1998).

Tropical cyclone disasters result mainly from huge gales, rainstorms and storm-surges caused by tropical cyclones. Grave disasters are mainly caused

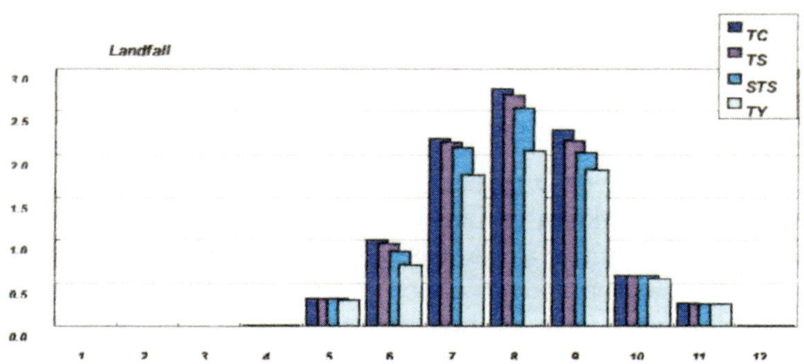

Fig. 8.1. Monthly distribution of number of TCs landing in China

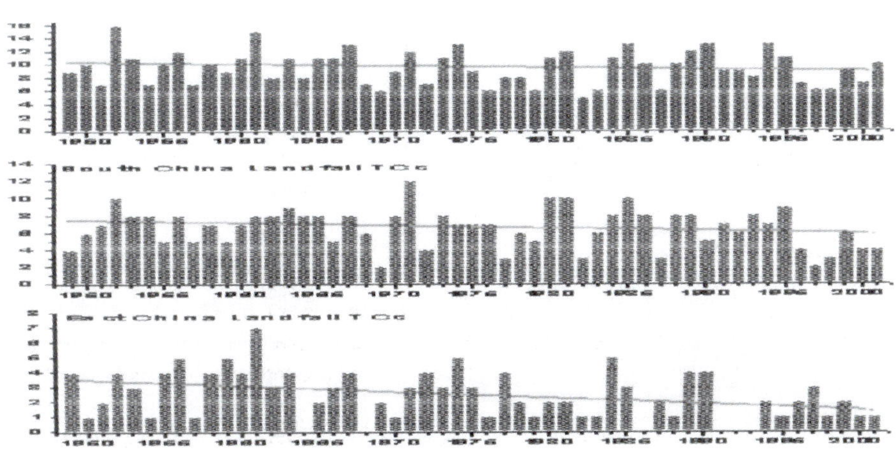

Fig. 8.2. Interannual variation of number of TCs landing in China

by landing tropical cyclones. Tropical cyclone disasters affecting China have the following features:

8.3.1 High Frequency in Occurrence

China is affected by typhoons throughout the whole year. The number of the tropical cyclones landing in China reaches a maximum of 15 and a minimum 4 per year, with an average of 9.2 per year. Grave typhoon disasters occur every year while ordinary typhoon disasters occur with an average of more than 9 times a year. The highest frequency of occurrence is from July to August.

8.3.2 Wide Range of Affecting Areas

From Liaoning province in Northeast China to Hainan province in South China, all the coastal provinces have been affected by tropical cyclones. Moreover, tropical cyclones could invade inland provinces occasionally. Except for six provinces and autonomous municipalities in west China, all the provinces in China have the records of being affected by tropical cyclones (Liang biqi et al. 1994). A tropical cyclone entering inland could bring grave disaster. For instance, No. 7,503 typhoon affected 8 provinces in the country and caused damages to more than 14,933,000 hectares farmland and direct economic losses more than 10 billion RMB *yuan*.

8.3.3 Violent in Sudden Occurence

Typhoon disasters are all caused by the sudden occurrence of typhoons. The disastrous weather, and storm surge brought by typhoons in particular, could often cause severe disasters in several hours. The severe typhoon disasters along the coast are all caused in one or two days, even in several hours. For instance, No. 8,309 typhoon caused disasters on about 134,000 hectares of farmland and turned many villages and towns into a vast expanse of water in only one or two hours after landing in the Pearl River.

8.3.4 Remarkable in Their Chain Effects

Typhoons affecting China not only cause disasters through huge gales, rainstorms, storm-surge and rough sea tides brought by them, but also result in a group of disasters from their chain effects. For example, the typhoon rainstorms can create flood, water-logging as well as mud-rock flow, avalanche and landslides and soil erosion. Moreover, the reverse irrigation of sea water caused by storm surges can form inner water-logging and cause land to be

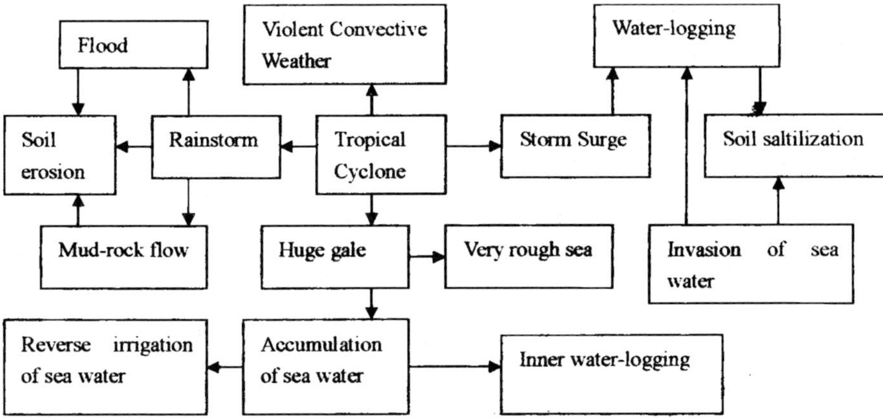

Fig. 8.3. Flow chart showing the chain effects of typhoon disasters

salted. The system of a group of disasters consisting of the chain effects of typhoon disasters can be seen in the flow chart (Fig. 8.3).

Typhoons and their related disasters often cause huge disasters and bring about grave losses in some areas. For instance, No. 8,607 and No. 9,107 typhoons caused great direct economic loss due to their multiple effects.

8.4 Impacts of Tropical Cyclones on Lowland Agriculture and Coastal Fisheries

Tropical cyclone disasters seriously affect lowland agriculture and coastal fisheries in particular. In China, the total area of lowland, swamp and wetland is 25 million hectares, about one quarter of area of arable land of the whole country. And the tropical cyclone prevailing time just meets the key period for crops' growth in coastal area. So every landing tropical cyclone causes disasters on large areas of cropland and decreases output of grains. From 1949 to 2000, the total area of farmland that was drowned by tropical cyclones attained 79,892.4 thousand hectares. Analysis of about 10 typhoons with the most disastrous effects in China showed that on average, every typhoon causes a disaster on more than 1,470,000 hectares of farmland.

Table 8.1 shows the area of damaged farmlands and number of affecting tropical cyclones in China during 52 years (1949–2000). Though the number of tropical cyclones landing in China decreases (Fig. 8.2), the number of tropical cyclones causing damages to agriculture increased quite obviously, while the area of damaged farmland increased too. Especially in 1991–2000, it is very impressing. Considering the quick development of China after 1980, it reflects the social factor of natural disaster influences.

Besides direct effects of tropical cyclones on agriculture, some indirect damages can also be made. These include serious destruction of the installations of water conservancy projects, soil-salanization, increased incidence

Table 8.1. Area of damaged farmlands and number of affecting tropical cyclones

Year	Area of damaged farmlands (Unit thousand hectares)	Number of TCs making agricultural damages
1949–1950	320	1
1951–1960	3,566.4	10
1961–1970	5,566.4	13
1971–1980	17,454.0	7
1981–1990	15,854.1	33
1991–2000	37,131.5	61

of insect pests after tropical cyclone invasion, etc. Because the relevant information is difficult to collect, we cannot provide concrete details on this topic here.

Coastal fisheries in China consist of three parts: off shore fisheries, sea water aquaculture and inland fresh water aquaculture. Tropical cyclones can cause sinking of fishboats, damage to aquafarms, and death of fishermen, thus leading to grave losses of fisheries. Tropical cyclone disasters also cut the piscatorial working time and thus reduce the production. Owing to the limited information we have collected, the fishery damages caused by tropical cyclones are represented by two indices: number of sunk fishboats and area of damaged fishponds.

To solidify the analysis of impacts of tropical cyclones on lowland agriculture and coastal fisheries, we choose Shanghai, Zhejiang province and Jiangsu province in East China to make further studies.

The coastal district of East China is on the delta regions of great rivers and the crossing belt of sea and land and is often invaded by sea tides, and the 70% of arable lands are frequently influenced by floods. Tropical cyclones, rainstorms and tides often bring about floods with severe influence on the agriculture and fisheries.

Table 8.2 shows seasonal distribution of frequency of tropical cyclones causing agricultural and fishery damages. From the third decade of July to the third decade of November, workers of agriculture and fisheries in this area should be cautious about tropical cyclone disasters.

Table 8.2. Seasonal distribution of frequency of tropical cyclones causing agricultural and fishery damages (1949–2000)

	Jun 1D	Jun 2D	Jun 3D	Jul 1D	Jul 2D	Jul 3D	Aug 1D	Aug 2D	Aug 3D	Nov 1D	Nov 2D	Nov 3D	Oct 1D	Total
Zhejiang	0	0	0	1	2	7	8	7	10	7	5	4	1	52
Shang-hai	1	0	1	1	2	7	5	7	8	2	5	3	1	43
Jiangsu	0	0	0	0	0	5	3	4	5	4	3	3	1	28
Total	1	0	1	2	4	19	16	18	23	13	13	10	3	123

Table 8.3 shows frequency of tropical cyclones causing agricultural and fishery damages every ten years. Increasing trend can be seen in this table, it is consistent with the fact exhibited in Table 8.1.

The losses to agriculture and fisheries during 1949 to 2000 are shown in Table 8.4. Zhejiang had heavy damages in fisheries, while agriculture in Jiangsu suffered great losses (Zikang and Weilun 1994).

Description of four particulars cases given below, provides an idea of the nature of damages caused by tropical cyclones.

Case #1: On August 1st, 1956, a severe typhoon landed in Xiangshan in Zhejiang province and 75 counties in Zhejiang encountered great disasters. 400,000 ha farmlands were flooded, more than 3,500 boats were damaged, and 4,925 persons were dead. In Shanghai, 18,000 ha of farmlands were flooded, while 2 million ha of farmlands were flooded in Jiangsu province.

Case #2: During 2–4 August, 1960, Shanghai and Jiangsu were affected by a typhoon. In Shanghai, 33.3 thousand ha of farmlands were flooded, 16.4 million fish were lost, and 67 boats were sunk or damaged. Strong rainstorms attacked east part of Jiangsu and 7.6 million ha of farmlands were flooded.

Case #3: On July 27, 1987, Zhejiang was hit by typhoon no. 8,707 (TY Alex), 619 boats were sunk, and 343 thousand ha of farmlands were flooded. 422 thousand ha of farmlands were flooded in Jiangsu.

Case #4: During 18–19 August, typhoon no. 9,711 (TY Winnie) landed in Wenling, Zhejiang province. 733,000 ha of farmlands were flooded, and 16,000 ha of fishponds were damaged in Zhejiang. In Shanghai, 30,000 ha of farmlands were flooded. In Jiangsu, 667,000 ha of farmlands were flooded.

Table 8.3. Frequency of tropical cyclones causing agricultural and fishery damages every ten years

Year	Zhejiang	Shanghai	Jiangsu
1949–1950	1	1	0
1951–1960	10	9	2
1961–1970	4	8	2
1971–1980	5	6	3
1981–1990	16	12	10
1991–2000	16	7	11

Table 8.4. Losses to agriculture and fisheries (1949–2000)

	Flooded Farmland (1,000 ha)	Damaged Fishponds (1,000 ha)	SunkFish-boats
Zhejiang	864.3	44.4	6,424
Shanghai	546.3	4.7	75
Jiangsu	10,846	7.5	221

Table 8.5. Water requirement of crops and rainfall in summer in Zhejiang

	Jiaxing	Huzhou	Shaoxing	Hangzhou	Ningbo	Wenzhou	Linhai
Total Rainfall [R] (mm)	333.1	355.6	426.5	452.2	431.1	564.5	593.4
Rainfall from Tropical cyclones [r] (mm)	112.3	104.8	154.6	141.6	188.3	275.6	294.8
R – r (mm)	220.8	250.8	271.9	310.6	242.8	288.9	298.6
Water Requirement Q (mm)	354.6	349.5	349.6	337.6	363.2	357.5	351.2
Q – (R – r) (mm)	133.8	98.7	77.7	27.0	120.4	68.6	52.6

The analysis above focused on the damages and losses induced by tropical cyclones. In fact, tropical cyclones also bring some beneficial effects on agriculture. Rainfall brought by tropical cyclones often reduces the drought in summer in Chinese coastal area. In some area the crop's growth depends on tropical cyclones' rainfall intensity. Table 8.5 shows the water requirement of crops and the rainfall in summer in Zhejiang. If there is no rainfall from tropical cyclones, crops would not grow well.

Though the beneficial effects of tropical cyclones on agriculture and fisheries are nontrivial, few studies have been conducted on this aspect. As an important aspect of tropical cyclone impact evaluation, it should be strengthened in the future.

References

Das HP (2003) Agrometeorological aspects of drought and desertification. In: Agrometeorology related to extreme Events. WMO No. 943, Switzerland, World Meteorological Organization

Henderson-Sellers A, Zhang H, Berz G, Emanuel K, Gray W, Landsea C, Holland G, Lighthill J, Shieh SL, Webster P, McGuffier K (1998) Tropical Cyclones and Global Climate Change: A Post-IPCC Assessment, Bulletin of the American Meteorological Society, vol 79(1). pp 19–38

Holland GJ, Elsberry RL (1993) Tropical cyclones a natural hazards: a challenge for the IDNDR. In: Lighthill J, Emanuel K, Holland GJ, Zhang Z (eds) Tropical cyclone disasters, pp 17–30

Liang Biqi, Liang Jingping, Wen Zhiping (1994) The Characteristics of Typhoon Disasters and Its Effects on Economic Development in Guangdong Province, Natural Disaster Reduction in China, vol 3(4). pp 145–150

Zikang Zhou, Weilun Liu (1994) Effects of Typhoon on the Agriculture of Zhejiang Province and Its Countermeasures (in Chinese), Journal of Hangzhou University, vol 2(1). pp 92–98

CHAPTER 9

Frost and High Temperature Injury in China

Du Yaodong

Abstract Frost and high temperature injury are two important agrometeorological disasters in China. In this paper, the concept of frost and its differences from other low temperature injuries are presented. Recent studies on injury mechanism of frost are introduced. Frost-prone areas, crops harmed, and statistical characteristics in China are analyzed and summarized. Frost preventive and preparedness measures in China are discussed. The detrimental impacts of high temperature injury on crops, vegetables, fruit trees, milk yield, livestock, and fish in China are also briefly reviewed.

9.1 Introduction

Frost and high temperature injury are two major agrometeorological disasters. Frost occurs in almost all parts of China and brings significant damage to many crops. Therefore, much attention has been paid to frost in the past in China (Zhang Yangcai, et al. 1991). There were many historical records on frost more than 3,000 years ago, but systematic frost study was not made until after 1949 when many meteorological stations were established and field studies started to carry out the observation and investigation of frost.

In this paper, the concept of frost, and injury mechanism of frost, frost-prone areas, crops harmed, and such statistical characteristics as frequency, distributions and inter-annual changes of autumn frost date, spring frost date, frost-free season, as well as the relationship between climate change and frost in China are analyzed and summarized. Preventive measures and proactive preparedness steps to reduce the impact of frost in China are given. High temperature injury will happen when temperature exceeds the upper limit of optimum temperature range for growth and development of plants and animals. The detrimental impacts of high temperature injury on crops, vegetables, fruit trees, milk yield, livestock and fish are introduced and discussed briefly.

9.2 Frost

9.2.1 Concept and Harmed Mechanism of Frost

In China, frost has different meanings in each of the subjects. Currently, the recognized definition of frost is a phenomenon that temperature on the soil surface or near the plant canopy drops to below 0 °C in a short time, and crops suffer injuries (Zhu Binghai, et al. 1985).

According to a recent study, injury mechanism of frost is concluded as follows (Zhang Yangcai, et al. 1991): frost does not result directly from low temperature, but from icing of plant tissue. The plant temperature below 0 °C is a necessary but not a sufficient condition for icing. Ice nucleation active bacteria (INA bacteria) is an inducement of icing. The density of INA bacteria is closely correlated with the icing temperature of plant tissues. In the past, it was generally considered that the injured degree of crop exposed to frost mainly was controlled by low temperature intensity. In the late 1970's, it was found that INA bacteria played an important role in the occurrence of frost (Lindow, et al. 1978). Recent research has indicated that there are a lot of INA bacteria in many plants of China (Sun Fuzai, et al. 1989). These INA bacteria can induce plants to form ice under the condition of relatively high temperature (Feng Yuxiang, 1990; Liu JiangHua, et al. 1990).

There are three types of icing: cell-to-cell icing, intracellular icing, and, plasmolysis. Only intracellular icing or cell-to-cell icing can cause cell to die thoroughly. If temperature ascends slowly in the course of defrosting, cell-to-cell icing or plasmolysis within the enduring ability does not cause cell to die.

Plant death is not the only result of frost. When some tissues have iced within their enduring ability, the cell is alive, but a film of the cell has been injured to a certain extent. This will adversely affect later growth of the plant.

Table 9.1. Difference among frost, freezing, cool, and chilling injuries

Types	Temperature (°C)	Occurring season	Physiological reaction	Harmed crop	Crop state
Frost	< 0	Comparative warm season (transition)	Dehydration, icing in short time	Winter crop, fruit tree and vegetable	Normal growth
Freezing	< 0	Winter, early spring or late autumn	Dehydration, icing	Winter crop, fruit tree	Wintering season
Cool	10–23	Warm season	Growth and development handicap	Thermophilic crop	Active growth
Chilling	0–10	Winter season	Physiological handicap	Tropical, subtropical crop	Slow growth

Recently, distinction between frost and freezing injury, cool injury, and chilling injury (Cui Duzhang, 1999) has been identified (Table 9.1).

9.2.2
Frost Area

9.2.2.1
Monthly Average Frost Area

Figure 9.1 depicts the monthly average frost area across China. Seen from Fig. 9.1, September, January, and May are the top three months in average frost area (Feng Yuxiang, et al. 1996). September is the mature season for maturity of autumn crops in the north of China and is also a period of sharply dropping temperatures in the continent. Therefore, September is a month that is prone to a larger injury range, higher frequency, and biggest injury area.

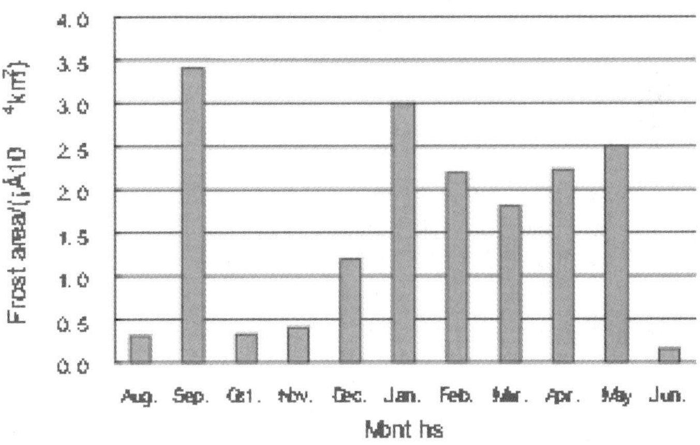

Fig. 9.1. Monthly average frost area across China

In January, because cold air is extremely severe, the range influenced by frost can extend to northern and central Hannan province and southern Yunnan province. Frost brings great damage to such crops as winter wheat, winter rape, and *Vicia faba* during the active growth period, as well as evergreen subtropical and tropical cash plants in the above areas.

In May, because spring-sowed crops have sprouted in northern China, frosts frequently pose severe injury to agriculture, so that average frost are in this month is maximum in the spring season.

9.2.2.2
Inter-annual Change of Frost Area

Annually, average frost area is 181.6 × 10^4 square kilometers, but there is a great difference from year to year. Frost area is as high as 568 × 10^4 square kilometers in 1977, but only 48.4 × 10^4 square kilometers in 1978 (Fig. 9.2) (Feng Yuxiang, et al. 1996).

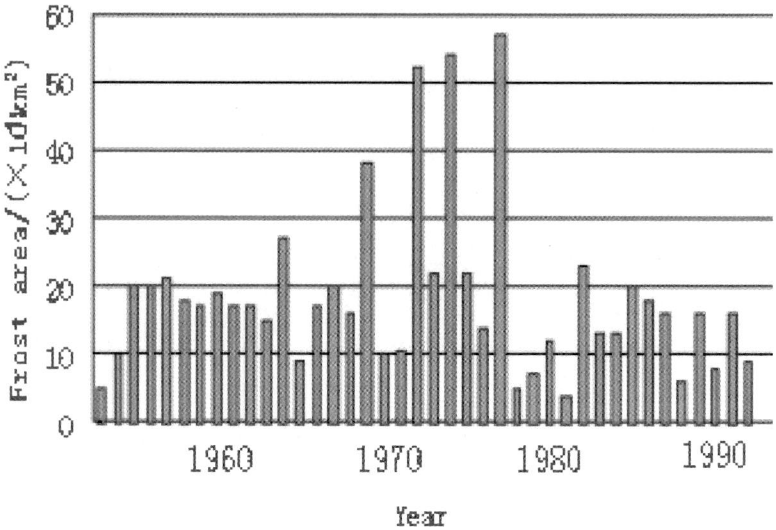

Fig. 9.2. Frost area in different years in China

9.2.3 Crops Harmed by Frost

Grain crops sown in the spring, such as maize, sorghum, millet, soybeans, buckwheat, naked oats, and potatoes, frequently suffer frost hazards in the northeast and northwest parts of China. Generally, frost brings hazards to the above crops during seedling and mature stages. Because of the short distance from the cold air source, the shorter growth season and preferred use of variety for long growth season in northeast and northwest parts of China, serious frost is more frequently seen. For instance, large scale frost between 20–24 May 1974, caused seedlings of more than 6,000 square kilometers to freeze to death only in Heilongjiang province, north China (Feng Yuxiang, et al. 1996). Cryophilic crops such as wheat, rape, *Vicia faba*, planted in cold areas and in the cold season in warm areas, easily suffer more frost hazards.

Wheat is an important grain crop in China and its sowing date is different in the northern and southern parts of the country. In spring wheat areas to the north of the Great Wall, the crop is sown in early spring. Frost during the seedling stage can destroy the leaf but not greatly impact the yield. In winter wheat areas to the south of the Great Wall, the crop is sown in autumn. Generally, wheat does not suffer frost hazards during the seedling stage. But, after stem elongation stages, decreased anti-frost ability of wheat increases risk of spring frost damage to the crop. For instance, ten large-scale frosts took place from 1953 to 1963 in the Huang-Huai-Hai area. Among these, severe frost between 19–20 April 1954, resulted in over 50 percent of wheat freeze in northern Jiangsu and Anhui provinces and yield decrease by more than 50 percent in Henan, Hebei, and Shangdong provinces (Feng Yuxiang, et al. 1996).

Because temperature in the main cotton area of China is unstable in spring and autumn, frost tends to bring hazards to cotton during seedling and ripening stages. Frost frequency during the seedling stage is higher than that during the ripening stage, and the harmed area mainly includes northern Tiangshan Mountain, Hexi Corridor, southern Shanxi, western Liaoning, and northern China plain. Frost during the late growth period of cotton in the boll-forming period greatly decreases both its yield and quality.

Such fruit trees as apple, pear, grape in the north of China and laquat, banana, pineapple, longan in the south of China have suffered frost disaster in the past. Apple trees during the flowering stage have the weakest anti-frost ability. If occurring during flowering stage, frost could make many flowers and young fruits fall off. For example, there were eight frosts in Etuoke county, Inner Mongolia Autonomous Region in the 1970s, with the apple yield decreasing by over 50 percent. Three frosts near Helanshan Mountain, Ningxia Hui Autonomous Region in 1971, 1972, and 1980 made apple yield almost decrease to zero (Feng Yuxiang, et al. 1996).

Because it has lower anti-frost ability than apple flower, pear flower easily suffers from spring frost. Four frosts in Ningxia Hui Autonomous Region from 1969 to 1973 decreased pear yield by over 70 percent.

Young bud and young leaf of grape have low resistance to frost, and the flower is sensitive to low temperature, therefore, frost could bring severe damage to grapes. For example, frost between 20–25 April 1986 caused the buds of grapevines to die by over 70 percent of Shanshan, eastern XinJiang.

Because of lowering anti-frost ability of flower, fruit, and leaf, the banana tree easily can partly and entirely die if it encounters frost. Frost in January 1992 brought about damage of 116 square kilometers in Maoming City and 50.7 square kilometers in Zhanjiang City, Guangdong province, and southern China, respectively.

Because of its low trunk, pineapple would easily be hurt if encountering advection-radiation frost. Frost between 11–13 January 1976, caused most pineapple to freeze to death, with yield decreasing by over 60 percent.

9.2.4 Statistical Characteristics of Frost

9.2.4.1 Frequency of Frost

Frost can occur in almost all parts of China. For spring frost, the highest frequency is recorded in west China, even as high as 64 percent. The frequency of frost is 47 percent in west Shangxi, north Jiangsu and north Anhui. The frequency of 25 percent is observed in southern parts of Jiangsu and Anhui.

For autumn frost, highest frequency of 39–46 percent is recorded in both eastern and western parts of China.

9.2.4.2 Distribution and Inter-annual Change of Autumn Frost Date

The higher the latitude, the earlier is the occurrence of average autumn frost date. The average autumn frost date is about the end of August to the north of Daxinganling mountain in the northeast China, but the average autumn frost date takes place in early January of the following year in the areas of central Taiwan, southern Fujiang, southern Guangdong, and southern Yunnan (Fig. 9.3) (Zhang Yangcai, et al. 1991).

The absolute variability of frost can be computed by equation (9.1):

$$V_a = \frac{1}{N} \sum_{i=1}^{N} |x_i - \overline{x}| \tag{9.1}$$

Where V_a is absolute variability, N, the number of years considered, x_i frost date in i year, $\overline{x}$, average frost date.

Absolute variability of autumn frost date is 4–7 days in northeast China, 6–8 days in north China plain, 8–10 days in middle and lower reaches of the Yangtse River, and 10–14 days in the Yungui plateau, respectively.

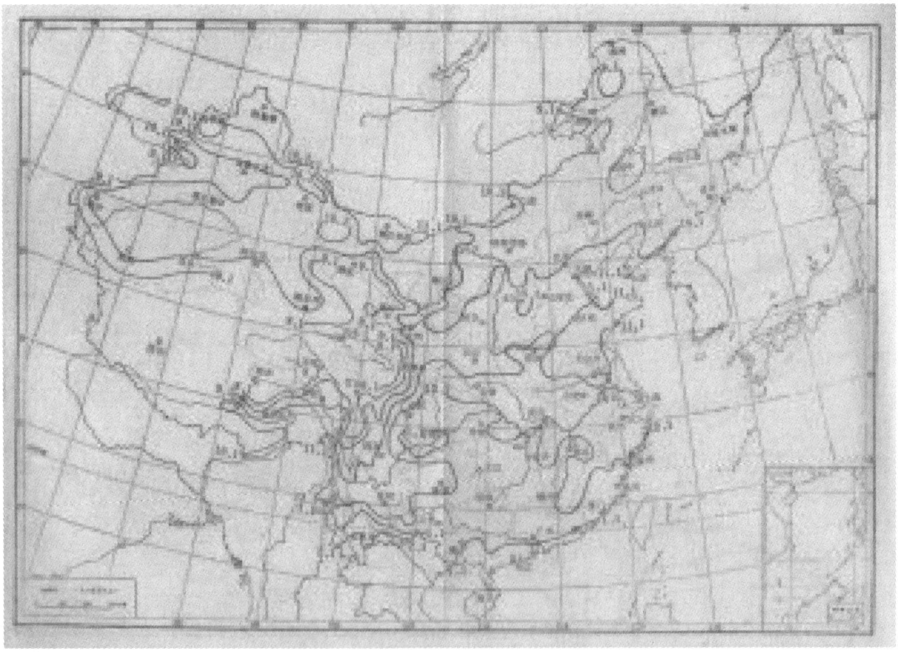

Fig. 9.3. Distribution of average autumn frost date in China

The earliest autumn frost date is 15 days ahead of the average autumn frost date in northeast China, about 20 days in the north China plain, and 30–45 days in the Yungui plateau. For areas with a shorter growth season, farmers like to plant late-maturing varieties for full use of frostfree season. If the autumn frost date occurs earlier, crops could suffer damages because they are not fully matured in time. Therefore, much attention must be given to the autumn frost.

9.2.4.3
Distribution and Inter-annual Change of Spring Frost Date

In contrast to the earlier autumn frost, higher latitudes lead to a later average spring frost date. The average spring frost dates occur in early March to the south of latitude 28° and between the end of March and early April to the south of Qinling Mountain, respectively. The average spring frost date does not occur until the middle ten days of June in the north of Daxinganling Mountain, northeastern China (Fig. 9.4) (Zhang Yangcai, et al. 1991).

According to equation (9.1), the absolute variability of spring frost date is 6–9 days in the northeast China plain, 8–13 days in the northern China plain, 6–13 days in the middle and lower reaches of Yangtse River, and 8–23 days in the Yungui plateau, respectively. Absolute variability of spring frost date also decreases with latitude's increase, similar to the distribution of autumn frost date.

Fig. 9.4. Distribution of average latest spring frost date in China

The latest spring frost date is 16–30 days behind average spring frost data in the east China plain, 22–31 days in the north China plain, and 22–36 days in the Yungui plateau. Because the absolute variability of spring frost date and the days of the latest spring frost date beyond the average spring frost date are comparatively large, spring frost is also of significant importance.

9.2.4.4 Distribution and Annual Change of Frostless Season

The average frost-free season is 100–150 days in northeast China, 150–200 days in the north China plain, about 250 days in the middle and lower reaches of Yangtse River, and more than 300 days in the south of Nanling Mountain, respectively. Frost does not occur in the south of Guangdong and Guangxi provinces all the year round (Fig. 9.5) (Zhang Yangcai, et al. 1991).

The absolute variability of frost-free season is 9–13 days in the northeast China plain, 14–16 days from the north China plain to the north of Yangtse River, 14–19 days from the south of Yangtse River to the north of Nanling Mountain, and more than 20 days in the Yungui plateau, respectively. The absolute variability of frost-free season is greater than that of autumn and spring frost date, indicating that frost-free season is more unstable than autumn and spring frost dates.

Fig. 9.5. Distribution of average frostless season in China

Difference of shortest frost-free season from average frost-free season is 20–42 days in the north China plain, 29–44 days in middle and lower reaches of the Yangtse River, and more than 40 days in the Yungui plateau, respectively. Seen from those figures, the shortest frost-free season is over 30 days less than average frost-free season in most parts of China. This issue must be considered in the prevention and preparedness of frost.

9.2.4.5
Climate Change and Frost in China

The climates in north and south China have gradually become warmer since the 1980s. Frost mitigation measures should be considered. In practice, frost still frequently takes place. There are two reasons. First, frost season begins to become more unstable. The second reason is an increased frost possibility with the use of late-maturing varieties and plastic film that is used widely in agriculture.

In order to explore the changing trends in frost areas, the 5-year moving average is computed (Fig. 9.6). Seen from Fig. 9.6, the average value in 1950s is slightly lower than that during 1953–2000. The average value in the 1960s is about equal to that during 1953–2000. The average value in the 1970s is significantly larger than that during 1953–2000. The average value in the 1980s is obviously lower than that during 1953–2000.

Although the 5-year moving average of spring frost date has become earlier since the 1980s, abnormal spring frost still occurs in some years. For example, spring frost on 5 June 1992 was 15–25 days later than average in eastern China. It brought about mass death of crops during seedling or transplanting stages. Spring frost on 2 May 1991, caused the death of 50 percent of cotton seedlings in Yuncheng district, Shanxi province of western China. According to the analysis of results from Beijing, since the 1980s autumn frost dates have become more stable, however, spring frost dates have become more unstable. This shows that there are abnormal spring frosts in obvious warming periods of climate.

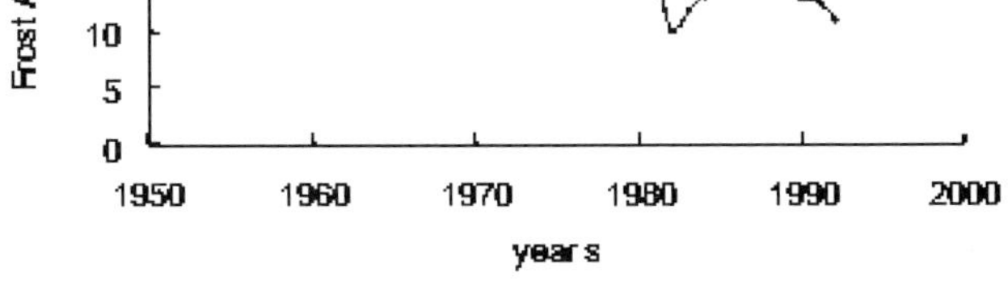

Fig. 9.6. Five-year moving average of frost area in China

9.2.5 Preventive Measures and Preparedness of Frost

Preventive measures and preparedness of frost in China are composed of long-term countermeasures and emergency measures. Long-term countermeasures include selection of favorable topography, breeding of anti-frost varieties, reasonable fertilizer supplies and timely sowing, etc. Emergency measures include artificial clouds of smoke, protective covering, irrigation, sprinkling crop with water and biological agents.

9.3 High Temperature Injury in China

9.3.1 Impact on Crops

Rice, cotton, and maize in the middle and lower reaches of the Yangtse River and the area of southern China and wheat and potato in the north of China frequently suffer from high temperature injury. From July to August, early and middle-season rice are at heading, flowering, milk-filling, or ripening stages in the middle and lower reaches of Yangtse River. Flowering and milk-filling stages for early-season rice also occur during June and July in southern China. It is reported that milk-filling for early-season rice is the fastest at 6–15 days after flowering and the optimum temperature is 25–28 °C in this period of time. If the temperature exceeds 32 °C, the crop has an adverse reaction to excessive heat. If the temperature exceeds 35 °C, the thousand-grain weight of rice can be decreased by as much as 0.42–3.2 grams. As the temperature increases further, the thousand-grain weight also continues to decline. Full-blossom stage of rice is also a stage sensitive to high temperature. At full-blossom stage, seed-setting percentage of rice decreases with the intensity and time of high temperature. According to the study from the Plant Physiological Institute, Chinese Academy of Sciences, seed-setting percentage of rice is 80.9 percent at 28 °C. As temperature ascends to 38 °C, seed-setting percentage of rice declines to zero.

In the middle and lower reaches of Yangtse River and southern China, there are a lot of days when maximum temperatures exceed 35 °C during July to August. For example, days of high temperature over 35 °C are more than 20 in Changsha, capital of Hunan province and Hangzhou, capital of Zhejiang province and 15 in Hefei, capital of Anhui province and Nanjing, capital of Jiangsu province. Longer days of high temperature have greater impact on yield in some years. For example, at milk-filling stage of early-season rice, days of high temperature over 35 °C are 21 in Haizhou, 17 in Nanchang, 16 in Changsha and Hefei, 15 in Wuhan and Nanjing, 13 in Shanghai, respectively. Continuous high temperature causes seed-setting percentage of rice to drop to less than 70 percent.

High temperature injury can shorten milk-filling time of wheat, thus, shortening accumulation of dry matter prematurely. High temperature injury can also accelerate grain's respiration that results in the rapid exhaustion of dry matter and the drop of thousand-grain weight. The harmful effect of high temperature is usually aggravated by the combined effect of low humidity and dry winds. Dry and hot wind is one of the major factors influencing yield of wheat in north and northwest China. High temperature or rapid ascent in temperature after rain can also result in the "green wilt" or death of wheat. Huang-Huai-Hai plain, Hexi corridor and Xinjiang are three severest areas for dry and hot wind in China. Their frequency of occurrence is as high as 30–40 percent.

High temperature is an important reason for the abscission of squares and bolls of cotton. According to research, as the daily mean temperature reaches over 30 °C and maximum temperature is more than 33 °C, abscission of squares and bolls of cotton is aggravated. As the temperature reaches over 36 °C, photosynthesis is close to zero and respiration is enhanced which accelerates the consumption of nutrients. In the cotton area of Yangtse River valley, the high temperature period mainly occurs in the last 10 days of July and first 10 days of August. In the Tulufang basin of Xinjiang cotton area, high temperature in summer often results in serious abscission of squares and bolls of cotton combined with atmospheric drought.

Rapeseed will be prompted to ripen when daily maximum temperature is higher than 30 °C. In China, the daily maximum temperature tends to exceed the optimum 22 °C, sometimes reaching over 30 °C at milk-filling and ripening stage of winter rape. Rapeseeds are frequently being prompted to ripen. The ripening date of spring rape is about 20 days later than winter rape. Therefore, frequency of high temperature is greater and yield is low for spring rape.

9.3.2 Impact on Vegetables

High temperature restrains photosynthesis of vegetable and affects development of budding, flowering, and pollination. With respect to tomatoes, its ability to resist high temperature is the weakest at 8–10 days before flowering. Excessively high temperatures can destroy meiosis preventing normal pollen and gastrula. High temperature at pollination stage can result in the split of the pollen tube and abnormal fertilization. High temperature injury is a significant disaster affecting vegetable production and the main reason for a slack season of vegetable supply in summer in northern China, Yangtse River basin and southern China.

9.3.3 Impact on Other Plants and Animals in Agriculture

High temperatures easily cause forest and fruit trees to scorch. Litchi, evergreens, and fruit trees in southern China need the inducement of relative low temperatures ranging from 0 °C to 10 °C to better flower. Warmer weather in the winter season tends to result in the bad flowering of litchi trees and, consequently low yield. High temperatures not only can accelerate respiration of flowering, but also can make the leaf be discolored and the flowering period be shortened.

High temperature is one of the factors limiting milk yield of dairy cows. As daily mean temperature is higher than 15 °C, milk yield of dairy cows evidently decreases. Under the conditions of excessively high temperatures, the amount of feed uptake and transformation rate of feed for livestock markedly decrease. High temperatures can bring about the shortage of oxygen and the degeneration of water quality that causes fish in pond to uplift head and even to die.

9.4 Conclusions

Frost is a damaging injury resulting from low temperature, generally in the fall and spring season. Frost does not cause injury directly from low temperature, but icing of plant tissue. That plant temperature drops to below 0 °C is a necessary, but not a sufficient condition, for icing. The density of INA bacteria is closely correlated with the icing temperature of plant tissues.

September, January, and May are the top three months in average injury area in China. Annual average frost area in China is 181.6×10^4 square kilometers, but there is a great difference from year to year. Average frost area in the 1970s is greatly larger than that during 1953–2000.

Frost brings injuries to not only thermophilic and cryophilic crops in north but also subtropical and tropical crops in the south.

The highest frequencies of autumn frosts occur in both eastern and western parts of China, while the highest frequency of spring frost occurs in west China. Because of their bigger inter-annual changes, autumn frost date, spring frost date and the frost-free season all must be given close attention to in agriculture. On the background of global climate warming, frost still frequently takes place in China.

High temperature injury has the serious detrimental effects on crops, vegetables, fruit trees, milk yield, livestock, and fish in China. Study of high temperature injury will be strengthened in the future.

References

Cui Duchang. 1999. Discuss on freezing injury, cool injury, chilling injury and frost. Chinese Agricultural Meteorology, 201 pp, 56–57 (in Chinese).

Feng Yuxiang. 1990. Relationship between cucumber frost and ice nucleation active bacteria. Chinese horticultural journal. (3):21–216.

Feng Yuxiang, He Weixun. 1996. Study on Frost. Beijing: Chinese Meteorological Press. 1–40.

Han Xiangling. 1999. Agroclimatology. Taiyuan: Science and Technology Press of Shanxi province, China. 233–238.

Lindow SE, Arny DC, Barchet WR, et al. 1978. The role of bacterial ice nuclei in frost injury to sensitive plants. In PH Li, A Saksi, eds., Plant cold hardiness and freezing stress mechanisms and crop implications. New York: Academic Press. 249–263.

Liu JiangHua, Tao Yufen, He Weixun, et al. 1990. A study on relationship between ice nucleation active bacteria and frost of maize and soybean. Chinese Agrometeorology. 11(9.1):1–6.

Sun Fuzai, Zhu Hong, He Weixun, et al. 1989. A preliminary study on types and distribution of ice nucleation active bacteria in China. Chinese Agricultural Sciences. 22(2):93–94.

Zhang Yangcai, He Weixun, Li Shikui. 1991. Outline of Agricultural Meteorology Disasters in China. Beijing: Chinese Meteorological Press, 129–199.

Zhu Binghai, Wang Pengfei, Shu Jiaxin. 1985. Meteorological Dictionary. Shanghai: Shanghai Dictionary Press, China. 992–993.

CHAPTER 10

Impacts of Sand Storms/ Dust Storms on Agriculture

Mannava V. K. Sivakumar

Abstract The drylands of the world are affected by moderate to severe land degradation from wind erosion and there is evidence that the frequency of sand storms/dust storms in increasing. Human induced change is by far the most significant factor in the alarming increase of sand storms in some regions. The definition of sand storms and dust storms were presented and the classifications currently in use were discussed. The occurrence of wind erosion at any place is a function of weather events interacting with soil and land management through its effects on soil structure, tilth and vegetation cover. The mechanics of sand and dust storms and their spatial and temporal distribution was described. The impacts of sand and dust storms were described in terms of crop damage, soil productivity losses, economic losses, mass migration, health impacts, and impacts on climate. Not all the impacts are negative and the positive impacts of sand and dust storms were described. Measures to combat sand and dust storms include reduction of the impact of wind speed through the use of live wind breaks or shelterbelts; protection of the loose soil particles through the use of crop residues or plastic sheets or chemical adhesives; and increasing the cohesion of soil particles by mechanical tillage operations or soil mulching. Future strategies for reducing the impacts of sand storms and dust storms must be based on community-based approaches that integrate effective crop and land management strategies as well as policy interventions to promote more effective soil conservation measures at the national and regional scales.

10.1 Introduction

Sand and dust storms are natural events that occur widely around the world, especially in the drylands which occupy half of the world's land surface. According to the measurements of deep-sea lithologic core and glacial cover sediments, sandstorms occurred before the end of the Cretaceous period, dating back 70 million years ago. Local chronicles in China documented sandstorms which had occurred in Wuwei of Gansu province in 351 AD and some collapsed houses and human and animal casualties had been recorded (Wang Shi-gong et al. 2001).

In Russia, the earliest events of wind erosion on sandy soils took place in the 12th to 14th centuries when nomadic tribes moved yearly to the North Caucasus sand lands for the winter period due to the absence of snowpack here from the vast terrain of the steppe zone between the Dnepr and the Volga rivers (Trushkovsky 1959).

It has been estimated that in the arid and semi-arid zones of the world, 24% of the cultivated land and 41% of the pasture land are affected by moderate to severe land degradation from wind erosion (Rozanov 1990). The major sources of present-day dust emissions are the subtropical desert regions and the semi-arid and sub-humid regions, where dry exposed soil is subject to severe winds at certain times of the year. The major world dust producing regions are located in the broad band of arid and semi-arid lands stretching from West Africa to North China, with the majority being in the Northern Hemisphere (Middleton 1986). The Sahara Desert is the world's most important source of dust. The world-wide total annual production of dust by deflation of soils and sediments was estimated to be 61 to 366 million tonnes. For Africa, it is estimated that more than 100 million tonnes of dust per annum is blown westward over the Atlantic (Middleton 1986).

In regions where long dry periods associated with strong seasonal winds occur regularly, the vegetative cover of the land does not sufficiently protect the soil, and the soil surface is disturbed due to inappropriate management practices, wind erosion usually is a serious problem.

In recent years, there is evidence that the frequency of sandstorms is increasing. For example, in China serious wind erosion occurs in the semiarid region, where overgrazing and ploughing of rangeland have dramatically increased since early in the 20th century (Zhu and Wang 1993). Every year desert encroachment caused by wind erosion buries 210,000 hectares of produce land in China (PRC 1994). Ci (1998) showed the annual changes of the frequency of strong and extremely strong sandstorms in China as follows: 5 times in the 1950s, 8 times in the 1960s, 13 times in the 1970s, 14 times in the 1980s, 20 times in the 1990s.

10.2 Anthropogenic Land Disturbances and Wind Erosion

Human-induced change is by far the most significant factor in the alarming increase in some regions in the scourge of dust storms. According to previous studies, wind erosion in the semi-arid regions of America, Africa, Australia, the Near East and many parts of Central Asia could only reach threatening proportions when man disturbed the ecosystem balance. Past policies on land-use and the promotion of farming systems that were unsustainable were the root cause of most disasters.

In the West African Sahel, rapid population growth, at annual rates of 3% during recent decades, has increased demand for food. Instead of intensifying farming systems, for instance by using mineral fertilizers, farmers have tried to enhance production by expanding the cropped area. The previously sustainable fallow system has broken down, yields have declined, and more marginal land, which used to be communal grazing land, is now cropped. Consequently, over-exploitation has resulted in land degradation, or desertification, on a large scale. The amount of dust arising from the Sahel zone has been reported to be around or above 270 million tons per year which corresponds to a loss of 30 mm per m^2 per year or a layer of 20 mm over the entire area (Stahr et al. 1996).

According to Khatteli (1984), the sand encroachment of olive crops in northern Africa resulted from excessive cultivation of soil with disk harrow that pulverised the soil and rendered it more vulnerable to wind erosion. This is manifested by the disappearance of olive crops where the deflation and formation of movable sand dunes occurred or where the sand deposition occurred.

In considerable parts of West Asia and North Africa, large areas of the traditional semi-nomadic rangelands, the steppe, are being opened for barley cultivation. The consequent removal of the natural vegetation cover has exposed the soil surface, leading to the loss of the fertile fine fraction of the shallow soils through wind erosion. This has led to a tremendous decline in soil productivity and quality of life. Hence the Government of Syria has forbidden the conversion of steppe for barley cultivation.

Even in the United States of America, the emergence of effective soil conservation strategies has its background in anthropogenic land disturbances over several decades. From 1850 to 1900, the population of the Great Plains of the United States increased 11 times, from about 700,000 to 8,000,000. Many of these settlers were farmers, who plowed the native grass sod and planted crops using tillage methods developed for more humid regions. The introduction of tractors allowed larger acreages to be farmed. These three events, coupled with the properties of the Great Plains soils, mostly aeolian derived, and the semi-arid climate of the region, set the stage for wind erosion during periods of drought.

Although dry spells are unavoidable in the Great Plains of the US, occurring roughly every 25 years (Warrick et al. 1975; Warrick 1980), it was the combination of drought and the misuse of land that led to the incredible devastation of the Dust Bowl years. Dust storms, known as "Black Blizzards", rolled across the region and deposited dust on ships at sea off the East Coast of the United States. The maximum number of dust storms occurred from 1935 to 1938. Between 1933 and 1937, the Prairie region of Canada experienced only 60% of its normal rainfall. Thousands of livestock were lost to starvation and suffocation, crops withered and a quarter of a million people across the region abandoned their land to seek better lives elsewhere.

According to Seifert (1936), changes in the climate of Middle Europe were caused mainly by rapid deforestation (from above 40% in the 16th century to 20% in the 19th century) and by intensive land reclamation works done in order to drain wet soils and marshes. New techniques of soil tillage reported in most countries in the 1960's resulted in the stimulation of the wind erosion process. As mechanization and effectiveness of cultivation increased, the field area expanded, and roadside and field trees were removed, the menace of wind erosion developed (Czarnowski 1956, Sommer 1984).

Since European settlement of New Zealand large areas of the country have been converted from forest and utilised for animal grazing or crop production.

As a result wind erosion became a major concern for soil conservation in New Zealand, and for a long time government subsidies were available to farmers for wind erosion control (Anon 1971; Salter 1984).

Wind erosion in Lithuania, as also the neighbouring Baltic countries, occurred mainly in areas where the natural plant cover has been replaced by agriculture (Racinskas 1997). In the years of Soviet power, the formation of vast continuous tilled fields (up to 1,000 ha) and extinction of agroplantations and individual farmsteads, contributed greatly to intensification of wind erosion.

Dolgilevich (2001) reported that in Ukraine wind erosion has become widespread in the forest zone. Selective soil blowing has caused degradation of light lands. For this reason, about 10% of the arable lands have been transformed into untilled lands.

10.3 Definition of Sand and Dust Storms

The transport, suspension and deposition of dust particles in the atmosphere mainly manifest itself as a dust storm. Squires (2001) presented a good description of the terminology of sand and dust storms. Sand describes soil particles in an approximate size range of 0.6–1 mm, while dust describes particles < 0.6 mm. In practice only those dust particles below 0.1 mm can be carried by suspension and be manifested in a dust storm.

Four definitions of the dust phenomena are the same as used by the Australian Bureau of Meteorology, which conforms to the worldwide standards of the World Meteorological Office (WMO). SYNOP present weather [WW] codes are included:

1. *Dust storms* (SYNOP WW code: 09) are the result of turbulent winds raising large quantities of dust into the air and reducing visibility to less than 1,000 m.
2. *Blowing dust* (SYNOP WW code: 07) is raised by winds to moderate heights above the ground reducing visibility at eye level (1.8 m), but not to less than 1,000 m.
3. *Dust haze* (SYNOP WW code: 06) is produced by dust particles in suspended transport which have been raised from the ground by a dust storm prior to the time of observation.
4. *Dust swirls* (or dust devils) (SYNOP WW code: 08) are whirling columns of dust moving with the wind and usually less than 30 m high (but may extend to 300 m or more). They usually dissipate after travelling a short distance.

At the southern fringe of the Sahara Desert, a special dry and hot wind, locally termed *Harmattan*, occurs. These NE or E winds normally occur in the winter season under a high atmospheric pressure system. When the wind force of

Harmattan is beyond the threshold value, sand particles and dust particles will be blown away from the land surface and transported for several hundreds kilometres to the Atlantic Ocean.

In the Northwest region of India, the convection sand-dust storm that occurs in the season preceding the monsoon is named *Andhi* (Joseph et al. 1980). It is called *Haboob* in Africa and Arabic countries. It is titled *"phantom"* in some regions: namely it means *"devil"* (Wolfson et al. 1986).

In general, two indicators, wind velocity and visibility, are adopted to classify the grade of intensity of sand-dust storms. For instance, Joseph has classified the sand-dust storms occurring in the Northwest part of India into three grades. Namely, the feeble sand-dust storm develops when wind velocity is at force 6 (Beaufort) degree and visibility varies between 500–1,000 m. The secondary strong sand-dust storm will occur when wind velocity is at force 8 and visibility varies 200–500 m. Strong sand-dust storms will take place when wind velocity is at force 9 and visibility is < 200 metres.

In China, a sand-dust storm is defined similarly to the above. The only difference is that the category of strong sand-dust storms is defined again into two grades, namely strong sand-dust storms and serious-strong sand-dust storms. When wind velocity is 50 metres per second (m/s) and visibility is < 200 metres, the sandstorm is called a strong sand-dust storm. When wind velocity is 25 m/s and visibility is 0–50 metres, the sandstorm is termed a serious sand-dust storm (some regions name it Black windstorm or Black Devil) (Xu Guochang et al. 1979).

10.4 Mechanics of Sand and Dust Storms

The occurrence of wind erosion at any place is a function of weather events interacting with soil (intrinsic properties) and land management (past and present practices) through its effects on soil structure, tilth and vegetation cover. As with water erosion, most wind erosion damage comes from relatively rare, severe events.

Wind erosivity is the main factor controlling the broad pattern of wind erosion. It has been defined as "that property of the wind which determines its ability to entrain and move bare, dry soil in fine tilth" (Painter 1978). It can be estimated from daily or hourly records of wind speed above a threshold related to the lowest speed at which soil particles are entrained (Skidmore and Woodruff 1968).

Soil erodibility, defined as "that property of dry, particulate soil which governs its entrainment by wind", determines locally whether erosive winds will entrain soil particles. It can be estimated from soil aggregate size distributions by the percentage of aggregates larger than 0.84 mm (Chepil 1950; Woodruff and Siddoway 1965).

When soil movement is sustained, the quantity of soil that can be transported by the wind varies as the cube of the velocity (Bagnold 1941; Chepil and Woodruff 1963). Soil roughness, soil erodibility, soil wetness, and the quantity and orientation of crop residues are a few of the parameters that impact the transport of eroded soils.

The erodibility of soils, which is an expression of their ability to resist wind erosion, is a result of soil evolution in an environment. For a given soil in a favorable environment, weathering such as wetting and drying and freezing and thawing, disturbs the original soil structure and makes the soil more erodible (Chepil 1954). Disturbing surface soils, by over-grazing that reduces surface coverage, over-cultivation, destruction of soil aggregates by mining, construction, off-road traffic, and military activities, stimulates and accelerates soil wind erosion in dryland regions, leading to land degradation and desertification (Fryrear and Lyles 1977; Gillette et al. 1980; 1982; Dregne, 1988; Saxton et al. 1996). Anthropogenic activities in wind erosion susceptible regions has made the problem more complex.

The most comprehensive summaries on the movement of surface material by wind action have been prepared by Bagnold (1941) for desert sands and by Chepil and Woodruff (1963) for agricultural lands. Wind erosion is possible when the wind velocity at the soil surface exceeds the threshold velocity required to move the least stable soil particle. The detached particle may move a few millimeters before finding a more protected site on the landscape. The wind velocity required to move this least stable particle is called the static threshold (Bagnold 1941). If the wind velocity increases, soil movement begins and if the velocity is sufficient, soil movement is sustained. This velocity is called the dynamic threshold.

When the wind force reaches the threshold value a number of particles will begin to vibrate. Increasing the wind speed still further, a number of particles will be ejected from the surface into the airflow. When these injected particles impact back on the surface, more particles are ejected, thus starting a chain reaction. Once ejected, these particles move in one of three modes of transport depending on particle size, shape and density of the particle. These three modes are designated suspension, saltation and creep. Its size and density determine movement pattern of sand-dust particles.

The suspension mode involving dust particles of less than 0.1 mm in diameter and clay particles of 0.002 mm in diameter are small in size and light in density. These fine dust particles may be transported at altitudes of up to 6 km and move over distances of up to 6,000 km.

Saltating particles (i.e. those between 0.01–0.5 mm in diameter) leave the surface, but are too large to be suspended. The remaining particles (i.e. above 0.5 mm) are transported in the creep mode. These particles are too large to be ejected from the surface and are therefore rolled along by the wind and impacting particles. Due to the nature of this mode the heights carried are rarely more than 30 cm and the distance travelled rarely exceeds a few metres.

During a storm, creep can move particles over distances from a few centimeters to several meters, saltating particles travel from a few meters to a few hundred meters, and suspension transport ranges from several tens of meters to thousands of kilometers. In the sand-dust wall, the up-lifting force produced by the rising air current is powerful. The sand grains at the lower stratum of the sand-dust wall are coarse particles, the sand particles in the middle stratum are the next in size and those in the upper stratum are mainly suspension dusts (Wang Shi-gong et al. 1993).

Sand particles, transported by saltation and by creep will accumulate to form new sand dunes when they are blown out, graded and transported for a distance.

Joseph (1980) studied the convection sandstorm of "*Andhi*" in Northwest India and results show that visibility can be reduced quickly from 1000 metres to 200 metres, and even 100 metres, while a strong sandstorm or dust storm sweeps across. Wind velocity can be increased up to 20 m/s from 4 m/s. Air temperature can be reduced about 5° and relative humidity can increase 10% or more. On May 20th 1976, a sandstorm took place in India and the visibility at New Delhi Airport was reduced to 280 metres from 4000 metres in a span of two minutes. Air temperature declined from 38–25 °C, relative humidity increased rapidly from 31–70% and wind speed was 73–80 km/hr.

10.5 Spatial and Temporal Distribution of Sand and Dust Storms

Sand-dust storms frequently occur in four regions throughout the world: the Sahel, Central Asia, North America, Central Africa, and Australia.

Mattson and Nilsen (1996) indicate that the Sahara region is the main source of aeolian dust in the world. Yaalon (1996) has indicated that North Africa is a source of dust for southern European dust deposition. Franzen et al (1995) made analyzed sand-dust storm processes that occurred in central and southern Europe and the north of Scandinavia in March 1991, which originated from the Sahara Desert, and they concluded that the sand-dust of the Sahara Desert was transported and accumulated in the northern region of Germany. The affected area of the sand-dust storm process of this sand-dust storm in March 1991 was at least $3.2 \times 105\,km^2$.

Several observations of dust from West Africa crossing the Atlantic were reported in the literature (e.g., Prospero et al. 1970; Carlson and Prospero 1972; Westphal et al. 1988). Prospero and Carlson (1972) estimated that between 25 million and 37 million Mg of wind-blown dust cross the Atlantic annually, with even greater amounts presumably deposited in the ocean (Lal 1988).

According to Swap's (1992) research, the sand-dust of the Sahara Desert was transported to the Amazon Plain of Brazil and approximately 4.8×10^5 tonnes of dust particles was brought to the northeastern part of the Amazon Plain during a one-time sand-dust storm that occurred in the Sahara Desert. The annual transport and accumulation was 1.3×10^7 tonnes ie., 190 kg of dust particles accumulated annually on one ha of land.

The Economic and Social Committee of Western Asia (ESCWA 1993) showed that about 18 million ha in Egypt are affected by different levels of wind erosion. Arroug (1995) used the wind prediction equation of Woodruff and Siddaway (1965) to estimate soil loss by wind erosion in El-Omayed area, Northwestern Coast. He showed that soil loss by wind erosion reached 100 mt/ha/yr. In South Sinai, Wassif et al. (1997) used Big Spring Number Eight (BSNE) dust sampler for measuring the amount of soil loss by wind erosion during 83 days from March, 1994. They found that the quantity was 3.07 ton/100 m width.

Pease et al. (1998) suggests that arid and semi-arid regions around the Arabian Sea are one of the principal sources of global dust. India, Pakistan, Iran and the Arabian Peninsular contribute to Arabian Sea dust deposition.

10.6 Impacts of Sand and Dust Storms

Sand and dust storms are hazardous weather and cause major agricultural and environmental problems in many parts of the world. There is a high on-site as well as off-site cost due to the sand and dust storms. They can move forward like an overwhelming tide and strong winds take along drifting sands to bury farmlands, blow out top soil, denude steppe, hurt animals, attack human settlements, reduce the temperature, fill up irrigation canals and road ditches with sediments, cover the railroads and roads, cause household dust damages, affect the quality of water in rivers and streams, affect air quality, pollute the atmosphere and destroy mining and communication facilities. They accelerate the process of land desertification and cause serious environment pollution and huge destruction to ecology and living environment (Wang Shigong et al. 2001). Atmospheric loading of dust caused by wind erosion also affects human health and environmental air quality.

10.6.1 Crop Damage

Wind erosion-induced damage includes direct damage to crops through in the loss of plant tissue and reduced photosynthetic activity as a result of sandblasting, burial of seedlings under sand deposits, and loss of topsoil (Fryrear 1971; Ambrust 1984; Fryrear 1990). The first two types of damage have an immediate effect on crop yields. However, the loss of topsoil is particularly worrying since it potentially affects the soil resource base and hence crop productivity on a long-term basis, by removing the layer of soil that is inherently rich in nutrients and organic matter.

10.6.1.1 Sand Blasting

Since damage to millet strongly depends on plant age, the risk posed by sandblasting and burial will to a large extend depend on the strength and timing of sand storms with respect to the time of sowing. Leaves are more sensitive to sandblast damage than are stems (Armbrust 1982; Fryrear 1971). Michels et al. (1995a) reported on the effect of wind speeds ranging from 8 to 14 $m\,s^{-1}$ and sand fluxes ranging from 0 to 42 $g\,m^{-1}\,s^{-1}$ on millet growth and photosynthetic activity in a wind tunnel experiment. Plants were exposed for 15 min. at 8 days after emergence (DAE), 16 DAE or on both dates. Sandblasted plants saw their viable leaf area reduced by an average of 19% across all treatments at 21 DAE compared to the control. In addition to the loss of viable leaf area, sandblasting affected photosynthesis in the viable parts of the leaves by an average of 55% 1 h after exposure at 8 DAE and was still reduced by 28% 4 days later. Exposure to sandblasting also reduced millet dry weight at the early stages of development. Brenner (1991) similarly reported reduced leaf area and dry matter production of young millet plants as a result of sandblasting during violent storms. However, at 46 days after sowing the crop had entirely recovered from these stresses. Exposure of seedlings to sandblasting prior to burial reduced dry matter production of millet at 70 days after emergence (DAE) by 47% compared to unexposed buried plants (Michels et al. 1995a).

10.6.1.2
Crop Burial

In the traditional sowing techniques of millet cropping systems of West Africa, small depressions are left after sowing the seed and burial of seedlings under deposited wind-blown sand frequently occurs during convective storms, sometimes necessitating partial or total resowing of the crop (Michels et al. 1993, Michels et al. 1995a; Sterk and Stein 1997). Burial of seedlings accentuates the effect of high surface temperatures by bringing the meristems of young seedlings directly into contact with the hot soil surface. Nutrient and water deficiencies at this critical stage may further hamper seedling recovery. In a 1990 field experiment sown in the traditional way, 90% of millet hills were covered by sand 23 DAE, which necessitated re-sowing and therefore increased the risk of exposure of the crop to an end-of-season drought (Michels et al. 1993).

10.6.1.3
Plant Development

In addition to the direct effects of partial burial on yield-determining factors such as number of tillers and panicle length, partially buried plants experience a delayed development compared to unburied plants. This may further increase the risk of an end of season drought stress to millet by extending the crop cycle beyond the usual growing season length. Michels et al. (1993) determined that panicle development was delayed by approximately 2 weeks following partial burial.

10.6.2
Soil Productivity Losses

Studies conducted at large regional scales as well as at field level demonstrated clearly that sand and dust storms result in loss of soil productivity and long-term soil degradation.

Fine airborne particles emitted into the atmosphere represent the loss of the most fertile fraction of the topsoil and can be transported over long distances from their sources of production (Péwé 1981). Calculations based on visibility and wind speed records for 100 km wide dust plumes, centered on eight climate stations around South Australia, indicated that dust transport mass was as high as 10 million tonnes (Butler, Davies and Leys 1995). Thus dust entrainment during dust events leads to long-term soil degradation, which is essentially irreversible (Yaping, Raupuch and Leys 1996). The cost to productivity is difficult to measure but is likely to be quite substantial.

At the field level, following intensive monitoring of sand fluxes in a 40 × 60 m plot, Sterk and Stein (1997) estimated soil losses from only four sand storms in 1993 at 45.9 t ha^{-1}. These measurements were made on a millet field with a 0.8 t ha^{-1} mulch of millet stover, which reflects on-farm conditions under favorable circumstances. Sterk et al. (1997) reported the only estimates of nutrient losses by wind erosion for Niger. They calculated nutrient fluxes based on the nutrient content of the trapped sediment for the two largest storms. The results reveal a total loss of 57.1 and 6.1 kg ha^{-1} of K and P, respectively. These amounts are roughly equivalent to the quantity of K and P required to produce a millet yield of 2,000 kg straw and 600 kg of grain ha^{-1}. They also correspond to approximately 3% of the nutrients contained in the top 10 cm of the soil.

On the basis of repeated measurements of surface elevation, Michels et al. (1995b) reported a decrease in surface elevation of 33 mm after one year on bare millet plots.

During dust transportation, many young plants are lost to the sand blasting nature of the process at ground level, resulting in a loss of productivity. However, major dust storms have most of their impact within the atmosphere. The most noticeable effect is the reduction of visibility. This is of course dependent on the severity of the dust event. It could range from a slight haze to a major dust cloud. In the worst cases, visibility can be reduced to only a few metres. This loss of visibility can be a major hazard to aircraft and in some cases to motorists.

10.6.3 Economic Losses

Wang Shigong et al. (1995) estimated the direct economic loss caused by the serious-strong sand-dust storm that occurred on May 5th 1993 that threatened 1.1 million km^2 of territory in China at 560 million RMB Yuan. Yang Gengsheng et al. (2001) reported that 373,000 ha of crops were destroyed, 16,300 ha of fruit trees were damaged and thousands of greenhouses and plastic mulching sheds were broken. About 120,000 animals were killed. More than 1,000 km of irrigation channels were buried by sand accumulation. Many water resource back-up facilities, such as reservoirs, dams, catchments, underground canals, and flood control installations were filled up with sand silts. It has brought about critical destruction of desert plants and ecological environment in Northwest China, promoting the desertification process in the affected areas, and its indirect economic loss was hard to assess (Wang Shigong et al. 1995).

Canada is plagued by localized dust storms and drought, especially in the Prairie Provinces of Manitoba, Saskatchewan and Alberta. At least 161 million tonnes of soil is lost each year because of wind erosion, and the annual on-farm costs of wind erosion in the Prairie Provinces are about USD$ 249 million. Sparrow (1984) argued that nearly 58% of the total annual soil loss due to wind and water may be ascribed to wind erosion which would account for a conservative estimate of 161 million t.

10.6.4 Mass Migration

Continued incidence of sand and dust storms over several years can result in total loss of soil productivity and force rural migration. Because of the Dust Bowl in 1930s, millions of hectares of farmland became useless, and hundreds of thousands of people were forced to leave their homes (Hurt 1981; Lee et al. 1999). The Dust Bowl exodus was the largest migration in American history (Bonnefield 1979). By 1940, 2.5 million people had moved out of the Plains states. Of those, 200,000 moved to California.

10.6.5 Health Impacts

Recently published data from South Australia show that wind erosion and dust storms are likely to cost millions of dollars each year. Costs such as house cleaning costs, days lost from work because of illness (or an inability to get to the job), the cost to the aviation industry of diverted flights and so on, are among the costly affects. The most significant cost is that associated with the onset of asthma in susceptible people. South Australia, for example, experiences 8.5 days per year on average when the dust load is sufficient to trigger an asthma attack (Williams and Young 1999). It was concluded that as much as 20% of the Province's asthma problem could be linked to the dust in the air.

10.6.6 Impacts on Climate

The very fine fraction of soil-derived dust has significant forcing effects on the radiative budget. Dust particles are thought to exert a radiative influence on climate directly through reflection and absorption of solar radiation and indirectly through modifying the optical properties and longevity of clouds. Depending on their properties and in what part of the atmosphere they are found, dust particles can reflect sunlight back into space and cause cooling in two ways. Directly, they reflect sunlight back into space, thus reducing the amount of energy reaching the surface. Indirectly, they act as condensation nuclei, resulting in cloud formation (Pease et al. 1998). Clouds act as an *"atmospheric blanket"*, trapping long wave radiation within the atmosphere that is emitted from the earth. Thus, dust storms have local, national and international implications concerning global warming. Climatic changes in turn can modify the location and strength of dust sources.

10.6.7 Positive Impacts

Mineral dust, it has been suggested, has an important role to play in the supply of nutrients and micronutrients to the oceans and to terrestrial ecosystems. Iron in the minerals composing the desert dust is a vital nutrient in oceanic regions that are deficient in iron. Further, more research has shown that the canopy of much of Central and South American rainforest derives much of its nutrient supply from dust transported over the Atlantic from the Sahara region of North Africay. Sahara dust occasionally reaches the State of Florida in the U.S., causing a high-altitude haziness that obscures the sun. Dust from China's deserts is transported to the waters near Hawaii in the south Pacific. As the dust settles in the waters around Hawaii, the primary productivity of the plankton in the water column increases (NOAA 1999). This research suggests that dust transport processes form an integral part of the global ecosystem.

According to Swap et al. (1992), it is indicated that each sand-dust storm in the Sahara Desert can blow up 480,000 tonnes of sand and dust into the Northeast part of the Amazon Valley. The annual sand transport is approximately 13 million tonnes, meaning sand-dust storms have brought about an accumulation of 190 kg of sand and dust particles per ha every year in the region. It was estimated that, along with the accumulation of sand-dust, 1–4 kg of phosphate has been transported and accumulated per ha per year. It can be assumed that the rate of production of the rain forest in the Amazon depends on the phosphorous and other elements transported along with sand-dust storms from the Sahara.

Sand-dust is partially alkali itself and can restrain certain harm from acid rain during its transportation in the affected area. Japanese scientists' research indicated that yellow sands and dust from Northwest China are the major component of coagulation tubercles cooling clouds in the sky of Japan and play an important role in precipitation in Japan. At the same time, ice crystals of yellow sand are alkali and play an active role in the neutralization of emerging of acid rain in Japan (Qu Zhang et al. 1994).

Apart from losses, the Sahel also receives nutrient-rich dust deposits. During the Harmattan season, dust is transported from the Sahara towards the Sahel, where it partly settles. Moreover, during the early rainy season, some of the dust raised by a convective storm is deposited by gravity or rain wash-out. Drees et al. (1993) measured 50% of the total annual dust input in southwest Niger in the early rainy season, whereas the Harmattan season contributed only 15% to the total. Dust deposits are particularly rich in sodium, potassium, magnesium, and calcium, but poor in phosphorus (Herrmann et al., 1996). At the field scale,

Sterk et al. (1997) measured nutrient inputs of 2.5 and 0.2 $kg\,ha^{-1}$ of K and P, respectively, from dust deposition.

10.7 Measures to Combat Sand and Dust Storms

Measures to combat sand and dust storms must concentrate on preventing the sand from being picked up in the source area as it may often be cheaper and more effective than to fix the dunes formed in the accumulation area. Basically these measures have concentrated on the following aspects:

- Reducing the impact of wind speed through the use of live windbreaks or shelterbelts
- Protecting the loose soil particles by use of crop residues or plastic sheets or chemical adhesives
- Increasing the cohesion of soil particles by mechanical tillage operations or soil mulching

10.7.1 Use of Wind Breaks or Shelterbelts

A "windbreak" is defined as any structure that reduces wind speed (Rosenberg 1974) and is commonly associated with a natural vegetative barrier against wind. A windbreak can be a single element or a system of elements that through its presence in the airflow reduces the effect of wind speed not only at the system itself but also at a certain windward and leeward distance. Wind barriers can control wind erosion by reducing the travel distance of wind across a field. The wind speed can be reduced by more than 50% at a leeward distance of 20 times barrier height (Skidmore 1986).

Planting and maintaining shelterbelts is an important conservation practice in the Great Plains region of the United States of America. Shelterbelts produce many benefits for farmers such as decreased soil erosion, increased crop yields, reduced livestock stress, control of drifting snow, building maintenance and energy savings, (e.g., Forman & Baudry 1984; Loucks 1984; USDA 1989). Shelterbelts also benefit many wildlife species by providing food, reproductive sites, escape cover and shelter from severe weather (e.g., Cable 1991). In addition, shelterbelts provide important opportunities for recreation, particularly hunting. (Cable and Cook 1990). This recreation in shelterbelts can have important economic impacts (Cook and Cable 1990). Moreover, shelterbelts have been shown to have a positive influence on landscape aesthetics (Cook and Cable 1995).

From meteorological observations, He et al. (1993) showed changes in wind velocity profile over the vegetation, which reduces wind shear at the ground surface and increases the friction velocity. Thus the drag coefficient is bigger compared with that for the shifting dune.

10.7.2 Use of Crop Residues

Surface crop residues help stabilize soils by reducing soil water loss and the erosive force of wind and by shielding the soil from a reduced number of saltating particles. Michels et al. (1995a) showed that covering the soil with 2,000 kg ha^{-1} of millet residues gives sufficient protection from the sand storms. Evaporation reduction by crop residues results from shading and protection from drying winds (Van Doren and Allmaras 1978). Standing stems reduce the wind energy available for momentum transfer at the soil surface (Hagan and Armbrust 1994; Lyles and Allison 1976; Raupach 1992).

The geometry of crop residues alters the surface microclimate, impacting the degree of soil protection and water conservation. Strips of partial mulch cover increased preplant soil warming (Bristow and Abrecht 1989), while standing stems increased crop water use by increasing the transpiration fraction of total evaporation (Lascano et al. 1994). Bilbro and Fryrear (1994) showed that vertical residues are much more effective that flat surface cover in controlling soil loss by wind. Vertical crop stems trapped more snow than horizontal stems (Nielsen and Hinkle 1994). Decomposition of surface residues and the persistence of standing stems change with climate, soil environment, and management conditions (Steiner et al. 1994; Vigil 1995), but impacts on soil water dynamics are less well understood.

Improved crop residue management is also an efficient and cost-effective practice to address land resource issues relating to soil conservation, water quality, sustainability, and enhanced nutrient cycling (Unger 1994; Steiner 1994).

10.7.3 Use of Mechanical Tillage

Increasing the cohesion of soil particles by mechanical tillage operations or soil mulching is an effective strategy, but it must be carried out with considerable care taking into account the weather and soil factors. Special farming implements, such as chisel-type ploughs, which permit the cultivation of vegetated surfaces and maintain a rough, well-textured surface could be used for this purpose.

10.8 Conclusions

As explained in some detail in the previous sections, there is an increasing frequency of occurrence of sand and dust storms in different parts of the world and they carry considerable impacts on agricultural productivity in the near term and on soil productivity losses in the long term. Despite the growing understanding of the problem and the need to counter it on an urgent basis, there are very few examples of comprehensive long-term implementation strategies to combat sand storms and dust storms. Past efforts in this regard have met with limited success because of lack of involvement of affected communities in planning and implementing the strategies to combat wind erosion. Given the complexity of the problem wind erosion control strategies need an integrated participatory and interdisciplinary approach that must be based on maintaining soil fertility; adoption of better crop and land management practices; development of policies to address land carrying capacities, land tenure and soil conservation; increased awareness of farmer perceptions, and a better understanding of local knowledge. Wind erosion is not just a local problem, but has regional and even global implications. Hence information should be collected from national and inter-governmental agencies concerning past successes at the national and regional levels, and the conditions and criteria conducive to ensure the success of future projects should be highlighted.

References

Anon (1971) Wind erosion control. Soil and Water 8: 23–27

Armbrust DV (1982) Physiological responses to wind and sandblast damage by grain sorghum plants. Agron. J. 74(3):133–135

Armbrust DV (1984) Wind and sandblast injury to field crops: Effects of plant age. Agron. J. 76:991–993

Armbrust DV, Bilbro JB (1997) Relating plant canopy characteristics to soil transport capacity by wind. Agron. J. 89: 157–162

Arroug SMM (1995) Relation of some soil properties to its erodibility. Ph.D. Thesis, Ain Shams University, Egypt

Bagnold RA (1941) The Physics of Blown Sand and Desert Dunes: Methuen and Ltd., London, United Kingdom

Bilbro JD, Fryrear, DW (1994) Wind erosion losses as related to plant silhouette and soil cover. Agron. J. 86:550–553

Bonnefield P (1979) The Dust Bowl: Men, Dirt, and Depression. University of New Mexico Press

Brenner AJ (1991) Tree-crop interactions within a Sahelian windbreak system. Unpublished Ph.D. thesis, University of Edinburgh

Bristow KL, Abrecht, DG (1989) The physical environment of two semi-arid tropical soils with partial surface mulch cover. Aust. J. Soil Res. 27:577–587

Butler, PR, Davies, WJ, Leys, JF (1994) Dust Storms in South Australia on 24–25th May, 1994. Technical Report Number 243. Primary Industries, South Australia

Cable TT (1991) Windbreaks, wildlife and hunters. In: Rodiek JE, Bolen EC (ed.) Wildlife and Habitats in Managed Landscapes, Island Press. Washington, DC, pp 34–55

Cable TT, Cook PS (1990) The use of windbreaks by hunters in Kansas. Journal of Soil & Water Conservation 45(5):575–577

Carlson TN, Prospero, JM (1972) The large-scale movement of Saharan air outbreaks over the northern equatorial Atlantic. J. Appl. Meteorol. 11: 283–297

Cook PS, Cable TT (1990) The economic value of windbreaks for hunting. Wildl. Soc. Bull. 18:337–342

Cook PS, Cable TT (1995) The scenic beauty of shelterbelts on the Great Plains. Landscape and Urban Planning 32:63–69

Chepil WS (1950) Properties of soil which influence wind erosion: II. Dry aggregate structure as an index of erodibility. Soil Science 69: 403–414

Chepil WS (1954) Seasonal fluctuation in soil structure and erodibility of soil by wind. Soil Sci. Soc. Am. Proc. 18: 13–16

Chepil WS, Woodruff, NP (1963) The physics of wind erosion and its control. Adv. Agron. 15:211–302

Ci Longjun (1998) Mechanism of Desertification and sustainable Strategies to Combat Desertification in China. Quaternary Sciences 5 (2)

Czarnowski M (1956) 0 mozliwosci odwrocenia procesu stepowienia w Wielkopolsce przy pomocy pasow przeciwwietrznych. *Zesz. Probl. Post. N. Roln.*, 7: 79–82, (in Polish).

Dolgilevich MJ (2001) Extent and Severity of Wind Erosion in the Ukraine. In: Yang Youlin, Squires V and Lu Qi (ed). Global alarm: dust and sandstorms from the World's Drylands. RCU of the United Nations Convention to Combat Desertification (UNCCD), Bangkok, Thailand

Drees, L.R., A. Manu, and L.P. Wilding. 1993. Characteristics of aeolian dusts in Niger, West Africa. Geoderma 59: 213–233

Dregne HE, Tucker CJ (1988) Desert encroachment. Desertification Control Bulletin 16: 16–19

Franzen LG (1995) The Saharan dust episode of South and Central Europe, and northern Scandinavia March. 1991. Weather 50 (9): 313–318

Forman RT, Baudry J (1984) Hedgerows and hedgerow networks in landscape ecology. Environmental Management 8:495–510

Fryrear DW (1984) Soil ridges-clods and wind erosion. Trans. ASAE. 27(2):445–448

Fryrear DW (1990) Wind Erosion: Mechanics, Prediction, and Control. Adv. in Soil Sci. 13:187–199

Fryrear DW, Lyles L (1977) Wind erosion research accomplishments and needs. Trans. ASAE. 20(5): 916–918..24

Fryrear DW (1971) Survival and growth of cotton plants damaged by windblown sand. Agron. J. 63(4):638–642

Gillette DA, Adams J, Endo A, Smith D, Kihl R (1980) Threshold velocities for input of soil particles into the air by desert soils. J. Geophysical Res. 85(C10):5621–5630

Gillette DA, Adams J, Muhs D, Kihl R (1982) Threshold traction velocities and rupture moduli for crusted desert soils for the input of soil particles into the air. J. Geophysical Res. 87(C11): 9003–9015

Hagan LJ, Armbrust DV (1994) Plant canopy effects on wind erosion saltation. Trans. ASAE. 37(2):461–465

He Hui-xia, Qian Zheng-an, Qu Zhang (1993) Example Registers of Partly Strong Sand-dust Storms in Northwest China. Journal of Atmospheric Information 30 (4): 14–18

Herrmann L, Stahr K, Sivakumar MVK (1996) Dust deposition on soils of southwest Niger. In: B. Buerkert et al. (ed) Wind erosion in West Africa: The problem and its control. Proc. Int. Symp., Stuttgart, Germany. 5–7 Dec. 1994. Margraf Verlag, Weikersheim, Germany. pp 35–47

Hurt RD (1981) The Dust Bowl: an agricultural and social history. Nelson-Hall, Chicago

Joseph PV, Raipal DK, Deka, SN (1980) "Andhi," the convective dust storms of Northwest India. Mausam, 31, 431–442

Khatteli H (1998) Review of major research on wind erosion in Arid and Desert Tunisia. In: Sivakumar MVK, Zobisch, MA, Koala, S, Maukonen, T (ed.) Wind Erosion in Africa and West Asia: Problems and Control Strategies. Proceedings of the ICARDA/ICRISAT/UNEP/WMO Expert Group Meeting, 22–25 April 1997, Cairo, Egypt. International

Center for Agricultural Research in the Dry Areas (ICARDA), Aleppo, Syria. pp 49–58

Lal R (1988) Soil degradation and the future of agriculture in sub-Saharan Africa. J. Soil Water Conserv. 43: 444–451

Lee JA, Gill TE, Mulligan KR (1999) The 1930s Dust Bowl: The Relative Roles of People and the Physical Environment. Geological Society of America Abstracts and Programmes 31(1): A11

Loucks WL (1984). Windbreaks Save Money! Cooperative Extension Publication C-645. Kansas State University, Manhattan

Lascano RJ, Baumhardt RL, Hicks SK, Heilman JL (1994) Soil and plant water evaporation from strip-tilled cotton: Measurement and simulation. Agron. J. 86:987–994

Lyles L, Allison, BE (1976) Wind erosion: The protective role of simulated standing stubble. Trans. ASAE 19:62–64

Mattson JO, Nilsen J (1996) The transport of Saharan dust to southern Europe: a scenario. Journal of Arid Environments 32: 111- 119

Michels K, Sivakumar MVK, Allison BE (1993) Wind erosion in the Southern Sahelian Zone and induced damage to pearl millet production. Agric. For. Meteorol. 67:65–77

Michels K, Armbrust DV, Allison BE, Sivakumar MVK (1995a) Wind and wind-blown sand damage to pearl millet. Agron. J. 87: 620–626

Michels K, Sivakumar MVK, Allison BE (1995b) Wind erosion control using crop residue: I. Effects on soil flux and soil properties. Field Crops Res. 40: 101–110

Middleton NJ (1986) Dust storms in the Middle East. Jour. Arid Env. 10:83–96

Nielsen, DC, Hinkle SE (1994) Wind velocity, snow and soil water measurements in sunflower residues. Great Plains Agric. Council Bull. 150:93–100

NOAA; As reported on the web-site on 09-12-99 (http://www.noaa.gov/); 1999

People's Republic of China (PRC) (1994) China 21st Agenda, China Environmental Science Press, Beijing, China

Péwé TL (1981) Desert dust: an overview. Geological Soc. of Amer., Spec. Pap., 186, 1–10

Painter DJ (1978) Wind erosiveness in New Zealand. New Zealand Journal of Science 21: 137–148

Pease P, Vatche P, Tchakerian N, Tindale N (1998) Aerosols over the Arabian Sea: geochemistry and source areas for aeolian desert dust. Journal of Arid Environments 39: 477–496

Prospero J, Carlson T (1972) Vertical and aerial distribution of Saharan dust over the western equatorial North Atlantic Ocean. J. Geophys. Res. 77: 5225–5265

Prospero JM, Bonatti E, Schubert C, Carlson TN (1970) Dust in the Caribbean atmosphere traced to an African dust storm. Earth Planetary Sci. Letters 9: 287–293

Qu Zhang, Xu Bao-yu, He Hui-xia (1994) Some Enlightenment from A Sandstorm Occurred in Northwest China. Arid land geography, 17 (1): 63–67

Racinskas A (1997) Assessment of Natural Conditions in Lithuania from the Point of View of Wind Erosion. In: Proceedings of the International Symposium/Workshop on Wind Erosion, 3–5 June 1997, Manhattan, Kansas, USA

Raupach MR (1992) Drag and drag partition on rough surfaces. Boundary-Layer Meteorol. 60:375–395

Rosenberg NJ (1974) Microclimate: the biological environment. John Wiley & Sons, New York

Ryabov EI (1974) The land asks protection. Stavropol. 160 p. (In Russian)

Salter RT (1984) Wind erosion. In: Speden I, Crozier MJ (Compilers) Natural hazards in New Zealand. New Zealand National Commission for UNESCO, Wellington, pp 206–248

Saxton K, Stetler L, Chandler D (1997) Simultaneous Wind Erosion and PM10 Fluxes. In: Proceedings of the International Symposium/Workshop on Wind Erosion, 3–5 June 1997, Manhattan, Kansas, USA

Sinclair HR Jr (1997) Soil Data for Wind Erosion Prediction System. In: Proceedings of the International Symposium/Workshop on Wind Erosion, 3–5 June 1997, Manhattan, Kansas, USA

Saxton KE, Stetler LD, Horning LB (1996) Principles of wind erosion: Principles to predict and control wind erosion and dust emissions from farm fields. In: Papendick R, Veseth R (ed.) Northwest Columbia Plateau Wind Erosion Air Quality Project: An Interim Report, Solutions in Progress, Washington State University, College of Agriculture and Home Economics, Miscellaneous Publication No. 0184: 7–12

Seifert A (1936) Die Versteppung Deutschlands. Deutsche Technik, Berlin (in German)

Skidmore EL (1986) Wind erosion control. Climatic Change 9:209–218

Skidmore EL, Woodruff NP (1968) Wind erosion forces in the United States and their use in predicting soil loss. Agriculture Handbook No. 346, United States Department of Agriculture Agricultural Research Service

Sparrow HO (1984) Soil at Risk, Canada's Eroding Future, A Report on Soil Conservation by the Standing Committee on Agriculture, Fisheries and Forestry to the Senate of Canada. Ottawa: Senate of Canada

Stahr K, Hermann L, Jahn R (1996) Long distance sand transport in the Sudano – sahelian zone and the consequences of sahel Soil properties. In: Buerkert B, Allison BE, von Oppen M (ed.). Proceedings of International symposium. "Wind erosion" in West Africa. The problem and its control. University of Hohenheim, 5–7th December 1994. Margraf Verlag, Welkersheim, Germany

Steiner JL (1994) Crop residue effects on water conservation. In: Unger PW (ed.) Managing Agricultural Residues, Lewis Publ., Boca Raton, Florida, USA. pp. 41–76

Steiner JL, Schomberg HH, Douglas, Jr. CL, Black, AL (1994) Standing stem persistence in no-tillage small-grain fields. Agron. J. 86:76–81

Sterk G, Stein A (1997) Mapping wind blown mass transport by modeling variability in space and time. Soil Sci. Soc. Am. J. 61: 232–239

Sterk G, Stroosnijder L, Raats, PAC (1997) Wind erosion processes and control techniques in the Sahelian zone of Niger. In: Proceedings of the International Symposium/Workshop on Wind Erosion, 3–5 June 1997, Manhattan, Kansas, USA

Sommer C (1984) Ziele und Aufgaben der Bodenschutzes heute. Mitt. Deutsch. Bodenkde Ges. 40:119–124 (in German)

Salter RT (1984) Wind erosion. In: Speden I, Crozier MJ (compilers) Natural Hazards in New Zealand, New Zealand National Commission for Unesco, Wellington. pp. 206–248

Squires VR (2001) Dust and Sandstorms: an early warning of impending disaster. In: Yang Youlin, Squires V, Lu Qi (ed.) Global alarm: dust and sandstorms from the World's Drylands. Asia RCU of the United Nations Convention to Combat Desertification (UNCCD), Bangkok

Swap R (1992) Saharan dust in the Amazon Basin. Tellus, 44B, 2: 133–149

Trushkovsky AA (1959) The stage of overgrowing Terek and Kuma sands. Botanical Journal 44(5):128 - 133. (In Russian)

Unger PW [ed.]. 1994. Managing Agricultural Residues. Lewis Publ., Boca Raton, Florida

U.S. Department of Agriculture (USDA) (1989). Soil erosion by wind. Agricultural Information Bulletin No. 555. U.S. Government Printing Office, Washington, DC

Van Doren DM Jr, Allmaras, RR (1978) Effect of residue management practices on the soil physical environment, microclimate, and plant growth. p. 49–83. In: Oschwald WR (ed).

Crop residue management systems. ASA Spec. Publ. 31. ASA, CSSA, and SSSA, Madison, WI

Vigil M, Sparks D (1995) Factors affecting the rate of crop residue decomposition under field conditions. Conservation Tillage Fact Sheet #3-95. USDA-ARS, USDA-NRCS

Wang Shi-gong, Dong Guangrong, Shang Kezheng, Chen Huizhong (2001) Progress of research on understanding sand and dust storms in the world. In: Yang Youlin, Squires V, Lu Qi (ed.) Global alarm: dust and sandstorms from the World's Drylands. RCU of the United Nations Convention to Combat Desertification (UNCCD), Bangkok

Wang Shi-gong, Yang De-bao, Meng Mei-zhi (1993) Analyses of Structure Characteristics and Formative Causes of the "5.5" Black Storm. Journal of Gansu Meteorology, 11(3): 28–31

Wang Shi-gong, Yang De-bao, Zhou Yu-su (1995) Analysis on the Formative Causes of Sand-dust Storms in the Northwest China during 3–12 April 1994. Journal of Desert Research 15 (4): 332–338

Warrick RA (1980) Drought in the Great Plains: A Case Study of Research on Climate and Society in the US. *In:* Jesse Ausubel, Biswas AK (ed.) Climate Constraints and Human Activities, IIASA Proceedings Series, Permagon Press, New York

Warrick RA, Trainer RB, Baker E J, Brinkman B (1975) Drought Hazard in the United States: A Research Assessment. Programme on Technology, Environment and Man; Monograph #NSF-RA-E-75-004, the University of Colorado, USA

Westphal DL, Toon OB, Carlson, TN (1988) A case study of mobilization and transport of Saharan dust. J. Atmos. Sci. 45: 2145–2175

Williams P, Young M (1999) Costing dust. How much does wind erosion cost the people of South Australia? CSIRO, Policy and Economics Research Unit, Land and Water, Adelaide

Woodruff NP, Siddoway FH (1965) A wind erosion equation. Soil Science Society of America Proceedings 29: 602–608

Xu Guochang, Cheng Minlian, Wu Guoxiong (1979) On an extraordinary heavy sandstorm on April 22nd in Gansu Province. Acta Meteorologica Sinica, 37 (4): 26–35 (in Chinese)

Yaalon D (1996) Comments on the source, transportation and deposition of Saharan Dust to Southern Europe. Journal of Arid Environments 36: 193–196

Yang Gengsheng, Xiao Honglang, Tuo Wanquan (2001) Black windstorm in northwest china: a case study of the strong sand-dust storm on may 5th 1993. In: Yang Youlin, Squires V, Lu Qi (ed.) Global alarm: dust and sandstorms from the World's Drylands. RCU of the United Nations Convention to Combat Desertification (UNCCD), Bangkok

Yaping S, Raupach MR, Leys, JF (1996) A model for predicting aeolian sand drift and dust entrainment on scales from paddock to region. Aust. J. Soil Res. 34: 309–42

Yang Youlin, Squires V, Lu Qi (2001) Global alarm: dust and sandstorms from the World's Drylands. Asia RCU of the United Nations Convention to Combat Desertification (UNCCD), Bangkok

Zhu Z, Wang T (1993) Trends of desertification and its rehabilitation in China. Desertification Bulletin, 22:27–30

CHAPTER 11

Disaster Reduction Planning and Response: The Example of National Drought Policy in USA

Raymond P. Motha

Abstract Severe droughts have occurred frequently in various parts of the United States throughout the 20th Century. These natural disasters caused severe hardship not only on the agriculture sector of society, but also on the entire socio-economic infrastructure. Until recently, governments and relief agencies responded to a drought disaster from a crisis perspective in a purely reactive approach. However, a comprehensive disaster management approach has emerged over the past decade, shifting from reactive relief to proactive risk management. Risk management is based on preparedness and mitigation measures. Interagency collaboration led to the development of a coordinated national drought policy.

11.1 Introduction

Natural hazards are inevitable. Floods, cyclones, wildfires, and droughts are normal occurrences in the natural environment. Disasters, however, result when human activity such as farming and its associated environment intersect with the natural hazards. The growing toll of disasters around the world presents a major challenge for public policy, scientific research, and their interactions. Policy seeks to reduce the human and economic effects of disasters; research seeks to provide the knowledge and tools that can contribute to the effectiveness of the policy. Challenges emerge in the face of societal trends that are converging to increase the likelihood, magnitude, and diversity of disasters. Growing populations, which increase economic and technological interdependence, enhance the increasing threat of disasters.

The severity and nature of the impact of a disaster depend on a range of factors including the type of hazard, the size of the economy and its socio-economic structure, and the sectors affected by the disaster. With respect to hazard types, droughts are often referred to as creeping disasters for their slow development and highly variable impacts on different sectors of society. Their lengthy duration creates significant hardship, especially on the agricultural sector. The loss of crops and livestock, increased wildfire potential, and reduced water supplies caused by drought can have a devastating effect on the economy. In contrast, sudden onset disasters such as tropical cyclones, floods, and blizzards can also have a direct impact on the physical infrastructure and productive resources as well as the socio-economic infrastructure.

Severe droughts have affected various parts of the United States frequently during the 20th Century. In the 1930's, a nearly decade-long drought affected more than 60 percent of the nation, turning millions of hectares of fertile agricultural land into the Dust Bowl across the Great Plains. Hensen et al. (1999) ranked this extreme event as the nation's top weather event of the 20th century. The drought caused a huge migration from the Southern Plains to California and resulted in a major federal agricultural policy. Drought returned to the Plains and the Southwest in the 1950's, claiming millions of cattle. Record drought created severe water shortages in the heavily populated Northeast in the early 1960's. Prolonged drought hit the Pacific Northwest from the mid-1980's to early 1990's, with a severe agricultural drought in the Midwest and the Southeast in 1988.

Drought is perhaps the most obstinate and detrimental of nature's extreme events. It can last much longer and extend across much larger areas than cyclones, tornadoes, floods, and earthquakes. At its most severe, drought creates vast, windblown dust bowls eroding the landscape, damaging terrestrial and aquatic wildlife habitat, contributing to widespread wildfires, and causing hundreds of millions of dollars in economic losses. It brings hardship to all water-dependent enterprises. In many small communities, downturns in farming and ranching have a crippling effect throughout the business and economic community. In large managed river basins and water systems, drought creates or exacerbates water conflicts, notably between agriculture against municipal populations, and water for direct human use against water for ecosystems.

11.2 Disaster Management: Shift from Reactive to Proactive Management

The paradigm in disaster management theory and practice has shifted over the last several decades. Until recent decades, disasters were viewed as on-off events and responded to by governments and relief agencies without taking into account the social and economic implications and causes of these events. It was a purely reactive approach to a crisis in response to a natural disaster.

Over the past few decades, a more comprehensive disaster risk management approach has emerged. Disasters are no longer viewed as extreme events created entirely by natural forces but as complicated unresolved problems of development influenced by physical, social, and economic factors as well. The evolution from reactive relief to proactive risk management has begun to influence the way disaster programs are now being planned and implemented. Risk management is based on preparedness and mitigation measures. Preparedness is designed to increase the level of pre-disaster readiness or improve operational and institutional capabilities for responding to a drought episode. Mitigation refers to both short-term and long-term actions, programs, or policies implemented during or in advance of drought that reduce the risk to life, property, and productive capacity.

Proactive disaster management, focusing on reduction of disaster risks, can contribute significantly toward protecting vulnerable communities, the lives of their inhabitants, the assets and livelihoods. Hazard mitigation, which is defined as sustained actions taken to reduce or eliminate long-term risk from hazards and their effect, is an integral part of a proactive drought management plan.

A significant trend, which has emerged during the shift to risk management, is the focus away from top-down and to a bottom-up approach. Wilhite, et al. (1986) noted that, due to the complex federal bureaucratic structure in the United States, successful drought planning efforts would most likely be initiated at the state level. This has been the emerging and successful trend. Communities are taking on a much stronger role in disaster management to reduce risk. Thus, aid and development agencies are finding new approaches to disaster management that attempt to merge the disaster reduction strategies defined by policymakers with the needs and resources of the local community, where eventually the success or failure of disaster management activities will be tested. Community-based disaster management can be seen as risk reduction programs designed primarily by and for the people in certain disaster-prone areas. The most effective way to ensure that appropriate actions are taken well before the hazardous event occurs is for the preparation and implementation of a comprehensive hazard mitigation plan.

Disaster mitigation using government and institutional interventions alone is insufficient because little attention is addressed to the community-level dynamics, perceptions, or priorities. At the same time, local communities are often either unaware of these formal disaster management interventions or find the interventions inappropriate to their specific needs. In fact, a recent WMO publication (WMO TD No. 201) provides an excellent overview of the agrometeorological aspects of drought and desertification, and addresses some of the latest literature on adaptation issues, drought management, and drought policy. This report notes that the reactive crisis management approach has been criticized by scientists, government officials, and many relief recipients as inefficient, ineffective, and untimely. This approach does not promote self-reliance, and in fact, it may increase vulnerability to drought as well to other natural hazards.

Another reason for implementing community-based approaches is that communities are knowledgeable about the disasters happening in their environment and, in some cases, able to anticipate them. They may not be scientific, but the insight of experience and indigenous knowledge are resources to be recognized and acknowledged. Hazard mitigation planning is, therefore, a collaborative process whereby hazards affecting the community are identified, vulnerability to the hazards is assessed, and consensus reached on how to minimize or eliminate the effects of these hazards.

A number of individual states and local governments include drought in their comprehensive water management, land-use, and long-term planning strategies. It is the local entities that know best how to communicate with their local populations to deal with specific risk management issues.

11.3 Rationale for a Coordinated National Drought Policy

Natural disasters such as droughts, pose severe threats to the social and economic well-being of the nation and their occurrence can cause severe economic hardship for many years after the event. Thus, it is imperative to shift to mitigation approaches to help reduce the impact of droughts. Such mitigation methods are based on scientific knowledge about the hazards and on technical means for countering their effects. Their implementation requires an effective partnership of all stakeholders. Thus, scientists and engineers must join with economists, insurers, politicians, and other decision-makers to make vulnerable communities more disaster resilient.

Disasters generally lead to immediate emergency responses but as time passes, the memory and need for preparedness fades, leaving communities just as unprepared for the next natural disaster. To develop safer, sustainable communities, it is necessary for these communities to become more knowledgeable about dealing with disasters. Given the increasing complexity of community living and the high-consequence risks that these communities are subjected to, the gathering and sharing of relevant disaster-related knowledge needs to be undertaken in a systematic and effective manner. An important role of scientists and natural disaster managers is to develop a culture of knowledge-based management within communities that effectively contributes to risk management and sustainability. The goal should be to help communities become aware of their own specific needs, so they can generate and share knowledge about natural disaster risk management. Equally as important, the community should not develop a dependence on external groups to always undertake this role.

With greater access to data bases, products and information on the Internet and the World Wide Web, a more integrated approach to data base development is needed at the national and regional levels to support risk assessment and natural disaster reduction. Sharing of data, resources, and knowledge-based tools are a cost-effective means of aiding community programs in natural disaster reduction and mitigation of extreme events. Access to these resources can also support the development of a knowledge management framework and standard for natural disaster reduction for communities.

Government agencies must provide these fundamental integrated data management systems as they are beyond the scope of local communities. While these agencies have the ability to manage data sets needed for risk management, to provide the analytical tools for assessment, and to deliver on-line telecommunications infrastructure, effective risk reduction and mitigation requires the active engagement and contribution of the local community. Natural disaster management agencies cannot achieve these results by themselves because the causes of risks are commonly in areas beyond the scope of practice or responsibility of the agencies. Thus, it is essential that a close working relationship is established and maintained for coordination of the disaster management plan between government agencies and local communities.

What is the plan of action for disaster management preparedness? Mitigation embraces all measures taken to reduce both the effect of the natural hazard itself and the vulnerable conditions to it, in order to reduce the scale of future disasters and their impacts. Mitigation also includes measures aimed at reducing physical, economic, and social vulnerability. Mitigation measures can be active or passive. Active measures are those that rely on providing incentives for disaster reduction. They are usually more effective than passive measures that are based on laws, controls, and responses to a disaster. The context within which disaster management and planning operates and is organized is referred to as structural mitigation measures. These decision-making activities include the preparation of preparedness plans, training and education, public awareness, evacuation plans, warning systems, and land use planning. Wilhite (1987) developed a well-coordinated plan of action for integrating various levels of government and various socio-economic and political entities into a drought policy, including management, mitigation, and preparedness strategies.

While drought is a normal part of climate for virtually all regions of the United States, it is of particular concern in the West. Any interruption of the region's already limited water supplies over extended periods of time can produce devastating impacts. Records indicate that drought occurs somewhere in the West almost every year. However, it is the multi-year drought events that are of the greatest concern to water planners, natural resource managers, and government policy makers.

Following initial action of the Western Governors' Association (WGA) in response to the 1995–1996 drought, the Western Drought Coordination Council (WDCC) was formed in February 1997 with federal agencies, the National Association of Counties and the WGA.

The goal of the WDCC was to improve the planning, communication, data, and response with regard to the management of droughts (The Western Drought Experience Report, 1999). Water scarcity continually defines and redefines the West. The steady growth that has been characteristic for much of the West created increased demands for agricultural, municipal, and industrial water supplies. Furthermore, such competing demands as the public's rising concern for meeting "quality of life" and environmental objectives create water supply management challenges in times of normal precipitation. Drought exasperates these challenges.

The WDCC worked to address a range of drought-related issues. It became increasingly clear that to achieve the goals, a national drought policy must be enacted. A comprehensive vision for future drought monitoring, mitigation, and response is essential. In the WDCC report (1999), it outlined a comprehensive, coordinated, and integrated approach that was needed to address future droughts. It included the following critical elements: monitoring/assessment/prediction; preparedness and mitigation; response; and communications. The monitoring/assessment/prediction element provided for the development of a comprehensive system to collect, analyze, and disseminate data, including onset, severity and termination of drought. Preparedness and mitigation provided a framework that assisted all levels of government and water agencies with assessing their vulnerabilities; provided incentives for a variety of preparedness actions, policies and mitigation options that can facilitate improved cooperation among all levels of government and promote individual responsibilities in planning for and mitigating drought impacts; and provided policy to promote drought contingency planning, emphasizing a more proactive, anticipatory approach to drought management. Response enhanced the current capability of the federal agencies, states, localities, and tribes through a variety of appropriate policies and programs to strengthen intergovernmental partnerships, response management, and customer service. Communications encouraged the use of a variety of communication tools to identify and use drought-related information networks to facilitate the exchange and dissemination of drought-related information.

This national framework must utilize risk-based analysis that provides for a greater opportunity to implement cost-effective policies.

The WDCC recognized that it was imperative to establish key partnerships among diverse sectors of society, including research institutions; local, regional, and national public-sector decision-making bodies; and public and private sector organizations. Active involvement, forward thinking policy, and coordinated action by multi-disciplinary partnerships were needed for a sustainable national plan.

11.4 National Drought Policy Commission

The United States has had no national drought policy. Over the years, from a national perspective, it has been a patchwork approach to risk management with the federal government responding to a drought emergency during a crisis. In fact, crisis management, rather than planning and proactive mitigation measures, often characterized the federal role to drought management. However, drought occurs frequently somewhere in the country with severe drought causing major economic hardship nearly every decade or two. Yet, each time drought occurs, many of the same issues are raised. How much damage is inflicted, on whom, and where? The response is to react to the crisis after the event and after evaluating the impact of the event on the economic sector of the nation. The impacts of drought were enormous. For example, the federal government spent $ 3.3 billion responding to the 1953–56 drought. The federal drought response cost at least $ 6.5 billion during the 1976–77 drought. Riebsame et al. (1991) estimated that the 1987–89 drought in the United States caused $ 20 billion in agricultural and forest production losses with an additional $ 10 billion loss associated with increased food costs. Thus, extraordinary federal expenses for drought along over the 1952–88 period averaged over half a billion dollars per year.

These were direct costs. There were other costs involved as documented by Riebsame, Changnon, and Karl (1991). The reductions in crop production and increases in food prices, associated with drought episodes, have a ripple effect across the local and regional economies. Low flows in the major river systems cause barge-shipping prices to substantially increase. Finally, poor crop harvests impact local and state employment, adversely affecting the economy.

Finally, in 1998, after several years of debate on national drought policy, recognition was given to the fact that the United States had no consistent, comprehensive policy driving the federal role to help reduce the impact of drought. The United States Congress passed the National Drought Policy Act in 1998 stating that this nation would benefit from a national drought policy based on preparedness and mitigation to reduce the need for emergency relief.

A national drought policy should be directed toward reducing risk by developing better awareness and understanding of the drought hazard and the underlying causes of society vulnerability. The principles of risk management can be promoted by encouraging the improvement and application of seasonal forecasts, developing integrated monitoring and drought early warning systems and associated information delivery systems, developing preparedness plans at various levels of government, adopting mitigation actions and programs, and creating a safety net of emergency response programs that ensure timely and targeted relief.

The Act created the National Drought Policy Commission, recognizing the need to prepare for and lessen the severe impacts of drought on the American people and the environment. The Commission was to advise Congress on formulation of national drought policy based on preparedness, mitigation, and risk management rather than on crisis management. It also directed the Commission to present a strategy to integrate federal programs with ongoing state, local, and tribal programs into a comprehensive national policy.

The missions of the Commission were: to determine what needs exist on the federal, state, local, and tribal levels to prepare for and respond to drought emergencies; to review all existing federal laws and programs relating to drought; to review pertinent state, local, and tribal laws and programs relating to drought; to determine what differences exist between the needs of those affected by drought nd federal laws and programs designed to mitigate the impacts of and respond to drought; to collaborate with the Western Drought Coordination Council and other appropriate entities to consider regional drought initiatives and the application of such initiatives at the national level; to recommend how federal drought laws and programs can be better integrated with on-going state, local, and tribal programs into a comprehensive national policy to mitigate the impacts of and respond to drought emergencies without diminishing the rights of states to control water through state law and considering the need to protect the environment; to recommend how to improve public awareness of the need for drought mitigation and response by governmental and nongovernmental entities, including academic, private, and nonprofit interests; and, to recommend whether all federal drought preparation and response programs should be consolidated under one existing federal agency and, if so, identify such agency.

The Commission conducted a number of public hearings and meetings to examine the far-reaching impact of drought. What makes this issue even more complex is the very definition of drought, or, more specifically, the lack of a single definition that is applicable to all segment of society. In the Commission's final report (National Drought Policy Commission, 2000), a generic definition states that "drought is a persistent and abnormal moisture efficiency having adverse impacts on vegetation, animals, or people." However, there are meteorological, hydrological, and agricultural definitions of drought that are specifically related to those disciplines. Declarations of drought are often triggered by specific and well-defined conditions, such as specific reservoir levels on a specific date. In some cases, there are well-defined exit points that trigger a resumption of normal activity. These drought triggers become a practical definition of drought for a particular region and for specific sectors of society. Defining these triggers is an inseparable part of planning for and responding to droughts. Once these triggers are defined, a region is much better able to estimate the costs, expected frequency, and risks of drought response.

However, in reality, drought is defined differently in different situations. For example, two months without rainfall during the growing season may result in serious drought conditions in sub-humid climates. This same period of dryness may be normal in semi-arid climates, where water reserves must be used for crops adapted to that region. National drought policy must therefore define drought so that it meets the needs of diverse water users and for diverse functions. It must be flexible enough to include a variety of drought situations. It must also be specific enough to distinguish between those situations that are true drought emergencies and those that are normal cyclical conditions. All of these factors must be accounted for in any definition but it clearly illustrates the need for careful evaluation at the regional level of the most appropriate triggers for a drought declaration and a drought termination.

The Commission reported a number of findings based on its extensive study (Final Report, 2000). Preparedness is a fundamental concept in a national drought policy. Preparedness includes drought planning, plan implementation, proactive mitigation measures, and public education. The Commission found that preparedness may well reduce the social, economic, and environmental impacts of drought and the need for federal emergency relief expenditures in drought-stricken areas. A variety of entities are engaged in some form of drought preparedness, ranging from individuals, citizen organizations, local and state governments, tribes, and regional bodies. This planning is often conducted within the framework of comprehensive water management planning by entities ranging from water districts and large multi-county urban areas to state water resources agencies and regional river basin commissions. The Commission, at the time, found that about 30 of the 50 states in the United States had drought plans, with most on oriented toward relief rather than toward preparedness (that number has now increased to 35). The assessment revealed that in most states, drought responsibilities are normally located in the agencies that are responsible for the functions of agriculture, natural resources, water management, environment, or emergency management. Fewer than five states have independent, designated drought coordinators, while more than 20 states have drought task forces.

Regional entities generally comprise several states with a common geographic boundary or water management jurisdiction such as a river basin. It is clearly evident that regional drought planning or incorporation of drought concerns into comprehensive regional water management plans is essential for any strategy to be successful. For example, in June 1965, during the peak of a serious drought in the northeastern United States, New York City stopped releases from its Delaware River reservoirs to maintain its withdrawal rate. With less fresh water flowing past the city of Philadelphia, there was a risk that salt water would be drawn into that city's water supply system. President Lyndon Johnson convened a special meeting of governors and majors from the Delaware Basin that led to emergency measures for managing the Delaware River.

While larger government entities can address the major issues related to drought, counties, towns, and rural areas must deal with the emergencies and respond to the disasters. These areas are facing suburban growth and development, which are increasing the demand for water and creating greater competition for available water. Local governments must be able to plan for future needs, but they need the technical data, tools, and resources to develop and implement these plans. Local governments must also inform and educate the local population about the need for drought planning, especially when an emergency is not imminent. It is at the local level where the most efficient and direct communication channels can be established to keep the population informed of drought emergencies that may be directly affecting a particular area.

Tribal lands in the western United States have experienced the vagaries of climate for many thousands of years, and the scope of tribal drought issues in current times is immense. There are 306 federally recognized tribes within the conterminous 48 states, with 289 of those west of the Mississippi River, where 95 percent of all tribal trust land is located. A total of six tribes were found to be developing drought contingency plans through cooperative agreements with the federal government. It is within these rural areas that the Commission found the least available information that is critical to basic drought planning. Some tribes lack access to basic weather data that is essential not only for planning but also for triggering emergency response efforts.

The Commission found that, in response to drought challenges over the years, Congress created a number of federal programs to lessen the impacts of drought. In fact, 88 drought-related federal programs were funded from 1988 to 1998 long, and were spread over a number of federal departments and agencies. The programs were classed into four broad program categories: 1) preparedness, including planning and mitigation; 2) information, including monitoring/prediction and research; 3) insurance; and 4) emergency response. Of these programs, 7 provide assistance for drought planning, 42 for drought mitigation, 22 for drought monitoring/prediction and research, and 47 for response. These total more than 88 because some programs cover more than one facet of drought. A major criticism heard at meetings and public hearings was that the federal action was primarily an ad hoc approach to drought. Moreover, limited authorities and funds as well as lack of coordination among and within federal agencies hindered planning efforts.

Based on guidance from the Commission findings, the development of a national drought policy included the following key principles. First, the policy favors preparedness over insurance, insurance over relief, and incentives over regulation. Second, research priorities need to be set based on the potential of the research results to reduce drought impacts. Finally, it is essential that there be coordination of delivery of federal services through collaboration with all non-federal entities.

This policy requires a shift from the traditional emphasis on drought relief following a natural disaster. Preparedness must become the cornerstone of the national drought policy. To achieve this objective, the pooling of non-federal and federal experience and the establishment of non-federal/federal partnerships must be nurtured to develop the tools needed to formulate drought preparedness strategies, including the incorporation of environmental concerns.

The Commission recommended that Congress pass the National Drought Preparedness Act that would establish the non-federal/federal partnerships through a national drought council. The primary function of the council would be to ensure that the goals of the national drought policy are achieved. The five goals are briefly summarized:

Goal 1: Incorporate planning, implementation of plans, and proactive mitigation measures, risk management, resource stewardship, environmental considerations, and public education as the key elements of effective national drought policy.

Goal 2: Improve collaboration among scientists and managers to enhance the effectiveness of observation networks, monitoring, prediction, information delivery, and applied research, and to foster pubic understanding of a preparedness for drought.

Goal 3: Develop and incorporate comprehensive insurance and financial strategies into drought preparedness plans.

Goal 4: Maintain a safety net of emergency relief that emphasizes sound stewardship of natural resources and self-help.

Goal 5: Coordinate drought programs and respond effectively, efficiently, and in customer-oriented manner.

A sound national drought policy can be derived from the development and implementation of these five goals. Testimony and expert analyses presented to the National Drought Policy Commission was convincing that most levels of government and most of the private sector were not adequately prepared for drought. The basic components of the first goal focused on coordinated drought preparedness programs to lessen the need for future emergency financial and other assistance. Basic components of preparedness include long-term planning, implementation of proactive mitigation measures, risk management, resource stewardship, environmental considerations, and public education.

The second goal acknowledged the essential value of observation networks, monitoring, prediction, and information gateways and delivery, and research to drought preparedness. A significant recommendation was the proposal for a drought data monitoring, prediction, and research summit of multidisciplinary, geographically diverse representatives to ascertain the needs and expectations of all concerned. Research priorities should address the impacts of drought non-irrigated systems, aquatic ecosystems, wildlife, and other aspects of the natural environment, including the potential negative impacts of drought mitigation measures.

The third goal promotes a risk management strategy to complement preparedness measures. Prolonged drought episodes cause risks that the best preparedness measures may not adequately address. The federal government offers a crop insurance program for some crops that does promote a financial strategy to cope with drought. However, the program does not cover all crops, nor does it cover livestock. Testimony presented to the Commission also indicted that programs were administered differently across the country. A major recommendation was to evaluate the crop insurance program to ensure that sound insurance and financial strategies will help the country move away from relying on emergency relief in response to widespread drought.

The fourth goal recognizes that over time, efforts at drought preparedness, including risk management, can greatly reduce, but not eliminate, drought-related emergencies. Response measures for drought emergencies can also be useful to respond to water shortages not caused by drought. Thus, the Commission acknowledged that a safety net of responsible emergency relief measures must be incorporated into the national drought plan.

Finally, the fifth goal recognized that federal drought programs are a collection of initiatives managed by different departments and agencies. Every analysis of past responses to major droughts notes that these programs need to be better coordinated and integrated. From this repeated lesson of past history, the Commission urgently recommended the permanent establishment of the National Drought Council that would formally oversee development of the national drought policy.

Several important points are to be emphasized from this major commission report. The enactment of legislation for a national drought policy recognizes the value and importance of observation networks and monitoring and prediction and information delivery systems as fundamental components of a drought preparedness strategy. Applied research that is focused toward operational applications at the regional and local levels must also be part of the strategy. Public education programs for effective drought preparedness at the community level is essential for current awareness and a proper understanding of the effort, scope, and purpose of the undertaking. Throughout the public hearings and meetings it was clear that there must be better coordination between non-federal entities and federal agencies to ensure that the key elements of an effective national drought policy are implemented. The commission endorsed the United States Department of Agriculture as the permanent federal co-chair of the National Drought Council. The fact that the Department of Agriculture was given the lead role recognizes the importance of agriculture in many facets of this natural disaster.

11.5 U.S. Drought Monitor

The National Drought Policy Commission found that about 22 federal programs have some responsibility for drought monitoring, prediction, and research. In relation to monitoring and prediction, these programs focus on weather patterns, climate, soil conditions, and streamflow measurements. Examples of three major networks are the Department of Agriculture's Soil Climate Analysis Network (SCAN)/Snow Telemetry Network (SNOTEL), the National Oceanic and Atmospheric Administration/National Weather Service's Cooperative Observer Network (COOP), and the U.S. Geological Survey's streamgaging and groundwater monitoring network. Federal programs often join with universities, private institutions, and other nonfederal entities to provide additional information. This is especially crucial for agriculture as data observation networks are often sparse in rural agricultural areas. It is well recognized that comprehensive weather, water, soil moisture, mountain snow amount, and climate observations are the foundation of the monitoring and assessment activity that alerts the nation to impending drought.

The vigorous debates and discussions during the National Drought Policy Commission meetings helped to formulate an important new operational drought product. This product was important to develop as it became the first prototype tool to integrate the basic data on current conditions and translate these data into meaningful information to the user community. The emergence of the U.S. Drought Monitor (Svoboda et al., 2002), established in 1999, was a major advancement in drought monitoring products. The Drought Monitor classifies drought severity into five categories. The category thresholds are determined from a number of indicators, or tools, blended with subjective interpretation. The United States Drought Monitor was developed as an operational tool for monitoring drought conditions, including aerial extent, severity, and type, around the country. The Drought Monitor has become a highly successful tool for assessing the development and duration of drought conditions. The U.S. Department of Agriculture, Department of Commerce, and National Drought Mitigation Center publish the map and text weekly and post them on the Internet (http://drought.unl.edu/dm). The product serves as an exemplary case of interagency cooperation. A major strength of the Drought Monitor is its inclusion of input from climate and water experts from around the country.

The Monitor requires a major collaborative effort to pull together the various sources of weather data and compile them in a single, comprehensive, operational, national report. The map not only delineates stages of drought but also specifies drought type when the impacts differ. For example, if severe drought affects wildfire danger and water supplies, but is not in a significant agriculture area, then the map would depict W (water) and F (wildfire danger) only. If drought affected a major crop area, that area would be denoted with "A" for agriculture. The map also reflects forecast trends. If the forecast of drought is expected to intensify, a "+" is depicted in affected area. Similarly, if the forecast calls for rain to diminish drought conditions, a "–" is depicted in the affected area. No change in the drought classification forecast is depicted by no sign. The text of the Monitor provides a detailed discussion of the map.

The Drought Monitor itself is not an index, nor is it based on a single index, but rather is a composite product developed from a rich information stream, including climate indices, numerical models, and the input of regional and local experts around the country. No single definition of drought works in all circumstances (Wilhite, 2000). Water planners and agricultural producers may rely on completely different sets of indicators. The Drought Monitor authors must rely on a number of key and ancillary indicators from different agencies. The map fuses these indicators, using human expertise from across the United States, into an easy-to-read image presenting a current status of drought conditions. The Drought Monitor process is an evolving one as new, or better, indicators and information sources become available.

Lead responsibility for preparing the Drought Monitor rotates among nine authors from four agencies who sequentially take 2 to 3 week shifts as the product's lead author. Nationwide experts respond to the lead author's first draft when it arrives by Internet and through an email list-server every Monday. An interactive process continues until the final product, both map and text, are released on Thursday morning.

Classification of drought magnitude in the Drought Monitor is based on farm levels using a percentile approach. The percentiles are standardized for the year rather than for all times of the year at once. They are not meant to imply an average areal extent value for the United States at any given time. The categories include: D0 (abnormally dry), 21 to 30 percent change of occurring in any given year at a given location; D1 (moderate drought), 11 to 20 percent chance; D2 (severe drought) 6 to 10 percent chance; D3 (extreme drought), 3 to 5 percent change; and D4 (exceptional drought), 2 percent or less change.

The Drought Monitor's severity categories are based on six key physical indicators and many supplementary indicators. The indicators are the Palmer Drought Severity Index (PDSI; Palmer 1965); CPC Soil Moisture Model Percentiles (CPC/SM); Huang et al. 1996); U.S. Geological Survey (USGS) Daily Streamflow Percentiles (http://water.usgs.gov.waterwatch/); Percent of Normal Precipitation (Willeke et al. 1994); Standardized Precipitation Index (SPI; McKee et al. 1993); and remotely sensed Satellite Vegetation Health Index (VT; Kogan 1995).

Ancillary indicators include the Palmer Crop Moisture Index; the Keetch–Byram Drought Index (KBDI; Keetch and Byram 1968); evaporation-related observations, reservoir and lake levels, and ground water levels; USDA field observations of surface soil moisture; and USDA snowpack and snow water equivalent measurements.

Classification of drought impact types is also included in the Drought Monitor. The categories include agriculture (crops, livestock, range, and pastures), water (streamflow, snowpack, groundwater, reservoirs), and fire (wildfire - forest and range fires). Crop stress is often the earliest indicator of a developing drought situation because of the plants need for moisture and moderate temperatures during critical phases of development. On the other hand, hydrological impacts of a major drought often linger for months or even years after agricultural concerns disappear. Thus, it is essential to monitor the evolution of drought types as well as overall conditions.

Finally, as mentioned earlier, a significant key to the outstanding success of the Drought Monitor is the process of gleaning information from many experts located across the country. Their input and verification of impacts at the regional and local levels is critical in both the production of the Drought Monitor and in establishing and maintaining the credibility of the product. These experts include regional and state climatologists, agricultural, and water resource managers, hydrologists, National Weather Service field office employees and others. The list of expert reviewers has grown to nearly 150. A Drought Monitor Workshop is held annually to allow all participants to meet and share ideas for improvement in the process and the product. The Drought Monitor is a dynamic product that is constantly searching for more timely and better indicators to assist the user community. This user community ranges from a farmer to a government policy maker.

11.6 Latest Developments

The Drought Monitor is an important step forward. However, many observations and information products on drought are currently being gathered, developed, and provided by a variety of federal, state, and local entities. The information is not comprehensive, not integrated, not easily accessed, and not readily usable to users. The integration of relevant information is important to customers of water/drought information in order to make proactive decisions on the impact of drought at state, regional, tribal, and local levels. The current Drought Monitor provides information and serves as an excellent prototype in displaying data and information for users but does not include all sources of information and does not provide for an interactive format for users to query data and analyze options for decisions at the regional and watershed scales.

It is envisioned that users will be able to easily access new a set of tools (databases, GIS products, maps, web-based educational material) and be able to use those tools in an interactive query-based process to analyze decision options for proactively responding to drought. This information system will promote scientific information for a variety of decision-makers that will enhance the adaptive capacity for the management of water resources, result in enhanced planning and preparedness for drought, and minimize the economic and ecosystem losses associated with drought. The National Drought Preparedness Act of 2003 includes a proposed National Integrated Drought Information System (NIDIS), which is defined as a comprehensive drought information system that collects and migrates information on the key indicators of drought, including forecasts, in order to make usable, reliable, and timely assessments of drought conditions.

While long overdue, the deliberations of the Commission and the successful implementation of the U.S. Drought Monitor have laid the foundation for active partnerships among key federal and nonfederal institutions involved in drought management. However, a comprehensive national drought policy still awaits congressional enactment.

References

Hensen R, Horstmeyer S, Rippey B, Koun P and Pinder E (1999) The 20th Century's Top Ten U.S. Weather and Climate Events. Weatherwise 52 (6): 14–19.

Huang J, VandenDool H and Georgakakos KP (1996) Analysis of Model-Calculated Soil Moisture Over the United States (1931–93) and Application to Long-Range Temperature Forecasts. J. Climate, 9, 1350–1362.

Keetch JJ and Byram GM (1968) A Drought Index For Forest Fire Control. Forest Service Research Paper SE-38, U.S. Department of Agriculture, 32 pp.

Kogan FN (1995) Droughts in the late 1980's in the United States as derived from NOAA polar-orbiting satellite data. Bull. Amer. Meteor. Soc., 76, 655–668.

McKee TB, Doesken NJ and Kleist J (1993) The Relationship of Drought Frequency and Duration to Time Scales. Preprints, Eighth Conference on Applied Climatology, Anaheim, CA. American Meteorological Society, 179–184.

National Drought Policy Commission. 2000. Final Report: Preparing for Drought in the 21st Century. United States Department of Agriculture, Washington, D.C.

Palmer WC (1965) Meteorological Drought. Office of Climatology, Research Paper 45, U.S. Weather Bureau, 58 pp.

Riebsame WE, Changnon SA and Karl TR (1991) Drought and Natural Resources Management in the United States. Westview Special Studies in Natural Resources and Energy Management, 174 pp.

Svoboda M, LeComte D, Hayes M, Heim R, Gleason K, Angel J, Rippey B, Tinker R, Palecki M, Stooksbury D, Miskus D and Stephens S (2002) The Drought Monitor. Bulletin of the American Meteorology Society, 83 (8): 1181–1190.

The Western Drought Experience, The Western Drought Coordination Council's Report to the National Drought Policy Commission. 1999. 600 17th Street, Denver, CO 80202.

Wilhite DA (2000) Drought as a Natural Hazard: Concepts and Definitions. Drought: A Global Assessment. D. Wilhite, Ed., Vol. 1, 3–18.

Wilhite DA and Easterling WE (1987) Drought Policy: Toward a plan of action. In: Wilhite DA and Easterling WE (eds.), Planning for Drought: Toward a Reduction in Societal Vulnerability. Westview Press, Boulder, Colorado, USA. Chapter 37.

Wilhite DA, Rosenberg NJ and Glantz MH (1986) Improving federal response to drought. Journal of Climate and Applied Meteorology 25: 332–342.

Wilhite DA, Sivakumar MVK and Woods DA (Eds.) (2000) Early Warning Systems for Drought Preparedness and Drought Management. Proceedings of an Expert Group Meeting held in Lisbon, Portugal, 5–7 September 2000. Geneva, Switzerland: World Meteorological Organization.

Willeke G, Hosking JRM, Wallis JR and Guttman NB (1994) The National Drought Atlas. Institute for Water Resources Report 94-NDS-4, U.S. Army Corps of Engineers, CD-ROM.

WMO (2003) Agrometeorology Related to Extreme Events. TD No. 201, WMO, Geneva.

CHAPTER 12

Agricultural Drought Policy and Practices in Australia

William J. Wright

Abstract This Chapter describes recent events in the development and application of a national drought policy in Australia. Prior to the 1990s, Government assistance to drought-affected farmers and pastoralists was based on the assumption that drought was an unpredictable natural disaster. A paradigm shift towards the present policy of "self reliance" occurred in the early 1990s. Under this policy, farmers were expected to be able to manage through most droughts, with Government assistance only available in the event of droughts of unusual length or severity – so-called "exceptional circumstances." Details of the policy, and its application in, and following, drought situations, are briefly outlined.

Illustrated by the experience of 2002–03, one of Australia's worst-ever droughts, we show that the development of drought policy in Australia is an ongoing process, in which attempts to apply aspects of the policy in practice generally reveal deficiencies, leading to subsequent modification of the policy. A review of National Drought Policy conducted by the Australian Government after the 2002–03 drought generated a number of recommendations. Among these were: the nearly universal endorsement by stakeholders that proactive drought preparedness measures should be encouraged by Governments; and that improvements were needed to the reliability of seasonal predictions, with calls for better-coordinated research into climate variability, and better presentation of prediction information.

The Chapter also describes changes in water management policy in Australia, which also underwent a fundamental change during the early 1990s. Case studies are provided of certain aspects of the policy, including the application of water entitlements in rural areas, and water restrictions in urban areas.

The Chapter discusses some likely future developments, including discussion of the possible implications for drought policy posed by future climate change.

12.1 Introduction – Past Policy

Australia is fortunate in that, unlike in many other countries, drought does not cause famine or economic collapse. Nevertheless, throughout Australia's history, drought has administered regular and often severe shocks to the country's economy, environment and socio-economic development. During the late 19th Century, and for much of the 20th Century, Australia's economy was dominated by agricultural and pastoral activities. Farm products accounted for approximately 75% of Australia's total export income during the first half of the 20th Century (Australian Bureau of Statistics, 1914 through 1918, and

1941 through 1950), therefore widespread drought exerted a major impact on Australia's economy. Although the relative importance of agricultural exports in the Australian economy has declined, major droughts (such as 1982–83 and 2002–03) still have the capacity to significantly impact on the Australian economy (Fig.12.1), and draw heavily on the resources of Government.

For most of the 20th Century, Australia's Federal (i.e., national) and State/Territory Governments and primary producers took the view that drought was a natural disaster, similar to bushfire, floods, etc, with land managers victims of a capricious climate. The responsibility for responding to natural disasters lay (and still lies) primarily with the various State and Territory Governments (Botterill, 2003), who provided financial support for those affected by drought. When a particular area was declared drought-affected, farmers became eligible for concessional rates on transport for feed and livestock, and for loans at reduced rates of interest. State/Territory Governments would typically provide emergency water supplies, subsidies on fodder for starving stock, and projects to provide employment in drought-affected areas.

Gradually, however, the Federal Government became increasingly involved, until it was contributing substantial sums to drought relief, through Natural Disaster Relief arrangements with the States/Territories (Botterill, 2003). In his history of drought in Queensland, Daly (1994) documented the political and

Fig. 12.1. Denuded earth and dry watercourses during drought near Gunnedah, in the normally well-watered Namoi Valley region of New South Wales, during sustained drought in the early 1990s (photo c/o the *NSW Dept of Land and Water Conservation*)

community factors involved in trying to develop an equitable, compassionate basis for Government support in drought. Relief was largely provided on a subjective and *ad hoc* basis.

There were two main problems with this approach:

1. In a country like Australia, with highly variable rainfall, and where much agriculture takes place in climatically-marginal areas, drought cannot justifiably be regarded as a "natural disaster", but rather, as a natural part of the environment.
2. Although the primary purpose of drought relief was to sustain land managers, by the late 1980s there was an increasing awareness that the existing policy often inadvertently rewarded poor managers, who hadn't adjusted their land use in accord with the limitations imposed by the climate, and whose activities led to serious environmental degradation. On the other hand, land users who had wisely and skilfully managed the drought conditions, and thereby conserved the land resources, were not eligible for support.

12.2 The 1990s – Development of a National Drought Policy

A paradigm shift in Australian Government policy took place in the early 1990s. It became officially accepted that drought was not a natural disaster, but should be viewed as a natural and cyclical part of the Australian environment. Drought came to be regarded as a "business risk" (like, for instance, changes in commodity prices), and subject to sound risk management practices. Under the new policy setting, drought relief for farmers and agricultural communities would be restricted to times of so-called "*Exceptional Circumstances*". In other words, the agricultural sector was expected to cope with the occasional drought, and relief would be available only for droughts of unusual length or severity. It took some time for many farmers, and much of the broader community, to accept this new risk management paradigm.

In July 1992, a National Drought Policy embodying the new paradigm was formally agreed to by the Federal, State and Territory Governments. This Policy had three main objectives:

- To encourage primary producers and other sections of rural Australia to adopt self-reliant approaches to managing the risks stemming from climatic variability;
- To maintain and protect Australia's agricultural and environmental resource base (i.e., to limit damage in these areas) during periods of extreme climate stress; and
- To ensure early recovery of agricultural and rural industries, but in a manner consistent with long-term sustainable levels.

A key part of implementing this policy was to provide farmers with the skills and tools needed to manage with climate variability. To this end the Australian Government, through agencies such as the Bureau of Rural Sciences, the Australian Bureau of Meteorology, the Australian Bureau of Agricultural and Resource Economics, the Department of Agriculture, Fisheries and Forestry (DAFF), and various other Federal and State/Territory agencies, undertook a capacity-building program. Activities included: research into climate variability and prediction, and the application of this knowledge to agriculture; the development of improved "tools" to aid decision-making (such as decision support software packages); and awareness-raising and education programs, with emphasis on the agricultural sector (Fig. 12.2). An important part of the capacity building program was in attempting to teach farmers how to employ seasonal climate predictions as an aid to on-farm decision making. Climate predictions are mainly based on the known and strong link between climate variability – including the occurrence of droughts – over much of Australia, and the state of the El Niño-Southern Oscillation (ENSO) phenomenon (e.g., McBride and Nicholls, 1983; Nicholls, 1985). This capacity-building program is described in more detail in a companion paper (Wright, 2005).

Between 1997 and November 2002 the Federal Government provided more than $ AU 800 million to programs that could be utilised by farmers to improve their general risk management skills, including those for managing drought (Truss, 2002). On top of this amount, payments were provided under Exceptional Circumstances (EC) criteria. A number of other measures were introduced to help manage drought. For instance, to help manage the sharp year to year fluctuations in income associated with climate variability and commodity price variations, a so-called *Farm Management Deposits* scheme was introduced, enabling farmers to put aside earnings from good years to enable them to manage through, and then recover from, subsequent income downturns, including those associated with drought. There are also community support services to help farm families manage through drought.

Fig. 12.2. The 2002–03 drought stimulated public debate about whether the Australian continent could be "drought-proofed." Suggestions such as turning northern Australian rivers inland were proposed. The illustration takes a light-hearted look at the issue (Cartoon copyright: Andrew Weldon)

As noted, for severe droughts, "Exceptional Circumstances" (EC) provisions exist, to support farmers who suffer financially from the drought. In its present form, EC relief consists of interest rate subsidies for two years, and income support for up to two years. To be eligible for EC relief, the following three criteria must be satisfied:

- the event must be rare and severe (a "rare" event is defined as one occurring, on average, once every 20 to 25 years; a "severe" event is one lasting more than 12 months, and of a significant scale);
- the event must produce a severe downturn in farm income over a prolonged period; and
- the event must not be predictable, or part of a process of structural adjustment.

The key point is that a severe income downturn should result from a specific, rare and severe event (usually, but not necessarily, drought) that is beyond normal risk management strategies employed by responsible farmers. "Exceptional Circumstances" can include various events that, in combination, could be considered rare, severe and destructive, such as frost, plagues of vertebrate and invertebrate pests, and drought-followed-by-flood. Many of these occur in combination with, or are exacerbated by, prolonged drought, however by far the greatest number of applications for assistance have been related solely to drought.

The various State/Territory Governments have also continued to provide assistance packages for drought relief (with, of course, different eligibility criteria). Thus, one aspect of the application of drought policy that remains unresolved is the relative contribution of the Federal and State/Territory Governments to drought assistance payments.

12.3 Climate Services in Support of the Drought Policy

A key point in applying National Drought Policy within Australia is in assessing the climatic background to the event. As noted above, the El Niño-Southern Oscillation (ENSO) phenomenon exerts a substantial influence on Australian rainfall, and the majority of serious droughts coincide with that phase when the eastern tropical Pacific is unusually warm (El Niño), and the so-called Southern Oscillation is negative (Nicholls, 1985). An extensive public education campaign by various agencies during the 1990s, targeted especially at the rural sector, has ensured that the link between ENSO and rainfall anomalies in Australia is widely appreciated (Wright, 2005). The Australian Bureau of Meteorology, along with some State Departments of Agriculture, regularly issue seasonal predictions. The Bureau, for instance, has been issuing seasonal climate predictions since 1989, and despite some changes in methodology, the ENSO pattern remains the primary predictor. Currently, the Bureau's seasonal predictions are based on broadscale sea surface temperature (SST) patterns, which exert a strong influence on the overlying atmosphere, yet themselves change only slowly. The dominant SST pattern (Fig. 12.3) reflects the ENSO phenomenon, characterised by strong SST anomalies over the eastern and central tropical Pacific Ocean. A second predictor based on SST patterns in the Indian Ocean is also used. The Bureau has been experimenting with seasonal predictions made by coupled atmosphere-ocean models for several years: it is expected that these dynamical model predictions will become fully operational in coming years. The Queensland Department of Primary Industries also issues seasonal predictions, based on historical links between rainfall and the "phase" of the Southern Oscillation Index (whether positive or negative, combined with whether the Index is rising or falling).

Such predictions have the potential to greatly assist on-farm management. Knowing, for instance, that a dry winter and spring is likely, farmers can take actions to reduce potential losses (such as planting reduced acreages, or more drought-hardy cultivars; or graziers could decide to sell or agist stock, or pre-purchase fodder). Seasonal climate predictions are also an important source of information for Government decision-makers, as they indicate the

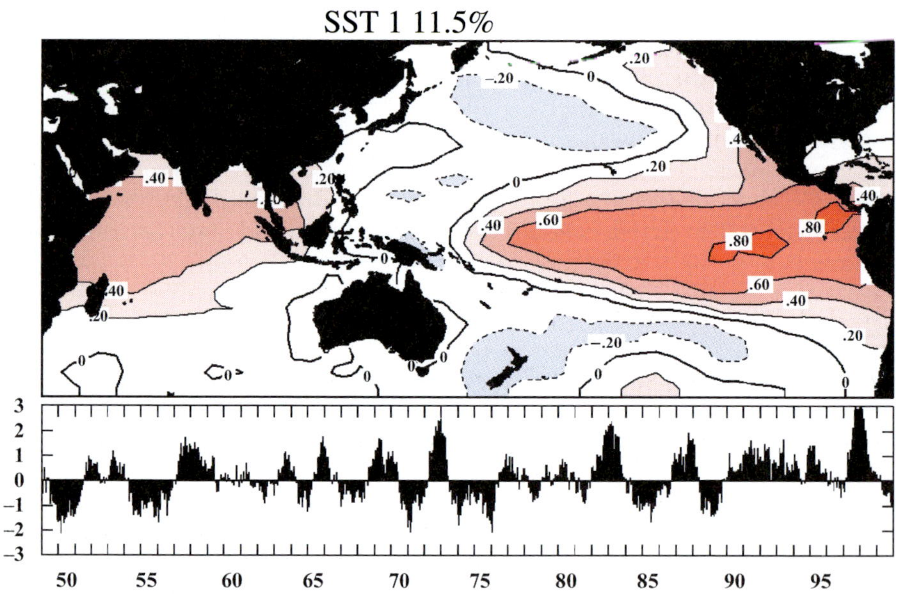

Fig. 12.3. The dominant pattern of sea surface temperature (SST) variability in the Indian-Pacific region. This pattern strongly reflects the so-called El Niño-Southern Oscillation phenomenon, and is closely related to rainfall variations over much of Australia. The time-series at bottom indicates the "strength" of the pattern in each month over the period 1949–2000. Strong positive values, when SSTs in the central and eastern Pacific are well above average (e.g., the early 1990s, and 1997), indicate El Niño conditions; strong negative (e.g., 1988–89) correspond to La Niña

likely duration of a drought, and therefore the probable "draw" on Government funds.

The Bureau also closely monitors recent rainfall patterns, and in times of drought, rainfall deficiency analyses – maps indicating where rainfall over specified periods is in the lowest 10% or 5% of recorded falls – are an important source of information for Government decision-makers. This is discussed in more detail in Sect. 12.5.

12.4 Application of the Exceptional Circumstances Policy

In applying the EC policy, the process is as follows. Applications are lodged by State/Territory Governments, on behalf of farmers, to the Australian Government. These applications are then assessed against the set of criteria described in Sect. 12.2. The assessment includes scientific and economic analyses, and involves a number of organisations. The coordination of this assessment process is handled by the Australian Government DAFF (see Sect. 12.2). If a *prima facie* case is established, an independent advisory body, the National Rural Advisory Council (NRAC: a group consisting of farmers, agribusiness professionals, and Government officials) is called on to assess the claim for full assistance. This assessment includes further scientific and economic analyses, and includes a visit to the region to assess conditions on the ground. NRAC then makes a recommendation to the Federal Government on the eligibility of the region for EC relief.

As mentioned above, rainfall deficiency maps are a crucial piece of information, as they show directly how rare an event is. It follows from the definition of percentiles that a 1 in 10 year dry spell over a given period is one in which rainfall lies within the lowest 10 percent of recorded falls, i.e., within the 10th percentile. A 1 in 20 year dry spell falls within the 5th percentile. During the 2002–03 drought, rainfall deficiencies over the period March to November 2002 were an important eligibility criterion for interim EC relief (see below). The map (Fig. 12.4) shows rainfall deficiencies over this nine month period, with hatching and cross-hatching indicating areas where rainfall was in the lowest 10th and 5th percentile respectively, and solid shading indicating "lowest on record" rainfall.

It is important to note, however, that in assessing EC claims, other measures that capture the effectiveness of rainfall are also important, such as the timing of the rainfall in relation to the growing seasons, the potential for plant growth, and the availability of water to industries relying on irrigation. The focus of these measures is to capture the impacts on agricultural industries of the rainfall deficiencies, rather than to rely on the rainfall deficiency alone as a suitable measure.

In attempting to apply these EC criteria, various issues and problems came to light, and these have driven ongoing modifications to the policy. A recurrent problem, in the 2002–03 and earlier droughts, has been how to define, fairly and objectively, those areas that qualify for support. Any definition will invariably

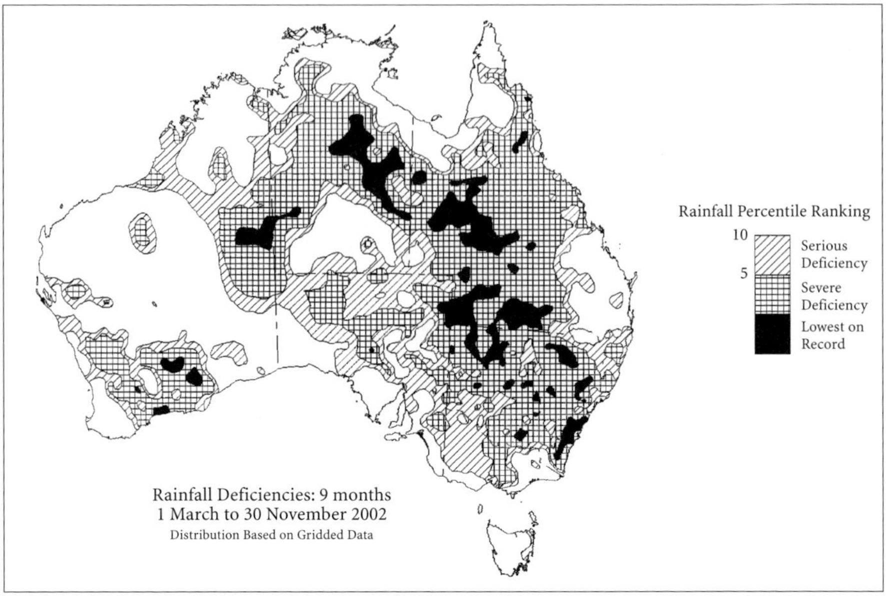

Fig. 12.4. Australian rainfall deficiencies for the 9 month period March–November 2002. Deficiencies are expressed as "Serious" where rainfall for the period is in the lowest 10 percent of recorded falls; "severe" if in the lowest 5 percent of recorded falls; and "lowest on record"

at some point exclude needy farmers whose situation is identical to those on the "right" side of a boundary. This is sometimes prosaically referred to as the "lines on a map" problem. In response, the concept of "buffer" zones was introduced in 2001, in which regions immediately adjacent to the application region could also be included. (Addition of buffer zones to application areas is not, however, automatic). Another problem was financial difficulties caused by delays in processing EC applications (due to the complexity of the process). During the 2002–03 drought, the Federal Government introduced "Commonwealth Interim Income Support" payments, which provided six months of income support, and interest rate relief, to farmers in areas where a *prima facie* case was established. (Even if the application for EC support was subsequently rejected, farmers would still receive a full six months support). Relief payments were also made available to non-farm people who lose their jobs, or suffer serious reduction in income, due to drought. However, various other issues remain unresolved at this stage – for instance, there is as yet no objective means of assessing the relative severity and impacts of long-term drought.

The current situation can be summed up by saying that, in Australia, drought policy and its applications continues to be an evolutionary process, with the assessment of some circumstances (e.g., the situation where a combination of factors lead to an EC application) requiring further development.

12.5
Water Management

12.5.1
Water, Irrigation and the Environment

Soon after the introduction of the National Drought Policy in 1992, a wide-ranging program of national water reform was also introduced. Commencing in 1994, the aim was that each Australian State/Territory Government became responsible for introducing urban and rural water reforms to increase efficiency of water use. (This initiative was broadly similar to a program of reforms that the Victorian State Government had first introduced in the late 1980s). An important aspect of the water reform program was the introduction of a system of water "entitlements" by various user groups. These entitlements – which were tradeable – effectively set limits on how much water was available to an enterprise, organization or other user in any one year, taking into consideration environmental needs.

Probably the most significant of these user groups is the agricultural sector. Much Australian agriculture takes place against a background of limited, and often highly variable rainfall, and therefore access to irrigation water is vital for many Australian farming enterprises. The issue of water entitlements is particularly important in the Murray River system, in Australia's southeast, where the bulk of the nation's agricultural activity takes place. Much of this activity is underpinned by irrigation water from the Murray River in Victoria, New South Wales and South Australia – the States through which the river flows. So important has been the need for irrigation that in the past, water from other rivers, notably the Snowy River, has been diverted into the Murray, causing adverse environmental effects downstream on these "donor" rivers. With the increased value placed on environmental issues, this became a contentious matter within Australian domestic politics.

To address this issue, the three State Governments reached an agreement about appropriate flow levels for the Snowy and Murray Rivers. One agreed outcome was the need to return to the Snowy at least a portion of the once-natural flow levels. The assumption was made that the extra water needed to meet the Snowy River's "environmental" flow levels could be obtained through a program of improved water use efficiency, involving measures such as replacing open water channels (which suffer huge evaporative losses) with enclosed pipelines, and through the sponsoring of research programs aimed at improving water use efficiency in certain key industries, such as viticulture and horticulture. A similar initiative to restore environmental flow levels to the Murray River has also been implemented. Incidentally, environmental and economic issues such as this that cross State boundaries have been a "driver" in the formation in Australia of inter-governmental bodies such as the Murray-Darling Basin Commission, with representation at the Ministerial and head-of-department level.

12.5.2 Urban Water Policy During Droughts

During severe droughts, the water storages supplying urban areas – including those of Australia's major capital cities – can become depleted. This means that care must be taken to protect remaining supplies. A common response to this situation has been for the imposition of mandatory water restrictions – supplemented as necessary by other measures, such as the installation of emergency ground-water bores. Water restrictions are generally imposed on a staged basis, with progressively tougher restrictions as conditions worsen. Lower level stages restrict such activities as washing cars and filling swimming pools. Higher level stages significantly restrict outdoor water use, and even impose limits for household use. (Should these higher level water restrictions not prove sufficient to conserve water supplies, rationing may be imposed: in the severe 1982–83 drought, residents of at least one Victorian country town were restricted to just 60 litres each a day). The development of formal drought response plans has led to changes in the way these restrictions, and other elements of water management policy, have been implemented. The situation in Victoria provides an interesting case-study.

Despite a reputation for generally reliable rainfall, Victoria has experienced many severe droughts. Recent examples including those of 1967–68, 1972–73, 1976–77, 1982–83, 1997 and 2002–03. In the first four cases (prior to 1990), water authorities imposed mandatory water restrictions to domestic and public use. These restrictions were imposed on the basis of subjective estimates of the overall water supply system, and the perceived likelihood of supplies falling to dangerously low levels. Restrictions became progressively tougher as drought continued and water supply levels fell. During the 1972–73 and 1982–83 droughts, restrictions in the Greater Melbourne area reached the point where hand-held hose watering of gardens and lawns could only be carried out between 7 pm and 9 pm three days a week, with odd-numbered and even-numbered properties alternating. Bucket watering – again for just two hours a day – was permissible on days when hose watering was not allowed. The situation attracted considerable media interest, and the popular press ran stories about such topics as "bucket back" (back injuries sustained due to carting water in buckets), "dob in a water cheat" campaigns, and the popular "save water – shower with a friend" bumper stickers, which appeared during the 1972 drought.

A problem with the restriction system was the subjective basis on which various stages were introduced. The situation was often politically charged, with sometimes vigorous exchanges between water supply authorities, State/Territory Governments, and other stakeholders, and pressure from various lobby groups about the need for restrictions (Keating, 1992).

In 1994 this reactive response to drought was replaced with a planned response. Each urban water authority was required to develop Drought Response Plans for the supply systems under their control. A centerpiece was the establishment of objective criteria, or 'trigger points', for restrictions. These were threshold water storage levels, and/or streamflow levels which, when reached, would automatically lead to the imposition of water restrictions and other measures to protect existing supplies. Such a system removed the subjectivity in the imposition of restrictions. 'Trigger points' are specific to each supply region, and were generally established from simulations using long-term records of rainfall and stream-flow. In the case of the Greater Melbourne area, a four stage set of restrictions currently applies.

The other historical response to drought in Victoria was that each water crisis led to the construction of higher capacity dams, mostly to supply the Greater Melbourne area. In 1984 the construction of a large dam on the Thomson River, well east of Melbourne, was completed, with the ambitious aim of "drought proofing" Melbourne (Keating, 1992). With a normal incidence of drought, this aim may well have been achieved. However the seven and a half years between late 1996 and June 2004 saw record low rainfall over most of the catchment area supplying the Greater Melbourne area, leading to a significant decline in storage levels (Fig. 12.5). Level One restrictions were imposed on 1 November 2002, advancing to Level Two restrictions in August 2003, where they remain at the time of writing (May 2004). Many towns in country Victoria, without access to such large water supplies, have suffered water restrictions - severe in some areas - from 1997 onward. This example underlines the need to consider longer-term climate variability and change in framing water (and drought) policy, an issue revisited in Sect. 12.7 below.

It is now accepted in most Australian States/Territories that there are limits to the amount of water that can be harvested from storages. For this reason, current planning emphasizes improving the efficiency of water storage, supply and use. In rural areas, typical measures are as suggested in Sect. 12.5.1 for preserving environmental flows; in urban areas, measures include attempting to develop a water conservation ethic, increased use of recycled water, and changes in pricing.

Fig. 12.5. The height of these two bridges above the dry, baked-mud bottom of Lake Eildon reservoir, in central Victoria, during the late 1990s is testament to the impact on water storages during extreme drought episodes in even relatively well watered areas such as southeastern Australia (photo c/o Catherine Beesley, Australian Bureau of Meteorology)

12.6
The 2002–03 Drought in Australia – a Case Study

The development of El Niño conditions in 2002 was accompanied by one of the worst droughts in recorded history in Australia. In the 11 month period March 2002 to January 2003, 56% of the country recorded rainfall in the lowest decile (i.e., totals were in the lowest 10% of recorded falls for that period). Since 1900, only the droughts of 1901–02 and 1982–83 have had (slightly) higher percentages of the country in decile 1 for a comparable period. The mean percentile in the March 2002–January 2003 period, averaged over the whole country, was 17%, second only to the 1982–83 drought.

The drought had severe impacts on many farmers and rural communities throughout Australia (Fig. 12.6). Farm incomes plummeted by around 70% from the previous (admittedly favorable) year, while agricultural exports fell more than 28%. Statistically this was the worst result in the last 40 years, and perhaps the worst since Federation in 1901 (Eslake, 2003). The price of feed for livestock rose strongly in the second half of 2002, strongly impacting

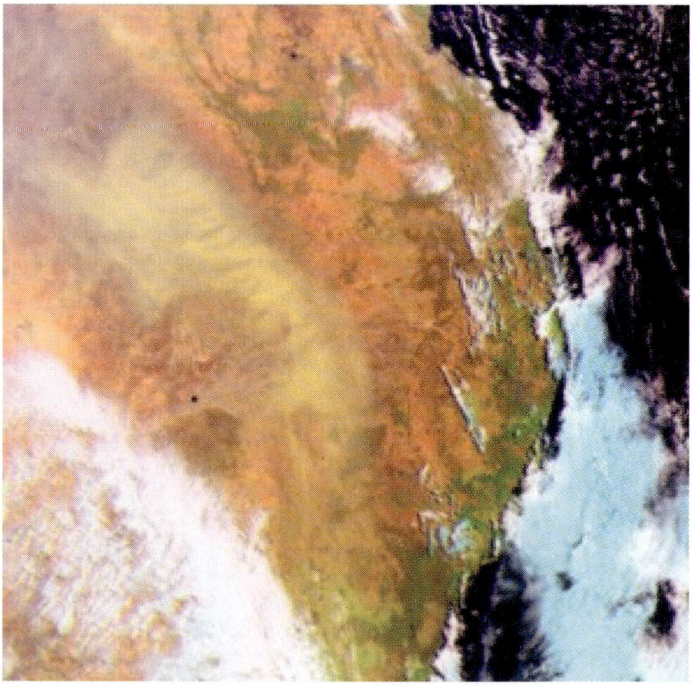

Fig. 12.6. A major dust-storm affects northern New South Wales and southern Queensland on 23 October 2003, as revealed in this satellite image showing the dust-cloud as a diffuse, buff covered area ahead of an eastward-moving frontal cloudband. The image clearly indicates the extensive nature of this dust-storm. Such storms are frequent during severe, and/or extended, drought episodes, and represent major degradation of potentially-arable land

upon the intensive livestock and dairy industries. Water restrictions were imposed in most of Australia's capital cities. The drought was exacerbated by extreme heat, with national average maximum temperatures in each of the austral calendar seasons autumn, winter and spring (March–November) 2002 exceeding by large margins the previous records (from a high quality dataset dating back to 1950). The result was that evaporation was enhanced, exacerbating the effects of the drought. There were severe bushfires: fires encroached close to Sydney, engulfed well over 1 million hectares of forested high country in northeastern and eastern Victoria, while in the nation's capital, Canberra, wildfires on 18 January destroyed 530 homes, with four lives also lost.

Despite the severity of the climatic conditions, the impacts on farmers were less severe than could have been the case, due to good antecedent conditions (i.e., abundant rainfall) over much of the country in preceding La Niña years, an overall sound economy, with low interest and inflation rates, and high export commodity prices. By contrast, some previous droughts in Australia (e.g., in 1914/15 and in the early 1940s) coincided with poor economic conditions (Australian Bureau of Meteorology, 2004). The 1901–02 drought, too, followed several years of dry conditions. Doubtless the impacts in 2002–03 would have been worse but for these inherently favourable mitigating circumstances.

With the drought worsening in the late winter and spring of 2002, the Director of the Bureau of Meteorology was required to provide as-needed briefings to appropriate State/Territory and Federal Government ministers, as well as to the heads of departments of agriculture and natural resources. These briefings consisted of updates on current conditions, including rainfall deficiencies, and the latest seasonal outlooks. This information was used as a guide to Government responses to the drought. The Bureau also provided data and regular briefings to the Bureau of Rural Sciences, whose primary involvement in the implementation of drought policy is to provide scientific advice on whether an EC application region has experienced an exceptional event, and the likely impact of this event on agricultural production. Their scientific advice was in turn provided to the Federal department of Agriculture, Fisheries and Forestry, and to the independent NRAC (see p. 201), who used it as a basis for a recommendation on eligibility to the Federal Government.

By the end of 2002, substantial areas of the country were receiving, or had applications pending, for EC relief. In December the Federal Government introduced its new Commonwealth Interim Income Support scheme (see Sect. 12.4), to support afflicted farmers pending resolution of EC claims. Additionally, in a one-off policy decision, areas suffering a one in 20 year rainfall deficiency over the nine months March–November 2002 were granted interim income support. On the basis that more than 80 per cent of farmers in the State of New South Wales fell into this category, it was decided to declare that entire State eligible for Interim Income Support, with the understanding that similar declarations would be made in other States reaching the 80 per cent "trigger".

While 2003 saw relief in many parts of Australia, drought persisted over much of the eastern States. It was not until late July/early August 2003 that the majority of the serious accumulated rainfall deficiencies finally began to be erased. However, the social, economic and environmental consequences of the drought continued to linger on, highlighting the long recovery period for agricultural activities following the end of a hydrological drought. Even after rainfall and water availability improve, it takes many months, even years, before these conditions translate into production and income, as new crops need to be sown and grown, herds rebuilt, and pasture rehabilitated. Moreover, at the time of writing (June 2004) some areas seemed to be sliding back into hydrological drought, with large areas of inland eastern Australia again experiencing severe water shortages.

Figure 12.7 shows the areas of Australia receiving EC relief as of the end of January 2004, including buffer zones. A small area in far southeastern Victoria was also eligible for Interim Income Support at that time. In February 2004 the Federal Government estimated that the cost of providing EC relief for the 2002–03 drought would amount to more than $ AU 1.2 billion – of this, the Federal Government would contribute around $ AU 1 billion up to 2006, with State/Territory Government assistance amounting to another $ AU 200 million (Australian Government, 2004).

12.6.1 Lessons Learned

The 2002–03 drought provided a stringent test for application of the drought policy, and revealed several areas where modifications were needed. To formalise this process of learning from the 2002–03 experience, the Australian Government established a Drought Review Panel to review national drought policy, taking into account the views of key stakeholders. The review looked at:

- The appropriateness of Australian and State/Territory Government drought assistance measures provided during the drought, viz:
  - the timeliness and accessibility of Australian and State/Territory assistance measures;
  - the role of financial instruments such as Farm Management Deposits and other risk management facilities.
- Key elements of future drought policy, including:
  - whether existing drought assistance arrangements should be retained, modified or replaced;
  - Alternative assistance measures, and/or drought preparedness policies, including preparations for "the next" drought (e.g., providing tax incentives for drought mitigation works, such as improving water and fodder storages);
  - Income-contingent loans for drought relief.

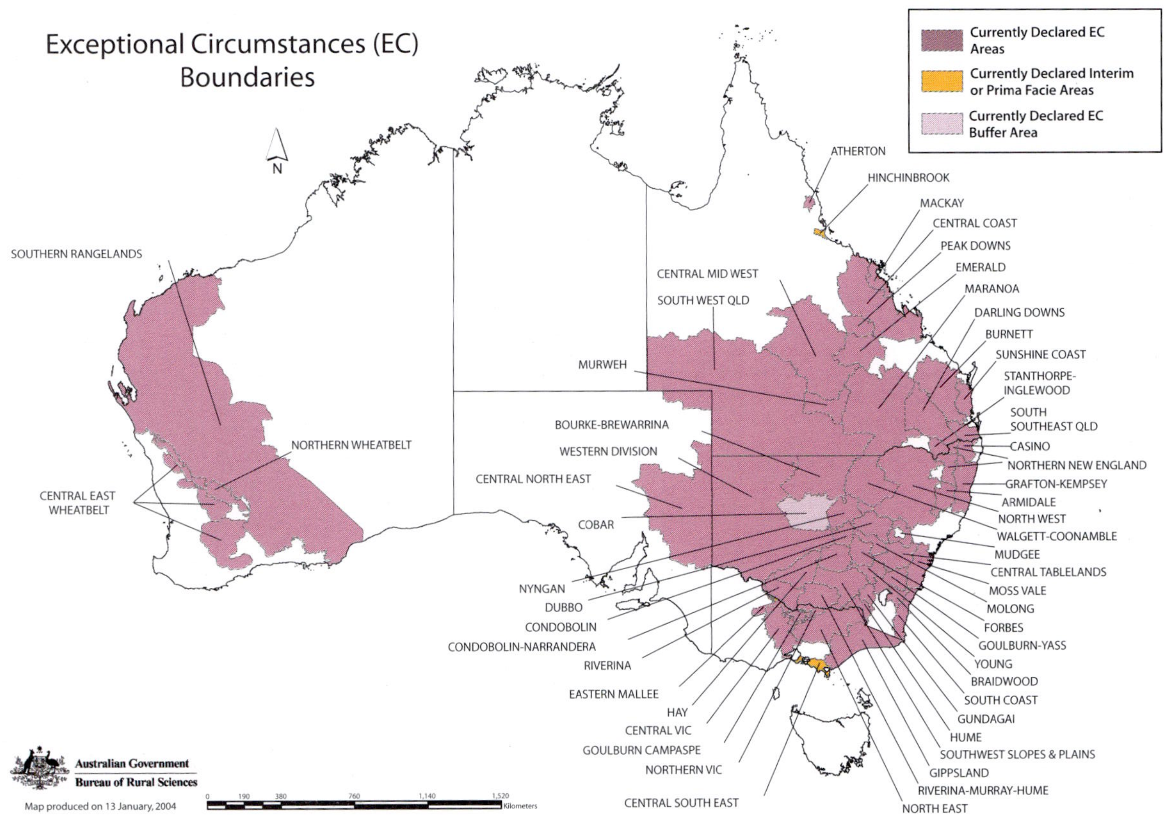

Fig. 12.7. Areas in receipt of Exceptional Circumstances relief, through declarations (*magenta*), buffer zones (*pale purple*), and Interim Income Support (*orange*), in January 2004. Courtesy of Australia's Bureau of Rural Sciences

- The adequacy of research on climate variability and its impacts on agriculture, including how seasonal predictions can be improved.

Some of the outcomes of this consultation process, which spanned the period November 2003–February 2004, are summarised in Box 12.1 (Australian Government, 2004). These reflect a need to emphasize drought preparedness, and when drought bites, for support to be directed towards welfare and community support and away from business support:

Box 12.1. Major issues arising out of a review of the 2002–03 drought in Australia. Adapted from: Australian Government, 2004: *Consultations on National Drought Policy – preparing for the future*

- Overwhelming support from key stakeholders for drought preparedness to be encouraged by Governments, and for appropriate drought preparedness measures to be a focal point of future drought policy;
- The need for clearer information about what drought assistance measures are available, and the eligibility requirements needing to be met (many stakeholders found existing information confusing, and added to tension among people already under stress);
- The need for greater consistency between States in the kinds of assistance measures applied (the measures implemented in some States were actually detrimental to farmers in other States), and also the need for agreement between States on a set of criteria for drought declarations;
- Drought assistance needed to be more timely, although it was emphasised by many that the need for assistance would be greatest during the recovery phase (e.g., for restocking);
- Equity issues, associated with both the "lines on a map" problem, and the equity issue that some industries within EC areas were eligible for assistance, while others were not;
- The need to support non-farm rural businesses and community support services, which were also severely affected by the drought. This reflects the fact that the impacts of drought are felt at all levels within a community, and have the potential to affect the future viability of the entire community (since rapid declines in population, and in access to services such as health and education, often occur in affected rural areas);
- Consideration of the special situations faced by young farmers, who could not be expected to have well-established coping procedures for managing drought, and for older farmers facing succession planning issues; and
- The need to improve the accuracy and usefulness of research into climate variability, and the reliability of seasonal predictions, for this research to be better coordinated, and for climate information to be presented in a form that could be readily understood and applied.

12.7 Relation to Long-term Climate Fluctuations

12.7.1 Cyclical Climate Fluctuations

Drought management, and EC responses to drought, are mostly aimed at managing relatively short-term droughts, associated with interannual climate variability. However Australia occasionally experiences extended periods of below average rainfall over extensive areas, on time-scales upward of five years (Fig. 12.8). Wright (2001) has described some of the more significant climate variations on decadal and multidecadal time-scales over Australia during the 20th Century, and their significant impacts on Australian agriculture and the environment.

Figure 12.9 summarises drought incidence in Australia, and indicates a number of extended episodes, including the so-called Federation drought (1895–1903); a long dry spell that affected much of the country between 1937–45; and more recently (early 1990s), an extended drought in Queensland and northern New South Wales that accompanied extended El Niño conditions in the early to mid-1990s. Each of these impacted severely on Australian agriculture, pastoral activities and the environment.

Climate scientists do not yet know why decadal and longer-term climate fluctuations occur. It may be that such periods arise simply by chance as, for example, when one throws eight heads in ten tosses of a coin. Alternatively, it has been hypothesized (e.g., Power et al., 1999) that such fluctuations could be

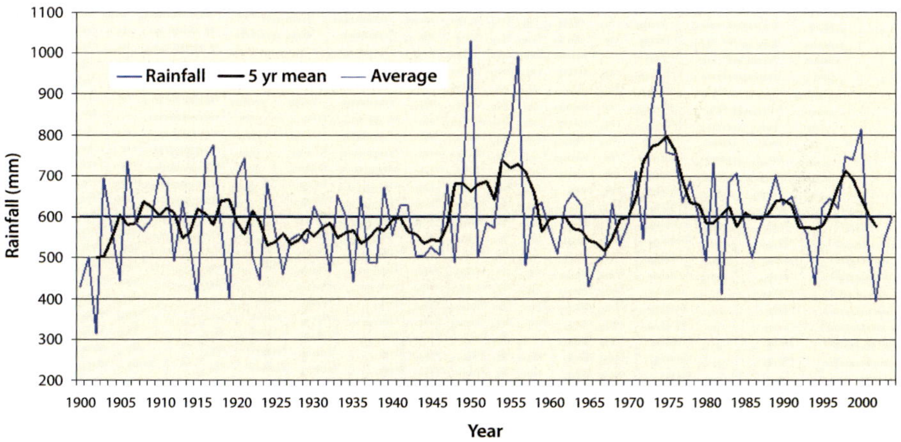

Fig. 12.8. Rainfall averaged over eastern Australia (i.e., land areas east of 130° E), 1900–2004. Rainfall in individual years is shown in *blue*. Also shown are the five-year running mean (to indicate the longer-term fluctuations) in *black*, and the long-term average (*thin blue horizontal line*). Dry conditions prior to 1905 represent the latter stages of the so-called "Federation drought", and other dry spells are evident in the 1920s through 1940s, and in the 1960s. By contrast, wet spells in the 1950s and 1970s stand out clearly

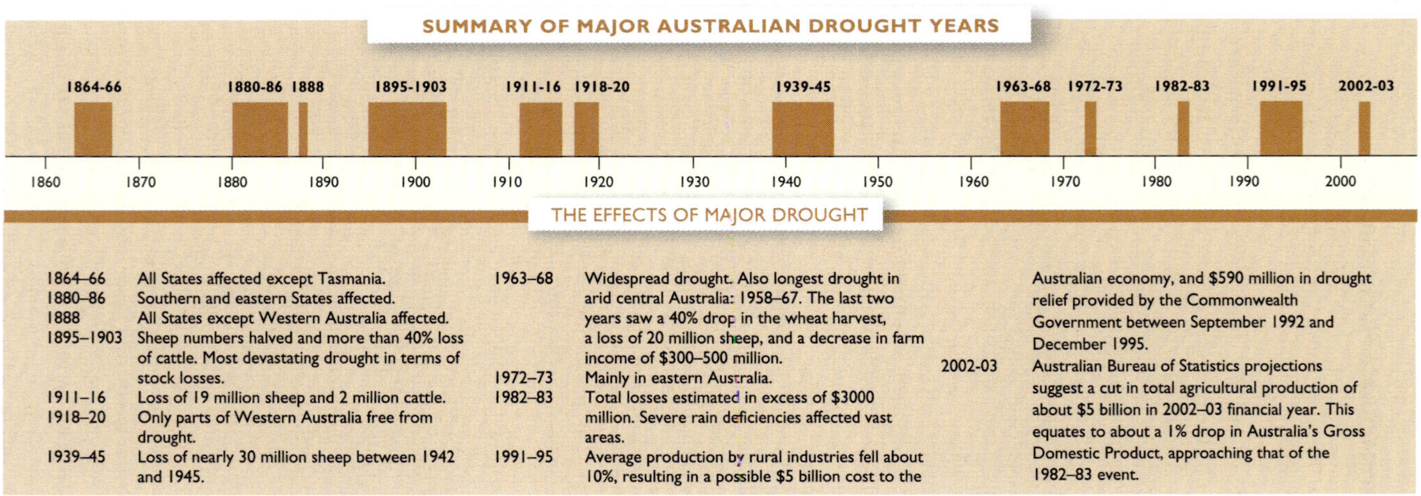

Fig. 12.9. Summary of the major droughts in Australia during the period 1850 to present

linked to long-term variations in Pacific SST. On decadal time-scales, Pacific Ocean SSTs vary in an essentially similar manner to the year to year fluctuations associated with ENSO. That is, there tend to be periods of around a decade and longer when the Pacific is more "El Niño-like", and other periods when it is more "La Niña-like". Such fluctuations go by the name of the ***Interdecadal Pacific Oscillation (IPO)***. The IPO was in its negative phase during the late 1940s–50s and 1970s, coinciding with wet eras over Australia. On the other hand, extended "warm" phases were evident throughout the late 1920s to early 1940s (a relatively dry era over Australia), weakly so in the 1960s, and again between the 1980s and late 1990s. During these phases – but particularly in the first – Australian rainfall was lower, and droughts more common.

However these phases – the El Niño-like "warm" Pacific phase and La Niña-like "cool" phase – do not follow each other in a systematic way, nor do they last for a consistent period of time. Therefore, predictability of the longer term fluctuations is not yet possible.

12.7.2 Climatic Trends and Discontinuities

Over the past 30 years southwestern Australia, and rather more recently, much of southeastern Australia, have been experiencing the effects of a rather abrupt and sustained drop in rainfall of some 15–20 per cent below their long term normals. Most of this decline has occurred in the normal wet season, in the (southern) winter half of the year, which is critical for agricultural activities and topping up water storages. The declines have been accompanied by substantial changes in Australian region circulation patterns (e.g., Allan and Haylock, 1993), with a tendency for mean sea level pressures to increase across southern Australia, as rain-bearing frontal systems and depressions embedded in the southern westerlies have retreated southwards. The decline in rainfall has had major implications for water resource management in both the southwest and the southeast (IOCI, 2002; Wright and Jones, 2003). For instance, Fig. 12.10 shows that, in southwestern Australia, inflow into storage reservoirs has dropped by approximately 40% since 1975, partly because of the rainfall decline, and also due to increasing temperatures (IOCI, 2002).

In response to the decline in rainfall and water availability over southwestern Australia, a research consortium called the Indian Ocean Climate Initiative, or IOCI, was established. This Initiative represents an excellent example of an effective cross-disciplinary approach to adaptation, essentially formalising collaboration between climate research groups and Western Australian water authorities. IOCI addresses questions such as: what is responsible for the rainfall decline, how likely is it to continue, and what are the likely implications for water management? Their findings to date (IOCI, 2002) indicate the likelihood that both natural variability and the enhanced Greenhouse Effect may have contributed to the decline, and therefore the most likely future scenario is for natural variability to continue against a backdrop of continued rainfall decreases and rising temperatures. Wright and Jones (2003) discussed some of the possible causes of the rainfall decline in southeastern Australia, consider-

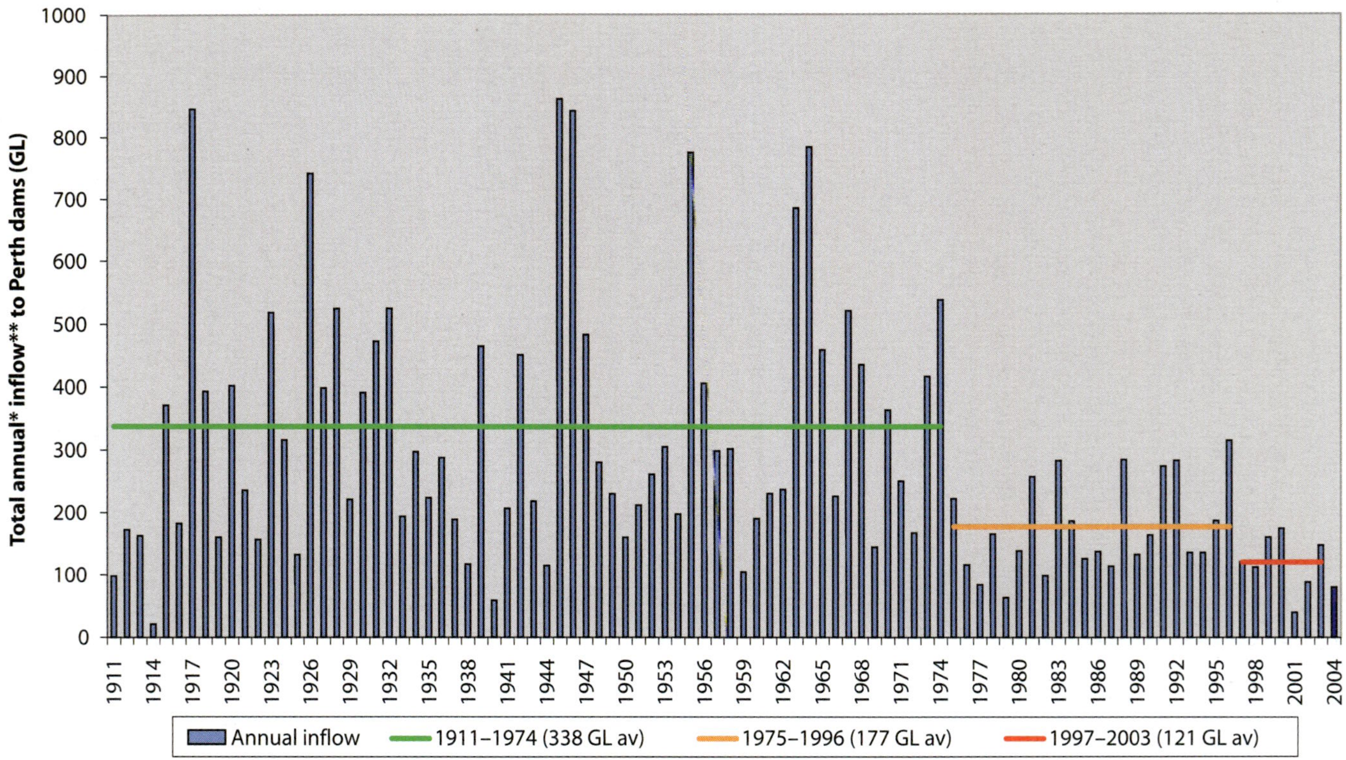

Notes: * year is taken as May to April and labelled year is beginning (winter) of year
** inflow is simulated based on Perth dams in 2001 and 2004 is an estimate (as at November 2004)

Fig. 12.10. Estimated inflow into Perth reservoirs, 1911–2000. Note the significant and sustained decline in inflows that commenced in the late 1960s, and accelerated after the mid-1970s, corresponding to the decline in rainfall over southwestern Australia. Source: IOCI (2002) ing decadal climate variability, and anthropogenic climate change associated with both the enhanced Greenhouse Effect (e.g., CSIRO, 2001) and/or changes to the so-called Southern Annular Mode (see Fig. 12.11), hypothesized to be linked to ozone depletion (e.g., Thompson and Solomon, 2002). Wright and Jones (2003) concluded that the causes of the rainfall declines could not yet be specified categorically, and that well-planned general circulation modelling studies were required to clarify this question. Without such clarification, it is not possible to specify whether the observed decline in rainfall would be maintained.

Such climate shifts and trends potentially pose problems for the application of drought policy in the future. Long-term droughts cannot be "managed through", nor can they be predicted using current techniques. The question must be asked: what implications could this have for the application of National Drought Policy, and in particular, the provision of EC support? These are difficult policy issues, made more difficult by the substantial uncertainty still surrounding climate change scenarios, and the fact that climate science is still far from being able to separate natural climate variability from anthropogenic climate change signals.

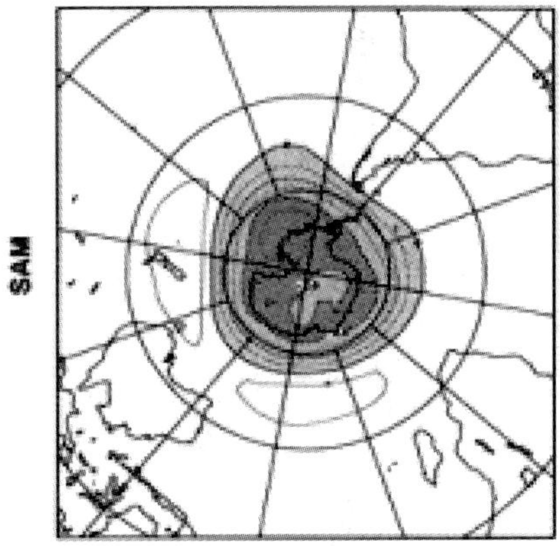

Fig. 12.11. The southern annular mode (SAM), encapsulating a tendency for Mean Sea Level Pressures (MSLPs) at high latitudes (around 60° S) to vary in the opposite sense to those in lower-middle latitudes (including over Australia). The SAM is represented by the so-called Antarctic Oscillation Index. A positive value of this Index implies lower pressures at high latitudes and raised pressures in the lower midlatitudes. A positive trend has been noted since the late 1960s, consistent with the observed circulation and rainfall shifts over Australia. Adapted from Boer et al. (2001)

12.8 Concluding Remarks

In Australia the process of refining drought policy, so that it can be fairly and equitably applied in support of productive and sustainable agriculture, will continue, as illustrated by a recent national Drought Policy review (Australian Government, 2004). In addition, there are other initiatives that could potentially improve the ability of farmers to manage drought. For instance, the Bureau of Meteorology has an internal El Niño alert system, with development work underway in the hope of eventually making it available to the public. And several agencies are collaborating in a bid to establish a Cooperative Research Centre for Climate Risk Technologies, which would ensure a steady funding commitment to drought-related climate research and its application to agriculture and other systems.

On the climatological side, one potential future issue is that, should the current warming trend in Australian temperatures continue (Della-Marta et al., 2004), as predicted by climate change projections (CSIRO, 2001), then future droughts in Australia may have more significant impacts. This is because (as demonstrated in 2002–03) the higher temperatures might be associated with higher levels of evaporation, exacerbating the effects of the rainfall deficiencies (Nicholls, 2004).

This has a number of implications for drought policy in the future. To begin with, if future droughts are indeed more severe, then the viability of at least some types of agricultural activity within Australia may be threatened. This has implications for sustainability, and future policy and planning will need to allow for such potential consequences of long-term climate variability and change.

The way in which monitoring of droughts is conducted will also need to be reviewed. The experience of 2002–03 has illustrated that rainfall deficiencies alone do not necessarily reflect the severity of drought. Therefore it may be desirable for future monitoring to take into account variables such as temperature or evaporation, as well as rainfall, in determining whether EC assistance is justified.

Acknowledgements

I would like to thank my colleagues Bruce Stewart, Neil Plummer, Michael Coughlan, and Scott Power within the Australian Bureau of Meteorology for helpful comments on earlier drafts of this article. The Australian Bureau of Meteorology is only one of a number of agencies which contribute towards drought and water management in Australia, and I would also like to thank Messrs Ashley Leedman, Kate Brinkley, Matthew Koval, and John Sims from the Bureau of Resource Sciences (Australian Government DAFF), and Ms Rae Moran from the Victorian Department of Sustainability and Environment, for their comments and advice.

References

Allan RJ and Haylock MR (1993) Circulation features associated with the winter rainfall decreases in southwestern Australia. *J. Climate*, 6, 1356–67.

Australian Bureau of Meteorology (2004) *Drought, dust and deluge: a century of climate extremes in Australia.* WJ Wright (Ed), 78 pp.

Australian Bureau of Statistics (1914 through 1918, and 1941 through 1950) Commonwealth Year Book, Canberra.

Australian Government (2004) *Consultations on National Drought Policy – preparing for the future.* Report of the Drought Review Panel, Australian Government Dept of Agriculture, Forestry and Fisheries, Canberra. 124 pp.

Boer GJ, Fourest S, and Yu B (2001) The Signature of the Annular Modes in the Moisture Budget. J. Climate, 14, 3655–65.

Botterill LC (2003) Government responses to drought in Australia. *In*: Botterill LC and Fisher M (Eds) *Beyond drought - People, Policy and Perspectives.* CSIRO Publishing, Collingwood, Vic, Australia. pp 49–66.

Botterill LC and Fisher M (2003) *Beyond drought - People, Policy and Perspectives.* CSIRO Publishing, Collingwood, Vic, Australia. 229pp

CSIRO (2001) *Climate Change Projections for Australia.* Climate Impact Group, CSIRO Atmospheric Research, Melbourne. 8pp.

Daly JJ (1994) *Wet as a shag. Dry as a bone. Drought in a variable climate.* QDPI Information series Q193028, Queensland Dept of Primary Industries, Brisbane.

Della-Marta PM, Collins DA and Braganza K (2004) Updating Australia's high-quality annual temperature dataset. *Aust. Meteorol. Mag., 53*, 75–93.

Eslake S (2003) media release issued Sept 2003.

IOCI (2002) *Climate variability and change in south west Western Australia.* Indian Ocean Climate Initiative Panel, Perth, September 2002. 34 pp.

Keating J (1992) *The drought walked through: a history of water shortage in Victoria.* Dept of Water Resources, Melbourne, 1992. 304pp

Mc Bride JL and Nicholls N (1983) Seasonal relationships between Australian rainfall and the Southern Oscillation. *Mon. Wea. Rev.*, 111, pp 1998–2004.

Nicholls N (1985) Towards the prediction of major Australian droughts. *Aust. Meteorol. Mag.*, 33, 161–66.

Nicholls N (2004) The changing nature of Australian droughts. *Climatic Change 63*, 323–336.

Power SB, Tseitkin F, Mehta V, Lavery B, Torok S, and Holbrook N (1999) Decadal climate variability in Australia during the 20th Century. *Int. Jnl Climatol, 19*, 169–84.

Thompson DWJ and Solomon S (2002) Interpretation of recent Southern Hemisphere climate change. *Science,* 296, 895–99.

Truss W (Minister for Agriculture, Fisheries and Forestry) (2002) *States lag in drought assistance*, media release, Parliament House, Canberra, 20 November 2002.

Wright WJ 2001: A review of Australian climate in the 20th century. *In*: Power SB et al (eds) Proceedings, *Cli-manage 2000 (a Conference on managing Australian climate variability)*, 23–25 October 2000, Albury-Wodonga, 127–30.

Wright WJ (2005) Significance of training, education and communication for awareness of potential hazards in managing natural disaster in Australia. *In*: Sivakumar MVK et al. (eds) *Natural Disasters and Extreme Events in Agriculture*, Springer-Verlag, Berlin, Heidelberg, New York, 2005, 219–239.

Wright WJ and Jones DA (2003) Long-term rainfall declines in southern Australia. In: *Science for Drought.* (*Proceedings of the National Drought Forum*, Brisbane, April 15–16, 2003, pp 35–40.

CHAPTER 13

Significance of Training, Education and Communication for Awareness of Potential Hazards in Managing Natural Disaster in Australia

William J. Wright

Abstract This Chapter describes the role of public education and training in mitigating the effects of natural disaster in Australia. Emphasis is on weather- and climate-related disasters, which have historically caused heavy economic losses, and sometimes significant loss of life. Three main themes are identified: (a) the continued development of a community-oriented approach to hazard minimisation, in which communities are encouraged and trained to take measures to protect themselves and their property when adverse weather or climate situations threaten; (b) issues involved in educating potential beneficiaries in the effective use of climate information to minimise impacts from climate extremes; (c) underpinning both the foregoing to some extent, the importance of developing a sound communication strategy between authorities responsible for disaster management, the people likely to be affected, and the media.

In describing the community involvement programs (Sect. 13.3), attention is focussed on responses to the threat posed by bushfires in southern Australia, and tropical cyclones in northern Australia. Public education campaigns, designed to minimise adverse effects of severe storms, skin damage due to solar radiation, and heatwaves are also described (the latter two elements are not high profile hazards, but have been responsible for many deaths in Australia). Sect. 13.4 briefly describes capacity-building activities designed to enable the effective use of seasonal climate predictions, a valuable tool in mitigating the effects of drought, both within Australia and over the South Pacific. It is shown that, while extensive educational and awareness campaigns over the past decade have led to increased use of seasonal climate predictions in decision-making by the agricultural sector, a number of barriers remain to more widespread acceptance of this tool. Section 13.5 highlights the importance of good two-way communication between service providers and end users to ensure information is relevant, adequately understood, and effectively applied. The role of the media in communicating information during times of major climate events, such as drought, is also discussed. We conclude with a summary of some promising avenues for improving the effectiveness of managing natural disaster through educational and training activities, along with some thoughts on the implications of future climate change.

13.1 Introduction

Natural disasters are responsible for loss of life, health hazards, loss of agricultural produce, interruption of economic development and creating a lack of confidence in the minds of planners and investors. The results are aggravated in an unprepared community which is unnecessarily exposed and not able to either provide or use warning signals. With the growth of science and technology, it is possible today to minimise these results. For this purpose, it is necessary to have well thought out pre-disaster and post-disaster planning in which training and education are important factors to mitigate the disasters. While Australia is mercifully spared the widespread loss of life from meteorological disasters that can afflict many other countries, events such as drought, fire, flood, cyclone and storm have nevertheless historically inflicted much suffering and heavy economic losses in this country (Australian Bureau of Meteorology, 2004). In the 20th Century, three bushfire events in southeastern Australia have each claimed over 60 lives, while many other fire events have claimed lives and caused enormous property losses. Floods and tropical cyclones have regularly inflicted losses of hundreds of millions of dollars and caused many deaths. Drought has an enduring and well-deserved reputation within Australia for inflicting hardship and loss, especially on farming communities, and has also played a key role in major land degradation episodes over the past century (McKeon et al. 2001).

Fortunately losses have been limited to some extent, especially in the latter half of the 20th century, by the development of robust warning systems, a capacity to respond rapidly to disaster, and improved community education. Each of these is linked to improvements in technology, but also to increasingly effective communication between those likely to be affected and disaster management authorities.

This paper describes some recent developments in awareness raising and community education within the Australian context that have contributed to the reduced losses. It does not attempt to provide a comprehensive review of training and education programs in particular sectors. In any case, there appears to be generally little in the way of formal training arrangements within disaster management organisations in Australia, and most of what there is concentrates on responses to disaster, rather than mitigation. Instead, the Chapter will assess developments in three key areas:

1. developing community awareness and self-management programs in relation to hazard mitigation. The emphasis here is on equipping community groups exposed to a certain type of hazard with the knowledge to manage that hazard;

2. training in the effective use of climate information, specifically climate predictions (which have the potential not only to help minimise heavy losses in "bad" years, but also to maximise returns in good seasons);

3. developing communication strategies between stakeholders, including effective use of the media. Good services and methodologies are ineffective

unless the user understands and applies the information provided. At the same time, it is important that users be able to feed back how the information can be improved.

13.2 Disaster Mitigation Through Education and Training Programmes by Government Agencies

In Australia, primary responsibility for the protection of life and property during most forms of disaster rests with State and Territory Governments, with overall coordination under the Federal (national) Government through the agency *Emergency Management Australia (EMA)*. This organisation promotes a national approach to emergency management, with emphasis on prevention, preparedness, response and post-disaster recovery activities. EMA also provide a wide range of education and training products to those involved in emergency management, and conducts public awareness and education campaigns aimed at advising on preparations for, and coping with, major hazards including severe storms, floods, bushfires, cyclones, earthquakes and heatwaves.

Towards the end of 2002 the Council of Australian Governments (COAG) completed a review of disaster relief arrangements in Australia, aimed at improving disaster mitigation (COAG 2003). Some of its main recommendations may be summarised as: (1) to further a nationally-consistent approach to disaster mitigation among the States, and integrated through local, State and national (Federal) Governments; and (2) to diverge further from the response-oriented approach of the past, by encouraging advance preparations aimed at reducing losses (via measures such as undertaking comprehensive disaster risk assessments, implementing statutory requirements for land use planning and building codes resilient to natural hazards, community awareness programs, and encouraging volunteers). The community awareness and self-reliance programs described in Sect. 13.3 below are a good example of the preparedness strategies mooted in the second of these recommendations.

13.3 Community Awareness and Self-help Programs in Australia

Australians have a long tradition of community involvement to meet the threats posed by natural hazards. For instance, most States and Territories have confronted the bushfire threat by establishing volunteer country fire services, where thousands of volunteers train in fire suppression techniques, and place themselves on standby on potentially hazardous days. The Country Fire Authority in Victoria is, indeed, one of the world's largest volunteer-based emergency services, with (in late 2003) around 59,000 volunteer members supported by over 400 career fire fighters and officers and more than 700 career support and administrative staff. This body was established in 1945, following investigations into major fatal bushfires in Victoria in 1939 and 1944 (Country Fire Authority, 2003).

Emphasis on community awareness and involvement has sharpened somewhat since the late 1980s, as governments and agencies responsible for managing natural disaster move towards imposing a greater sense of self-reliance on those likely to be impacted by severe events. Official government policy for drought and water management has also shifted, with drought now regarded as a natural and inevitable part of Australia's farming environment. The relevant government agencies are now focussing on providing farmers with training on, for instance, how to manage with climate variability. With regard to other forms of natural disaster, community education and training efforts have been best coordinated in areas where significant losses are common. We describe here the community awareness activities conducted in southern Australia in relation to bushfires, and in northern Australia in relation to tropical cyclones.

The essential element of the community-based approach is involving the public, at a local community level, in decision-making. Community members are encouraged to take actions relevant to their own situation, based on a knowledge of the hazard, an acceptance of its likelihood of occurring, and a knowledge of preventive management strategies. This clearly involves an education/training campaign for the communities exposed to risk. Open consultation processes are also critical, since community attitudes and values need to be taken into account in deciding the most appropriate strategies.

Campaigns such as this need to be built around a thorough knowledge of the nature of each type of natural hazard, a process that is more advanced with some kinds of natural hazards than others. Knowledge bases of this type require detailed historical inventories of past disasters, including information on the meteorological and other relevant conditions leading up to and accompanying the event; climatological frequency analyses of certain phenomena (e.g., lightning incidence, tropical cyclone frequency); analysis of broadscale influences (e.g., phase of the El Niño – Southern Oscillation phenomenon) likely to influence variability of the phenomena. These should be supported by statistics on, and analyses of, the impacts of the phenomena (e.g., financial losses, fatalities, etc; assessments of how/why people died). This in turn requires integrated data management systems, for collecting, quality controlling, analysing and presenting data, including metadata. Presentation of information to users (including "intermediaries", such as disaster management agencies) should make use of the best available technology, particularly GIS and the Internet.

13.3.1 Bushfires

As indicated earlier, bushfires are a major hazard in southern Australia during the warmer months, especially in the southeast where, over the years, they have resulted in hundreds of deaths, devastating property losses, and the destruction of large tracts of valuable forests. Fig. 13.1 is a photograph taken during the infamous "Black Friday" fires of January 1939, which claimed 71 lives and destroyed millions of hectares of forest. Much of the loss of life and property

Fig. 13.1. Buildings ablaze during the January 1939 "Black Friday" bushfires in Victoria's forest country, southeastern Australia. (*Courtesy of the Victorian Dept of Sustainability & Environment, Victoria, Australia*)

destruction has occurred in residential areas set among bushland on the fringes of cities. Between 2002 and 2004 major fires have occurred in Canberra (on 18 January 2003, fires destroyed over 500 homes in Canberra's suburbs, and caused four deaths), in outer suburban Sydney, and in eastern Victoria and southeastern New South Wales. Moreover, the forested areas east of Melbourne are regarded as one of the most fire-prone areas on earth. An important part of the strategy for managing bushfires in such fire-prone areas is community education and consultation.

A community-based bushfire risk management program, "*Community Fireguard*", has been initiated in the States of Victoria, South Australia, New South Wales, and Tasmania (essentially the southeastern corner of Australia). This program stands as an excellent example of highly effective risk management in communities regularly exposed to climate-related hazards. In the lead-up to the fire season, fire authorities facilitate community meetings in bushfire-prone areas. In these meetings, the authorities outline the nature of the fire threat, and advise residents on how to protect their property against it. Typical recommended preparatory and preventive measures include: clearance or reduction of flammable vegetation; covering or removing flammable material (e.g. exposed wood); and assembling fire fighting equipment. Homeowners are also encouraged to consider in advance whether on potentially severe fire days they will evacuate early, or prepare to stay and defend their properties. Reports from past severe fires, in Hobart in 1967, and in Victoria/South Australia in 1983, have indicated that well-maintained properties attended by able-bodied people offer a refuge from fire, and are much more likely to be saved than unattended buildings (Ramsay et al. 1986; Packham 1992). It is also argued (Krusel and Petris 1992) that the most effective bushfire safety strategies are developed when communities work together to develop arrangements most appropriate to their specific situation. In some cases, residents are encouraged to operate neighbourhood 'phone trees, so that, in the event of an emergency, residents keep each other informed of their plans (e.g., evacuating, staying and defending, ability to take elderly and infirm to safety).

In the State of Victoria, seasonal climate predictions can provide an indication of the likelihood of severe and less severe fire seasons (Williams and Karoly 1999), and this information can be used by fire authorities and community managers in making decisions on, for example, allocating fire suppression resources, planning personnel deployment, and preparing communities. Specifically, El Niño summers tend to be associated with an increased fire risk in many parts of southeastern Australia (Fig. 13.2). Dangerous conditions, with serious bushfires, are much more likely in southeastern Australia during the summers of El Niño years (negative SOI) than La Niña summers (positive SOI). The association arises because the lead-up to the fire season is likely to have experienced below average rainfall, while there is evidence that extremely hot days and strong winds are more likely in El Niño summers. Prior

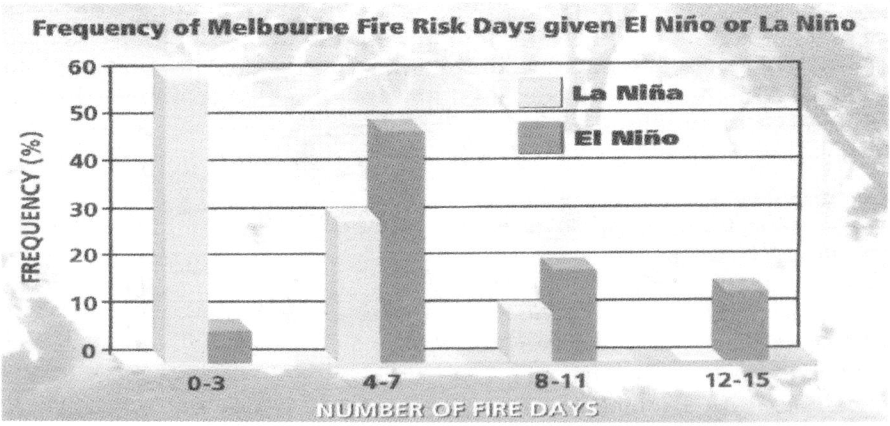

Fig. 13.2. Relationship between El Niño/La Niña and extreme fire danger days in the area around Melbourne, Victoria. *(Courtesy of Harvey Stern & Mark Williams, Australian Commonwealth Bureau of Meteorology, 1989)*

to the severe 2002–03 fire season in Victoria, advice about increasingly severe rainfall deficiencies, and an adverse seasonal climate outlook, led to a greater allocation of fire fighting resources (including fire-bombing aircraft), and public awareness campaigns highlighting the increased risk (Billing 2004). This may have been a factor in the subsequent relatively low level of fatalities and property losses in what was potentially a devastating season, with many large fires.

In some fire-prone regions, local Government bodies enact strict planning regulations, aimed at ensuring that new houses are built in a style that reduces the risk of destruction in wildfire, including provisions for an adequate water supply, and design features aimed at limiting the risk of ignition by airborne burning embers.

13.3.2 Tropical Cyclones

In the lead-up to the tropical cyclone season in northern Australia (normally December-April), the Bureau of Meteorology, in partnership with emergency management and local government authorities, issues detailed information to the public about the risks posed by cyclones, and effective steps to manage this risk. Bureau and emergency management staff also visit communities to inform residents about tropical cyclones, provide general information on cyclone survival preparations, and provide expert advice for specialised users and managers of sensitive infrastructure. Many of these communities are in remote localities, so it is not practicable to visit each community every year. However special arrangements are being made to improve communications through, e.g., improved Internet access and radio.

In the event that a tropical cyclone threatens communities, a comprehensive and regularly-updated series of advices and warnings is provided by the Bureau of Meteorology, to alert communities in areas that may be affected about the likelihood and probable timing of cyclone-related hazards such as destructive winds, flooding and storm tides. The advices also alert residents to the need to activate staged precautionary and defensive measures aimed at reducing losses. An example of such an advice, for residents in northwestern Australia facing an approaching severe tropical cyclone is given below (Box 13.1).

Broader mitigation strategies have also been developed, and supported by legislation and various relevant Codes of Practice. In response to cyclone "Tracy", which devastated Darwin on Christmas Day 1974 (Fig. 13.3), authorities in the Northern Territory imposed strict building regulations for all new buildings. Under these regulations, new buildings must be strong enough structurally to withstand wind-speeds equivalent to a Category 4 cyclone (i.e., with mean wind-speeds up to 211 km/h and gusts to 285 km/h). Elsewhere around northern Australia, similar construction codes have been imposed: buildings along the cyclone-prone Pilbara coast of northwestern Australia are constructed to withstand Category 5 cyclones (wind-gusts in excess of 285 km/h).

Box 13.1. Warning/advice in relation to TC "Fay", a Category 3 cyclone (i.e., with sustained wind-speeds of 64 knots (118 km/h) or more) that affected northwestern Australia in March 2004

SES advises of the following community alerts:

YELLOW ALERT: People in or near coastal communities between Cape Leveque and Bidyadanga including Broome, One Arm Point, Lombadina, Djarandjin, Beagle Bay and Bidyadanga should commence action in readiness for the cyclone's arrival.

People in Broome and Bidyadanga are advised that later this morning they will go on RED ALERT.

BLUE ALERT: People in or near coastal communities between Bidyadanga and Wallal should start taking precautions.

The next warning will be issued at 10 am. Cyclone advices and State Emergency Service Community Alerts are available by dialling 1300 659 210

Rest of message:

TROPICAL CYCLONE ADVICE NUMBER 56

Issued at 7:05 am WST on Thursday, 25 March 2004

BY THE BUREAU OF METEOROLOGY

TROPICAL CYCLONE WARNING CENTRE PERTH

A CYCLONE WARNING for a CATEGORY 3 CYCLONE is now current for coastal communities between Cape Leveque and Pardoo.

A CYCLONE WATCH extends southwest to Roebourne.

At 6 am WST SEVERE TROPICAL CYCLONE FAY was estimated to be

150 kilometres west southwest of Cape Leveque and

110 kilometres north northwest of Broome

and moving south at 8 kilometres per hour.

Severe Tropical Cyclone Fay is moving on a southerly track but is expected to turn towards the south southwest during today. Destructive winds with gusts to 150 kilometres per hour are possible on the Kimberley coast between Cape Leveque and Bidyadanga today, gradually extending to the eastern Pilbara tonight.

Tides between Cape Leveque and Pardoo are likely to rise above the normal high tide mark with very rough seas and flooding of low-lying coastal areas. Residents of Bidyadanga are specifically warned of a dangerous storm tide as the cyclone centre passes to the west. Tides are likely to rise significantly above the normal high tide mark with damaging waves and dangerous flooding.

Fig. 13.3. The near-complete devastation inflicted on Darwin by cyclone "Tracy", which struck early on Christmas Day, 1974. Some 65 people died, and within days, three quarters of the remaining population had been evacuated from the devastated town. *(Photo c/o Australian Department of Foreign Affairs and Trade)*

13.3.3 Public Weather Advice and Alerts

A range of lower-profile climatic threats exist which are increasingly recognised as being responsible for the death of significant numbers of Australians. As such, they are becoming the subject of public safety campaigns, with attendant warnings often carrying advice about preventive actions. Some examples are:

Severe weather advices and warnings, issued through the media, warn of the risk of severe thunderstorms, and provide advice on action to be taken to minimise damage and loss from strong wind squalls, flash flooding and large hail. The Bureau warns graziers of the likely onset of conditions conducive to hypothermia in livestock, and also issues a variety of warnings about conditions likely to cause crop disease or losses.

Australia has a very high incidence of **sunburn and skin cancer**, and for many years now, health authorities have conducted public education campaigns advising the public on how to prevent skin damage, which has potentially fatal consequences. In recent years the Bureau of Meteorology has issued daily Ultraviolet Radiation forecasts, indicating the timing and severity of conditions conducive to skin damage (Fig. 13.4). Index levels higher than 2–3 require preventive actions (e.g., sunscreen, hats, protective clothing) – themselves a subject of an intense community education program – by people facing prolonged sun exposure

It is a fact that **heatwaves** kill more Australians annually than any other meteorological phenomenon (Coates 1996), and this is of increasing concern to health authorities in particular. The elderly and chronically ill are the most likely victims, and the threat appears greatest in areas with otherwise mod-

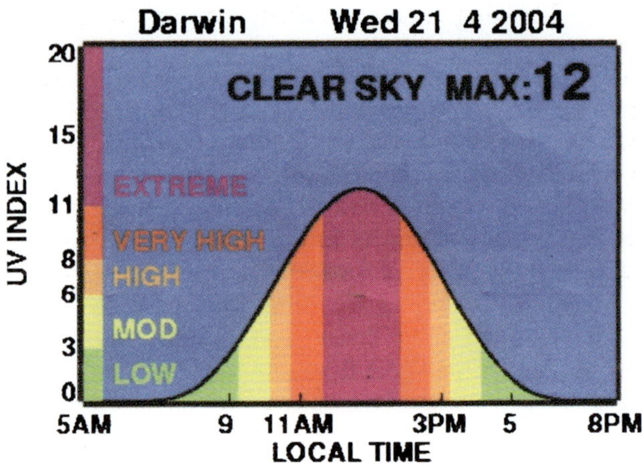

Fig. 13.4. Daily Ultraviolet Radiation forecast for Darwin, Australia, showing variation in intensity of the radiation throughout the day

erate climates subject to occasional hot periods (Kalkstein and Davis 1989). In response, various public authorities are now publishing community information guidelines about avoiding heat stress, and occupational health and safety organisations, and sporting bodies such as Sports Medicine Australia, are providing safety tips for physical activity in hot weather. Increasingly, the Bureau of Meteorology is issuing media releases when unusually hot periods are forecast, and is currently developing the scientific basis for a heat stress warning.[1]

13.4 Droughts and Flooding Rains – the Effective Use of Climate Information in Managing Extremes of Climate Variability

Drought not only causes crop failure and stock losses, but also sets the scene for extreme phenomena such as fires, dust-storms, and general land degradation. This section describes the issues associated with educating potential beneficiaries, particularly farmers, in the effective use of climate information to reduce impacts from drought and other extreme climate events. Broader drought policy issues are described in detail in Botterill and Fisher (2003).

There is a strong link between the incidence of severe drought in Australia and the El Niño-Southern Oscillation (ENSO) phenomenon (Nicholls 1985). Conversely, La Niña events are usually accompanied by above average rainfall, an increased risk of flooding, and reduced frost incidence. Following a series of papers in the 1980s about the impacts of ENSO in Australia, and the potential for seasonal prediction based on indicators of ENSO (McBride and Nicholls 1983; Nicholls 1985), the Australian Bureau of Meteorology com-

[1] As this book went to press, the Bureau had just commenced publicly displaying on the Web heat stress indices for selected locations.

menced issuing seasonal climate predictions in 1989. Predictions were first based on the Southern Oscillation Index, and in recent years, the Bureau has introduced sea surface temperatures (SSTs) as predictors (with the dominant pattern still representing ENSO, and a second predictor based on SST patterns in the Indian Ocean). In addition to these statistical schemes, experimental seasonal predictions are also made using coupled atmosphere-ocean models, and it is expected that dynamical model predictions will become fully operational in coming years. For reasons to be described later, seasonal climate outlooks are expressed in terms of probabilities; however the concept of probabilities has proven a difficult one to describe to potential users.

13.4.1 Seasonal Climate Predictions – Educating the User

Seasonal climate predictions are presented to the users and general public via a number of different media. These include publications such as the Seasonal Climate Outlook (a subscription-based service), on the Internet, via face-to-face briefings, and through the mass media. The information is presented in a variety of different ways, one of which is illustrated via the Table 13.1. In this example, Sydney has an enhanced risk (40%) of being comparatively dry (with seasonal rainfall less than 257 mm), and a reduced chance of being in the upper tercile (377 mm or more; 27%). The "climatological" probability – i.e., in the absence of any climate forcing influence such as ENSO – is 33% in each of the three categories.

Prediction services are only useful if they can be understood, applied, and provide value for decision-making. During the 1990s a huge campaign was mounted to demonstrate to primary producers (farmers, pastoralists, horticulturalists, etc) the links between ENSO and rainfall variations (especially drought) in Australia, and the value of seasonal climate predictions in aiding on-farm decision-making, and minimising losses and environmental degradation during drought periods. Government bodies such as the Bureau of Meteorology, Federal and State Departments of Agriculture and Natural Resource Management (SDAs), and other scientific and research organisations, have collaborated closely in this effort.

Table 13.1. Probabilities of being in the lower and upper terciles (cols 3 and 5 respectively), together with the upper bound of tercile 1 rainfall (col 2), and the lower bound of tercile 3 rainfall (col 4)

Town	33rd Percentile (mm)	Prob. "dry" (%)	67th Percentile (mm)	Prob. "wet" (%)
Perth	21	35	44	32
Melbourne	109	29	177	38
Sydney	257	40	377	27

The Bureau's education/training activities have included:

1. workshops designed to bring together the producers and users of climate information, e.g., the *Cli-Manage* and *DroughtCom* workshops (Australian Bureau of Meteorology 2001; Plummer et al. 2004). During these workshops, the producers of climate information describe what information is available to potential users of the information, while the users provided feedback on their needs for climate information, and the limitations of existing information, especially seasonal predictions;
2. information about ENSO and seasonal prediction via web-sites, educational videos, publications and articles in the media (updates on El Niño and of the seasonal prediction are also routinely provided via the above media); and
3. awareness-raising, via e.g., attendance at agricultural shows, and talks to user groups.

The SDAs are generally in more direct contact with the agricultural sector, and in addition to the above activities (often conducted in collaboration with the Bureau), carry out the following training/awareness-raising activities:

1. workshops/seminars designed for farmers, and usually held in country centres;
2. contact with users via Extension programs (in which Extension officers "package" climate information for farmers in a way that directly assists on-farm decision-making). Extension personnel also provide feedback to research organisations on what information best meets the needs of the end-user;
3. some States run climate education courses designed for farmers at tertiary institutions; and
4. development and demonstration of Decision Support software as an aid in assessing likely outcomes of various management strategies, such as changing planting time, stocking rates, etc. Such programs can be run for a range of possible seasonal conditions (including climate scenarios). For instance, there are packages to simulate growth and potential yield for a range of crops, pasture growth, even economic outcomes.

In addition, while there is no formal coordination of education/training activities between organisations, there is much informal communication and information-sharing at conferences and workshops attended by representatives of SDAs, the Bureau, and other organisations. This has the advantage of encouraging a cross-fertilisation of ideas. Surveys of farmers have led to many improvements in the way information has been presented – for instance, the Australian farming community made it clear that they want information "localised", i.e., what is the prediction for the nearest town?

13.4.2 Responses from Farming Community

In response to these activities, the level of uptake of seasonal predictions (and other climate information) is generally high within stakeholder government agencies, but lower in farming communities. Surveys (White 2001) show that, nationally, about 32% of farmers take seasonal climate predictions into account (with a higher proportion – 44% – in New South Wales and Queensland, where active education campaigns have been in place for many years). However farmers' preparedness to change major decisions is not necessarily influenced by this information. This is true even in extreme situations such as El Niño events, and is despite the fact that, because of the extensive educational efforts in the 1990s, the El Niño – Drought link is now widely known and accepted within the Australian agricultural community.

Feedback from users has indicated a number of factors that have contributed to the limited uptake and use of seasonal climate predictions by farmers:

1. Perceptions of limited accuracy. ENSO-based predictions sometimes suggest an outcome inconsistent with that observed. During the 1997 El Niño, timely rainfall in areas where the El Niño influence on rainfall is normally strong led to some highly critical comments about seasonal outlooks from some sections of the community;

2. Difficulties in understanding, or applying, probabilistic predictions. Because chaos limits the deterministic accuracy of predictions, there are good reasons for expressing seasonal predictions in probabilistic terms. However the communication of probabilistic information remains a serious challenge. Many farmers do not understand or like probabilistic information. The problem is compounded by the fact that, in practice, the probability of rainfall being in a particular tercile range (lower, upper or middle tercile) seldom exceeds 50%. That is, climate prediction input to decision-making often takes the form of relatively minor shifts in probabilities;

3. Farmers' reticence to change traditional strategies/adopt new approaches;

4. Farmers' inability to apply predictions directly in decision-making. However, the development of Decision Support software and ongoing education/training services to farmers by SDAs is reducing this as an issue.

5. Sometimes-conflicting opinions in the media from so-called "experts", including some from outside Australia, leading to confusion and distrust ("if even the experts can't agree ... "); and

6. Inappropriate use of predictions, again leading to a loss of confidence in the system. For instance, using Decision Support software without a clear understanding of the underlying principles, it is possible to generate a whole range of predictions, from which one can then be biased either towards those that appear to give the best results, or to particular outcomes by chance – clearly a scientifically-unsound and misleading practice.

In response to these issues, the means by which seasonal climate information is presented to users in Australia is constantly evolving. For instance, a probability of 70% of above-normal rainfall would now be presented as "in years like the present, 7 out of 10 years would have above average rainfall, and 3 of 10 below".

Where climate predictions are utilised, they are generally seen to be part of a suite of decision-making tools, and should rarely, if ever, be used in isolation. This is reasonable, because climate is generally only one of several factors that influence farm profitability and sustainability.

13.4.3 Other Climate Extremes

Frost is another serious threat to agriculture. In particular, crops flowering in spring can be seriously damaged by late frosts. In some parts of Australia, notably Queensland, research suggests that late frosts are more likely during El Niño events (Stone et al. 1996). Again, education programs by SDAs have drawn attention to this link, and provided strategies for minimising the risk. Examples of effective responses to increased frost risk include planting cultivars that are more frost-hardy, or which will flower later.

Mitigation of the effects of **flooding** is linked to awareness building within the community, and is also effected through specific projects aimed at reducing impacts in problem areas identified in risk assessment studies (COAG 2003). In Australia, hydrology units within the Bureau of Meteorology oversee a highly effective public warning system for river flooding. These warnings, which include recommended defensive actions agreed by various emergency services, ensure that when warnings of moderate to major flooding along river valleys are issued, action is triggered within communities likely to be affected, as are response strategies by emergency services personnel. In local areas, partnerships have become established between flood warning personnel and vulnerable residents and business owners in flood-prone situations. For instance, the Bureau of Meteorology's South Australian Regional Office has worked with residents in identified flash-flood-prone areas in urban Adelaide and surrounding areas (Dickins, *pers comm.*, 2004). To date, despite the strong statistical links between La Niña and above average rainfall and flooding over much of Australia, seasonal climate predictions tend not be used, in any direct sense, as indicators of an increased or reduced likelihood of flooding.

As indicated above, specific mitigation activities are carried out to reduce the threat from flooding, under a program supported by all levels of government known as the Flood Mitigation Programme. Of the various natural disaster types, flooding is the only one for which a targetted mitigation program of this sort currently exists (COAG 2003). Typical mitigation activities include: risk assessment studies; the establishment of planning regulations prohibiting development in flood-prone areas; building resilient infrastructure; community awareness, and a number of others. A variety of engineering solutions have also been developed to constrain floods on Australian rivers, including retarding basins and dams, levee banks, and river diversions. For example, after disastrous flooding in Brisbane during January 1974 (in which large sections of the city were inundated and 14 people died), the Wivenhoe dam was built on the Brisbane River in 1985.

Finally, awareness-raising by SDAs and others has also led to an improved understanding of actions (e.g., application of fungicides) for preventing certain wet weather-related crop diseases.

13.4.4 Training in the Application of Seasonal Predictions in the South Pacific

The ENSO phenomenon is a primary influence on the climate of the South Pacific region, as it is in Australia. Building on this fact, the Australian Bureau of Meteorology is undertaking a project aimed at providing Pacific Island Countries' National Meteorological Services (NMSs) with the capacity to provide a seasonal prediction service for people in their countries. The need for such a service was identified in The Pacific Meteorological Services Needs Analysis Project, which in May 2001 reported (South Pacific Regional Environmental Program 2001) that most NMSs in the Pacific region were struggling, and often failing, to provide basic services for the citizens and industries of their countries. The report identified the need for assistance programs in two clear priority areas:

1. improved severe weather warning services; and
2. seasonal and climate prediction services, especially for droughts.

The Project described here tackles the latter issue. It partly involves the provision of computer software, based on the Bureau's operational probability-based seasonal prediction system. Equally important, however, is the planned provision of an extensive training program, aimed at both the National Meteorological Service (NMS) staff who are to provide the service, and the likely users of the predictions. This training program covers topics related to the "prudent" use of climate predictions in supporting decision-making. Some aspects of the project are occurring in collaboration with other countries active in climate service provision in the region, notably New Zealand and the United States. The project has strong parallels with the objectives of WMO's CLIPS program, which has the stated aim of developing the NMSs' abilities to take advantage of recent advances in climate science, with the potential to improve climate services to the user community.

The emphasis in the training program is on providing users with information about how to interpret and apply probability-based seasonal predictions in decision-making, in areas such as agriculture, fisheries, forestry and water management. NMS staff are being trained in how to provide a seasonal prediction service, and potential users are being trained in how to apply probabilistic forecasts in their decision-making. Fig. 13.5 illustrates the use of pie-charts plotted on a map (a feature of the software developed for this project) as an aid in demonstrating probability shifts and how they vary with location.

The above capacity-building exercise is one example of how developed countries in the Asia-Pacific region (including Australia, New Zealand, the US,

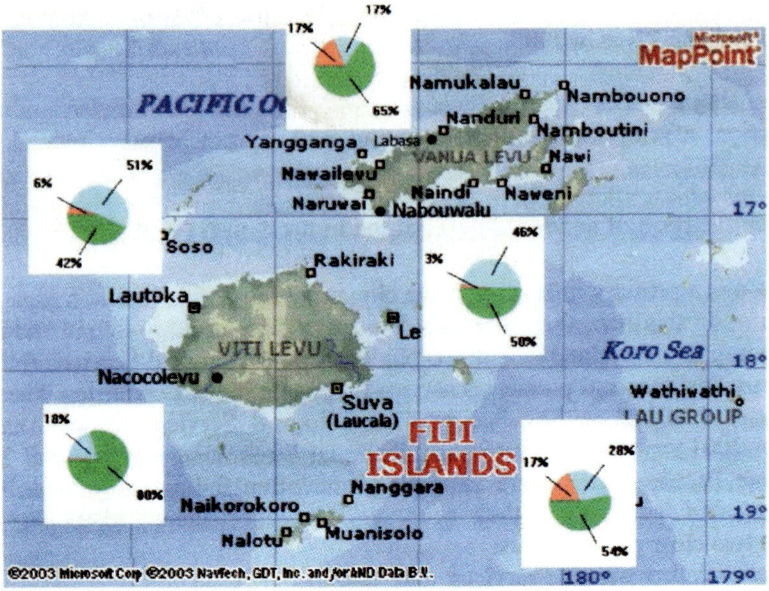

Fig. 13.5. Rainfall tercile probabilities, represented as pie-charts, for the Fiji Island group. Hayman (*pers. comm.*, 2003) has demonstrated that pie-charts are an effective way of presenting probabilistic information

and several other countries) actively participate in aiding the development of meteorological and climate services in developing countries in the region. In addition, in-house meteorological training at the Australian Bureau of Meteorology is regularly provided to staff from Pacific Island NMSs, while Australia and New Zealand have also contributed personnel and resources to train climatologists from the Asia-Pacific region in seasonal outlook development and verification techniques. Workshops on analysing climate extremes have also been held, and some valuable results obtained (Manton et al. 2001). Such assistance is vital in helping these countries minimise the heavy losses associated with weather- and climate-related disasters.

13.5 Towards Improved Lines of Communication

Much of the foregoing illustrates that, at the heart of effective risk management is the need for sound communication between agencies responsible for disaster mitigation and the people potentially affected. Effective information dissemination, carried out via a variety of communication methods, is necessary for educating the public about how they should mitigate, prepare for, and recover from the impacts of hazards, and in providing essential information once an event is underway. At the same time, effective public relations, in this case between agency and public (or user group), is probably best served in the current context through two-way communication in which the "public" have the opportunity to influence the way in which information is conveyed to them.

While communication between the users and producers of climate information has improved substantially in recent years, there is still room for improvement. In an analysis of the communication of climate information during the 2002/03 drought, which drew on a wide variety of stakeholders, Plummer et al. (2004) identified the following areas where increased effort was needed:

1. Better national coordination (so that information is consistent and complementary);
2. Better understanding of user needs;
3. More capacity building (i.e. more information about climate products and their application);
4. More information about products and services;
5. Improved media interactions;
6. Integration and value adding (e.g., combining climate information with other types of information); and
7. More information on current research and its implications.

13.5.1 Role of the Media

Dealing effectively with the media is crucial for any organisation involved in managing natural disasters. Mass media can reach a large audience at minimal cost, and is a popular avenue through which the public receive information. An effective media strategy ensures that information aimed at reducing losses from meteorological hazards reaches the public unambiguously. Also, in reporting on the operational, political and social aspects of events, the media help galvanise national support for victims of adverse events. The latter was emphatically illustrated during the 1994–95 drought, which was the culmination of several very dry years over much of northeastern Australia. The severe drought was causing major hardship in rural communities in the affected areas, with many families, unable to cope any longer from years of adversity and no income, simply abandoning their land. Many in the cities had little knowledge of, or sympathy for, the plight of farmers. However the media drew attention to the farmers' often grave situation, and according to the Federal Primary Industries Minister in the Australian Government (the Honourable Bob Collins), were instrumental in turning around the negative image of farmers (Wahlquist 2003).

It is important to note that the aims of the mass media and service organisations can sometimes be at odds. Media articles in general need to be fairly generalised, timely, brief, and catchy to best capture their audience, whereas scientific organisations (such as meteorological agencies), are more accustomed to dealing with objective information, sometimes with lengthy detail (Mullen, *pers. comm.*, 2004). Add to this the combination of possibly poor explanation by the organisation spokesperson, lack of expertise in the subject area by journalists, and space/time/editorial constraints, and it is not surprising that misrepresentations occur.

Nevertheless, strategies to minimise confusion of the message, as identified in a media analysis study of the 1997/98 drought commissioned by the Bureau of Meteorology's National Climate Centre (Walsh 1998) include:

1. Update, clarify and document clear communication objectives, and specify a consistent message strategy;
2. Train staff in dealing with the media;
3. Direct resources towards the most aware or active "publics" (i.e., particular user groups);
4. Research the information needs of "publics"; and
5. Host education seminars for the media's awareness[2].

13.6 Concluding Remarks

The past 15 or so years have seen considerable advances in the level of coordination between different government agencies responsible for minimising losses due to weather and climate extremes in Australia, and also in the level of coordination between these organisations and community groups likely to be affected by the extremes. Progress in the latter area has been achieved through targeted training and education campaigns, aided by changes in government policy, and better communication strategies.

Drawing on recent experience in Australia, the following policy objectives, from the education/training perspective, appear to hold promise for improving the effectiveness of managing meteorological disasters, and mitigating the impacts of extreme events:

1. Continual fostering, via government agency engagement with community groups, of a sense of self-reliance, extending to the local community level;
2. Further consultation with community groups exposed to natural hazards (e.g., the agricultural sector), resulting in improved awareness of climate information by the users, and better presentation of this information by the producers;

[2] A fine illustration of how this can pay off in times of crisis is provided by the Victorian Country Fire Authority (CFA) who, prior to the 2002–03 fire season, sponsored more than 600 media personnel through a formal course in fire knowledge and safety. They (CFA, 2003) attribute the subsequent "excellent" coverage and timely information dissemination provided by the media during the severe 2003 bushfires in part at least to the training course.

3. Effective communication and media strategies for educating the public and disseminating information;
4. Technological advances in computing and communications, particularly in view of the likely benefits for distance education and the development of Decision Support and other learning tools;
5. Collation and effective management of information about natural disasters, including the meteorological circumstances that precede and accompany an extreme event, and the impacts of the event. The aim of compiling such databases is to better understand the nature of particular types of event, in turn facilitating improved monitoring, assessment, prediction and preparedness. This requires free and unlimited access to such databases, and collaborative links for the exchange of information.

Finally, the development of effective risk management strategies, to which education and training contribute a fundamental role, is likely to assume even greater importance in the future. This is a likely consequence of two parallel trends, which have emerged globally and are reflected to some extent in Australia (Wright et al. 2003). These are:

1. Global warming, which the Intergovernmental Panel on Climate Change argues is at least partially due to an increase in greenhouse gas concentrations in the atmosphere; and
2. A rise in the social, environmental and economic impacts of natural disasters.

With regard to the first of these trends, there has been speculation (e.g., Nicholls, 2004) that the severity of the 2002–03 drought in Australia, and of related phenomena such as bushfires, may be manifestations of global climate change. Moreover, the potential exists for future climate change to further increase the frequency and severity of extreme events like droughts, floods and bushfires.

Given the severe impacts of natural disasters, increased precautionary and defensive planning and action to mitigate known hazards is needed, and National Meteorological Services are likely to play an important role in these activities. Finally, the communication of information and uncertainty about the potential impacts of climate change will pose new challenges for NMSs in the future.

Acknowledgements

The author gratefully acknowledges the assistance of colleagues within the Bureau of Meteorology, including Mr Neil Plummer, Dr Linda Anderson-Berry, Mr Bruce Stewart and Ms Clare Mullen, and Dr Peter Hayman (New South Wales Dept of Agriculture), for providing helpful comments on earlier drafts of this article. Ms Carol Skinner provided the information on heatwaves.

References

Australian Bureau of Meteorology (2001) In: Wright WJ, Power S, Della-Marta P (eds) Proc Cli-Manage 2000: A conference on managing Australian climate variability, 23–25 October, 2000, Albury, NSW, 133 pages.

Australian Bureau of Meteorology (2004): *Drought, dust and deluge: a century of climate extremes in Australia.* Wright WJ (Ed), Melbourne. 78 pp. (avbl from Bureau of Meteorology, GPO Box 1289K, Melbourne 3001 Australia).

Billing P (2004) Impact of the drought on forest fires. In: Plummer N, Flannery M, Mullen C, Trewin B, Watkins A, Wright W, Powell T, Power S (eds) DroughtCom Workshop: improving the communication of climate information, Vol 2. Extended Abstracts, Bureau of Meteorology, Melbourne, pp 1–6.

Botterill LC, Fisher M (2003) Beyond drought – People, Policy and Perspectives. CSIRO Publishing, Oxford St, Collingwood, Vic, Australia, 229 pages.

Coates L (1996) An overview of fatalities from some natural hazards in Australia. In: Heathcote RL, Cutler C, Coetz J (eds) NDR 96, Conference on natural disaster reduction, Institution of Engineers Australia, Barton, ACT, pp 49–54.

Council of Australian Governments (2003) Natural disasters in Australia: Reforming mitigation, relief and recovery arrangements. Commonwealth Dept of Transport & Regional services, 2003, 201 pages.

Country Fire Authority, 2003: CFA – at a glance (Website). http://www.cfa.vic.gov.au/about/documents/cfa_ataglance_nov03.pdf.

Kalkstein LS, Davis RE (1989) Weather and human mortality: An evaluation of demographic and inter-regional responses in the United States. Annals of the Association of American Geographers, Vol. 79. pp 44–64.

Krusel N, Petris SN (1992) A study of civilian deaths in the 1983 Ash Wednesday bushfires, Victoria, Australia. CFA Occasional paper No. 1, Country Fire Authority, Victoria. 28 pages.

Manton MJ, Della-Marta PM, Haylock MR, Hennessy KJ, Nicholls N, Chambers LE, Collins DA, Daw G, Finet A, Gunawan D, Inape K, Isobe H, Kestin TS, Lefale P, Leyu CH, Lewin T, Maitrepierre L, Ouprasitwong N, Page CM, Pahalad J, Plummer N, Salinger MJ, Suppiah R, Tran VL, Trewin BC, Tibig I, and Yee D (2001) Trends in extreme daily rainfall and temperature in southeast Asia and the South Pacific: 1961–1998. Int. J. Climatol, Vol. 21. pp 269–284.

Mc Bride JL, Nicholls N (1983) Seasonal relationships between Australian rainfall and the Southern Oscillation. Mon. Wea. Rev, Vol. 111. pp 1998–2004.

McKeon GM, Hall WB, Yee JY, Stone GS, Crimp SJ, Peacock A, Richards R, Tynan RW, Watson IW, Power SB (2001) Learning from history: land and pasture degradation episodes in Australia's rangelands. In: Wright WJ, Power S, Della-Marta P (eds) Proc Cli-Manage 2000: a conference on managing Australian climate variability, 23–25 October, 2000, Albury, NSW, pp 64–67.

Nicholls N (1985) Towards the prediction of major Australian droughts. Aust. Meteorol. Mag., Vol. 33. pp 161–166.

Nicholls, N, 2004: The changing nature of Australian droughts. *Climatic Change*, Vol. 63, 323–336.

Packham D (1992) Bushfires in Australia, Australian Planner, Vol. 30. pp 8–12.

Plummer N, Flannery M, Mullen C, Trewin B, Watkins A, Wright W, Powell T, Power S (2004) DroughtCom Workshop: improving the communication of climate information, 22–23 July 2003. Bureau of Meteorology, Melbourne, Vols 1 (38 pages) and 2 (52 pages).

Ramsay GC, Macarthur NA, Dowling VP (1986) Building survival in bushfires. Paper presented at Fourth Biennial Conference, Institute of Fire engineers, Perth, 21–24 October, 1996.

SPREP (2001) Pacific Meteorological Services: Meeting the Challenges. SPREP and Australian Bureau of Meteorology, 103 pages.

Stone R, Nicholls N, Hammer G (1996) Frost in northeast Australia: trends and influences of phases of the Southern Oscillation. J. Climate, Vol. 9. pp 1896–1909.

Wahlquist A (2003) Media representations and public perceptions of drought. In: Botterill LC, Fisher M (eds) Beyond drought - people, policy and perspectives. CSIRO Publishing, Oxford St, Collingwood, Vic, Australia, pp 67–86.

Walsh ME (1998) El Niño: Climates of Uncertainty: Media Analysis Report for the National Climate Centre, Bureau of Meteorology. RMIT University, Melbourne.

White BJ (2001) Survey shows forecast use up. Climag Newsletter of the Climate Variability in Agriculture Program, Vol. 5. p 4.

Williams AJ, Karoly D (1999) Extreme Bushfires in Australia and the Impact of ENSO. Aust. Meteorol. Mag, vol 48. pp 15–22.

Wright WJ, Coughlan M, Plummer N, Watkins A, James R, Drosdowsky L, Southern B (2003) The 2002 Australian drought: meteorological services to rural and regional communities. Outlook 2003 Conference Proc on weather & farming session papers, Canberra, March 2003, pp 1–13.

CHAPTER 14

Agrometeorological Disaster Risk Management in China*

Wang Shili, Huo Zhiguo, Guo Jianping, Wang Chunyi

Abstract Efficient management of risks due to meteorological disasters is very important in China because of their increasing occurrence and limited capability for prevention and preparedness in China. Losses due to meteorological and related disasters account for more than 70% of the natural disaster losses in China. Early warning and prediction of agrometeorological disasters includes improvements in statistical prediction models, predictions based on climate prediction and agrometeorological models, and early warning for agrometeorological disasters using GIS. Risk assessments of agrometeorological disasters are based on identification of risk and use of models to characterize disaster intensity, disaster loss and the capability to combat disasters. Several ways of popularizing the preventive measures of agrometeorological disasters were described with suitable examples.

14.1 Introduction

Agricultural production is highly dependent on weather, climate and water availability, and is adversely affected by the weather and climate related disasters. Abnormality of rains and unfavorable temperature could lead to crop failure, food insecurity, famine, loss of property and life and negative national economic growth. Due to frequent occurrence of disasters and limited capability of prevention and preparedness, meteorological disasters have serious impacts on sustainable development of agricultural production in China, therefore the risk management of meteorological disasters is more important.

China is located in monsoon climate region. Monsoon and its manifestations in terms of onset time, intensity and its impact on various sectors in China are different every year. Some meteorological disasters and extreme climate events such as drought, flood, heavy rain, low temperature damage, frost, hail and heat wave occur frequently. In this situation the meteorological disasters strongly influence national economy, especially agricultural production. According to statistics the losses due to meteorological and related disasters account for more than 70% of natural disasters losses in China. The average damaged acreage of cultivated land caused by natural disasters is 30,660,000 hectares from year 1950 to 1979, and the averaged reduction of crop yield in each

* Supported by national key project "Early warning and control techniques of desease and pest in agriculture and agrometeorological disasters".

year is about 20,000,000 t. In recent years the unfavorable impacts of climate variation and meteorological disasters on agricultural production seem to be increasing, for example, the mean drought-damaged acreage per year from 1950 to 1990 was 20,426,000 hectares, while in 1991–2000 it was 27,147,000 hectares; the flooded areas from 1950 to1990 were 7,982,000 hectares, in recent 10 years it reached to 15,066,000 hectares, nearly doubled. As such, it is very important to strengthen risk management of meteorological disasters so as to mitigate unfavorable impacts on agricultural production and to ensure sustainable development of agriculture. From agrometeorological point of view the major approaches of disasters risk management should be focused on early prediction, assessment and preparedness of agrometeorological disasters and on providing service of disasters information and of practical techniques of disaster-preventing measures.

Chinese meteorologists pay considerable attention to assessing the impacts of meteorological disasters on agricultural production, apart from strengthening and improving the accuracy of weather forecasting and climate prediction. In recent years, agrometeorologists have made good progress in study and service of agrometeorological disasters, which helped in reducing disaster risk, minimizing disaster losses and ensuring stable development of agriculture. In this paper an attempt has been made to highlight the efforts and experience in reducing risk of agrometeorlogical disasters through prediction and extending preventive measures to mitigate agrometeorlogical disasters in China.

14.2 Early Warning and Prediction of Agrometeorological Disasters

Agrometeorological disasters have serious influence on agricultural production due to frequent occurrence of agrometeorlogical disasters and in the absence of capacity building in managing these disasters. Timely and accurate prediction of agrometeorological disasters is very useful for reducing losses of disasters and ensuring sustainable development of agricultural production in China. However, agrometeorological disasters are different from common meteorological disasters, as their occurrence, extent and impacts are determined not only by the variations in weather elements, but also by crop growth status, soil and management level. In recent years keeping in mind the demand of sustainable development of agriculture and mitigating disasters, studies on early warning and prediction of agrometeorological disasters were carried out (Wang Shili 2003).

14.2.1 Improvement in Statistical Prediction Models

Multi-regression analysis, time series analysis, analogue analysis and cadence are often used to develop prediction models for agrometeorological disasters. In multi-regression prediction models the predictors are selected from parameters related to atmospheric circulation, sea surface temperature and other meteorological elements. In the study on prediction of low temperature damage in Northeast China, low temperature damage patterns during the summer period are firstly determined by crop yield index and anomaly of sum of degree-days (daily temperature above 10 °C) during the period from May to September, and subsequently regression analysis, clustering analysis, distinguishing analysis are used to develop cold injury prediction models in which the statistical values of atmospheric circulation are selected as predictors (Zhou Lihong et al. 2001). Some previous signals in atmospheric circulation, along with synoptic and climatic characteristics are revealed to forecast the disasters. As for example sea surface temperature in winter in warm pool of tropical west Pacific is detected as strong signal of low temperature in summer in Northeast China based on EOF (empirical orthogonal function) and, SVD (Singular Value Decomposition) analysis (Zheng Weizhong and Ni Yunqi 1999). SNR (signal noise ratio) is applied to identify previous stronger signal from atmospheric circulation values and sea surface temperature for probable occurrence of drought in North China. It is found out that the abnormal changes in 500 hPa height of Urals and Lake Baikai as well as abnormality in Kuroshio warm pool and Nino 4 are the strong signals of probable severe drought in North China (Wei Fengying and Zhang Jingjiang 2003).

In some models time series analysis is combined with regression analysis. In low temperature prediction during anthesis of rice, the time series of low temperature index are firstly generated as mean generated functions, and then auto-regression prediction models where time periods are independent variables are developed. Other prediction models are multiple regression models with period variation (Na Jiafeng 1998; Xu Bingnan et al.1999). Damaged acreage due to disaster is estimated by GM (1,1) model (Wang Weiming 1998).

Phenological phase reflects past and present weather and climate condition, sometimes which are used to predict abnormality of climate or agrometeorological disasters. In Northeast China the phenological phases of almond tree, poplar tree and willow trees as well as phenological phases of maize crop are used to predict mean temperature of spring and accumulated temperature during growing season. The time of jointing stage of maize is used to forecast the occurrence of crop cold damage in Northeast China (Bi Bojun 2000).

14.2.2 Prediction Models Based on Climate Prediction and Agrometeorological Models

14.2.2.1 Irrigation Scheme for Winter Wheat in North China

In irrigation scheme for winter wheat in North China the regional climate models are connected with soil water model. Based on forecast of daily weather elements from regional climate model and soil water balance model, the soil water contents are forecasted dynamically. The optimal irrigation date and amount are calculated according to the threshold of water requirement of wheat in different development stages (Fig. 14.1). The water deficit and yield

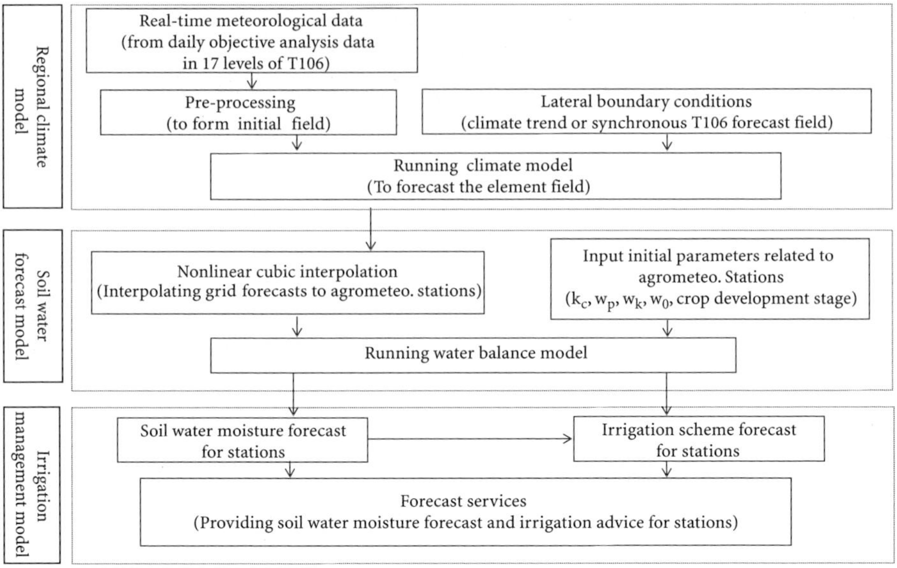

Fig. 14.1. Predicted irrigation scheme for winter wheat in North China

reduction are also estimated. The relative error is found to be 8.6% for soil water content during next 10 days (Zhang Guangzhi et al. 2001). The irrigation scheme information service was shown on China Central Television Station.

14.2.2.2
Prediction Based on Climate Prediction and Crop Growth Models

Statistical prediction models have their limitation, such as the mechanism is not clear and the predictions are often beyond the record data, and so on. On the other hand, presently the process-based crop growth simulation models have been applied to crop yield estimation, climate risks evaluation and agricultural resources utilization (Gerrit Hoogenboom 1999). Therefore it may be a way of agrometeorological disaster prediction to apply the seasonal climate prediction combined with crop growth simulation model. Some research on this aspect was carried out in recent years.

Wheat Drought Prediction in North China

For identifying and forecasting winter wheat drought, the predicted meteorological elements (decaded values) are put into soil water balance sub-model and the changes in water content and ratio of supply to demand in water are estimated. According to the index expressed with the supply/demand ratio, drought with four grades was identified, assessed and forecasted (Zhao Yanxia et al. 2001). In another study, a winter wheat growth simulation model in North China is developed based on field experiments, where crop parameters such as

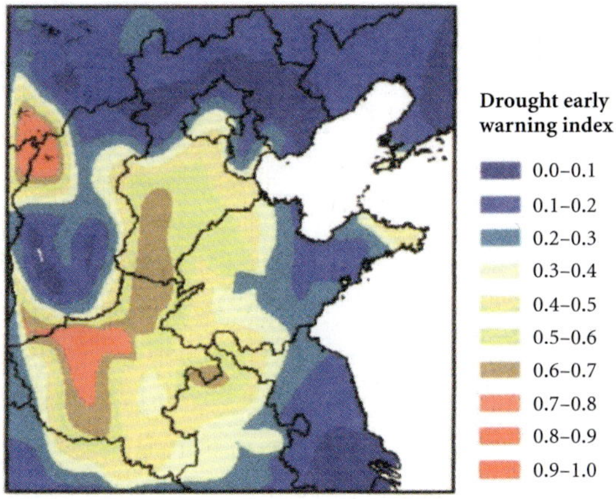

Fig. 14.2. Simulated wheat drought in middle of May, 2000 based on crop model (early warning index, higher value denotes severe drought)

maximum photosynthesis rate and initial water use efficiency for local variety in North China under various water conditions are measured and determined, and the cumulated value of relative transpiration ratio is defined in terms of crop drought stress index and drought early warning index. The model was validated for 1998–2000 with satisfactory results. The spatial distribution of simulated drought stress index was drawn using geographic information system (GIS) for historical drought cases (Fig. 14.2) (Liu Jiandong et al. 2003).

Prediction of Low Temperature Damage of Maize in Northeast China

A regional maize growth model was developed and applied to predict low temperature damage in Northeast China. Based on the determination of crop parameters in different regions and the spatial interpolation of daily weather elements, the maize growth status for 12 stations in 1961–2000 and growth processes in grids with $0.25° \times 0.25°$ for the typical low temperature damage year were simulated, respectively (Fig. 14.3). Further-more, experiment of low temperature disaster prediction was made using maize model and regional climate model reGCM-NCC (National Climate Center) by down-scaling from China-GCM (Liu Buchun et al. 2003). It is an attempt to combine the down-scaled climate model with a scaled up crop model.

Prediction of Wet Damage for Wheat in the Middle and Lower Reaches of the Yangtze River

A wheat model for wet damage in middle and lower reaches of Yangtze River has been developed, which includes influences of water-logging days on photosynthesis rate, dry matter allocation and leaf decline. When prediction of regional climate models output or real-time weather forecast is put in, the impact of water-logging on wheat yield can be determined by running wheat water-logging model and the model without water-logging, and by comparing two results, the early warning for water logging can be made (Shi Chunlin and Jin Zhiqing, 2003) .

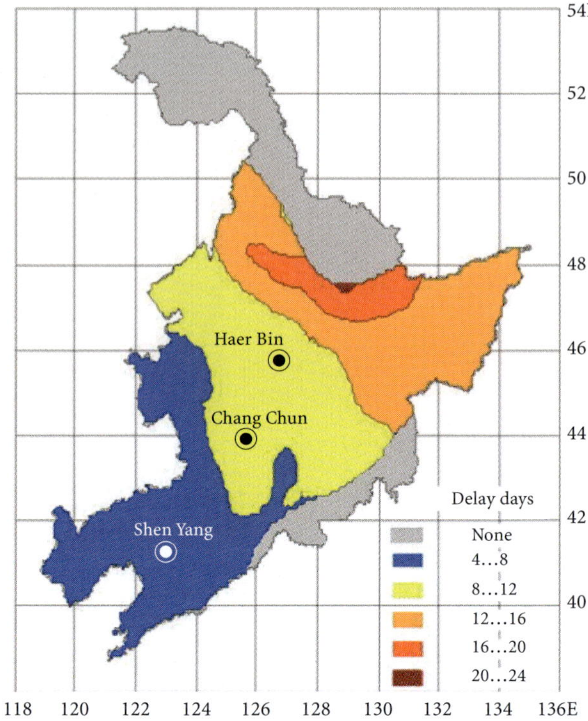

Fig. 14.3. Simulated cold damage in Northeast China in 1969 based on regional maize model (difference of tasseling stage between 1969 and average, unit: d, no cultivation)

It must be pointed out that till now prediction of agrometeorological disasters is not successful due to its particularity and complexity, so it is difficult for any single method or model to get perfect results. The better way is a comprehensive analysis based on various prediction methods and approaches.

14.2.3 Early Warning for Agrometeorological Disasters Using GIS

Due to the great spatial differences in occurrence and impact of meteorological disasters on agricultural production, users often require more specific prediction. It requires application of GIS to prediction of meteorological disasters. Here is a case study of predicting chilling injury of banana and litchi trees using GIS in Gunagdong Province, China (Wang Chunlin et al. 2003).

By using T213 numerical weather forecast products, MOS (model output statistics) prognostic equation is established, and the future 24, 48 and 72 hours temperatures (mean and minimum temperatures) for 86 weather stations are forecasted. Considering that topographic differences will result in different spatial distributions of temperature and impacts on chilling injury, the values of temperature prediction on grids with $0.5° \times 0.5°$ are estimated by GIS, which are corrected by longitude, latitude, elevation, slope gradient, and slope orientation. Furthermore according to the chilling injury index of banana and

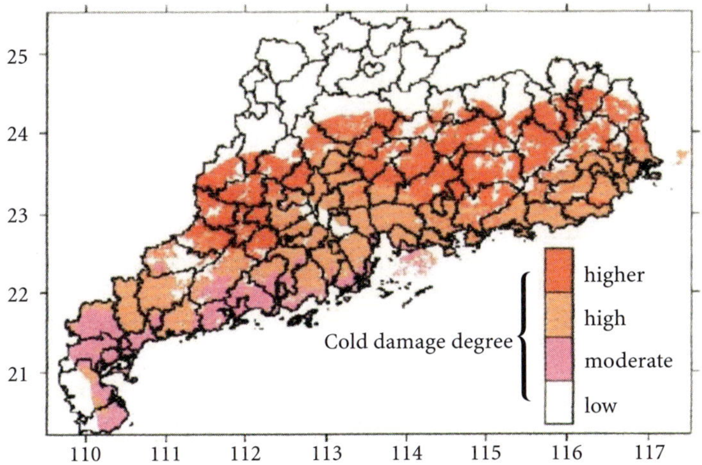

Fig. 14.4. Simulated cold damage degree of litchi in the third-decade, Dec. 1999 in Guangdong Province by GIS

litchi trees and their actual planted distribution, the predicted degree of chilling injury of banana trees or litchi trees on grids could be obtained (Fig. 14.4). The distribution map of early warning and prediction is visual and detailed, which is useful and beneficial to managers and farmers.

14.3 Risk Assessments of Agro-meteorological Disasters

Meteorological disasters risk assessment is helpful for managers of agriculture, government officials and farmers to make decision in agricultural planning, to formulate preventive measures, to rescue and to provide relief after occurrence of disasters, as well as to make insurance so as to minimize the unfavorable impact on agriculture. Up till now it is still lack of quantitative assessment and comprehensive risk analysis of agro-meteorological disasters in China. Studies on risk analysis system including risk identification, risk assessment and types of agro-meteorological disasters were carried out for four major agro-meteorological disasters e.g. winter wheat drought in North China, wet damage of wheat and rape in middle and lower reaches of Yangtze River and Huai River basin, low temperature damage in summer of maize and rice in Northeast China and cold damage of litchi and banana trees in South China (Huo Zhiguo, et al. 2003).

14.3.1 Risk Identification of Major Agro-meteorological Disasters

Risk identification refers to analyzing the locations and seasons of disaster occurrence, characteristics of weather and climate and their effects on agricultural production. For example, drought of wheat in winter and spring in North China is mainly caused by less precipitation in winter and spring and less available water in soil before sowing, which results in severe reduction of wheat yield. The wet damage for winter wheat and rape in middle and lower reaches of Yangtze River and Huai River basin is due to a long period of more precipitation in winter and saturated soil water as well as weak sunshine. Wet damage shortens the duration of grain filling and reduces the crop yield. Low temperature in summer in Northeast China prolongs the crop development stages and crops such as maize and rice are not able to mature before frost in autumn. In some situation, low minimum temperature may damage rice during its productive stage. When a sharp decrease in minimum temperature due to cold wave in winter results in temperatures falling below critical temperatures for sub-tropical fruit trees such as banana and litchi, there will be total damage of these fruit trees (Li Shikui et al. 2004).

14.3.2 Risk Assessment of Major Agro-meteorological Disasters

Risk assessment is the core component of the risk analysis system. Risk assessment of agro-meteorological disasters includes risk assessment models of intensity, losses and capability of combating disasters. Risk assessment models of intensity describe the various grades of intensity and their probabilities while risk assessment models of losses are used to estimate the direct and indirect losses at different intensities. Risk assessment models of capability of combating disasters reflect social and economic level. Risk assessment models of intensity, losses and capability of combating four agro-meteorological disasters mentioned above are established, and spatial distribution are plotted in maps (Wang Suyan et al. 2003; Liu Ronghua et al. 2003a; Xue Changying et al. 2003a; Ma Xiaoqun et al. 2003; Ma Shuqing et al.2003; Liu Jinluan and Du Yaodong 2003).

14.3.2.1 Risk Assessment Models of Disaster Intensity

In the study carried out by Li Shikui et al. 2004 risk index of disaster intensity (I) is a function of different intensity (G) and their probabilities (P) expressed as follows:

$$I = F(G, P) = \sum_{i=1}^{n} G_i P_i \,. \tag{14.1}$$

The risk indices of disaster intensity for different grades of 4 agro-meteorological disasters were determined. For an example, risk indices of water stress under natural water supply during whole growing season in North China (L) consists of 5 grades, which include light drought, moderate drought, drought, severe drought, extreme drought.

$$L = \left(1 - \frac{E_T}{E_{TM}}\right) \times 100\% \tag{14.2}$$

where E_T is soil water available under natural water supply during whole growing season, which is sum of effective precipitation and water available in soil before sowing, E_{TM} represents maximum water requirement of crop during whole growing season, which is calculated by multiplying reference crop evapotranspiration with the crop coefficient (Richard et al.1998).

In order to calculate probabilities of disasters intensities, the probability distribution density functions are fitted using 6 different probability distribution models such as normal distribution based on historical data (Du Yaodong et al. 2003; Huo Zhiguo et al. 2003).

14.3.2.2 Risk Assessment Models of Disaster Loss

The loss due to disasters is expressed by reduction of yield caused by unfavorable meteorological conditions, which is a negative value of the climatic component of the weather-crop yield model. Similarly, risk index of disaster loss is the sum of different yield reduction percentages multiplied by their probabilities.

14.3.2.3 Risk Assessment Models of Capability of Combating Disasters

The capability of combating disasters is related to the resistance of crops and the level of agricultural production. The latter is considered in the study related to agrometeorology. Some models and indices are adopted, such as the ratio of actual yield to maximum yield, the slope of weather yield regression equation, and so on.

14.3.3 Comprehensive Risk Division of Major Agro-meteorological Disasters

Combined with disaster intensity, losses and capability of combating disasters, the comprehensive risk indices are calculated, and risk divisions for four disasters are prepared and plotted. (Liu Ronghua et al. 2003b; Xi Zhuxiang et al.2003; Liu Jinluan 2003 et al.). There is an example of a case study on comprehensive risk division for water stress under natural water supply during growing season. The comprehensive risk index (M) is expressed as:

$$M = \frac{1}{a}\sum_{i=1}^{n}\left(f_i \times P + R_i \times D_i\right) \tag{14.3}$$

were f is water stress, P is its probability, R and D are yield reduction and its probability respectively, a is the coefficient reflecting change trend of capability

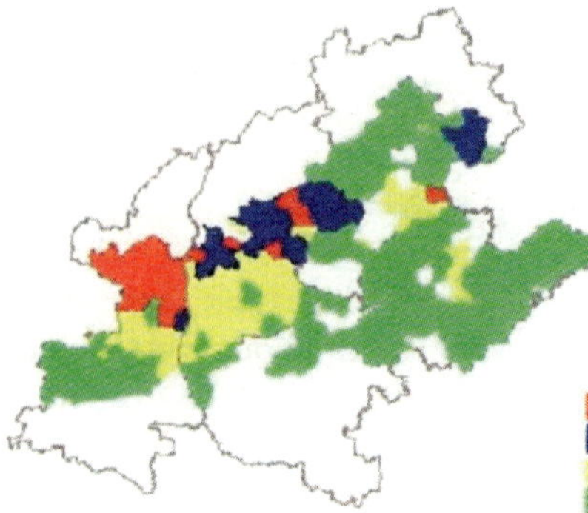

Fig. 14.5. Comprehensive drought risk zoning of wheat in North China

of combating drought and M is normalized as value of 0–1. The division result is shown in Fig. 14.5.

Risk assessment is an effective risk management approach of agro-meteorological disasters. The results of risk assessment are used in guiding management of agricultural production. For example, the drought risk assessment under different irrigation system scheme shows that the drought risk could be effectively reduced under 1–3 times irrigation in North China (traditionally 4–5 irrigation), and the benefit of three-time irrigation in severe drought year is the greatest (Xue Changying et al. 2003b). In 2003 the result was demonstrated in 100,000 hectares farmland in Henan Province. The rational arrangement of tropical fruit trees and adjustment of cropping system are proposed according to the risk division distribution. In 2002–2003, when the results of risk assessment for 4 agro-meteorological disasters were popularized in 4 provinces, the disasters losses were significantly reduced.

14.4 Popularizing Preventive Measures of Agro-meteorological Disasters

It is practical and helpful for farmers to understand and know well the techniques of disasters-preventing measures based on agro-meteorology. Chinese agro-meteorologists carried out studies on techniques of agro-meteorological disasters-preventing measures in the past five years, and popularized the techniques to farmers by various ways.

14.4.1 Rational Use of Agro-climatic Resources to Mitigate Disasters

14.4.1.1 Arranging Proper Crop Varieties According to Climate Conditions

The temperature condition is reflected in a major agro-climatic index determining regional distribution and yield of crops in Northeast China. Because of large variations in climatic conditions in Northeast China and longer growing period and unstable yield for late maturing crop varieties, it is an important counter measure of alleviating low temperature damage to rationally arrange and adjust crop varieties requiring different thermal conditions so as to use thermal resources and avoid yield reduction. It is an objective way to prevent and mitigate low temperature disasters. Yield reduction percent (r) of maize due to lack of thermal adjustments, the relation between yield of different matured varieties and temperature condition is given below:

$$r = 1 - \frac{1}{1 + e^{(a + b\Delta\Sigma T / \Delta\Sigma T_{\max})}} \tag{14.4}$$

where $\Delta\Sigma T$ is difference between the accumulated temperature from sowing date to harvest and the one required from sowing date to complete maturity ($\Delta\Sigma T_{\max}$), a and b are empirical constants. The benefits of each variety in region and the climatic probabilities of each benefit were estimated, and the proper variety which could use thermal resources and get maximum benefit is chosen as the dominant crop variety by probability decision method. In addition, decision-making in matching and adjusting proportions of various varieties under different heat patterns by multi-goal linear program method was made. As a result the spatial distribution of proportion of dominant variety and other varieties under each heat pattern year are determined and plotted. (Liu and Sun 2000).

14.4.1.2
Sowing Crop According to Predicted Proper Sowing Date

Sowing on proper date is one of the effective counter measures to avoid low temperature injury of crops in Northeast China, which could use thermal condition in spring and ensure crop maturity before frost. Normally, indirect and direct prediction methods are adopted (Ma Shuqing et al. 2000).

The numerical weather forecast model and mean function are used to predict mean ten-day temperature of April and May. Based on the relationship between mean temperature in April and sowing date the predicted model of maize sowing date is developed. The accuracy of sowing date prediction is up to 79%. The suitable sowing date for maize in Northeast China is normally determined as starting day of mean daily temperature $\geq 7\,^{\circ}$C. There is a model to forecast starting date of temperature $\geq 7\,^{\circ}$C using co-regression analysis between beginning date and departure field of 500 hPa geopotential height with accuracy of 80%. In routine forecasts, the weighting integration with historical fit accuracy and validation accuracy for different forecast models is adopted.

14.4.2
Adopting Practical Agricultural Measures to Mitigate Disasters

Some practical agricultural measures to mitigate disasters were studied and popularized to farmers.

14.4.2.1
Effect of Plastic Film Mulching and Nursery Transplanting

Film mulching has the significant effect of increasing temperature and preserving soil moisture. Some experimental studies on effects of mulching and nursery transplanting on crop growth in Northeast China were conducted. In Northeast China late maturing varieties could be cultivated by using mulching. Its growing period is 15 days longer than local major variety and gains more than 200 °C of accumulated temperature, and it could be planted at 5 °C of mean daily temperature. When using film mulching the yield could increase 45% (Ma Shuqing et al. 2004). The nursery transplanting prolongs 20 days of maize growing period and increases accumulated temperature up to 220 °C. The yield is enhanced by 40% (Guo Jianping et al. 2003).

Table 14.1. Effect of plastic film mulching and nursery transplanting

Technique treatments	Advanced days of sowing date (d)	Prolonged days of growing period (d)	Increment in sum of temp. during growing period (degree day)	Increment in yield (%)	Increment of weight per 100 grains (%)
Mulching	−5	+15	+200	+45	+40.5
Nursery transplanting	−7 ~ −10	+20	+220	+40	+6.9

14.4.2.2
Effect of Straw Mulching

With straw mulching, the thermodynamic and dynamic characteristics of soil surface are changed, which results in sensitive heat flux increasing and latent heat flux reducing, more water is stored in soil. It is favorable for soil water accumulation in early growing period and promotes transpiration and dry matter accumulation in later growing period (after jointing stage). The results indicate that straw mulching is an effective measure for preserving soil water and raising crop yield. Different amount of straw, different mulching time and different soil water content have different influences on the dry matter weight, leaf area, intensity of transpiration, yield and water use efficiency of winter wheat. The optimum mulching date in North China plain is the end of December when winter wheat stops growing. The optimum straw amount used for mulching is about 4,500–6,000 kg/hm^2, and the optimum soil water content is 55%–70% of relative soil humidity. Under the combination of above conditions, the yield of winter wheat increased by 18.5% and water use efficiency raised by 22.2% (Zhu Zixi et al.2000; An Shunqing et al.2001).

14.4.3 Preparing and Applying Reagents to Mitigate Agrometeorological Disasters

14.4.3.1 Low Temperature-resistant Growth Promoting Regulator

A new low temperature-resistant growth promoting regulator without pollution and hormone was prepared for combating low temperature damage in Northeast China. It consists of various nutrients required by crops, improves chlorophyll content of leaf and promotes photosynthesis and anti-adversity performance. The results of field experiments and demonstration on farm land show that spraying growth promoting regulators on leaf surface of crops twice during vegetative growth stage and spraying again when low temperatures occurs, helps in advancing the crop maturity by 5–6 days and increasing seed weight by 9%–10% and in production increases of 6%–11% under normal climatic conditions, and 10% in low temperature years. The ratio of input to output is about 1:20 (Wang Chunyi, et al. 2000).

14.4.3.2 Growth- and Tiller-promoting Regulators of Rice

In order to mitigate low temperature injury during sowing period of rice in Northeast China, a new regulator of promoting growth and tillers of rice seedlings was prepared. It consists of multi-organic and mineral substances. The results show that by using the regulators in seedling period, the time of tillering shifts earlier. Effective tiller percent is 25%–27% compared with contrast of 2.2%–6.7%, proline content increases 10%–46% and electrical conductivity decreases 23%, which reflects the feature of low temperature resistance (in the year 2001). The rice yield raises 2.5%–6.7% in four site experiments (in 2002), and 8%–12% in farmland with 7,500 hectares (in 2003) and the ratio of input to output is 1:10 (Wang Chunyan et al. 2003).

14.4.3.3 Drought-preventing Reagent

A multi-functional drought-preventing reagent restraining evaporation and promoting crop growth was prepared in laboratory and tested in field in North China. It plays a role in combating drought and saving water. Its use may enhance stomatal resistance of leaf by 24%–51%, depress plant transpiration by 2%–9%, thus reduce water consumption by 27.0–28.1 mm during the stage of ear formation and 10.2–11.3 mm in grain-filling stage, and improve water use efficiency by 2%–10%. At the same time the multi-functional drought preventing reagent may retard the senescence of green leaves and therefore prolong the wheat grain-filling process. The leaf area index increased by 0.24–1.12, the rate of grain filling increased 22%–44%, and photosynthetic intensity was enhanced by 18%–28%, as compared with increase in 1,000-grain weight by 1.6%–3.3%, and yield increased by 6.7%–8.4% (Zhao Guoqiang et al. 2000).

14.4.4 Popularizing Agrometeorological Disaster-preventing Techniques

In order to make more farmers learn and operate the techniques of meteorological disaster preventing so as to reduce disaster risk, Chinese agrometeorologists actively popularize agro-meteorological disaster preventing techniques through demonstration, education and training. The handbooks of countermeasures against low temperature damage in Northeast China and drought in North China were published and issued to farmers. The training courses on agro-meteorological disaster-reduction techniques are held in disaster prone regions. In some areas the agro-meteorological disaster preventing techniques are shown on TV station. During the past 2–3 years the growth and tiller promoting regulators of rice were popularized in 7,500 hectares and average increase in yield was 8%–12% while the increase in net rice production was 2,400 tons. The comprehensive techniques of preventing drought of winter wheat has been popularized on about 50,000 hectares.

14.5 Agrometeorological Information Services

Timely information on crop growth status, agro-meteorological conditions as well as occurrence and effects of agro-meteorological disasters is important and necessary for decision-makers, managers of agricultural production and farmers as it helps them to manage agricultural production and avoid or mitigate adverse impacts of meteorological disasters on agricultural production. In China Meteorological Administration (CMA) there are various levels of operational agro-meteorological information service.

14.5.1 National Level

There is a huge and effective agrometeorological information service system in CMA, it relies on an agrometeorological observation network with 584 stations and advanced communication network. It consists of data collection and their transmission, processing and analysis, and subsequent product preparation as well as delivery system. The agrometeorological information bulletins are issued every ten-days to users by mail, e-mail, internet, and the daily agrometeorological conditions along with review and assessment is also broadcasted on the agricultural channel of China Central TV. Drought and flood are monitored by using remote sensing techniques and provided to the public. When the meteorological disasters occur or are predicted, the possible unfavorable effects and practical counter measures are proposed and suggested through bulletin or TV channel. The yield forecasting for major crops all over the China is provided to governments in advance.

14.5.2 Provincial Level

In all the Provincial Meteorological Bureaus, agrometeorological information services are carried out routinely. Many information services are integrated systems including prediction and assessment of agrometeorological disasters followed by decision-making, crop yield forecasting and consultation on agricultural production management. The service products are issued by bulletin, Internet, TV program, radio and newspaper. There are also some professional information service systems, such as special information service for cotton crop. (Wei Li and Lu Shuming 1997; Ge Huiyan and Zhang Yong-hong 2001). The role of guidance and consultation in provincial level is more detailed and directive, and benefits obtained are also more pronounced.

14.6 Conclusions

With increasing incidence of natural disasters around the world, timely and accurate prediction of disasters, comprehensive assessment of their impacts on agriculture, forestry, and fisheries as well as relevant strategies for mitigation of natural disasters are critical for sustainable development, especially in the developing countries. It may not be possible to prevent the occurrence of these natural disasters, but effects of the resultant disasters and vulnerability associated with the hazards of natural disasters could be reduced considerably through proper planning and effective preparedness and also can be controlled to some extent by accurate and timely prediction and by taking counter-measures.

It is very important to strengthen and to improve risk management of natural disasters. Disasters should not be viewed as extreme events created only by natural forces but complicated unsolved problems influenced by physical, social, and economic forces as well. An effective approach for risk management is to put forward timely and accurate prediction of agrometeorological disasters, comprehensive risk assessment and practical disasters preventing measures. At the same time transferring agrometeorological disaster information to users and popularizing disasters counter-measures to farmers have the desired effects on sustainable development of agricultural production in China.

In order to better make disaster risk management, mitigate extreme events and reduce loss due to agrometeorological disasters, it is necessary to further strengthen monitoring of agrometeorological information, establish a comprehensive database in accordance with the users needs, including adequate collection, quality control and presentation of data specific to certain kinds of disaster, and make good use of modern available technology, e.g., remote sensing, geographic information system and Internet. There is an urgent need to improve the forecasting skills for agrometeorological disasters. Lack of good forecast skill is a constraint to improved adaptation, management, and mitigation. Prediction models based on improved statistical models and combination between seasonal climate prediction and crop growth simulation models should be integrated. Agrometeorological disasters risk zoning is an essential component of disaster mitigation and preparedness strategies. The study on models for disaster loss assessment should be further promoted. More practical and useful techniques of preventing or reducing loss of disasters on agricultural production should be extended and popularized to farmers by imparting education and training. Agrometeorological information service system should be established with more capabilities to deliver information of agrometeorological disasters to public and users more easy and faster.

It is essential to promote and foster the use of international and regional programs to enhance collaboration and building of partnerships on issues related to improved management of and preparedness for meteorological disasters. In addition, the cooperation between agrometeorologists and users, institutes and government agencies is important.

References

An Shunqing, Liu Gengshan, Guo Anhong (2001) The integrated drought defending technique system for winter wheat in North China and its application. Journal of Natural Disasters, Vol. 10(4). pp 237–245.

Bi Bojun (2000) Diagnose and prediction of temperature abnormality using plants phenomena. Meteorology, Vol 26(3). pp 56–57.

Du Yaodong, Mao Huiqin, Liu Jinluan (2003) Study on probability distribution models of cold damage in South China. Journal of Natural Disasters, Vol 12(2). pp 103–107.

Ge Huiyan, Zhang Yonghong (2001) The forecast system of cotton meteorological disasters. China Cotton, Vol. 28(3). pp 21–22.

Gerrit Hoogenboom (1999) The state-of-the art in crop modeling. In: Sivakumar M.V.K (ed) Climate prediction and agriculture. Proc the START/WMO International Workshop held in Geneva, Switzerland, 27–29 September. Washington, DC, USA, International START Secretariat, 2000, pp 69–75.

Guo Jianping, Wang Chunyi, Ma Shuqing (2003) Epxerimental study on defending corn from cold injury and spring drought by growing and transplanting seedlings. Journal of Natural Disasters, Vol. 12(1). pp116–120.

Huo Zhiguo, Li Shikui, Wang Suyan (2003) Study on the risk evaluation technologies of main agrometeorological disasters and their application. Journal of Natural Resources, Vol. 18(6). pp 692–703.

Li Shikui, Huo Zhiguo, Wang Suyan (2004) Risk evaluation system and models of agrometeorological disasters. Journal of Natural Disasters, Vol. 13(1). pp 77–87.

Liu Buchun, Wang Shili, Zhuang Liwei (2003) Study of low temperature damage prediction applications in NE China based on a scaling-up maize dynamic model. Quarterly Journal of Applied Meteorology, Vol. 14(4). pp 616–625.

Liu Jiandong, Wang Futang, Yu Qiang (2003) A drought prediction model in North China Plan and its application. Quarterly Journal of Applied Meteorology, Vol. 14(4). pp 593–604.

Liu Jinluan, Du Yaodong (2003) Cold-damage influencing factors on litchi and their economy-loss evaluation model. Journal of Natural Disasters, Vol. 12(2). pp 188–191.

Liu Jinluan, Du Yaodong, Mao Huiqin. (2003) Risk analysis and zonation of cold damage to litchi in South China. Journal of Natural Disasters, Vol. 12(3). pp 126–130.

Liu Ronghua, Zhu Zixi, Fang Wensong, (2003a) Risk and losses assessment of drought disaster for winter wheat in North China Plain. Journal of Natural Disasters. Vol 12(2), pp 170–174.

Liu Ronghua, Zhu Zixi, Fang Wensong, (2003b) Preliminary study on drought zonation of winter wheat in North China Plain. Journal of Natural Disasters, Vol 12(1). pp 140–144.

Liu Xinan, Sun Yuting. (2000) Study on reasonable arrangement techniques of corn variety in Northeast China. ACTA Meteorologica Sinica, Vol. 58. (Supplement), pp 889–898.

Ma Shuqing, Wang Qi, Guo Jianping (2004) Study on cold and drought preventing of corn using mulching, Journal of Natural Disaster (in press).

Ma Shuqing, Wang Qi, Zheng Qinglin (2000) Study on Differentiate Methods of Corn Seeding Date in Northeast China. ACTA Meteorologica Sinica, Vol. 58. (Supplement), pp 881–888.

Ma Shuqing, Xi Zhuxiang, Wang Qi (2003) Risk evaluation of cold damage to corn in Northeast China. Journal of Natural Disasters, Vol. 12(3). pp 137–141.

Ma Xiaoqun, Chen Xiaoyi, Sheng Shaoxue(2003) Models of evaluating water-logging loss of winter wheat in Anhui Province. Journal of Natural Disasters, Vol. 12(1). pp 158–162.

Na Jiafeng (1998) EOF analysis and forecast model based on periodic function of rice chilling damage weather. Chinese agrometeorlogy, Vol. 19(4). pp 50–52.

Richard G. Allen, Luis S. Pereira and Dirk Raes (1998) Crop evapotranspiration guidelines for computing crop water requirements. FAO Irrigation and Drainage Paper 56, Rome.

Shi Chunlin, Jin Zhiqing (2003) A WCSODS-based model for simulating wet damage for winter wheat in the middle and lower reaches of the Yangtze River. Quarterly Journal of Applied Meteorology, Vol. 14(4). pp 462–468.

Wang Chunlin, Liu Jinluan, Zhou Guoyi (2003) Research on real-time cold-disaster watching and prediction in Guangdong Province based on GIS technology. Quarterly Journal of Applied Meteorology, Vol. 14(4). pp 487–495.

Wang Chunyan, Wang Lizhi, Li Zhongjie (2003) Primary study on effect of rice seedlings and tillering-promoting regulator on cold-resistant ability. Journal of Natural Disasters, Vol. 12(2). pp 108–112.

Wang Chunyi, Guo Jianping, Ma Shuqing(2000) The experiment and example study on a low-temperature resistance medicine for defending crop cooling injury. Acta Meteorologica Sinica, Vol. 58. (Suppl.), pp 961–968.

Wang Shili(2003) Review of the progress in methods of agrometeorlogical disaster prediction in China. Quarterly Journal of Applied Meteorology, Vol. 14(5). pp 574–582.

Wang Suyan, Huo Zhiguo, Li Shikui, (2003) Risk assessment of drought effect on yield of winter wheat in northern China. Journal of Natural Disasters, Vol 12(3). pp 118–125.

Wang Weiming (1998) The topological prediction model on agricultural meteorological calamity in Gansu Province. Journal of Gansu agricultural university, Vol. 33(3). pp 276–282.

Wei Fengying, Zhang Jingjiang (2003) Climatological background and previous stronger signal of anomaly drought over North China. Acta Meteorologica Sinica, Vol. 61(3). pp 354–363.

Wei Li, Lu Shumming (1997) A study of the prediction of agroclimatic harvest and decision service system. Quarterly Journal of Applied Meteorology, Vol. 8(2). pp 224–229.

Xi Zhuxiang, Ma Shuqing, Wang Qi (2003) Risk evaluation and zonation of the low temperature and cold damage in Northeast China. Journal of Natural Disasters, Vol. 12(2). pp 98–102.

Xu Bingnan, Zhang Yufa, Luo Gang (1999) Predicting trend of meteorological disasters in Guizhaou using multi-regression models with periods components, Guizhou Meteorology, Vol. 23(3). pp 29–33.

Xue Changying,, Huo Zhiguo, Li Shikui, (2003a) Risk assessment of drought and yield losses of winter wheat in northern part of North China. Journal of Natural Disasters, Vol. 12(1). pp 131–139.

Xue Changying, Huo Zhiguo, Li Shikui, (2003b) Action of irrigation on decreasing yield reduction due to drought-a risk assessment of winter wheat in North China Plain. Journal of Natural Disasters, Vol. 12(3). pp 131–136.

Zhang Guangzhi, Xu Xiangde, Mao Fei (2001) A new approach of irrigation management for winter wheat by the climatic-agrometeorological model consensus system. Quarterly Journal of Applied Meteorology, Vol. 12(3). pp 307–316.

Zhao Guoqiang, Zhu Zixi, Fang Wensong (2000) Application study of multi-functional drought-resistant reagents. Quarterly Journal of applied Meteorology, Vol. 11. (Suppl.), pp 186–191.

Zhao Yanxia, Wang Futang, Qiu Guowang (2001) A study of assessing and forecasting models of drought in agriculture. Quarterly Journal of Applied Meteorology, Vol. 12(2). pp 235–241.

Zheng Weizhong, Ni Yunqi (1999) Diagnostic study for impact of sea surface temperature anomalies over tropical and mid-latitude Pacific on summer low temperature cold damage in Northeast China. In: Studies on integrated preventing technology of summer low temperature cold damage, Meteorological Press, pp 33–40.

Zhou Lihong, Liu Xinan, Zhou Yuhui (2001) Research of general circulation features and forecast of summer low temperature and cold damage in the Northeast of China. Journal of Shenyang Agricultural University, Vol. 32(1). pp 22–25.

Zhu Zixi, Zhao Guoqiang, Deng Tianhong (2000) Characteristics of Microclimate in Mulched Wheat Field. Quarterly Journal of Applied Meteorology, Vol. 11. (Suppl.), pp 112–118.

CHAPTER 15

Degradation of Vegetation and Agricultural Productivity due to Natural Disasters and Land Use Strategies to Mitigate Their Impacts on Agriculture, Rangelands and Forestry

Bernard Gomez

Abstract Natural disasters are known to cause severe economic losses and in many developing countries in the form of decline in agricultural production, destruction of food reserves, damage to water supplies, etc. This report focuses on land use strategies to mitigate natural disasters including tropical storms, floods, droughts, dust storms and sand storms, frosts, volcanoes, landslides, bush, and forest fires. It is important that disaster preparedness and prevention projects are built into a wider development strategy. Mitigation efforts need an integrated and concerted action by all the agencies concerned. Long term disaster reduction efforts should aim at promoting appropriate land-use in the disaster-prone areas, by harmonizing land suitability with agricultural development strategies.

15.1 Introduction

Natural disasters affect human society and cause suffering, damage to infrastructure worth billions of dollars each year, loss of life and deterioration of the environment, substantially exceeding normal expectations. In recent decades, people throughout the world have become increasingly concerned by extreme meteorological and hydrological events, which are becoming more frequent and more destructive. Natural disasters that affect agriculture are mainly those that are related to weather and climate such as drought, floods, frost, excessive rainfall, tropical cyclones, storm surges, high winds, hailstorms, forest fires etc.

In the developing world, some of the impacts that can accompany natural disasters include the decline in agricultural production and destruction of food reserves and damage to or loss of water supplies through drought or through pollution of traditional water sources during floods. The extension of cultivation into less suitable climates also increases the risk of damage.

A major attempt has been made to focus this work on land use strategies to mitigate natural disasters. With this case studies an attempt to analyse the strengths and weakness of the strategies, would have been most useful, as we firmly believe that the ultimate objective of this work is to define effective and efficient ways of mitigating the negative effects of natural disasters. We also recognise that the types of extreme events have not been fully exhausted.

15.2 Definitions of Terminologies

Unless stated otherwise in this study, the following terms are construed to have the following meanings:

- degradation – reduce to a lower organic type or a simpler structure
- vegetation – plants collectively
- agricultural productivity – efficiency in producing agricultural goods
- natural disasters – a serious disruption in the functioning of a society, causing widespread human, material, or environmental losses which exceed the ability of the affected society to cope using their own resources.
- land use strategies – plan(s) or policy(ies) to achieve the purpose an area of the Earth is put to (e.g., agriculture, forestry, urban dwellings, or transportation corridors) or its character (e.g., swamp, grassland, or desert)
- mitigate – the process of taking action in advance to minimize the effects of future hazards.
- impacts – the detrimental and beneficial consequences of a phenomenon on natural and human systems.
- agriculture – the process of cultivating land and rearing livestock.
- rangelands – Unimproved grasslands, shrublands, savannas, and tundra.
- forestry – the science or practice of planting and caring for forests.

15.3 Natural Disasters

To combat the situation arising from natural disasters and take effective measures and draw up appropriate plans to minimise the loss particularly in agricultural sector, it is essential to know about them. Some of these major natural disasters in the context of agricultural activities are outlined as follows:

15.3.1 Tropical Storms (Cyclones, Hurricanes, Typhoons)

Tropical cyclones are amongst the most destructive natural disasters to affect many regions of the globe, causing tremendous loss of life, property, agriculture etc. the impact of tropical cyclones is greatest over coastal areas, which bear the brunt of the strong surface winds and flooding from rainfall at the time of landfall. Apart from the strong winds and heavy rainfall, cyclones are also associated with devastating storm surges, which inundate vast areas of coastlines.

15.3.1.1 Process of Damage

The principal causes of destruction by tropical storms are:

1. Fierce winds;
2. Torrential rain and associated flooding;
3. High storm tides.

Most casualties are caused by coastal inundation from the resulting storm tides. The collapse of buildings, the falling of trees, flying debris, electrocution and disease from contaminated food and water in the post-cyclone period contribute substantially to loss of life and destruction of property (Das 2003).

The effects of strong winds in coastal areas are seen in stunted and often very sculpted trees providing unmistakable evidence of the direction of the strong winds. In addition to the battering effects of winds, there is the additional damage caused by airborne sea salt which occurs within a few hundred meters of the coast. Winds which blow from the coastal seas spray a lot of salt on coastal areas, making it impossible to grow crops sensitive to excessive salt.

Moreover, a rise in the ecstatic sea level results in the territorial extension of coastal salinity under the direct or indirect influence of sea water. Fields inundated by the storm surge suffer a loss of fertility due to salt deposition, even after the sea water has receded. The affected land takes a few years to regain its original fertility. The period of high water can last from 6 hours to several days, if the drainage is poor, and may leave the soil saline and unfit for crops. Saline soils are predominantly observed in the coastal areas of India. A rise in the sea level would adversely affect the 7,000 km coastal belt of India, comprising 20 million ha of coastal ecosystem, increasing coastal salinity and reducing crop productivity.

15.3.1.2 Impacts on Vegetation and Livestock

Impacts of tropical storms to agriculture, rangelands and forests can be characterised as follows:

1. Physical destruction of vegetation and livestock;
2. Damage to irrigation infrastructure such as canals, wells and tanks; and
3. Long-term loss of soil fertility from saline deposits over land flooded by sea water.

15.3.1.3 Mitigation Measures

In any effective tropical storm mitigation measure, public awareness plays the most important role. People in affected areas must understand the physical nature of the threat and its disastrous potential. Aside from early warning actions aimed at saving lives and property from impending danger, the following land-use strategies could be very useful:

- Hazard-prone areas should be subject to land reform. Harvests should be phased such that they coincide with the safe period or periods of low risk.
- Protective forests (mangrove) should be encouraged, both to reduce the frequency of inundation and to initiate reclamation.
- Risk zone mapping and analysis of land use pattern should be undertaken to guide growth and development away from hazard-prone areas.
- Land-use legislation and building regulations should be established and strictly enforced in hazard-prone areas.
- Raised platforms for livestock, emergency food grain storage facilities, drinking water storage and wells with covers to avoid pollution and silting during inundation should be built in storm-risk areas and properly maintained.

15.3.2 Floods

Floods are produced by tropical cyclone, severe thunderstorm or prolonged heavy (including monsoonal) rainfall. These floods occur mainly in areas where the soil is not well drained or are clayey and hence non-drainable. Heavy rains and floods result in certain constraints in agricultural activities but their impacts may vary greatly from one area to the other depending on the rate of drainage seepage and drying. Examples of negative effect, either direct or indirect of heavy rain and flood on agriculture are the damage to fragile plant organs like flowers and buds, soil erosion, water logging and conditions favourable to crop and livestock pest development as well as on pollution and pollinators.

In monsoon area such as southern Asia and coastal areas of West Africa, flashfloods are common, often caused by heavy rains associated with the monsoons. These heavy rains, apart from causing damage to economic structures such as roads, dams and bridges, affect agricultural products in the areas.

Flooding caused by flash floods is not always linked to bodies of water. On sloping terrain the water masses can turn the smallest of channels into raging watercourses; on even terrain the water does not run off fast enough and accumulates on the surface. Flash floods are a great danger to life and limb. It also causes landslides with resultant damage to life, property, loss of soil etc.

15.3.2.1 Impacts on Vegetation and Livestock

Flooding often has significant, deleterious effects on agricultural production, rangelands and forestry. The impacts can be wide ranging, both temporally and spatially. Its economical and political impacts are also far reaching. The following effects can be observed:

Erosion: The movement of water at considerable speed, in some parts of a flooded area, particularly during flash floods causes erosion of the topsoil, resulting in the loss of the soil nutrients.

Sediment transport: Flowing water carries sediments along with it. The amount of sediment that can be transported depends on the flow velocity and the size of the sediment particles. Sediments enhance the fertility of the soil and beneficial for agriculture.

Deposits and debris: Sediments are often deposited outside the course of a river. Debris and refuse (trees, structural components, cars, ice floes, etc) may become wedged at bridge passages and cause backwater which can lead to very high water levels even when there is only a small increase in discharge. Backwater can be used for aquaculture.

Debris flows: This is usually associated with torrential rain. They are a mix of water and solids that advance at high speeds of 15 m/s and over and carry almost everything along which get in their way.

Contamination: Flowing water transports impurities, which can lead to very unpleasant pollution. Leaking oil tanks and flooded sewage plants in particular, generate significant hazard to natural systems. Freshwater fisheries are also affected by contamination of water

Landslides: It is usually triggered by torrential rain. The infiltrating water activates slip planes and at the same time increases the weight of the upper soil layer by filling up the pores in the soil. Road transportation can be interrupted significantly.

Waterlogging: Crops, nurseries, pastures can be completely destroyed due to asphyxiation, if flood waters stay for a significant period of time. Flood water causes interruption to tillage, planting, crop management and harvesting. It also causes permanent damage to perennial crops, trees, livestock, building and machinery. Soil temperature reduction and retardation due to flood is harmful for agricultural production. Stagnant water due to flood could be favourable breeding ground for insects and diseases.

15.3.2.2
Protection Against Flood

Various devices may be employed in order to control an excess flow of water so that a flow may be prevented, or at least, the worst effect reduced. These devices include engineering works, embankments, detention reservoirs, the adaptation of river channels and facilities for flood diversion. Short- and long-term protective measures, which can be classified as conventional, and non-conventional methods of flood control are as follows:

1. Conventional Methods

Depending on the cause of flooding, two conventional approaches to flood control can be distinguished. When the cause is excessive rainfall over the basin, then levees and floodwalls along the river are used to confine the water to the river channel and thus accelerate the floodwater flow to the sea or storage reservoir. When the cause of flooding includes tidal influence, then control gates at the river mouth are used to regulate the tidal effect.

2. Non-conventional Methods

Other possibilities for minimising flood include the following:

- On-farm storage in lowland ricefields: Practised in ASEAN (Association of South East Asian Nations) countries, this involves the storage of excessive waters in ricefields, especially during the vegetative phase of the crop. With properly maintained ricefield bund spillways, the amount of rainfall that can be trapped as on-farm storage is equal to the height of the bund. This practice can also be conducted in upland farms or fairly flat areas, with 2- to 3-m deep dug-out farm reservoirs, can also trap some runoff which would otherwise flow immediately to the river.
- Low earth dams: The construction of a series of low earth dams (up to 3 high) on shallow waterways or gullies leading to the rivers, can be done by farmers with technical assistance from the appropriate national agency.
- Reforestation and terracing of hilly lands constitute long term approaches to minimising flash flooding.
- In urban areas, flooding can be minimised by constructing appropriate drainage channels based on the estimated peak floods. Allowance should be made to cater for clogging of drainage systems by garbage, particularly non-bio-degradable materials.

15.3.3 Drought

Dryness and the more serious drought conditions occur in a regional climate system when the supply of water, that is available for plants, humans and livestock cannot satisfy the demand adequately. Demand may differ significantly from one region to another and thus as a rule adjusts to the long-term supply, which, in turn, is determined in essence by the average annual precipitation and the local evaporation rates. Regions in which the availability of moisture is more or less constant may be severely affected by a short dry period, whereas arid and semi-arid regions, have no major problems enduring relatively long periods without any precipitation at all.

Dryness and drought differ from the majority of natural hazards in that it is not possible to pinpoint inception date, at least not at the time. The last precipitation event observed is not synonymous with the beginning of a drought because the important role played by the amount of antecedent precipitation and the ground moisture. Therefore, the transition from a dry period to a drought is gradual.

In addition to precipitation and evaporation, such factors as temperature, wind, type of soil and its capacity to store water, depth and availability of groundwater, vegetation growth, and a few other factors play a role in the making of a drought. On account of this complexity it is extremely difficult to predict a drought and to react in good time. It is usually not recognisable until it is already well-advanced.

The vulnerability of a region to drought is a function of the social, political, and economic conditions there. A drought event can quickly change a basic vulnerability due to an unstable supply system into a famine if external assistance is not provided on time.

15.3.3.1 Response of Plants to Drought Conditions

Not all parts of a plant react to drought in the same way. According to Molga (1962) the upper leaves of the plant maintain their physiological activity the longest by drawing water from lower leaves. The latter are first to wilt or dry up during a drought.

It is not unusual for plants under such conditions to develop short lateral shoots with tiny leaves from the axils of the dropped old leaves. These new small thickish leaves can be kept alive and turgid even with a much reduced water supply from the roots. If the drought occurs during the seed formation in grain the seed will be small, deprived of adequate amounts of nutrient substances, its size being inversely proportional to the area of the assimilating surface of the plant.

Drying out of the tilled layer of the soil at any time during the vegetative period has an adverse effect upon yields. However, with respect to grains , the greatest yield reduction takes place when drought occurs during or immediately after the stem elongation stage, i.e. at the beginning of the heading stage.

The increased sensitivity of grains to droughts during the period from rooting to heading stage is well known. The development of spikelets coincides most frequently with appearance of the fourth leaf. If the tilled layer dries out during this period the number of spikelets is greatly reduced. Particularly adverse in this respect is the combination of drought and high temperature. These conditions also reduce the number of heads which are formed at this time. This leads to the reduction of yield even though moisture conditions improve during later stages of development (Kulik 1957).

Prolonged dry periods coupled with the absence of effective water in the one-metre soil layer reduce yields to zero. If there is some effective water in the 20–100 cm soil layer, dry periods reduce the yields but do not ruin them completely. Any additional water accumulated during the fallow period prior to seeding becomes highly significant under these conditions.

15.3.3.2 Impacts on Vegetation and Livestock

Drought may result in the following phenomena:

Subsidence: Certain types of soil (clays in particular) shrink tremendously when they dry out. Subsidence occurs especially when these types of soil are subjected to stress (e.g. by a building).

Desertification: In prolonged or repeated periods of drought, the ground and the flora are so badly damaged that their resilience, is impaired or even destroyed permanently. Die-back of forests occurs due to drop in water tables. Mangrove forests are threatened by increased level of salinity.

Famine: When a drought is prolonged, the entire food supply system in a region for both human and livestock may break down. International aid is usually required to restore it.

Forest, bush, brush, and grass fires: Vegetation that is dry or parched is extremely combustible. A fire may be caused by a stroke of lightning, a spark, or even an object that just needs to act as a burning glass (e.g. a piece of broken glass).

Conflicts: Shortage of fodder and drinking water for cattle may lead to hostile confrontations between farmers and pastoralists, or between neighbouring states sharing common water resources.

15.3.3.3 Mitigation Measures

Drought is a recurring phenomenon and its occurrence cannot be avoided. Furthermore, the consequences of future drought events may be considerably aggravated, because the progressive evolution of the ecosystems tends to worsen the imbalance between population and resources. Generally, there are always some areas which are not affected by drought while some other areas may be reeling under drought. Therefore, there is a need to develop infrastructure for mitigation of drought (Das 1995). The impacts of drought can be minimised through the application of science and technology in developing suitable drought management plans as mentioned below:

1. Current or Short-term plans

The availability of accurate agricultural, meteorological, hydrological and/or other environmental monitoring data/information on the occurrence of drought in any area cannot be over emphasized. Since the onset and cessation of a drought is generally difficult to determine, continuous monitoring of the environmental variables is the only way to track the occurrence of the phenomenon.

A recently introduced technique in the CILSS (Permanent Interstate Committee for the Control of Drought in the Sahel – French acronym) member countries of the West African Sahel, is that in addition to monitoring the environmental variables during the crop growing season, Geographical Information System techniques are used to locate "zones-at-risk" of drought (famine).

Although the technique is not operational yet in most of the countries, it goes without saying that given the least developed status of the countries of the Sahel, timely information on the risk of famine/water scarcity in any area is a major achievement, if not the most important tool to withstand the negative impacts of drought. Using the information generated from this planning tool, policy/decision-makers can intervene (either shift national priorities or appeal for external support) on time to alleviate a catastrophe.

2. Medium-term plan

- Water control: During drought periods, the water factor becomes of primary importance and often constitutes an element of survival. Availability of sufficient potable water supply for both humans and livestock therefore becomes urgent. This can be achieved through the use of ground water (deepening of old wells, digging of new wells, drilling of boreholes, etc.) resources, where advantageous and the creation of water reservoirs as appropriate in order to maximise the use of surface water resources.
- Moisture Conservation: In fragmented and/or sand soils, the ability to retain moisture for proper crop growth is not always guaranteed, especially during low rainfall periods. This is all the more pronounced in sloppy terrain and where rainfall intensities are significant, e.g., most parts of the West African Sahel. Upland vegetation does not benefit much from the incoming rainfall, which drains as runoff to the lower lands. Placing obstacles (stones in general) on waterways in farms encourages infiltration of the water into the soil for use by crops.
- Contour Bunding: In areas receiving seasonal rainfall, another device of water conservation is to build small bunds on the small perennial rivulets. During period of rainfall water gets stored in such check-dams which can be used by the village folks during dry periods. It also recharges the adjoining soils. In some areas of rural India, this technique has found wide acceptability.

3. Long-term plan

One of the objectives of long-term planning to enhance resilience to drought is to organize and equip vulnerable countries/regions in a way as to make their economies less vulnerable to drought.

Irrigation: Since water is the major limiting factor in drought environments, the possibility of irrigating crops should be explored. This will certainly need the assessment of various factors such as topography, availability of water, costs and benefits analysis, etc. In food insecure regions, this option should be examined seriously, as drought is a recurrent phenomenon and self-sufficiency in food production is more secure than reaching out for imports or food aid.

Drought planning and water crisis management need to be proactive. The overall policy, legislation and specific mitigation strategies should be in place well before a drought or water crisis affects the regular supply of the country's water resources.

15.3.4 Dust Storms and Sandstorms

Dust storms occur all over the world but have large frequency over arid and semi-arid lands. Their spatial extent varies markedly from several hundred meters to thousands of kilometers horizontally and from several meters to several kilometers vertically.

The largest dust storms occur in sandy deserts. Here they are responsible for putting large volumes of dust into the atmosphere and moving great masses of sand on the earth's surface. As a result of selective blowing, soil with a high sand particle content is created and the organic element (the main source of nutrients for plants) decreases. Thus in these places the dust storms cause loss of soil fertility and alter the chemical structure of the soil. The physical properties of the soil are changed also with changes in soil structure. In dry areas, steppes and deserts, dust storms are observed usually in spring and early autumn in connection with dry soil.

Soil structure, moisture conditions, presence of snow cover and relief are important factors in the formation of dust storms. The drying of the upper soil horizon, absence of crop and vegetation cover, low relative air humidity (below 30 percent) in winter as well as a poor cementation of soil, etc. increases the probability of dust storms. Dust storms mostly occur in spring and summer, when the wind strengthens, and fields are in a ploughed state or contain poorly developed vegetation.

When the soil moisture content in the ploughed layer is 25 mm or less, dust storms may be formed at wind speeds above 15 m/s. When this content is below 10 mm, dust storms occur even at wind speeds of 8–10 m/s. In stable air conditions particles of sand and dust may be lifted up to great heights under the influence of convective mixing. The light soil particles move as dust in great distances; the relatively heavier particles fall out and dislodge other soil particles, which are then taken up by wind. Thus, there is a chain reaction.

The intensity of the blowing of soil is proportional to the speed of the wind raised to a third power. For example, when the wind speed changes from 12 to 15 m/s, the intensity of the blowing of soil is increased approximately twofold (Adamenko 2003).

15.3.4.1 Impact on Agriculture and Rangeland

Dust storms cause great damage to agriculture and vegetation cover, if any. They destroy and damage crops. High winds may remove significant volumes of earth, leading to reduced soil fertility. Dust storms are often associated with droughts. Such a combination causes major losses to a national economy. Crops are ruined and poor developed seedling being blown away; fields, roads, canals, water reservoirs, irrigation systems are covered by sand and dust; all kinds of infrastructure and communication networks are adversely affected.

15.3.4.2
Prevention/Mitigation Measures

In most countries, field afforestation is the main measure to protect the soil from dust storms. Improving soil resistance to erosion can be achieved by careful selection of cultivation methods, applying mineral and organic fertilizers, sowing grasses and spraying various substances, which enhance soil structure. It is important to reduce the areas where dust can gather, especially in tracts in areas characterized by erosion. One major protection strategy is to establish well-developed plant cover before the dust storms period. This can encourage a reduction in the wind speed in the layer next to the ground by forming an effective buffer.

When looking at the conditions in which dust storms develop and data on storm-induced damage it is evident that measures to reduce the wind speed at the soil surface and to increase the hooking of soil particles are both crucial. Such measures include the establishment of tree belts and windbreaks. Leaving stubble in field, non mouldboard ploughing application of chemical substances promoting the hooking of soil particles, soil protective crop rotation using perennial grasses and seed of annual crops are also important (Adamenko 2003).

In planning and implementing protection measures it is necessary to take into account the direction of prevailing winds, relief, microclimatic details and soil properties.

In regions with intensive wind erosion, especially on wind shock slopes or on light soils, strip cultivation may be used. On fallow lands bare fallow strips of 50–100 m can be alternated with strips of grain crops or perennial grasses; spring crops can be alternated with winter crops. The direction of strips should be perpendicular to the damaging winds.

Cultivating without turning over furrow and retaining a cover of stubble from previous crops are two important measures to reduce the oscillations of soil temperature and protect soil from wind erosion during the winter–spring period.

Barriers, such as tree belts and buildings, cause a slackening of the wind and the heavier particles to fall out, forming land-drifts. The lightest particles of soil can remain suspended in the atmosphere for a long time. And so, during a dust storm visibility and light penetration worsen.

15.3.5
Frost

Frost is a condition in which the temperature of the air close to the ground surface is below 0 °C. An important role is also played by regional circulation systems in the oceans (e.g. Gulf Stream) and in the atmosphere (e.g. monsoons). A second decisive factor is the elevation, as the mean temperature decreases by 0.6 °C on average per 100 m of elevation, although this rate is influenced by the humidity. The frost risk is not synonymous with the level and duration of the minus temperatures recorded.

During the winter, there is a constant decrease in the temperature of both the ground and ocean surfaces in higher latitudes and of the air above these areas. Since cold air streams into areas which would not cool down to below freezing on account of the reduced radiation alone, the cooling effect can also be "transported" into these areas.

On clear nights there is increased radiation from the warmer surface of the earth to the colder atmosphere, as a result of which the earth's surface cools (as well as objects, vegetation, etc.). For this reason ground frost may occur even at air temperatures above freezing.

15.3.5.1 Impact of Frosts on Vegetation

Losses due to frost are primarily to be found in the agricultural sector. At temperatures below 0 °C the water contained in plant cells freezes. On account of the expansion of the ice, the cells burst and are destroyed. Although some plants have adjusted to low temperatures, others, are extremely susceptible to frost and cannot even survive short periods at temperatures slightly below freezing without being damaged; sometimes they are injured even at temperatures above freezing. Damage may also occur if the blossom period is too cold for insects to fly and pollinate the flowers. Although this does not lead to direct plant damage, yields may be reduced or the crop may even fail altogether. Freezing rain is a special case in which the temperature is not the critical parameter but the extreme weight of the ice under which the branches break and entire trees collapse.

Damage to crops by frost and freezing temperatures causes serious loss to farmers in many parts of the world. In the middle-latitude, the incidence of killing forest determines the length of the growing season, which is practically limited to the time-interval between the last killing frost of spring and the first killing of autumn. The shorter this period is, the earlier maturing will be the varieties that can be grown and lower their potential yielding ability. In warmer regions occasional and exceptional frosts may do considerable damage because the crops grown in such regions are usually very susceptible to low temperature. Frosts are, on the other hand, common in the temperate region and in the subtropical areas which suffer occasional incursions of cold air masses.

15.3.5.2 Mitigation Measures

Water pipes and other installations exposed to frost must be adequately protected against the cold. Pipes in the open freeze more quickly if the water in them does not move. Insulation is a good alternative. This can be done by covering the pipes (with earth) or wrapping them with a thermal insulating material. Water pipes in the open must be laid at a sufficient depth.

Standing water and consequently ice too are almost as poor thermal conductors as air and are thus good insulating materials. Wine-growers often exploit this property by watering vines when there is a risk of frost.

Precautions can be taken against radiation frost by covering the objects that require protecting. In most cases, canvas sheet stretched out like a tent will do the job. Another possibility is to churn up the often shallow layer of cold air with the warmer air above it using fans.

15.3.6 Volcanic Eruption

Major volcanic eruptions may be even more destructive than earthquakes, but their occurrence probability is much lower than that of strong earthquakes. The ash that is thrown out in large eruptions may reach into the earth's upper atmosphere, where it may stay for a relatively long time (months to even years) and affect the global climate by blocking out the sun's rays. The mean global temperature in the year after the eruption of Pinatubo in the Philippines in 1991 was reduced by approx. 0.6 °C.

15.3.6.1 Impact of Volcanic Eruption on Vegetation and Livestock

The hazard derives from various features of volcanic eruption, in particular, ash fall, tsunamis, lava flows and lahars, gases, glowing clouds and volcanic earthquakes.

Volcanic gases (sulphur oxides and nitrous oxide) threaten the entire living world in the surroundings. Plant life could be poisoned by volcanic gases, resulting in the loss of rangeland, and death of livestock. Property is also threatened by the corrosive effect of volcanic gases. Volcanoes as natural springs of new rock have many advantages. It contains numerous minerals formed at high temperatures, which quickly break down when attacked by water, providing essential nutrients for plants. In the aftermath of a major eruption of volcanic ash, farmers (at a distance) often enjoy greatly improved harvests.

15.3.6.2 Mitigation Measures

For many volcanoes throughout the world there are hazard maps that represent a good basis for land use regulations, engineering measures, and emergency response planning with a view to minimization or prevention of loss. Engineering measures are viable in the case of lahars and even at short notice in the case of lava flows due to their slow speeds. Barriers may be erected either to check or to channel the progress of the flows, as is the practice in connection with the Mount Etna eruptions.

15.3.7 Landslides

Landslides involve the movement of the topmost portion of the earth caused by internal earth forces (relief and exogenic geological processes) and external actions (climatic and human activities). They constitute a major problem during the rainy season along roads with steep embankments, on mountain slopes and on hillsides. The manifestation of these processes is often catastrophic, causing destruction and havoc over large areas, transforming lush valleys and agricultural lands into lifeless wastes, and maiming and killing human beings.

Exogenic geological processes can occur independently of human activity. In denuded mountain slopes, some soils crack in the dry season due to excessive drying. These cracks can be as wide as 10 cm or more, and run to a few metres in depth. With the arrival of rains, landslides of these cracked soils, often occur.

At the zone of occurrence, landslides displace objects sitting on the earth to a lower hypsometric level, therefore making them loose their grip and thus tumble and fall.

15.3.7.1 Impacts on Vegetation and Frost

Losses due to the negative impacts of landslides include both direct and indirect losses. Direct costs involve the replacement, repair, or maintenance due to damage to installations or property (industrial plant, houses, buildings, and equipment, including tractors, automobiles, agricultural machinery, etc), whilst indirect costs related to this work include:

1. Loss of productivity of agricultural or forest lands;
2. Temporary closure of industrial plants due to interruption of transport systems;
3. Loss of agricultural productivity due to the destruction of irrigation systems, etc.

15.3.7.2 Control of Landslides

There exists a general consensus that the incidence of landslides has gone up due to hydrological problems arising from widespread denudation of hill slopes. Majority of slope stability problems in the Western Himalayan areas have their origin in a cumulative erosion of hill slopes. Effective system of surface drainage is needed to intercept the surface runoff, to minimise percolation of water into the slope material and to establish a stable moisture regime. Drainage enables a reduction in the weight of the mass tending to slide, as well as, leads to increase in the strength of the slope forming material. Plantation of grasses and shrubs to restore the vegetation cover has been found to be successful in arresting this type of mass movement. Since a high proportion of landslides are of erosional type, the need for providing a vegetation cover on denuded slopes is an urgent task.

15.3.8
Bush/Forest Fires

The extensive forest fires of the late nineties, not only in Latin America and the Caribbean, but also in the rest of the world, have moved this issue off the back burner to the forefront of the world's attention. Fires are changing in their frequency, intensity and pattern. For example, if fire incidence stays at current levels or increases in frequency, then many rain forests will be replaced with vegetation that is less diverse and more combustible, contributing further to the trend. People are now concerned not only with the disappearance of the forest, but also the extensive and widespread impacts of fires on human health and the economy.

Weather plays a significant role in the fire environment (Pyne et al. 1996). Federal fire mangers in the United States recommend that fire danger rating indices, based in part on weather data, be considered part of a comprehensive set of criteria used to decide whether to control naturally occurring forest fires or to let them burn as part of a management strategy for reducing fuel load and/or ecological purposes (Andrews and Bradshaw 1995).

15.3.8.1
Favorable Conditions for Fire

Prevailing weather conditions, fuel composition and structure of forest cover and topography are highly influential to fire behavior. In dry and hot weather conditions, especially over an extended period, there will be more fuel available across the landscape. Fires spread when adjacent fuel catches alight before fuel that are already burning go out; this can continue until all fuel is used up. Some favourable conditions have been mentioned below:

- Both heat and dryness serve to reduce the required amount of heat to cause ignition.
- Wind aerates a fire and increases oxygen availability, enhancing the rate and intensity of the combustion and fire spread.
- Chemical composition affects the temperature of ignition and also the energy release from combustion.
- Fuel structure determines the availability of oxygen and effectiveness of heat transfer.
- Topography affects the relative geometry of a fire's flames, thereby affecting the efficiency of heat transfer from the fire to adjacent fuels, and hence, the likelihood and rate of fire spread.

15.3.8.2
Impacts of Forest Fire

Forest fires have seriously affected human health, the economy and the environment. The extent of the damage to human health from smoke inhalation depends on the constituents of the smoke, its concentration, and exposure time. Smoke from fires causes thousands of respiratory, cardiovascular and eye problems (e.g. conjunctivitis). In addition to numerous cases of constrictive or obstructive lung disorders, the number of cases of asthma, pneumonia bronchitis, acute laryngitis, and bronchiectasis rise dramatically. It also can kill, as in 1998, when 70 Mexican firefighters were killed, and there were 700 smoke-related deaths in the Brazilian Amazon.

The environmental impacts from forest/bush fires ranges from local to global. Local impacts include soil degradation, increased risks of wet-season flooding and dry-season drought, reduced number of animals and plants, and increased risk of recurrent fires. Global effects of these fires include the release of large amounts of various greenhouse gases, reduced rainfall and increased dry lightning as well as contributing to the reduction of world biodiversity through the extinction of populations or species.

15.3.8.3 Prevention/Mitigation Measures

These can be considered at several levels: fire prevention, fire management, fire prediction, fire detection and monitoring and fire fighting.

1. The importance of fire prevention in any fire management strategy cannot be over-emphasized. Prevention is less expensive than suppression and obviously has the benefit of reducing fire damage.
   Fire prevention methods are quite straightforward and well-known. They include:
   - The creation of firebreaks, by clearing the ground of all fuels, around any area that is to be burned
   - The use of back or perimeter fires so that firebreaks are created by burning towards the centre instead of toward an edge
   - The cutting of dead trees which could span firebreaks as they fall and burn
   - Consulting with neighbours before burning lands

   Mopping-up operations to prevent fires in tropical/subtropical forests are particularly necessary and time-consuming. Falling foliage from trees destroyed in a fire can begin to blanket the ground with a new fuel layer within a few days. This process has resulted in as many fires in a single area within a given year. To prevent this, a detailed search of the entire affected area is necessary to extinguish anything still smoldering. If a fire is rapidly contained, the mop-up operation will be minimal, but if it is allowed to burn over substantial areas of forest, time and manpower needed to deal with this increase substantially. Monitoring the progression and spread of fires or all these fire prevention efforts can be expensive for local owners.
   Zoning of land use across tropical landscapes can be effective. For example, the disassociation of fire-maintained agricultural area from selectively-logged forest would reduce the probability of forest fires and the severity

of fire damage from accidental fires. Easily-flammable fuel sources should not be exposed to frequent ignition events (e.g. pasture maintenance or new slash fires). Land use intensification has been cited as a way to increase land values and possibly decrease deforestation. This would also help the fire problem because greater productivity and value of property would make fire prevention and protection efforts more cost-effective.

2. Fire management at a national level involves the establishment of the necessary infrastructure, equipment and trained personnel to be able to detect, monitor and respond to forest/bush fires. Special emphasis needs to be placed on international cooperation, since very few nations have the human, material, or financial resources to cope alone with severe fire situations. Interaction between nations and personnel exchange/training programmes should be encouraged and fostered by international agencies.
   Many different strategies can be adopted to tackle the fire problem, from better education and fire management to economic incentives and land use planning. But these will not be successful without the support and involvement of local people.

3. Fire prediction or early warning systems need to integrate information about weather, vegetation dryness, fire detection and fire spread to provide a simple measure of the fire situation. It is therefore important to know the status and distribution of land cover. Reasonably accurate and current maps of the changing landscape are therefore needed. Local populations must be involved in supportive efforts against fire.

4. Patrols, towers and aircraft are integral parts of the fire detection and monitoring process but satellite detection is a necessity. Trade-offs exist for using various satellite platforms for fire detection and monitoring and mapping. New satellite sensors will enhance capability but aerial detection and suppression of fire in tropical forests, although critical, can still be problematic. The forest canopy disperses smoke and obscures vision, making it difficult to locate fire lines. It intercepts much of the water and fire-suppression agents that are dumped on fires, making them much less effective.

5. Fire fighting should make use of aerial and ground suppression forces. They should be well trained, equipped and coordinated through a well-defined command structure.

15.4 Conclusion

The recent spurt in the periodicity and intensity of disasters calls for a systematic mitigation strategy, which is comprehensive as well as contextual. More so, because the ever changing countenance of disaster cannot be totally attributed to nature's wrath. Vulnerability to natural disasters is also increasing because of unmindful human actions and insensitivity. We are completely out of tune with our surroundings. The heedless pace of development has left us ecologically barren. Developing countries face the most debilitating consequences in the form of economic and social disruption caused by disasters. Only when disaster preparedness and prevention projects are built into a wider development strategy or when mitigation efforts go beyond the individual action of a single department, can there be any lasting impact on the lives of the affected. Long-term disaster reduction efforts should aim at promoting appropriate land-use in the disaster-prone areas, by harmonizing land suitability with agricultural development strategies. The measures to promote proper land-use should include both legislative and economic inducements and creation of public awareness of proper land practices. Formulation of land-use policies for long-term sustainable development is, therefore, imperative.

References

Adamenko TI (2003) Hail, Highwinds and Cold injury. In: Agrometeorology related to Extreme Events. WMO No. 943, Geneva, Switzerland: World Meteorological Organization.

Andrews PL, Bradshaw LS (1995) Fire danger rating and the go/no-go decision for prescribed natural fire. Symposium on Fire in Wilderness and Park Management: Past lessons and Future Opportunities, 1993 March 30-April 11, Missoula MT, General Technical Report, INT-GIR-320. Ogden, UT: US Department of Agriculture, Forest Service, Intermountain Research Station, pp 149–151.

Das HP (1995) Incidence, Impact, monitoring and mitigation of large scale droughts in India. A review paper submitted to the Expert Meeting on Drought Monitoring in WMO, Geneva.

Das HP (2003) Incidence, predication, monitoring and mitigation measures of tropical cyclones and storm surges. In: Agrometeorology related to Extreme Events, WMO No. 943, Geneva, Switzerland: World Meteorological Organization.

Kulik MS (1957) Estimation of the Effects of Droughts. Collected Volume of instructions for the analysis and evaluation of the past and expected agrometeorological conditions. Hydromet Publ House, Moscow.

Molga M (1962) Agricultural Met Centralny Instytut, Informacji Nankowo-Techniczneji Ekonomicznej, Warszawa.

Pyne SJ, Andrews PL, Laven RD (1996) Introduction to Wildland Fire, John Wiley and Sons, p 769.

CHAPTER 16

Agricultural Drought Mitigation and Management of Sustained Agricultural Development in India

Haripada P. Das

Abstract Drought is a many faceted natural disaster that leads to serious socio-economic impacts particularly affecting agricultural production and water supplies. There are two distinct phases in which the application of the knowledge of weather and climate can reduce the impact of drought on the communities. The first is the long term planning in which strategies can be devised, and precautions taken to reduce impact. The second phase is the action taken during the onset of the event to reduce adverse effect. Efforts were made to stabilize dryland agriculture by evolving contingent crop production strategies in rainfed areas of India. Drought management policies included agricultural planning and practices with consideration of overall water requirement within the individual agroclimatic zones.

The ill effects of drought, to a considerable extent, can be alleviated by adopting proper crop management strategies. These strategies may vary from moisture conservation to manipulation of plant population, and even mid-season corrections. Rainfall also can be harvested in either farm ponds or in village tanks and can be recycled. In case of drought mitigation, it was recommended that economies diversify to include agro-industry or various tertiary products, which could create new forms of income. The focus in mitigation should be on measures like improvement in agriculture, management of rangeland, development of water resource and animal husbandry. There is an urgent need to develop appropriate policies and strengthen institutional mechanisms for drought preparedness and mitigation accompanied by concrete programs.

16.1 Introduction

Among the various hazards of nature, drought is one of the most disastrous because it brings innumerable miseries to mankind. Few areas of the world can boast of not having faced this calamity one time or the other in their history. A few thousand years ago people were aware of how adverse weather conditions affected food production and water supplies. The idea of a genesis strategy was suggested in biblical times, i.e., to store surplus grain produced in the good rainfall years for use in the years of poor rainfall. This idea reminded societies that droughts do recur and that the people must learn from history, or they will repeat their mistakes.

Drought and famines have occurred in India for centuries and have even been mentioned in folklore. No precise data of these events are available,

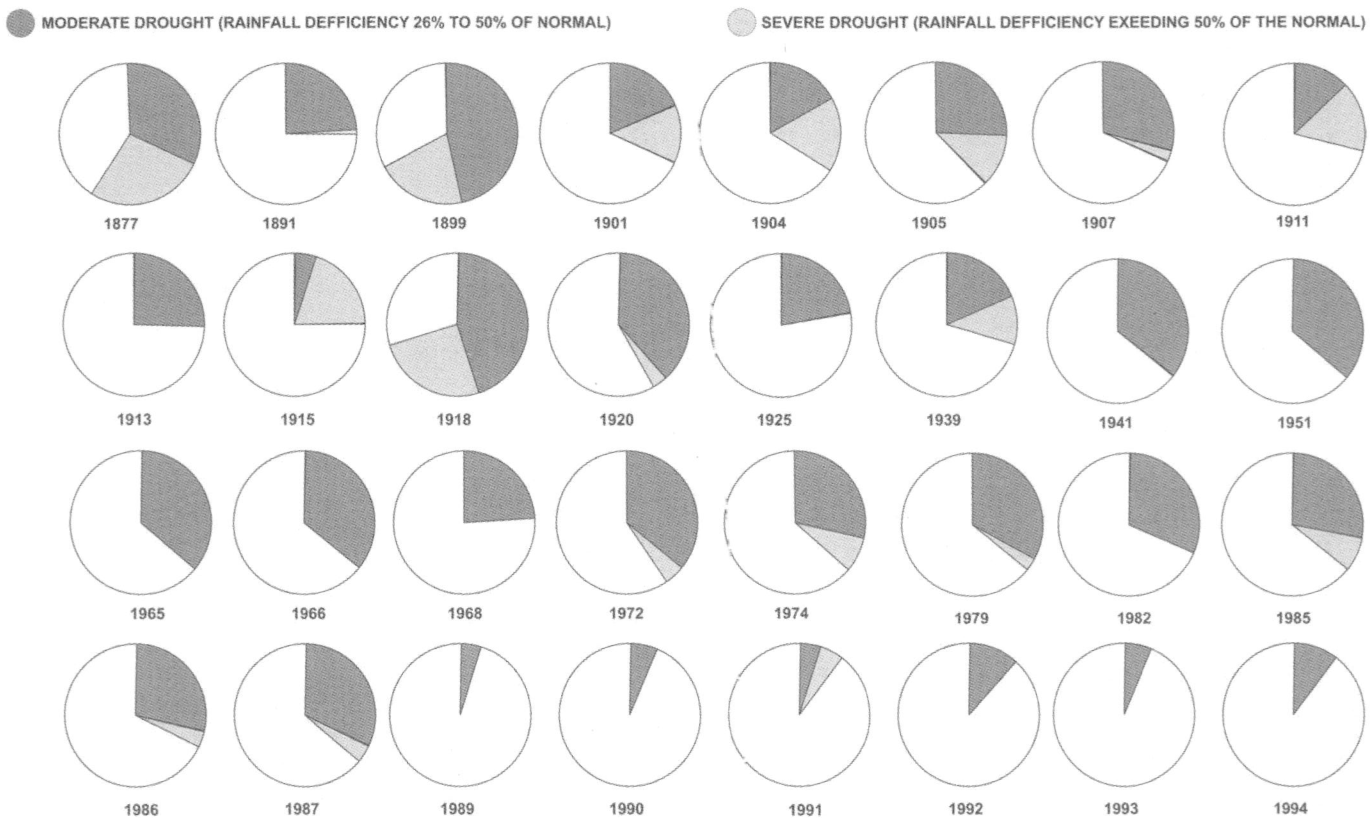

Fig. 16.1. Drought years in India, with percentage of the area affected since 1875

however. Since the establishment of the India Meteorological Department in 1875 and systematic data generation, it has been possible to demarcate areas affected by droughts in each year. Some typical cases of significant droughts in India are depicted in Fig. 16.1. While drought by itself may not appear to be a major cause of societal dislocation, it can combine with underlying societal problems to initiate new changes that are already under-way. Often, the impacts of droughts linger long after a drought has ended, thereby dissociating the drought itself from many of its impacts (Das 1995).

Poor people living in drought-prone environments are consummate risk managers. They often have successful track records in managing drought, even though the effectiveness of indigenous coping strategies to protect food and livelihood security has been undermined by successive droughts and impoverishment. Although crops and cropping patterns are adjusted to given climate and physiographic conditions, intermittent drought spells or recurring drought conditions in a given area can be managed with the help of different strategies. A range of policy tools has been developed to mitigate the effects of drought, but their effective implementation has proved elusive. National organisations to manage and plan for drought relief have proliferated in sub-Saharan Africa (SSA) often in alliance with donor agencies, and have met with varying degrees of success. Many have drafted drought contingency plans, but these tend to remain at the level of theory. Wilhite (1993) outlined in considerable detail a generic process with meteorological steps that may be adopted by governments to develop comprehensive drought planning and management.

The basic objective of drought risk management is to minimise possible adverse outcomes within the constraints of the costs involved. Given the certainty that drought will occur but the uncertainty about the timing of such events, the management task encompasses anticipation of and preparation for the drought, and management of the various phases of the drought event, including drought onset, occurrence, and recovery.

The effective management of drought risk is therefore a matter of concern not only to those countries subject to significant climate uncertainty, but to the international community as a whole.

16.2 Drought Management

Drought is indeed a many faceted natural disaster that leads to serious socio-economic impacts, which have long-term implications. Associated human misery underscores our vulnerability to this natural hazard and makes it imperative to improve our knowledge of the nature and causes of drought and to innovate strategies for reducing its adverse impacts.

It is important to note that drought does not descend all of a sudden. It results from a set of weather sequences that require an extended period to develop. Thus it takes a good period of time for a drought situation to begin, expand and decay. This allows us some times to adopt contingency plans to reduce the adverse effects of drought. There are two distinct phases in which the application of the knowledge of weather and climate can reduce the impact of drought on the communities. The first is the long term planning in which strategies can be devised, and precautions taken to reduce impact. The second phase is the action that is taken during the onset of the event to reduce adverse effect. It is in this context that development of appropriate drought management strategy is of great importance, for individual countries as well as through internationally initiatives. Drought management is currently addressed by the following mechanisms and sectors:

Governmental

- Policy issues at national, regional and district level
- Rural development infrastructure
- Input supply, marketing and farm advisory services

Non-governmental

- NGOs
- Rural institutions, local self-governments
- Private sector
- Philanthropic organizations
- Community codes (tribes, herders)
- International aid agencies
- Alternate land use systems

Research and development institutions

- Best practices for rainwater and soil management through linking on-station and on-farm research
- Weather forecasts
- Contingency crop planning/mid season corrections

16.2.1 Risk Management

Risk is the probability of loss due to drought event. The probability is the likelihood of occurrence of an event assessment based on the historic data. For the purpose of decision making, it is adjusted to take into account currently available information and referred mainly as subjective probability. Risk aversion refers to an individual's attitude towards risk. Risk assessment takes into account the consequences of event and the likelihood of its occurrence. Risk management refers to actions taken to reduce the consequences or probability of unfavourable event. It provides the basis for an effective warning system at any level of responsibility. It identifies potential threats from hazards and establishes the degree of local exposure or vulnerability to hazardous conditions. This knowledge is essential for policy decisions that translate warning information into effective preventive action.

16.2.2 Drought Management Strategy

Research and development initiatives especially in areas like understanding the monsoon behaviour, agrometeorology, arid/dryland farming systems and hydrology have since contributed substantially to the knowledge base on drought management. These advancements have contributed in the development of useful technological outputs and also infused dynamism in agricultural production strategies and development of appropriate farming systems. These efforts thus enable the farmers to tide over difficulties created by the drought situations. However, the technology adoption by small holders who constitute the bulk of stakeholders in drought prone areas has not been to the desired extent. This shortfall is not only due to inherent difficulties in developing appropriate technologies for extremely complex rainfed-drought prone environment, but also due to the poor socio-economic conditions of the farmers and lack of adequate development infrastructure. Of-late, the drought management approach has shifted significantly in the region from crisis response to risk management through early warning systems, advance planning for emergency response and better preparedness. This paradigm shift in research and development strategy has since made some impact, but the sustainability in rainfed areas is yet to be achieved (Singh and Sharma 2002). The quality of life of small holders in rainfed areas therefore is yet to attain the minimum level of the desirable standards of living to reduce the gap with high productivity zones i.e. the irrigated areas. This is a national priority.

In India, having gone through the unnerving experience of suffering widespread droughts due to high variability of rainfall in time and space, a fairly organized drought management system has evolved in last 150 years. The problems caused by drought were faced with fair amount of competence and dedication. However, the approach was essentially that of starting adhoc relief works, as per famine code, to provide employment to the distressed population rather than having the long-term perspective for improving the conditions in the region and to reduce the probability of drought. The approach did not prove effective in mitigating the drought conditions. It was only since 1970s, with the launch of Drought Prone Area Programme (DPAP), that a long-term view was taken for evolving technological and organizational innovations, which were accepted as the strategy for integrated development of drought prone areas. This policy improvement has paid rich dividends while dealing with the infamous droughts of 1987 and 1999–2000. At present, the main objectives of drought management strategy are (Anonymous 1994):

- Distribution of essential commodities such as water, fodder, and food at subsidized rates, sometimes under Food for Work Programmes;
- Optimum utilization of resources in the affected areas with emphasis on primary resources viz. soil, water, vegetation, livestock, manpower, etc. Relief works undertaken for providing employment to drought stricken population are mainly for drought proofing; and,
- Improvement in terms of living conditions of the rural poor who suffer most due to scarcity and drought in particular, and the community in general by creating direct and indirect wage employment and taking up short gestation programmes of development.

In India, major research efforts on improving the productivity of rainfed areas with focus on reducing the adverse effects of drought have been underway for the last 2–3 decades. Packages of practices involving appropriate crops, improved varieties, tillage and seedling practices, soil and rainwater conservation/harvesting, use of fertilizers, weed control, alternate land use systems and plant protection have been developed through intensive and location specific research efforts. However, except for improved crop varieties, other components of the technology-particularly those aimed at conservation of natural resources (soil and water) have not been adopted by the farmers, more so the small holders who dominate the agriculture sector in rainfed areas. The integrated watershed management programmc which forms the central strategy for the national programme of development of drought prone areas in India, initiated in mid-seventies has not made the desired impact in terms of replicability of pilot project despite the huge investment of public funds. Among other reasons, inadequate rural infrastructure and lack of efforts in rain water harvesting and water shade management may be attributed to this set back.

In arid, semi-arid and marginal areas with a probability of droughts incidence of say, at least once in ten years, it is important for those responsible for planning of land use, including agricultural programmes, to seek expert climatological advice regarding rainfall expectations. Drought is the result of the interaction of human pattern of land use and the rainfall regimes. There is thus an urgent need for a detailed examination of rainfall records of these regions. In this regard, the development of methods of predicting many weeks/months in advance, the occurrence of rainfall deserves high priority.

Since the technological impacts quickly reach an optimum level, more emphasis should be placed on the drought management policies, i.e., land and water management policies to minimize the food grain loss. This is especially important in dry-land farming areas. The agricultural planning and practices need to be worked out with consideration of overall water requirement within the individual agroclimatic zones. Crops that need shorter duration to mature and require less water need to be encouraged in the drought prone areas. Irrigation through canal and ground water resources needs to be monitored with optimum utilization avoiding soil salinity and excessive evaporation loss. Food reserves to meet the emergency of maximum up to two consecutive droughts must be planned. A variety of policy decisions on farming, human migration, population dynamics, livestock survival, ecology, etc. must be formulated.

16.2.2.1
Crop Adjustment for Alleviation of Drought

Efforts were made to stabilize dryland agriculture by evolving contingent crop production strategies in rain-fed areas of India. One of the best methods evolved was the adjustment of a suitable crop to the rainfall quantum and distribution in a given area. Using years of experience, crops have been chosen to suit the zones, based on rainfall and physiographic features. However, certain crops often are grown more for convenience, by convention, or for family needs. With more and more commercialization of agriculture, the attitude of the farmers is changing. As for example in Madhya Pradesh state of India, during the last decade, soybean area has increased from a mere thousand ha to more than 1 million ha. This is one of the best examples of crop choice to suit the rainfall pattern. Thus selection of a crop suitable for the rainfall quantum and distribution can contribute to sustainable agriculture.

In attempts to deal more effectively with problems of complex, marginal, diverse, risk prone agriculture and disadvantaged farmers operating in harsh environment, the farming system, as a whole should be regarded as a part of the larger system, such as the rural livelihood situations and administrative domains. The dryland farming system comprises several subsystems such as natural resources, the commodities, external inputs, internal inputs and marketing.

Small scale farming in upland areas of India is generally rainfed and on marginal lands and thus risks are high. Rather than depending on annual crops alone, it would be desirable to adopt the strategy of spreading the risks by using as wide a range of options as possible, including tree crops and livestock. The use of perennial tree crops in upland areas has a long history in India. Major types of trees grown include fruit (as inter cropping), paulowni, and white mulberry. Agroforestry is being practiced on a large scale in the region only for the past 20 years. In drought prone eastern areas, growing period of rice is shortened for late transplanting due to delayed onset of rain. Further, in jute-rice cropping pattern, harvest of jute extends to planting time of rice. The risk of water stress during the grain filling stage of rice is thus increased. As a cropping strategy, farmers use older seedlings in many situations. Changes in cropping patterns, varieties and input uses, reduction in rice area by replacement with other remunerative crops, are the major components of the recommended strategies when drought is of milder intensity and occurs early in the growing season (Pandey et al. 1999). Most of these innovations based on decades of research and largely adopted by farmers are centered to crops and cropping systems. Some of these are also based on the indigenous technical knowledge. Growing trees, and raising livestock, fish and poultry could be the important components of drought proofing strategies in the low land drought prone areas, and hence need concerted research efforts. There have been success stories to amply demonstrate that the more diverse the production system, lesser is the risk of failure of farm level economic security, inherent to rainfed drought prone regions.

16.2.2.2
Drought Resistance Varieties of Crop

Drought resistant plants are those varieties that are able to grow and yield satisfactorily in areas liable to periodic drought (Hounam et al. 1975). May and Milthorpe (1962) identified three types of drought resistance. They are:

- drought 'escapability' – the ability of a plant to complete its life cycle before being subjected to serious water stress,
- drought endurance with high internal water content, enabling a plant to survive drought by virtue of a deep root system or reduced transpiration; and
- drought endurance with low internal water content during the period of drought but the ability to recover and grow when soil water is replenished.

In India, the third type of drought tolerance is needed in the rice variety at the seedling stage because the root has not penetrated to deeper layers, the first type (drought-escaping) is needed in the rice variety, which completes its reproductive stage by the end of August.

In the areas where prolonged dry spells occur, varieties of the second type of drought tolerance are needed. For example, sorghum is more drought resistant than corn because of differences of root concentration. Greater drought resistance is shown by these plants with deeper and more extensively branched root systems. Root extension virtually ceases in most species during flowering and podding stages and hence the plants with faster maturity can survive during a drought spell.

16.2.2.3
Crop Management to Alleviate Drought Conditions

In the case of drought incidence during the growing season, the ill effects of drought, to a considerable extent, can be alleviated by adopting proper crop management strategies. These strategies may vary from moisture conservation to manipulation of plant population, fertilization, and even some midseason corrections. For example, recurring drought conditions can be managed primarily by crop variety adjustments, either by planting early maturing varieties that can escape drought or by planting drought resistant varieties. If drought occurs during the crop growing season, it can be alleviated with either soil water conservation techniques or supplemental irrigation. In high rainfall areas where there are a series of wet and dry spells, rainfall can be harvested in either farm ponds or in village tanks and can be recycled as lifesaving irrigation during a prolonged dry spell. The remaining water can also be used to provide irrigation for second crop with a lower water requirement, like chickpea. There have also been significant improvements in the ability to predict the consequences of drought on people's access to food, through the establishment of famine early warning systems (EWS). Much less progress has been made in using this information to improve preparedness and to trigger timely and appropriate responses, at either national government or donor level. Some of the strategies that can be adopted are discussed below. However, one point to be noted is that no one strategy can be applied to every situation. In fact, every strategy is specific to a given location, crop, time, and socioeconomic conditions.

16.2.2.4
Plant Population Management

If water stress conditions occur at the time of or immediately after germination, it would be best to resow the crop, preferably with some early-maturing variety. On the other hand, if water stress occurs around forty to fifty days after crop emergence, ratooning and thinning need to be considered. According to Venkateswarulu (1992), ratooning of a drought-affected sorghum crop with a subsequent rain resulted in a grain yield of 0.8 t/ha, compared to 0.2 t/ha for the same crop without ratooning. Similarly, pearl millet had a yield of 2.51 t/ha with ratooning, whereas the non-ratooned crop yielded 1.6 t/ha. Thinning the plant population by removing every third row at the time of drought increased the productivity of sorghum from 1.55 t/ha to 2.11 t/ha at Bellary in south India. When the plants started to recover after the return of rainfall, a 2–3 per cent urea spray was also found to be beneficial.

16.2.2.5
Rainfall Distribution Modeling and Contingency Planning

The corrective measures that can be taken in the event of moisture stress depend on the period during which drought occurs. If the rain falls immediately after sowing, resulting in a heavy mortality of seedlings, resowing has to be done. Mid season droughts usually occur when the crop is 40–50 days old with maximum leaf area which results in fast depletion of soil moisture. Therefore, reduction of leaf area either by water rationing or thinning can mitigate the drought effects to some extent. Weed control and mulching are other short-term measures that can mitigate drought by conserving the scarce moisture. Long-term measures for improving the soil moisture storage through appropriate land configurations also contribute to minimise the mid-season drought effect. However, these measures are to be planned well in advance at least in areas vulnerable to recurring droughts, as stated earlier. In short, the main strategy of mid-season correction revolves around extending the moisture availability period in soil. Such contingency plans can be prepared only if rainfall distribution models are prepared for different situations and are made available.

16.2.2.6
Weed Management

During a dry spell or water stress conditions, weed competition is a problem for crops because weeds also use the little moisture that is available. It was found that when a prolonged dry spell occurs immediately after the germination of the crop under rain-fed conditions, weeds infest the entire field. Bushening, one of the traditional practices of weed control and a means of creating semi-puddled conditions (a condition of partially pulverized soils), cannot be done under dry conditions.

Under such conditions, at thirty or thirty-five days after sowing, harvesting the entire field (weeds and rice together) can economically reduce the weed problem (Sastri et al.1990). Even in dryland crops sown in line, weed control through interculture operations was found to be beneficial under water stress conditions.

16.2.2.7
Management of Pastures and Range Lands

Among the different land use units in the rainfed agro-ecosystem, the permanent pastures are the most degraded and neglected. In many instances these are devoid of even the basal vegetation cover. Since the land does not belong to any individual, there is no concern for it. Overgrazing and other forms of destruction of natural vegetation, as well as encroachment, have increased so much that other poor communities have largely replaced the original plant communities. In order to reverse the process, a serious and sustained effort (perhaps even legislation) is warranted. In the arid areas, pelletted seeds of grasses such as *Lasiurus sindicus, Cenchrus ciliaris and Cenchrus setigerus* may be planted. Such simple interventions can increase the carrying capacity of an average permanent pasture from 2.5 sheep ha^{-1} to as much as 4.5–6.9 ha^{-1} sheep in sandy and 9.0–13.8 ha^{-1} sheep in loamy sand soils (Abrol and Venkateswarlu 1995). Multipurpose trees and shrubs such as *Prosopis cineraria, Ziziphus nummularia, Capparis decidua, and Acacia nilotica* are suitable for silvipasture. The areas marked for development need to be protected from grazing for at least two years. The controlled rotational grazing can then be introduced. Community mobilisation is thus the central issue. Farmers may be encouraged to establish pastures on part of their holdings (short and long term/degraded fallow lands) to decrease the pressure on village commons. The size of pasture may depend on the individual requirement and holding size. The approach, however, should be integrated with research and development in complimentary sectors. This may include replacing the large herds of poor quality animals with improved quality livestock of smaller number, which can be sustained in arid and semi-arid rainfed environments, improving grazing land management/stall feeding, improving the fuel energy situation, people's willing participation, and even legislation, if absolutely required and practical. Dynamic models need to be developed that help to identify the appropriate balance between the resilience of vegetation to short-term overgrazing, the variability in the weather, the indebtedness of herdsmen and their attitude to risk. Mention may be made here of an outstanding example of efforts made by community participation to involve the people in the management of their land in 'Banaskantha', a poor tribal district of Gujarat state of India. Banaskantha is one of the 18 districts in the North of Gujarat state of India, which is a drought prone and arid area. The main problem here has been lack of water and unpredictability of the rainfall, with the result that there has been a drought in the district in almost every other year. Livestock forms an essential part of the farming system. Droughts led to migration, fall in agricultural production, food scarcity, and so on.

An initiative of the Self-Employed Women's Association (SEWA), the National Dairy Development Board (NDDB), and the Disaster Mitigation institute (DMI) evolved into a Fodder Security System. This was aimed at increasing the ability of the people and the environment to cope with the adverse impacts of drought. Its immediate objective was to provide fodders to the members of the local dairy cooperatives in times of scarcity so that their animals would continue to be fed, the milk yield stabilized and their income sustained. The fodder was purchased at the lowest prices by the local cooperatives by pooling their resources during the earliest periods. Then, the fodder was stored at convenient locations and distributed at the right time in required quantities, thereby minimizing wastage. The final aim of the Food Security System was gaining access to ownership, production and distribution of fodder. It was felt that too much money was being spent on purchase of fodder and if they could collectively grow the fodder, the amount of fodder that needed to be purchased would be reduced (Bhatt 1997).

Wastelands are common property resources that are individually used but owned by the village community. These are important sources of fuelwood and fodder. The village community including the Panchayat was convinced by SEWA about the acquisition of wasteland for a fodder farm. Fodder was grown collectively and the yields were distributed among the members according to the number of cattle they owned and quantity of milk they produced. The Fodder Security System had a good impact on the community. It prevented death of cattle to a great extent caused by drought conditions, enabled continuous milk supply to cooperatives, decreased migration levels and produced positive ecological consequences. This initiative made the people realize as to how an organized effort could help them cope with the drought (Bhatt 1987).

16.2.2.8
Water Conservation Practices

The water conservation practices require improved irrigation (reduce conveyance loss, trickle irrigation), reduced evaporation from water surface, conservation of water in soil, reduced transpiration and evapotranspiration, increased underground recharge, conjunctive use of surface and sub-surface ware, etc. The storage of scanty rainfall through simple as well as extensive types of surface and ground water has been an important source of water in arid and semi-arid regions. These practices have often saved the drought-affected regions from water famine. One among the reasons of large-scale drought in the country is due to higher consumption of water and discontinued use of traditional water harvesting practices.

Monsoon dependent regions require sustainable surface and sub-surface water conservation practices. Such conservation practices should have social binding/responsibility in maintaining it. Conservation practices that were followed in the past are irrespective of rainfall amount in the region. In arid regions, precipitation is a basic element responsible for the water resources influencing development of the area. There is need of rain water conservation systems like Nadir (pond) harvesting catchment system of harnessing precipitation for human and livestock consumption; Tank – a covered underground tank for collection and storage of surface run-off (rainwater); Toba (pond catchment system of water harvesting) for human and livestock uses and Khadins (a water harvesting method where run-off water is stored) for irrigation. Conservation of both surface and ground water has been an integral part in most of the drought-affected areas of India. Ponds and tanks represent an important community resource for drinking water in rural India (Narain et al. 2000). The main attributes to their success are the sound scientific knowledge and methods on which they were built (Barah 1996). Destruction of the catchment area has resulted in excess water runoff away from the ponds leading to water storage. The boring of the large number of tube wells mainly by rich farmers has resulted in the steady decline in the level of ground water in these areas. The modern lift technology withdraws water exceeding the recharge rate and pushes the water table below the natural depth of water. Large-scale lift technology had left 30 to 40% of wells dry up annually in Anantapur district. In the coastal alluvial plains of Kutch and Saurashtra regions of Gujarat, high pumping of potable ground water along the sea has resulted in intrusion of saline water into freshwater aquifers. The water shortage is also due to the misuse of traditional water systems that could partially cope with the problem to meet the needs of cattle and irrigation. Increase in the population of livestock and humans and their fodder and fuel requirements have led to depletion of natural resources.

16.2.2.9
Lifesaving Irrigation

Drought prone areas suffer from inadequacy of soil moisture resulting into partial or total failure of crops. One life saving irrigation can transform the situation from failure to productivity. The concept of life saving irrigation received greater impetus with the concept of water harvesting. It is already reported that 15–20% rainfall goes as run off from the cultivated areas. The run off can be stored into farm pond and used at the critical stages. Generally for cereals, grain formation stage is critical. However, in dryland, the stage is decided by the soil moisture content. Under normal situations, one irrigation at the critical stage would boost yield by 50–60 per cent (Randhwa and Venkateswarlu 1979). One presoaking irrigation may make difference between crop versus no crop. When taken as a second crop, there may not be enough moisture below seeding zone due to which crop can not be established on the seed zone. Irrigation of 6 cm depth can saturate seed zone to establish the crop stand. Whatever may be the situation; the life saving irrigation is important. Use of available water judiciously for such kind of situation would be the best way to manage the drought. Even under certain situations life saving irrigations can be managed from regular irrigation projects.

In drought-prone areas of India, although there are no irrigation facilities, technological developments have been made to harvest, conserve, and recycle excess rainwater for use during water stress period. Such techniques are in vogue in almost every dryland area of the world. However, in recommending a particular practice of harvesting rainwater for providing lifesaving irrigation, care should be taken to match the practice to the available natural resources. Land, soil types, physiography, rainfall quantum, rainfall distribution and intensity, and cropping patterns must all be considered in this process.

16.2.2.10
Fertilizer Management

The crops in drought prone region suffer two kinds of stresses viz. moisture stress and nutrients stress. Nutrient stress is more pronounced due to low nitrogen content of soils. By providing moderate fertilizer the crop growth can be enhanced and drought can be escaped. It is, therefore, rightly said that ‘beat the drought by fertilizer application’. Numerous large-scale trials have been conducted on farmers’ fields to prove the worth of fertilizer use in dry farming.

Use of fertilizer, particularly nitrogen accelerates the growth and develops profuse root system thus making plants more capable to face the drought. The time and method of application are important. Placing the fertilizer 10 cm below the soil surface in the vicinity of moist soil at seeding is the most effective. Usually 25 to 50 kg N/ha dose is recommended. For kharif sorghum the level can be increased by 25% more due to alternate wetting caused by intermittent rains.

16.2.2.11
Contingency Crop Planning

Under rainfed conditions, the farmers have learned to keep their cropping practices flexible so that corrective measures can be introduced, depending on the type of weather aberrations (Subbiah 2000). Normally, the following kinds of aberrations are observed under rain-fed conditions:

- delayed onset of monsoon rains;
- long break in rainfall during the middle of the rainy season;
- lack of rainfall during the post-rainy season; and
- high soil temperature at sowing time (in case of post-rainy season crops).

In response to late onset of rainfall conditions, farmers change from long-duration high-yielding crop varieties to short-duration low-yielding varieties. The mid-season correction for each aberration varies from place to place, depending on the rainfall pattern, soil type, choice of crop, and so forth. Corrective measures include reducing the plant population by thinning crops to reduce crop competition for available moisture and providing supplemental irrigation in case there is a long break in rainfall during the crop season. However there are very few options to manage terminal droughts due to early withdrawal of monsoon. Harvesting the crop for fodder is the only way to slightly reduce the damage. Nevertheless, most of the drought management strategies need to be adopted from the beginning of the season in order to be effective. In other words, "one has to plan for drought management, before planting the crops even if the forecast is for a good year". Therefore, the farmers sow a mixture of crops and thin out the cropstand as the intensity and duration of rainfall becomes clearer over the passage of time.

16.2.2.12
Drought Watch System

There is a need to have a Drought Watch System at district and state levels, which should be developed, implemented, and managed by experts in meteorology, agriculture, irrigation, public health, food supplies etc. (Das 2000).

The pre-requisites for the operation of such a drought watch system are:

- a network of rainfall stations, with reliable records of good quality, that are homogeneous and extend over a period of at least 20 and preferably more than 50 years;
- weekly/monthly rainfall records that are in computer compatible form;
- weekly/monthly rainfall totals that are available at the drought watch center within two or three days at the end of the week/month; and,
- the drought watch centers should have the capability of issuing weekly/ monthly drought watch statements whenever rainfall situation demands.

16.2.2.13
Climate Forecast for Drought Management

As drought is a phenomena associated with scarcity of water, the period during which the scarcity of water is likely to be experienced, the extent of scarcity of water and the areas/regions that are likely to be affected by drought have to be known in advance. Information about the onset of drought conditions in a timely and reliable manner is useful in many ways to organize corrective steps in mitigating droughts for farmers. The information will facilitate:

- decisions at the time of sowing about the choice of crops/varieties;

- adoption of management practices related to soil moisture conservation, fertilizer application and manipulation of plant population during the crop growing period;
- Improved long range forecasts would diminish uncertainty in long-term planning and facilitate;
- development and sectoral planning – especially for water, an increasingly scarce regional resource;
- drought preparedness and mitigation strategies – governments need justification and confidence to embark on long-term approaches; and
- capital investment projects – through a more certain environment for investors throughout the economy.

Reliable long-term forecasts would also enable improvements in a larger number of strategic planning decisions in the public and private agricultural support services. For example, the availability of reliable seasonal forecasts should vastly extend the scope of response farming, allowing researchers to develop a limited range of different agronomic packages; in any particular year, the most appropriate extension packages could be promoted in accordance with the actual forecast for the area (Williams 2000). In India, demand for water – urban, rural, industrial, agricultural, and ecological – is increasingly outstripping supply. Options to satisfy future requirements increasingly demand international collaboration. Potential benefits from better regional water resource planning and management would justify major investment in improving the reliability and precision of long-lead forecasts.

The meteorological community in India has been contributing its experience to manage the drought in the following manner:

- Provision of improved long-range forecast of the seasonal rainfall before the beginning of the season. This provides time to the planners to adopt different strategies;
- Close monitoring of the rainfall over different parts of the country on daily, weekly and monthly scales within the rainy season;
- Delineation of different agro-climatological zones which helps in specific measures for agricultural planning on climatological basis; and,
- Continued research efforts to enhance our capabilities of forecasting monsoon rain over local, regional, and all India based on different temporal scales.

16.3 Drought Planning in Relation to Climate Change

Drought planners should also take steps to plan for the possibility of climate change or climate variability. They must plan for the change, since they are already considering changes in population and technology that will affect the demand. To address climatic uncertainty, drought planners should assess the sensitivity of water resource systems to climatic change. One could conduct a simple analysis by examining for example how a 2 °C or 4 °C increase in temperature and a 10% increase or decrease in rainfall affects a system's vulnerability to drought. Alternatively, one could use a more sophisticated scenario based on climate models. The sensitivity of the system to these kinds of changes could be compared to the impacts of other factors such as population growth and technological change. Drought planners in sensitive regions can take steps to improve the ability of water resource systems to recover from drought (Das 1999). Among the steps that can be taken are:

Water conservation promotion

- Water conservation can be encouraged by reducing or eliminating wastage. Agriculture is the greatest user of water in the country and has the greatest potential for conservation.

System optimization

- Better use of existing infrastructure would reduce vulnerability to regional droughts. Water resource managers should explore ways to transfer water between neighbouring systems during droughts.

Water quality protection

- Water pollution, control and abatement programs can increase the amount of water available for consumption during droughts. States and river basins should have comprehensive drought contingency plans that will reduce the impacts of drought quickly, such as automatic short-term water rationing.

All of these measures should have benefits even if climate does change or does not change.

16.4 Mitigation of Drought

Drought is a recurring phenomena and its relatively frequent occurrence cannot be avoided. However, its impact can be minimised through application of science and technology in developing suitable drought management plans. Generally there are always some areas that are not affected by drought while some other areas may be reeling under drought. Therefore, there is a need to develop an infrastructure for drought mitigation. Mitigation measures specific to drought can be defined as short and long term actions programmes or policies implemented in advance of drought that reduce the degree of risk to people, property, and productive capacity (Das 1999).

16.4.1 Drought Mitigation Strategies

Droughts are caused not merely by failure of rains but also due to lack of proper dispersal. Some areas get scanty rainfall, and therefore the strategy has to be built around water conservation and a cropping pattern, which should be supported by a judicious cropping pattern. Only those crops should be permitted which can be supported by the rain water. This should be worked out on the basis of the recorded rainfall data and its dispersal. There should be a proper regulation of ground water use. Apart from proper geographical dispersal of shallow and deep tube wells and dug wells through an appropriate legislation, sinking of tube wells should not be permitted beyond a certain depth. This would ensure availability of ground water in times of water scarcity. The legislation should also provide for regulation of flowing water in the rivers, streams and rivulets so that it is not used for irrigation in the years of scanty rainfall and water scarcity. Restoration of village ponds, tanks, streams and rivulets should be undertaken after village-wide survey. The work of survey and restoration should be completed in a period of three years. Fodder is a big problem at the time of droughts. A combination of pasture development and afforestation programme will provide not only the fodder during normal and drought years but also lead to soil and water conservation, and employment generation on a sustainable basis. Funding of these programmes can be done from the funds under the rural development and employment generation programmes under the various centrally sponsored schemes.

Some of the measures that would help in effective response and mitigate the hardship of the people are as under:

- Arrangement for reasonable buffer stock of food grain and fodder;
- Ensure supply of good drinking water in rural areas for human and livestock in drought affected areas;
- Assess fodder requirements in drought affected districts and locate areas where shortage are likely to occur and arrange supplies from outside;
- Fodder cultivation to be encouraged wherever feasible;
- Rejuvenation of traditional rainwater systems viz., Nadis, Tankas, Khadins, etc;
- Rainwater harvesting for both the drinking and cropping purposes;
- Management of human livestock population to reduce pressure on fragile arid ecosystem;
- Timely availability of credit, postponement of revenue collection, and re-payment of short-term agricultural loans;
- Appropriate land-use planning (Inter-cropping system), discouraging the cultivation of water-intensive crops, and encouraging sprinkler and drip irrigation systems;

- Creation of local task force in each district to initiate relief measures immediately after the drought takes place;
- Implementation of crop and livestock insurance schemes;
- Provisions for cattle camp in drought affected areas; and,
- Early warning and drought monitoring should be carried out on the basis of long, medium and short term forecasts.

Typically, rural communities in drought-prone areas have managed their vulnerability by storing harvests and diversifying their livelihoods (e.g., crops, cattle, and off-farm employment). Population growth and the sometimes resulting degradation of natural resources are circumstances that can undermine the ability of communities to cope with drought. This is particularly true in dryland areas. It is often recommended that economies diversify to include agro-industry or various tertiary products, which could move some of the population out of direct farm employment and create new forms of income.

16.4.2 Government Action to Combat Drought

People of India from ancient times, have been concerned about the occurrence of this natural calamity. Governments from the historical past to the present, have tried methods to combat drought situations in different parts of the country so as to mitigate the sufferings of the people and reduce losses to the country's economy. These measures included: the reduction/waiver of the land revenue; distribution of free food from the government stock/granaries; provision of employment to the poor, marginal agriculturists, and landless labour force to provide purchasing power; migration of labour from the part of the country under drought to another part where rainfall is more normal, and food and employment are available; and, digging of water reservoirs in the form of tanks, wells and canals to provide irrigation facilities to avoid or reduce drought impacts in future events, etc.

The government also initiated the policy of maintaining adequate food reserves. Grain storage facilities have been improved and enhanced and a network of fair-price shops have been established in different parts of the country under a Public Distribution System. This has helped to instill confidence in the people and avoid panic purchase by the affluent section of the society. Agricultural research and other inputs like, the use of fertilizers, provision of power for tapping underground resources for irrigation and additional irrigation resources have also contributed to the 'green revolution' in the country. Different types of 'food for work' programmes have been initiated by the central and state governments to augment purchasing power to the poor. Loans granted under the Land Improvements Act have stimulated the construction of wells and other minor irrigation works.

The commonly adopted drought mitigation strategies (Sahni 2003) in India include: (i) Construction of check dams (ii) Water rationing (iii) Cattle management (iv) Proper selection of crop for drought affected areas (v) Levelling, soil conservation techniques (vi) Reducing deforestation and fire wood cutting in the affected areas (vii) Checking of migration and providing alternate employment for people in government sponsored relief schemes (viii) Education and training to the people (ix) Participation in community programmes.

16.4.3 Development of New Agronomic Practices and Modern Technology

Attempts to induce drought resistance by various pre-sowing hardening treatments of seeds, such as pre-sowing exposure to radiation, are being made. Attempts have also been made to evolve drought resistant cultivars and shorter duration crop varieties. Agronomic practices are being developed so that the active growth of the crop, during peak water demand, does not coincide with rainless period.

The focus in mitigation should be on the following measures:

- Improvement in agriculture through modifying cropping patterns and introducing drought-resistant varieties of crops;
- Management of rangelands with improvement of grazing lands, improved grazing patterns, introduction of feed and protection of shrubs and trees;
- Development of water resource system with improved irrigation, development of improved storage facilities, protection of surface water from evaporation, and introduction of drip irrigation system; and,
- Animal husbandry activities helping in mitigation with the use of improved and scientific methods, increasing outputs without destroying the eco-system.

What we need today is change over from traditional farming methods to more modern dynamic, flexible practices, making greater use of modern technology. This is necessary for farmers to respond rapidly to the changing demand for agricultural produce, replacing and adjusting varieties, crops and technologies to cope with new situations. What is needed is that the farmer is actually able to apply the new technologies. This calls for intensive training for all those connected with agricultural production and planning. An efficient agrometeorological training service consisting of professional scientists of high caliber is needed. This entails advanced level of training both in theory and practice.

16.4.4 Education and Training Programmes for Mitigation of Drought

Educational and training programmes should concentrate on several points. First, a greater level of understanding must be established to heighten public awareness of drought and water conservation and the ways in which individual citizen and the public and private sectors can help to mitigate impacts in the short and long term. The educational process might begin with the development of a media awareness programme. This programme would include provisions to improve the media's understanding of the drought problem and the complexity of the management issue involved, as well as a mechanism to ensure the timely and reliable flow of information to all members of the media (e.g., via news conferences). Second, the drought planner should initiate an information programme aimed at educating the general population about drought and water management and what they can do as individuals to conserve water in the short run. Educational programmes must be long-term in design, concentrating on achieving a better understanding of water conservation issues among all age groups and economic sectors. If such programmes are not developed, governmental and public interest in and support for drought planning and water conservation will wane during periods of non-drought conditions.

16.4.5 Farming Systems Approach for Drought Mitigation

As discussed earlier, extension and development programmes have failed to inspire farmers to adopt technologies in drought prone areas. Besides lacking a farmer participatory approach, these failures partly stem from the inherent difficulty of the task. Of late, it has been increasingly recognized how difficult it is to develop a sustainable technology for heterogeneous agro-ecological and socio-economic conditions of small holders in drought prone areas, unlike that for irrigated areas. The problems are complex, characterized by a host of environmental and socio-economic issues. Addressing not only a component of the farming system, e.g. crop variety, fertilizer use, or even crop husbandry per se, does not generally result in a dramatic increase in productivity as witnessed in irrigated areas. Mixed farming, consisting of crop production and animal husbandry for risk aversion, has been the mainstay of subsistence for such farmers. The farming systems perspective, therefore, can only be the proper management strategy for such regions.

Planning on-farm farming system research requires an understanding of what farmers are already doing. Four sets of information are needed to initiate farming system research in real field situations (i) socioeconomic conditions of people, their perceptions, priorities, requirements, and indigenous technical knowledge; (ii) natural resource conditions; (iii) technology (research information) developed at the research stations; and (iv) infrastructure (market and input issues) (Singh 2001). These data are collected and then carefully analyzed to balance resource conditions and farmers requirements for selecting technologies to apply and test as treatments. To begin with, more emphasis should be placed in refining indigenous technical knowledge. Strict prescriptions of component technologies developed at the research stations should be avoided. Required adjustments can be incorporated to offer baskets of options that suit the client's requirements and priorities.

Sustainability of rainfed agriculture is anticipated to be one of the most crucial problems of the next two decades. In this context sustainable agriculture refers to a farming system, which on a continuous basis enhances the productivity and economic returns, protects the environment, conserves the natural resources, and finally leads to an improved quality of life of people. The burgeoning human and livestock population regime in the region demands a continuous increase in productivity to meet the increasing requirements. Economic growth associated with development further adds to it. The fragile resource base in these regions is, therefore, required to be balanced with production activities in such a way that it continuously responds to the increased requirements without compromising its quality.

In irrigated areas the synergy among components of the recommended production technology (viz. high yielding variety, water management, fertilizer use and plant protection) resulted in a quantum jump in productivity. Farmers got impressed and readily accepted the technology, which led to the green revolution. However, in rainfed region, such synergy could at best operate at a very low level. Water, the most important input is uncertain. Moreover, farmers are poor and cannot cope with the risk factor. Inputs such as fertilizer are thus seldom used. In this setting, it is difficult to get farmers interested in improved technology as willing participants. We cannot do so until productivity substantially increases. For this reason farmers in rainfed areas have not accepted technologies (in isolation) despite vigorous extension efforts during the past two decades. Crops alone cannot meet the formidable task of sustainable dryland farming. We therefore, have to look into the whole production/farming system for synergy among its components, e.g., arable cropping, livestock management, alternative land use systems, and management of village commons/degraded lands. With such a synergistic integration the farming systems approach can meet the objective of productivity increase to the desired extent in the rainfed areas.

16.4.6 Strategy to Induce Farmer's Participation

Technology development for rainfed areas in the past has by and large ignored the farmer. It is known that people's participation is hard to come by unless some tangible benefits come through in the immediate future.

The following methodology may hitherto be appropriate to obtain interactive participation of farmers in drought mitigation efforts:

- Participatory resource appraisal (PRA);
- Demonstration of land degradation processes and water and soil losses with tools such as portable rainfall simulator;
- Focus group interactions to clearly obtain farmer's perceptions on: (i) drought prone issues; (ii) watershed and its relevance; (iii) sustainability of rainfed farming; and,
- Identification/listing of indigenous technical knowledge aimed at its utilization for on-farm assessment and refinement.

The success in interactive participation if focused in a mission-mode may eventually lead to voluntary participation or self-mobilization, which must be the national strategy for technology transfer in rainfed areas in the next decade. This methodology should invariably be adopted for planning on-farm drought prone research in a farming systems perspective.

16.5 Strategic Planning for On-farm Research

Strategic planning for developing best agricultural practices and farming systems is crucial and central to attain the goal of sustainable management of drought prone areas. In this context it would be permitted to divide the on-farm program into short-term (5 years), medium-term (10 years) and long-term (20 years) technological interventions (Singh and Sharma 2002).

16.5.1 Short-term Measures

Technologies to address the problems on a short-term basis aim at immediate benefit to farmers. An ideal approach would be to start from refinements in indigenous technical knowledge based practices. These interventions would not need much monetary inputs, but at the same time may lead to tangible benefits from improved crop productivity, thus in resulting in indigenous capacity building. This would build the confidence of the farmer in the programme and motivate them to adopt medium and long-term measures in due course, which will obviously need more of his contribution both in terms of labour and cash input. These consist of 'doable' technologies, which the farmers can adopt at their farm level (irrespective of farm size) and of course, as far as possible on a watershed basis with a little or no help from external agencies. Drought tolerant varieties, inter-bund treatments like key-line and appropriate tillage, summer tillage, ploughing/sowing across the slope, green-loppings-land cover-cum-manure treatment, vegetative barriers, ridge furrow configuration for planting, opening of conservation furrows, etc. are some of the examples for short-term measures. These technologies aim at creating an impact within a short span of 2–5 years. Nevertheless, such technologies act as "starters" or "stepping stones" for the longer-term task of managing water stress by developing sustainable water resources. These interventions aim at ensuring utilization of the in situ conserved moisture integrated with improved crop management for enhancing the crop yield under erratic and uncertain water supply.

16.5.2 Medium-term Measures

Once the farmers begin to realize the benefits of adopting technologies (through short-term measures), they can better understand the goals for control of drought and respond to participate in the adoption of medium and long-term measures. The medium-term measures address the problems in a time frame of 5–10 years. Integrated watershed management for development of farming systems and best practices is central to the medium-term measures. Regularization of runoff and storage in medium size reservoirs, renovation of existing tanks, adoption of alternate land uses systems and improvement of common pool resources are examples of medium-term interventions. Construction of such structures needs major initiative from the farmer with support from the government. Best practices are those which are appropriate both from biophysical and socio-economic angles.

16.5.3
Long-term Measures

These measures have a time frame of 10–20 years. Besides generation of large-scale surface and groundwater resources, development of wastelands that have gone out of cultivation should receive special attention in such interventions to reduce pressure on existing arable land resources for ease in drought mitigation. The emphasis on rehabilitation of wastelands should be to develop alternate land use systems with silvi-pasture integrated to livestock production as the main production enterprise. Extremely degraded and rocky terrains can be utilized as catchments for generation of surface water resource for utilization in adjoining areas. Such long-term measures thus can restore the land back to the production chain.

Long-term measures also include structures, which are erected to regulate overland flow and reduce peak flow. These structures, aim at improvement of relief, physiography and drainage of watersheds on macro scale, say 2,000–5,000 ha.

16.6
Drought Management and Remote Sensing

By virtue of a synoptic and repetitive coverage at regular intervals, space-borne measurements offer immense potential for early warning of and addressing most of the issues related to drought management on a near-real time basis. India has developed a space-based strategy for all the three phases of drought management, namely the preparedness phase, the prevention phase and the relief phase. Since 1989, the National Agricultural Drought Assessment and Monitoring system (NADAMS) has been providing biweekly drought bulletins which describe prevalence, relative severity level, and persistence through the crop growing season in most of the peninsular and northern India. The drought interpretation takes into account rainfall and aridity anomaly trends (Fig. 16.2). Plans are afoot for a comprehensive assessment of drought by accounting for variations in rainfall, surface water availability and soil moisture, in addition to vegetation condition. Apart from general crop condition assessment, crop-specific damage assessment is also contemplated using high spatial resolution

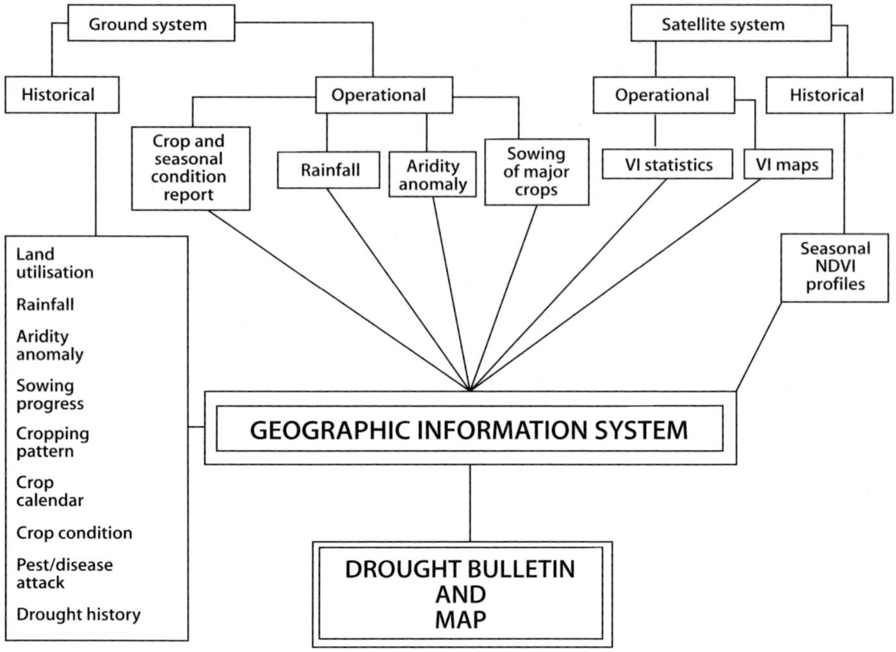

Fig. 16.2. National Agricultural Drought Assessment and Monitoring System (NADAMS)

satellite data. Future earth observation missions with improved spatial spectral and temporal resolution may enable the realization of our goals especially in the relief phase of drought management, by way of enabling the refinement of the locale-specific action plan for land and water resources development aimed at providing long-term relief.

16.7 Risk Transfer Through Insurance

In keeping with the impact of natural disasters, especially drought in India, disaster insurance could be a critical instrument of development in the field of crop production, providing financial support to the farmers in the event of crop failure. It can encourage farmers to adopt progressive farming practices, better technology in agriculture, besides providing significant benefits not merely to the insured farmers, but to the entire community directly and indirectly through spill over and multiplier effects in terms of maintaining production and employment generation of market fees, taxes, etc., and net assertion of economic growth. Crop insurance could also streamline loss assessment procedures and help in building up huge and accurate statistical base for crop production (Sahni 2003).

It envisages coverage of all the food crops (cereals, millets and pulses), oil seeds and annual commercial/horticultural crops.

The basic objectives of the crop insurance are to:

- Provide insurance coverage and financial support to the farmers in the event of failure of any of the notified crop as a result of natural calamities, pests and diseases;
- Encourage the farmers to adopt progressive farming practices, high value inputs and higher technology in agriculture; and,
- Help stabilise farm incomes, particularly in disaster years. Comprehensive risk insurance is provided to cover yield losses due to non-preventable risks namely natural fire and lightening, storm, hailstorm, cyclone, flood, inundation, drought, dry spells, pests, diseases, etc.

16.8 Conclusion

Drought inflicts considerable pain and hardship on society. The impacts of contemporary droughts have demonstrated this fact again and again over the past several decades. Drought illustrates in innumerable ways the vulnerability of economic, social, political, and environmental systems to a variable climate. It also illustrates the dependencies that exist between systems, reinforcing the need for improved coordination within and between levels of government.

Extended periods of normal or benign weather conceal the vulnerability of societies to climate variability, while drought exposes these sensitivities. Projected changes in climate because of increased concentrations of CO_2 and other atmospheric trace gases suggest a possible increase in the frequency and intensity of severe drought in the future. Any increase in the incidence of drought will aggravate drought management problems further. Coupled with increasing population and the associated rise in demand for water and other shared natural resources, there is a sense of urgency for reducing the personal hardships and economic and environmental impacts of drought.

The task of reducing the impacts of drought in developing countries poses tremendous challenges in regions where large segments of the population are poor and food-insecure and poverty levels are expected to increase in the future. The development of appropriate policies and strengthening of institutional mechanisms for drought preparedness and mitigation need to be accompanied by concrete programmes to promote sustainable livelihoods, as well as safety net to protect lives and livelihoods in the event of major drought episodes. Ultimately, development strategies for dryland areas need to help create increased opportunities for the populations and reduce marginalisation in order to contribute significantly to reducing vulnerabilities to drought.

Acknowledgement

The author expresses his gratefulness to Dr. S.K. Srivastav, Director General of Meteorology, New Delhi for his nomination and participation at this expert meeting.

References

Abrol IP, Venkateswarlu J (1995) Sustainable development of arid areas in India with particular reference to Western Rajasthan. In: Sen AK, Kar A (eds) Land degradation and desertification in Asia and the Pacific region, Jodhpur: Scientific publisher (India), pp 135–153.

Anonymous (1994) Report of the Technical Committee on Drought Prone Areas Programme and Desert Development. Ministry of Rural development, New Delhi.

Barah BC (1996) Traditional water harvesting systems. New Age International Publishers, New Delhi.

Bhatt CP (1987) The Chipko Anadolan: Forest conservation based on people's power. In: Agarwal A, Darvy D'Monte, Samarth U. The fight for survival – People's action for environment, Centre for Science and Environment, New Delhi, pp 45–55.

Bhatt M (1997) Maintaining families in drought India: The fodder security system of the Banaskhantha Women. In: Fernando P, Fernando V (eds) South Asian Women: Facing disasters, securing life, Duryog Nivaran, pp 35–44.

Das HP (1995) Incidence, impact, monitoring and mitigation of large scale droughts in India. Presented to the Expert Meeting on Drought Monitoring in WMO, Geneva.

Das HP (1999) Management and mitigation of adverse effects of drought phenomenon. In: Sinha DK, Rahim MB (eds) Natural disasters – some issues and concerns. Natural Disasters Management Cell, Visva Bharati, Shantiniketan, Calcutta, India pp 87–103

Das HP (2000) Monitoring the incidence of large scale droughts in India. In: Wilhite DA (ed) Drought A Global Assessment, Vol. 1. Rauteldge London and New York, pp 181–195.

Hounam CE, Burgos JJ, Kalok MS, Palmer WC, Rodda J (1975) 'Drought and agriculture', WMO Technical note 138, World Meteorological Organization, Geneva, p127.

May LH, Milthorpe FL (1962) 'Drought resistance of crop plants', Commonwealth Bureau of Pastures and Field Crops, UK Field Crops Abstracts, pp 171–179.

Narain P, Sharma KD, Rao AS, Singh DV, Mathur BK, Ahuja UR (2000) Strategy to Combat Drought and Famine in the Indian Arid Zone. Central Arid Zone Research Institute, Jodhpur.

Pandey S, Singh HN, Villano R (1999) Rainfed rice and risk coping strategies: some micro economic evidences from eastern India. Annual Meeting of the American Agricultural Economics Association, Nashville.

Randhwa NS, Venkateswarlu J (1979) Indian Experiences in Semi-Arid Tropics, Prospects and Retrospects. In: Proc. Intl. Symp. on development and transfer of technology. ICRISAT, Patancheru, Hyderabad, A.P.-502324.

Sahni P (2003) Drought profile, management and risk reduction in India. In: Sahni P, Ariyabandu MM (eds) Disaster Risk Reduction in South Asia. Prentice-Hall of India Private Limited, New Delhi, pp 299–326.

Sastri ASRAS, Siddique MRH, Urkurkar JS (1990) 'New approaches for drought management of rainfed rice in Central India', Drought Network News 2, 1: pp 12–13.

Singh HP (2001) Developing farming systems and best practices for drought prone areas. In: FAO Asia-Pacific conference on Early Warning, Prevention, Preparedness and Management of Disasters in Food and Agriculture, Food and Agriculture Organization of the United Nations, Regional office for Asia and the Pacific, Bangkok, pp 289–319.

Singh HP, Sharma KD (2002) Drought mitigation and farming system research in India. In: Proc. National Workshop on recent trends in drought assessment, monitoring and management, CSRE, IIT Bombay, pp 30–37.

Subbiah AR (2000) Response Strategies of local farmers in India. In: Wilhite DA (ed) Drought A Global Assessment. Vol. 1. Rauteldge London and New York, pp 29–34.

Venkateswarulu J (1992) ' Steps to mitigate moisture stress in crop production', lecture notes of First SERC School on Agrometeorology, Indira Gandhi Agricultural University, Raipur (India), 14 September - 3 October.

Wilhite DA (1993) Drought assessment, management and planning: theory and case studies. Kluwer Academic Press, Boston, MA.

Williams J (2000) Drought risk management in Southern Africa. Developing institutions to transform 'belated disaster response' into informed preparedness. In: Wilhite DA (ed) Drought, A Global assessment, Vol. II. Routledge, London and New York, pp 168–177.

CHAPTER 17

Early Detection and Monitoring of Drought and Flood in China Using Remote Sensing and GIS

Yanxia Zhao, Sanmei Li, Yeping Zhang

Abstract Drought and flood are the most frequent and severe natural disaster. With the development of remote sensing and GIS technology, early detection and monitoring of drought and flood using remote sensing and GIS has become a reality in China. The paper summaries the status of the field including the methods and data used. It especially presents the methods and products for early detection and monitoring of drought and flood developed by National Satellite Meteorological Center (NSMC), and gives an example of Flood monitoring.

17.1 Introduction

Drought and flood are worldwide problems, causing enormous destruction and creating human suffering and producing negative impacts on national economies. Due to geo-climatic condition, China is one of the world's most disaster prone countries (Li, et al. 1996, 2000). Statistics show that among all the natural disasters that country faces, drought and river floods are the most frequent and often devastating (see Fig. 17.1, Fig. 17.2). They averagely account for 80 percent in terms of the area struck by disasters, drought is simply responsible even for 55 percent. Especially in the North China Plain, drought is the most serious natural disaster to winter wheat, and it usually accounts for half of the total agricultural loss. The east part of China is under the influence of the continental monsoon climate, and the rainfall is high in intensity and concentration in time. The flood season in the east region starts from June through September. During the flood season the maximum rainfall in one month can make up 25 percent to 50 percent of the total annual rainfall. Based on analysis of climate data (see Fig. 17.3 and Fig. 17.4), we can see that temperature generally increases and rainfall decreases. Drought becomes very severe. Meantime, due to rainfall distribution unevenly in China in recent decades, devastating floods frequently hit China, especially in 1990s. Many of them were the most severe in the last century (i.e. the 1998 flood in China).

Drought/flood affects few million hectares of crop area, take to a heavy toll of human and animal lives, and have damaged millions of houses annually during the last decade. Drought and flood early warning and monitoring are

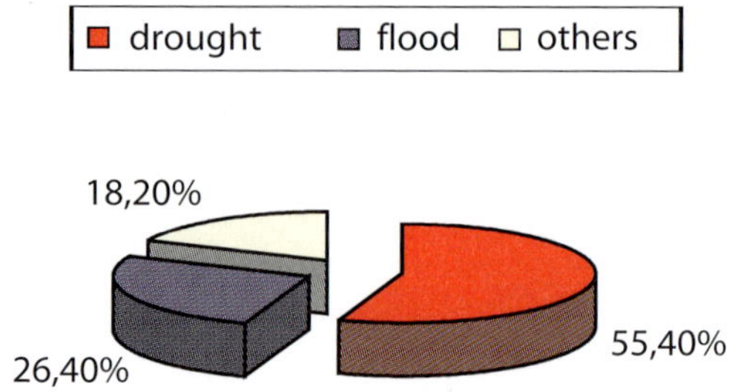

Fig. 17.1. Percent of drought and flood area in the total area suffering from natural disasters (based on statistics during 1951–2000)

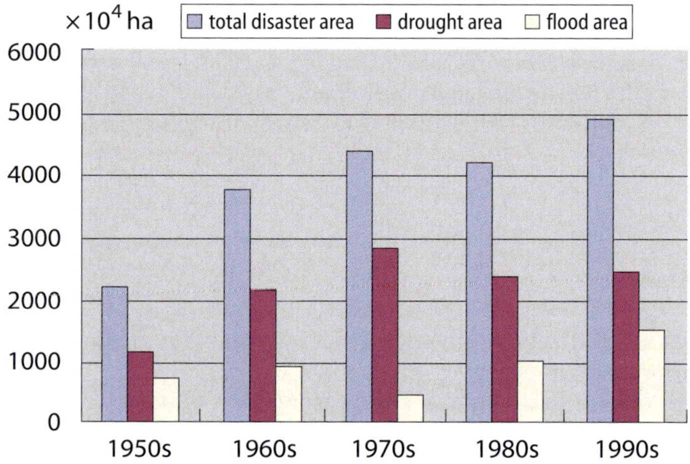

Fig. 17.2. Area affected by drought and flood in the total area suffering from natural disasters (based on statistics during 1951–2000)

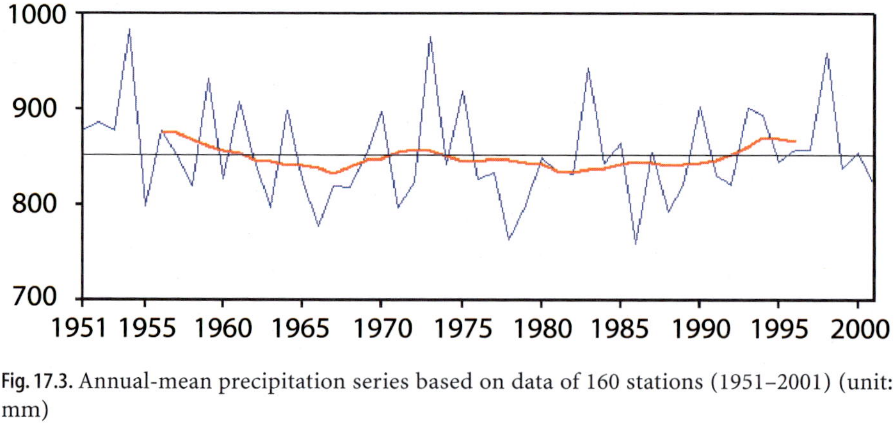

Fig. 17.3. Annual-mean precipitation series based on data of 160 stations (1951–2001) (unit: mm)

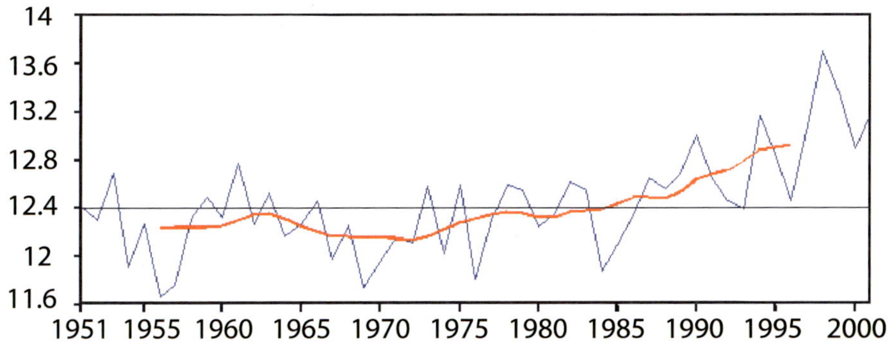

Fig. 17.4. Annual-mean temperature series based on data of 160 stations (1951–2001) (unit: °C)

crucial components of drought preparedness and mitigation plans. Recent advances in operational space technology have improved our ability to address many issues of early drought warning and efficient monitoring. With help from satellites, drought, for instance, can be detected 4–6 weeks earlier than before and delineated more accurately, and its impact on agriculture can be diagnosed far in advance of harvest. Though it is not possible to completely avoid the disasters, but the sufferings can be minimized by creating proper warning system, disaster preparedness and management of disasters through application of information technology tools such as GIS and remote sensing. The emergence of China as an advanced country in the area of remote sensing with its own satellite in orbit has led to better management of drought/flood.

17.2 Status of Drought and Flood Early Detection and Monitoring in China

Remote sensing and Earth Observation System (EOS) are available to disaster management professionals today, which make the effective preparedness very much possible and more accurate now than ever before. Although none of the existing satellites and their sensors has been designed solely for the purpose of observing natural hazards, the variety of spectral bands in VIS (visible), NIR (near infrared), IR (infrared), SWIR (short wave infrared), TIR (thermal infrared) and SAR (Synthetic Aperture Radar) provide adequate spectral coverage and allow computer enhancement of the data for this purpose. Repetitive or multi-temporal coverage is justified on the basis of the need to study various dynamic phenomena whose changes can be identified over time. Typical polar orbiting satellites are the NOAA-AVHRR, the French SPOT, the US LANDSAT, TM and FY-1 series. The data received from polar-orbiting satellites is very useful in understanding and monitoring the natural hazards. Also, the effect of the space and time distribution of water accompanied by phase changes on the water and energy budget at the earth surface from the point of view of understanding and control of the mechanism of environmental changes can be better evaluated with satellite data.

GIS are computer-based systems that efficiently store, retrieve, manipulate, analyze and display spatial data according to user specifications. It is used as a decision support system. When it is integrated with remote sensing systems that collect vast amounts of biological and physical data, it can realize the full potential of both systems, thus monitoring of the environment.

Drought and flood monitoring using remote sensing and GIS began in the middle of 1980's in China. Several sectors such as National Satellite Meteorological Center (NSMC) of China Meteorological Administration (CMA) and provincial meteorological bureaus, Ministry of Water Resource, Ministry of Agriculture, some research institutions in Chinese Academy of Science (i.e. Institute of Remote Sensing Application, Institute of Geographical and Natural Resources Research) and some research institutions in the universities have been engaged in this work and provide services for government and public. So far, many methods and products suitable to different regions and different seasons have been developed and some of them have been put into practice, providing service. In terms of scientific research, many kinds of satellite remote sensing are used to develop the methods of early detection and monitoring of drought and flood (Tian 1993, Xiao 1994, Tian et al. 1998). However, for the operational systems, such as those developed by NSMC, FY-1 series and NOAA series satellite data are mainly source of information. This paper will take National Satellite Meteorological Center (NSMC) of China Meteorological Administration (CMA) as an example to detail operational early detection and monitoring system of drought and floods in China using remote sensing and GIS (NSMC, 2003).

17.2.1 Drought

One of the missions of NSMC is to employ satellite data to monitor the natural disaster and environment, including drought/flood monitoring. Based on the two types (polar-orbiting and geostationary) of satellite data receiving and processing system, NSMC has been carrying out drought/flood monitoring since the middle of 1980s. In order to monitor the occurrence and development of droughts efficiently and provide information on the strength and range of drought, several operational systems for drought monitoring were developed. National Satellite Meteorological Center (NSMC) of China Meteorological Administration (CMA) has developed practical products of drought monitoring suitable for different seasons using weather satellite data and GIS. They can dynamically monitor the occurrence and evolution of droughts over the whole country and provide early warning of droughts. NSMC and some of provincial meteorological bureau usually distribute monitoring information once every ten days. But once the drought occurs, they are needed to provide their products as soon as possible in unfixed time manner.

Table 17.1 lists the features of some drought monitoring operational products. An improved thermal-inertia model has been developed for monitoring

Table 17.1. Main features of methods

Models	Data source	Condition	Period	Quantitative	Resolution
Thermal inertial	NOAA	Bare soil; less vegetated	Spring, Autumn	Soil moisture	1.1 km
WSVI	FY-1, NOAA	Fully vegetated	Summer	Relative drought level	1.1 km
Energy and water balance	GMS	Bare soil; vegetated	All seasons	Relative drought level	5.0 km

soil moisture in areas without vegetation cover. For vegetated areas, Water Supplying Vegetation Index (WSVI) is adopted. In addition, a method based on energy and water balance was developed to estimate evapotranspiration (ET) and soil moisture (SM), thus monitoring drought.

17.2.1.1
Thermal Inertial

Thermal inertial is defined as:

$$I = (k\varrho C)^{1/2} \quad \left[\mathrm{Jm^{-2}K^{-1}s}\right] \tag{17.1}$$

K: bulk thermal conductivity $[\mathrm{Wm^{-1}\,K^{-1}}]$
ϱ: bulk density $[\mathrm{kg\,m^{-3}}]$
C: specific heat capacity $[\mathrm{J\,kg^{-1}\,K^{-1}}]$.

Thermal inertia represents the ability of a material to conduct and store heat.

$$I = C \cdot (1 - A) / \left(T_{\mathrm{day}} - T_{\mathrm{night}}\right) \tag{17.2}$$

A : albedo
$T_{\mathrm{day}}, T_{\mathrm{night}}$: temperature of day and night

$$SM = f(I) \tag{17.3}$$

SM: soil moisture

One of the limitations of the method is to need two images one day with cloud free; the other is just applicable to bare soil or less vegetated.

17.2.1.2 Water Supplying Vegetation Index (WSVI)

The Water Supplying Vegetation Index (WSVI) is a newer method to detect drought information by using meteorological satellite data.

$$\text{WSVI} = \text{NDVI}/T_s \tag{17.4}$$

NDVI: normalized difference vegetation index
T_s : surface temperature derived from AVHRR channel 4.

When vegetation suffers from drought, the NDVI decreases and the temperature of the canopy increases, WSVI decreases. Therefore, WSVI can reflect drought effectively. Using this method, vegetation growth can be monitored timely and the regional effects of summer drought detected quickly. The method is adopted when ground is completely covered by vegetation.

17.2.1.3 Energy and Water Balance

Figure 17.5 is the chart of elements of energy and water balance.
The above can be expressed by the following equations:

$$\text{Energy balance:} \quad In = H + LE + G \tag{17.5}$$

$$\text{Water balance:} \quad P = E + I + R \tag{17.6}$$

LE: latent heat flux ($\text{J/m}^2\,\text{s} = W/\text{m}^2$)
E : evapotranspiration, moisture flux ($\text{kg/m}^2\,\text{s}$)
L : heat of evaporation (J/kg).

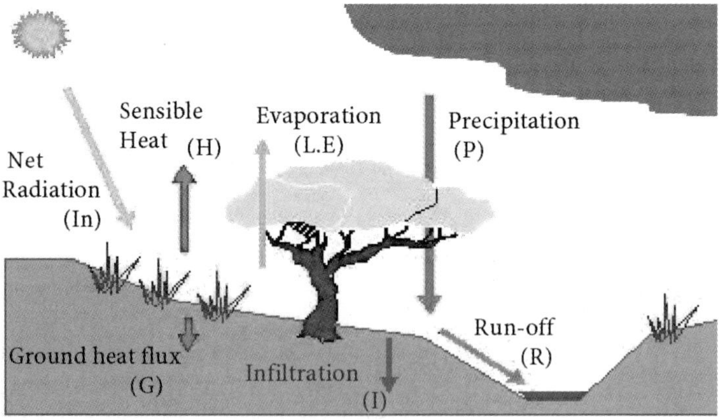

Fig. 17.5. Elements of energy and water balance

17.2.2
Flood

In the summer, from May to September, NSMC uses FY-1 series and NOAA series satellite data for daily water body monitoring of seven major rivers' basins in China including NenJiang, LiaoHe, HaiHe, Yellow River, HuaiHe River, HanShui River, Dongting Lake, Poyang Lake, Taihu Lake, Zhujiang River, Yongjiang River, and North of Xinjiang (See Fig. 17.6). The monitored area is divided into several regions, and each region has a range of 5° × 5°. In recent two year, EOS/MODIS data has been extensively used in water body monitoring and played an important role for its higher resolution. Generally, three types of efforts should be made in monitoring flood: water body identification, determination floodwater body and flood area mapping. Fig. 17.7 is the flow chart of NSMC monitoring flood.

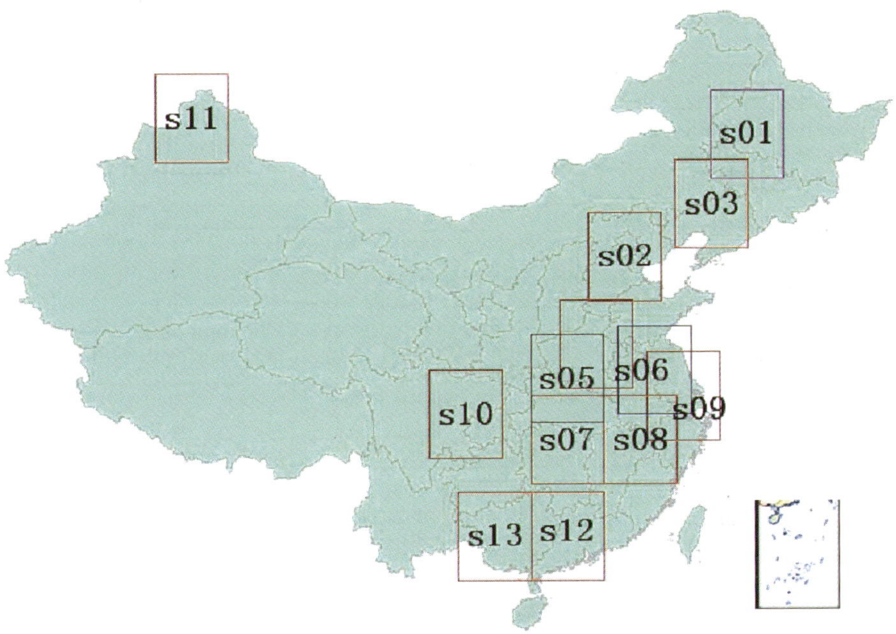

Fig. 17.6. The sketch of water body monitoring regions in China

17.2.2.1
Water body identification

Water body identification is the base of water monitoring. In visible and near infrared channels, the reflectance of water body, vegetation and bare soil are quite different. Water body has a strong absorption in near infrared channel, leading to a very low reflectance in this channel and a lower reflectance than

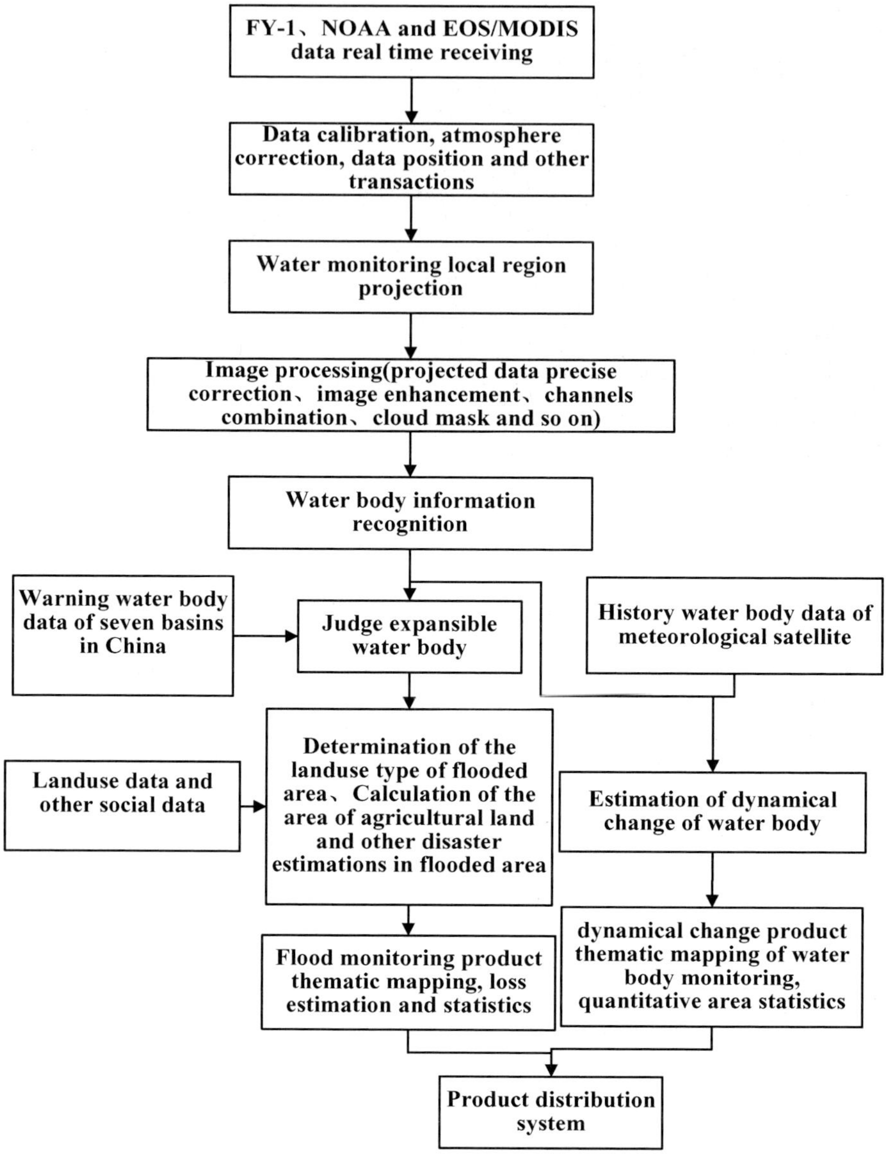

Fig. 17.7. Flow chart of flood monitoring by NSMC

in visible channel. On the contrary, vegetation has a low reflectance in visible channel but a much higher reflectance in near infrared channel. The reflectance of bare soil in visible channel is much higher than both water body and vegetation, and in near infrared channel, the reflectance is higher than water body but lower than vegetation. So based on the difference of the ratio of reflectance in near infrared channel and visible channel for the three objects,

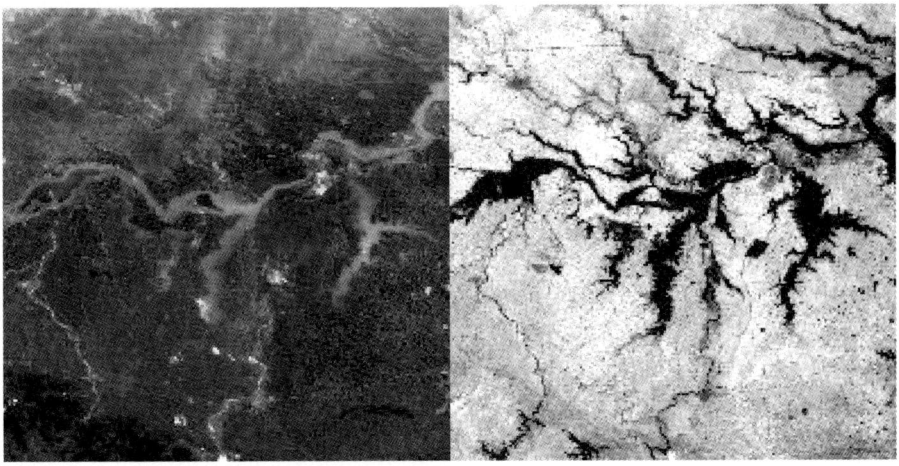

Fig. 17.8. Ch.1 and Ch.2 of EOS/MODIS is sensitive to water body (*left* is Ch.1 (visible) image, *right* is Ch.2 (near infrared) image)

it is possible to recognize water body by using FY-1, NOAA and EOS/MODIS satellite data (see Fig. 17.8)

In cloud free area, water body recognition can be processed based on the feature of the clear boundary between water and land in near infrared channel data, using threshold by interactive mode in a limited area.

In the area covered by thin cloud, the reflectance observed by satellite includes cloud and land surface.

$$V_{si} = V_{ci} + V_{gi} \tag{17.7}$$

V_{si}: reflectance from channel I,
V_{ci}: reflectance from cloud,
V_{gi}: reflectance from land.

The ratio of Ch.2 and Ch.1 is:

$$R_{21} = \left(Vc_2 + Vg_2\right) / \left(Vc_1 + Vg_1\right) . \tag{17.8}$$

For water body,

$$R_{21}(W) = \left(Vc_2 + Vg_2(W)\right) / \left(Vc_1 + Vg_1(W)\right) . \tag{17.9}$$

For land surface:

$$R_{21}(L) = \left(Vc_2 + Vg_2(L)\right) / \left(Vc_1 + Vg_1(L)\right) . \tag{17.10}$$

According to the characteristics of water body and land surface:

$$Vg_1(L) < Vg_1(W) , \; Vg_2(W) < Vg_2(L) . \tag{17.11}$$

Therefore,

$$R_{21}(L) > R_{21}(W) . \tag{17.12}$$

By selecting proper threshold, the water body covered by thin cloud can be calculated, then the influence of thin cloud can be deleted.

17.2.2.2 Determination of Unusual Expanding Water Body

Unusual expanding water body is the water body beyond the normal, namely floodwater body. In China, the water body area of most of lakes and rivers is changed constantly with seasons and precipitation. In certain period, the relative expansion of water body is not always floodwater body. So it is necessary to build up warning water body data to discern the expanding water body. Warning water body data can be obtained by extracting lakes and rivers water body information when the water surface is on the warning level. Compared with warning water body, the expanding part of water body is the flood area. Similarly, compared with the maximum water body in flood period, the reduction area is the shrunk water body.

17.2.2.3 Flood Mapping

Flood mapping is by overlaying a pre-flood image and a peak flood image to delineate the inundated area. Damages, in terms of properties and crops, are assessed with the help of existing land use map and GIS. Post-flood image could be an effective tool to evaluate impact of flood on environmental concerns, under the support of GIS, of course.

17.3 Examples of Flood Monitoring by NMSC in 2003

In the summer of 2003, a severe flood happened in Huaihe River. NSMC monitored its evolution. Fig. 17.9 is a product of flood monitoring in Huaihe River Basin on July 18,2003, using EOS/MODIS data. From it, we can learn the evolution of the flood.

The following is an example of flood product for a county's loss evaluation (Fig. 17.10, Fig. 17.11, Table 17.2). Land use data and other social data are extensively used by the support of GIS technique to estimate the area of the flood. Here flooded size, non-irrigated farmland inundated size, irrigated land inundated size and percentage of flooded size to the county size was precisely calculated for quantitative loss estimation.

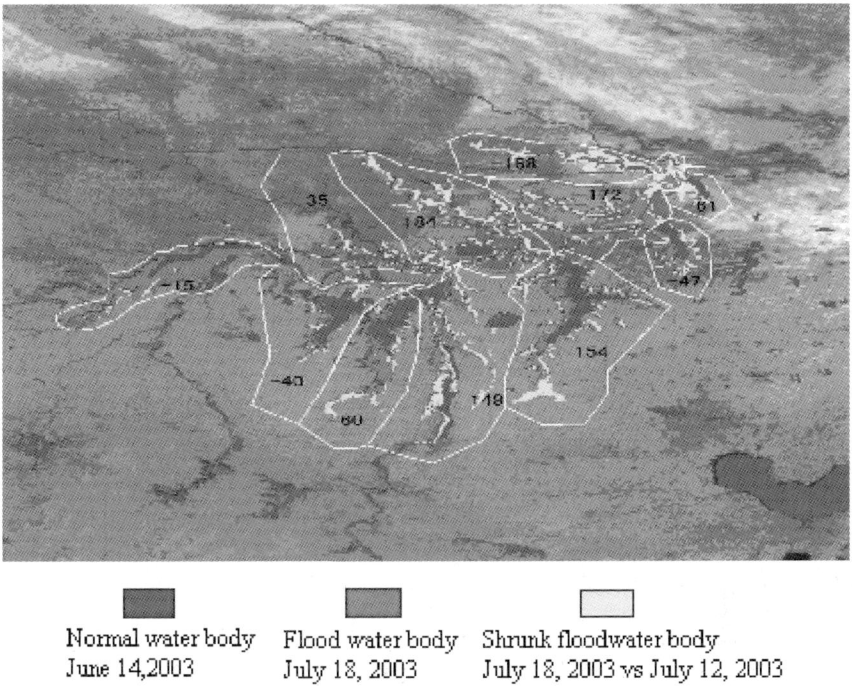

Fig. 17.9. The evolution of flood in Huihe River in 2003

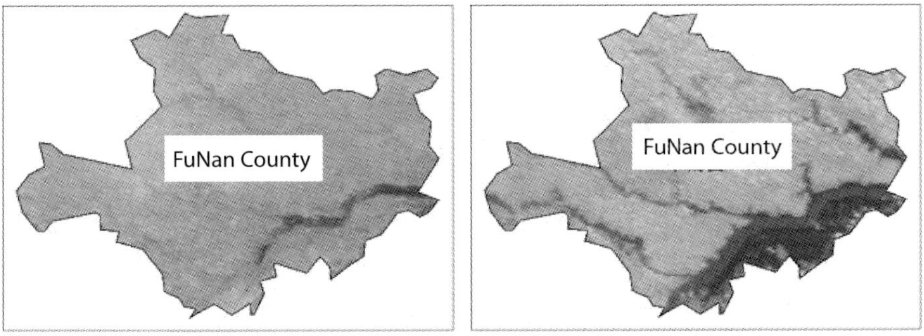

Fig. 17.10. Water conditions before and after flood (left is before flood, June 28, 2003; right is after flood, July 23, 2003

17.4 Conclusions

(1) Drought and flood are China's severest natural disaster. Early detection and monitoring of drought and flood are particularly useful for drought planning and mitigation.

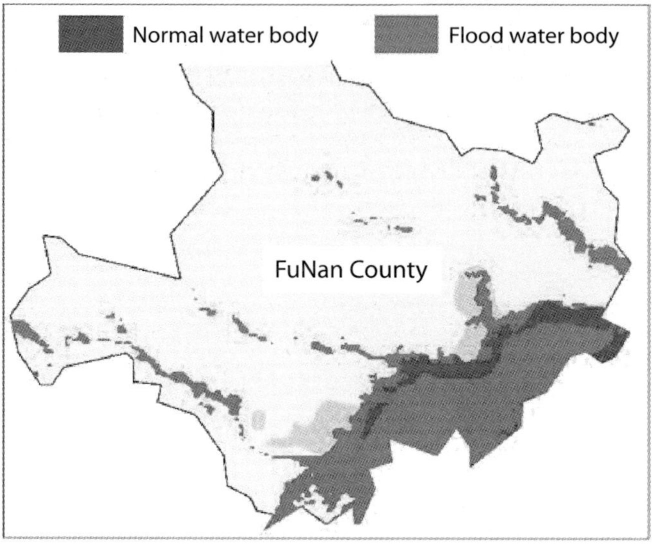

Fig. 17.11. Output by comparison of before and after flood images

Table 17.2. Assessment of the flood area for Funan county

County name	Funan
Total size	1,927 km^2
Water body size (June 28)	50 km^2
Water body size (July 23)	370 km^2
Flooded size	320 km^2
Nonirrigated farmland inundated size	271 km^2
Irrigated farmland undated size	27 km^2
Percentage of flooded size to the county size	16.6%

(2) NSMC and other organizations provide the central government, all national economic sectors and public with real time information through letter, internet, and other applicable way.

(3) Substantial progress in both academic and operational aspects of the subject has been achieved in China. But we have to see remote sensing and GIS are not perfect for disaster monitoring. There are a variety of limitations.
   a) Monitoring methodology: techniques for processing remotely sensed data, i. e. geometric and radiometric calibration; development of models and algorithms to extract the physical variables that are needed for disaster monitoring;
   b) Database: inadequate observation stations in a so large territory and inadequate data sharing between different sectors; lack of geographical, economic and social data, especially in less developed regions.

(4) Anyway, natural disasters monitoring by remote sensing technology was one of the main contributions in last century. Integration of remote sensing with GIS and web technology makes it a powerful tool to identify indicators of potential disasters. The future is promising with the new generation of very high-resolution satellites. They will provide high spatial and temporal resolution images of the world to track natural and human-made disasters.

Acknowledgements

Many thanks to Professor Cheng Liu at the National Satellite Meteorological Center, China Meteorological Administration, for providing many of the figures and giving helpful advice.

References

Li K, Yin S and Sha W (1996) Characters of time-space of recent drought in China. *Geographical Research*. 15(3):6–14.

Li K (2000) Y. Proceedings of an Expert Group Meeting held September 5–7, in Lisbon, Portugal.

NSMC, Report of satellite monitoring drought and flood of China in 2003.

Tian G (1993) Estimation of evapotranspiration and soil moisture and drought monitoring using remote sensing in North China Plain. Pages 23–31 *in* Space and Environment, Special Plenary Session, 44th lnternational Astronautical Federation, IAF Congress, Graz, Austria.

Tian G, Jupp DLB, Qin Y, McVicar TR and Li F (1998) Monitoring Soil Moisture and Drought Using AVHRR Satellite Data II: Applications. CSIRO Earth Observation Centre, Canberra, ACT.

Xiao Qianguang, Chen Weiying, et al. (1994) Soil moisture monitoring by meteorology satellite, Quarterly Journal of Applied Meteorology, Vol. 5, No. 3, (8).

CHAPTER 18

The Decision of the Center of a Tropical Cyclone over China Coastal Waters Using a Doppler Radar

Xu Yinglong, Jiao Meiyan, Bi Baogui, Chang Tao

Abstract Using the Doppler radar radial velocity data simulated by Rankine combined vortex and observed by a Doppler radar and typhoon yearbooks, the applied study about the objective positioning methods of a typhoon was made. Two methods had been established. A series of simulated and actual tests were made for the two methods. The results show that the errors of the two methods in simulated tests are below 0.5 ∼ 1 km and those of radial maximum wind are below 0.12 km. Compared to the best operational track data, the errors of the two algorithms in actual tests are below 3 ∼ 13 km, sometimes those are below 3 km. So it is feasible to decide the center of a tropical cyclone with the data of a single Doppler radar radial velocity in operational tropical cyclone positioning. The positioning results of the two methods are of important reference and guidance in operational tropical cyclone forecasting and warning.

18.1 Introduction

As we know, China is one of the most severely countries which suffers from tropical cyclones. There are about 7 tropical cyclones make landfalls over China every year and a slight increasing trend in the number of landing tropical cyclones (Fig. 18.1). Tropical cyclones are the most important disastrous weather system over Southeast China coastal area from summer to autumn every year and often bring great casualties and damages. In total there are US$ 33.528 billion economic losses resulting from tropical cyclones since 1993 and US$ 335.28 million every year (Fig. 18.2). In total 7,158 persons were killed by tropical cyclones since 1988 and 477 persons were killed every year (Fig. 18.3). Because of the effective forecasting and warning service, economic and life losses have some decreasing trends in the past.

In August 1994, Typhoon Fred (9417) hit Zhejiang province which killed 1,126 persons and economic loses were estimated over US$ 2,167 million.

In August 1996, Typhoon Herb (9608) landed over Fujian province, over 700 persons were killed and economic loses were estimated over US$ 8,025 million.

In June 2001, Typhoon Chebi (0102) hit Fujian province (Fig. 18.4a), 122 persons were killed and economic loses were estimated over US$ 547 million.

In September 2002, Typhoon Sinlaku (0216) made landfall over Zhejiang province (Figs. 18.4b–d), 29 persons were killed and economic loses were estimated over US$ 984 million.

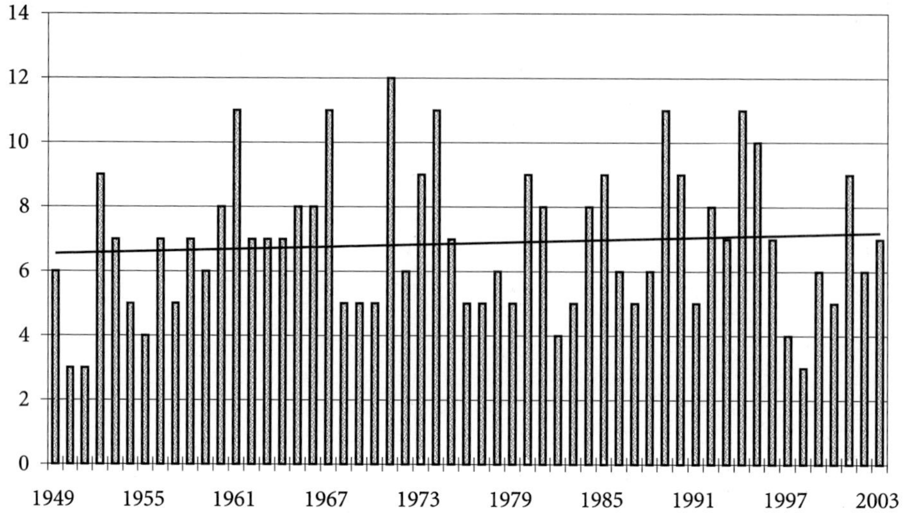

Fig. 18.1. The number of tropical cyclones landing over China since 1949

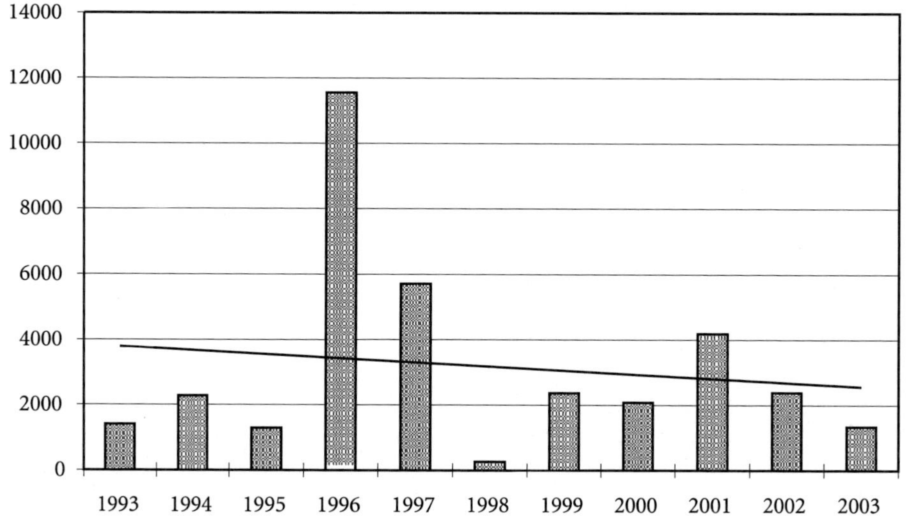

Fig. 18.2. Economic loss of tropical cyclone disaster since 1993 (unit: million US$)

Therefore, the operational tropical cyclone forecasting and warning is of crucial importance in all operational forecasting and warning of CMA. In the course of actual operational forecasting, exact positioning is the basis for successful forecasting, timely and effective prevention and fight against tropical cyclones.

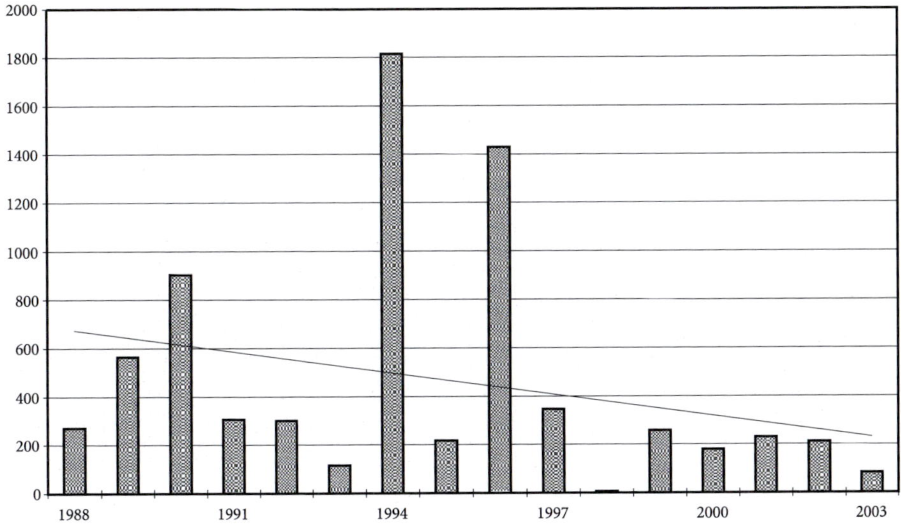

Fig. 18.3. Life loss of tropical cyclone disaster statistics since 1988

Fig. 18.4. A large ship was pulled up to the sea bank by Typhoon Chebi in 2001 (**a**). The storm surge in the Qiantang River (**b**), the flooded area in Zhijiang Street in Hangzhou City (**c**) and the big trees fell to the ground in Cangnan County (**d**) during the period of Typhoon Sinlaku in Zhejiang province in 2002

In the past, the tropical cyclone center was commonly decided by satellite images. However, when tropical cyclones approach land, their structures became more complicated in the influence of islands and continental terrain. So, it is hard to get exactly centers of tropical cyclones only by satellite images. At present, the radar is the best detecting tool. The positioning ways using radar echo include spiral rain belt (Senn and Hiser, 1959; Anthes, 1982), geometric center of eye wall (Griffin, 1992), etc. But its positioning accuracy is poor when the spiral rain belt is not evident, or when the structure of radar echo is in disorder, or when spiral rain belt or eye wall rain belt is not observed in the influence of terrain after landing, etc. The reconnaissance mission observations show that the center position of radar echo is same as the center position of typhoon circulation, and the circulation center often slants to the side of the most intensive radar echo region (Marks, 1992).

With the development of Weather Surveillance Radar-1988 Doppler (WSR-88D) during the 1990s, it is possible to detect tropical cyclone center using the pattern distribution of Doppler radar velocity. The research of Wood and Brown (1983) and Chang Paoliang (2000) indicated that Doppler radar velocity of the circulation near tropical cyclone center, which is characteristic of approximately axisymmetric distribution or Rankine vortex, has a particular dipole pattern distribution feature. A lot of methods were designed to decide a tropical cyclone center on the basis of the distribution feature and the hypothesis about the radius maximum wind (RMW). Wood and Brown (1992) first developed the Geometric Axisymmetric Center Positioning Method (WB92) to decide a tropical cyclone center using Doppler radar radial velocity and its operational positioning accuracy is below 4 km compared to aircraft reconnaissance. Ben Jong-Dao Jou et al. (1996) advanced the Velocity Distance Azimuth Display Positioning Method (VDAD) to detect a tropical cyclone center using the product ($V_r \times D$) of the Doppler radar radial velocity value (V_r) and the distance from observation site to radar center (D) and obtained better results. However, when the asymmetric component or environmental mean flow is evident, big errors are introduced.

18.2 Pattern Recognition Technique

Because Doppler radar only can observe radial velocity along the radar beam direction and can't get the structure of real three-dimensional flow, the analysis and the research about the flow structure observed by Doppler radar are made by the particular pattern distribution of Doppler radial velocity according to the early research about the aspect by Donaldson (1970), Wilson and Wilk (1982) and Wood and Brown (1986). Pattern recognition technique mainly is an analytical technique with which the real circulation of atmospheric system (e.g. the thunderstorm mesoscale cyclone and typhoon circulation etc.) can be identified by the particular pattern distribution of Doppler radial velocity.

For a run of increasing velocities in any radial direction, a shear segment is defined as the beginning and ending velocity data points of the run. It consists

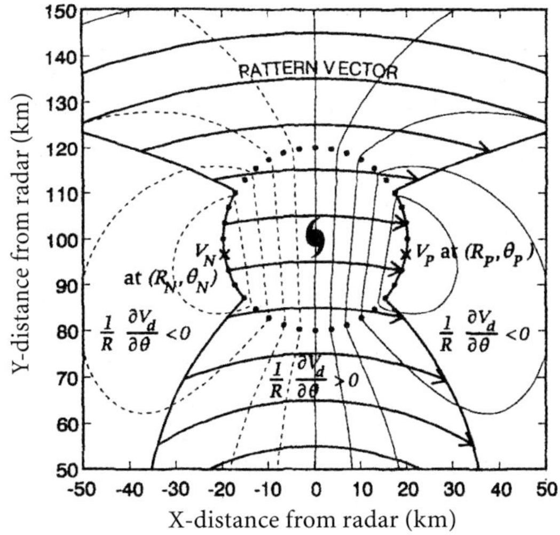

Fig. 18.5. The single Doppler velocity pattern an axisymmetric Typhoon Votex. (Source: Vincent T. Wood, 1994 RSSL of NOAA)

of the slant range (R), the beginning and ending azimuth angles (θ_b, θ_e), the beginning and ending Doppler velocity values (V_b, V_e), and the beginning and ending times (t_b, t_e) at both ends of the shear segment (Fig. 18.5).

Here, for identifying the tropical cyclone vortex, the azimuthal shear is defined as the shear of Doppler velocities in the azimuthal direction for a pattern vector, and is given by

$$S = \frac{\Delta V}{R\left(\theta_e - \theta_b\right)} .$$

Where $\Delta V = V_e - V_b$ is the velocity difference and $R(\theta_e - \theta_b)$ is the azimuthal distance across the shear segment.

If there is $S > 0$ in any pattern vector, a cyclonic shear will be identified. The shape of the region of the pattern vectors of all cyclonic shears will be detected by a cyclonic circulation of a real atmospheric system.

18.3 The Geometric Axisymmetric Center Positioning Method

18.3.1 The Detection of a Tropical Cyclone Vortex

1. Searching for Cyclonic Shear (i.e., $S > 0$).
2. The Determination of Detection Thresholds. In order to prevent the influence of non-typhoon circulation, some detection thresholds will be given in advance. And these thresholds will be given on the base of the intensity of the typhoon. In the following typhoon case, the detection thresholds are

given as:

$$\Delta V = V_\mathrm{e} - V_\mathrm{b} = 60 \sim 70\,\mathrm{m\,s^{-1}}$$
$$S = \Delta V / R\left(\theta_\mathrm{e} - \theta_\mathrm{b}\right) = 1\,\mathrm{m\,s^{-1}\,km^{-1}}\,.$$

18.3.2 The Determination of the Locations of Extreme Doppler Velocity

1. The determination of the extreme negative Doppler velocity value (V_N) and its range (R_N) and azimuth (θ_N) are given as:

$$V_\mathrm{N} = \frac{\sum_k \left(V_\mathrm{b}\right)_k \left(W_\mathrm{b}\right)_k}{\sum_k \left(W_\mathrm{b}\right)_k}\,,$$
$$R_\mathrm{N} = \frac{\sum_k \left(R_\mathrm{b}\right)_k \left(W_\mathrm{b}\right)_k}{\sum_k \left(W_\mathrm{b}\right)_k}\,, \tag{18.1}$$
$$\theta_\mathrm{N} = \frac{\sum_k \left(\theta_\mathrm{b}\right)_k \left(W_\mathrm{b}\right)_k}{\sum_k \left(W_\mathrm{b}\right)_k}\,.$$

2. The determination of the extreme positive Doppler velocity value (V_P) and its range (R_P) and azimuth (θ_P) are given as:

$$V_\mathrm{P} = \frac{\sum_k \left(V_\mathrm{e}\right)_k \left(W_\mathrm{e}\right)_k}{\sum_k \left(W_\mathrm{e}\right)_k}\,,$$
$$R_\mathrm{P} = \frac{\sum_k \left(R_\mathrm{e}\right)_k \left(W_\mathrm{e}\right)_k}{\sum_k \left(W_\mathrm{e}\right)_k}\,, \tag{18.2}$$
$$\theta_\mathrm{P} = \frac{\sum_k \left(\theta_\mathrm{e}\right)_k \left(W_\mathrm{e}\right)_k}{\sum_k \left(W_\mathrm{e}\right)_k}\,.$$

3. The determination of the weight $(W_\mathrm{b})_k$ and $(W_\mathrm{e})_k$.

Here the weight $(W_\mathrm{b})_k$ and $(W_\mathrm{e})_k$ are decided with the technique of Desrochers and Donaldson (1992), and $(W_\mathrm{b})_k$ and $(W_\mathrm{e})_k$ are confined to $0 < (W_\mathrm{b})_k \le \Delta W$ and $0 < (W_\mathrm{e})_k \le \Delta W$, and ΔW is given as:

$$\Delta W = 3\,\mathrm{m\,s^{-1}}\,, \quad \Delta\overline{V} < 35\,\mathrm{m\,s^{-1}}$$
$$\Delta W = 5\,\mathrm{m\,s^{-1}}\,, \quad \Delta\overline{V} \ge 35\,\mathrm{m\,s^{-1}}\,.$$

Where $\Delta\overline{V} = (V_\mathrm{max} - V_\mathrm{min})/2$ is the average of peak Doppler velocity values obtained readily from all the pattern vectors. Then the $(W_\mathrm{b})_k$ and $(W_\mathrm{e})_k$ are given as:

$$\left(W_\mathrm{b}\right)_k = \Delta W - \left|\left(V_\mathrm{b}\right)_k - V_\mathrm{min}\right|\,,$$
$$\left(W_\mathrm{e}\right)_k = \Delta W - \left|\left(V_\mathrm{e}\right)_k - V_\mathrm{max}\right|\,.$$

18.3.3 The Decision of the Center of a Tropical Cyclone

After (18.1) and (18.2) are computed, the estimated range (R_E) and azimuthal (θ_E) of the tropical cyclone center are given by the following:

$$R_E = F\left(\frac{R_N + R_P}{2}\right),$$

$$\theta_E = \left(\theta_N + \theta_P\right)/2 .$$

Where R_E is the range of the center; θ_E is the azimuth of center; F is the correction factor and is given by

$$F = \sec\left(\frac{\Delta\theta}{2}\right)$$

$$\Delta\theta = \theta_P - \theta_N ,$$

$$0^\circ \geq |\Delta\theta| < 180^\circ .$$

Where the correction factor F is made to avoid the large difference between the estimated and real range from typhoon center to radar while the maximum wind region of typhoon circulation is approaching the radar.

18.3.4 The Decision of the Radius Maximum Wind (RMW)

After the locations of the extreme negative and positive Doppler velocity have been computed, we can use a half of the distance between the two points as the radius maximum wind. Supposing (X_{min}, Y_{min}), (X_{max}, Y_{max}) are the locations in the Cartesian coordinate of the extreme negative and positive Doppler velocity, RMW is given as:

$$X_{min} = R_N \sin\theta_N ,$$

$$Y_{min} = R_N \cos\theta_N ,$$

$$X_{max} = R_P \sin\theta_P ,$$

$$Y_{max} = R_P \cos\theta_P ,$$

$$\mathrm{RMW} = \frac{1}{2}\sqrt{\left(X_{max} - X_{min}\right)^2 + \left(Y_{max} - Y_{min}\right)^2} .$$

18.4 The Velocity Distance Azimuth Display Positioning Method

When a tropical cyclone is very close to the radar, Doppler radial velocity observed by radar will become deformed and the big errors of estimated center will occur by directly using Doppler radial velocity. Ben Jong-Dao Jou etc. (1996) developed a method to detect a tropical cyclone center using the product ($V_r \times D$) of the Doppler radar radial velocity value (V_r) and the distance from observation site to radar center (D). The flow of detecting the center of a tropical cyclone is given as:

1. Compute $V_r \times D$ in constant level and get its space distribution chart.
2. Compute the values and locations of the extreme negative and positive of $(V_r \times D)/D_{CR}$. In the course of computing, the handling way is the same as the Geometric Axisymmetric Center Positioning Method. D_{CR} is the distance from a reference center of a tropical cyclone to radar, so the value of $(V_r \times D)/D_{CR}$ will be eaqual to the Doppler radial velocity value (V_r). The formula are given by:

$$V_N = \frac{\sum_k \left[\left(\frac{V_r \times D}{D_{CR}}\right)_b\right]_k (W_b)_k}{\sum_k (W_b)_k},$$

$$R_N = \frac{\sum_k (R_b)_k (W_b)_k}{\sum_k (W_b)_k},$$

$$\theta_N = \frac{\sum_k (\theta_b)_k (W_b)_k}{\sum_k (W_b)_k},$$

$$V_P = \frac{\sum_k \left[\left(\frac{V_r \times D}{D_{CR}}\right)_e\right]_k (W_e)_k}{\sum_k (W_e)_k},$$

$$R_P = \frac{\sum_k (R_e)_k (W_e)_k}{\sum_k (W_e)_k},$$

$$\theta_P = \frac{\sum_k (\theta_e)_k (W_e)_k}{\sum_k (W_e)_k},$$

$$(W_b)_k = \Delta W - \left| \left[\left(\frac{V_r \times D}{D_{CR}}\right)_b\right]_k - \left(\frac{V_r \times D}{D_{CR}}\right)_{min} \right|$$

$$(W_e)_k = \Delta W - \left| \left[\left(\frac{V_r \times D}{D_{CR}}\right)_e\right]_k - \left(\frac{V_r \times D}{D_{CR}}\right)_{max} \right|$$

$$0 < (W_b)_k \le \Delta W$$

$$0 < (W_e)_k \le \Delta W$$

$$\Delta W = 3\,\mathrm{m\,s^{-1}}, \quad \Delta\overline{V} < 35\,\mathrm{m\,s^{-1}}$$

$$\Delta W = 5\,\mathrm{m\,s^{-1}}, \quad \Delta\overline{V} \ge 35\,\mathrm{m\,s^{-1}}$$

$$\Delta\overline{V} = \frac{\left[\left(\frac{V_r \times D}{D_{CR}}\right)_{max} - \left(\frac{V_r \times D}{D_{CR}}\right)_{min}\right]}{2} .$$

3. After computing the coordinate of the extreme negative and positive of $(V_r \times D)/D_{CR}$, the center of a tropical cyclone and the radius maximum wind of typhoon circulation are given by

$$X_C = \frac{X_{min} + X_{max}}{2} \qquad Y_C = \frac{Y_{min} + Y_{max}}{2}$$

$$\mathrm{RMW} = \frac{1}{2}\sqrt{(X_{max} - X_{min})^2 + (Y_{max} - Y_{min})^2} .$$

Where (X_C, Y_C) is the coordinate of a tropical cyclone center, (X_{max}, Y_{max}) is the coordinate of the maximum extreme value of $(V_r \times D)/D_{CR}$, (X_{min}, Y_{min}) is the coordinate of the minimum extreme value of $(V_r \times D)/D_{CR}$, and RMW is the radius maximum wind of a tropical cyclone circulation.

18.5 Simulated Test

18.5.1 The Building of the Simulated Doppler Radial Velocity

The analytic dataset is based on a single-level, idealized Rankine combined vortex at sea level. The rotational and convergence/divergence flows are constructed as

$$V(R) = V_{max}\left(R/R_{max}\right) \quad R \le R_{max} ,$$

$$V(R) = V_{max}\left(R_{max}/R\right) \quad R > R_{max} .$$

Where $V_{max} = 40$ m/s, $R_{max} = 10$ km, the Cartesian coordinate of the vortex center is (100 km, 100 km).

The environmental horizontal mean flow added in a vortex is constructed as

$$\vec{V}_M = -V_M \sin\left(\theta - \theta_M\right) + V_M \cos\left(\theta - \theta_M\right) .$$

Where V_M is the mean velocity of environmental horizontal flow, θ_M is the included angle between mean flow and eastward.

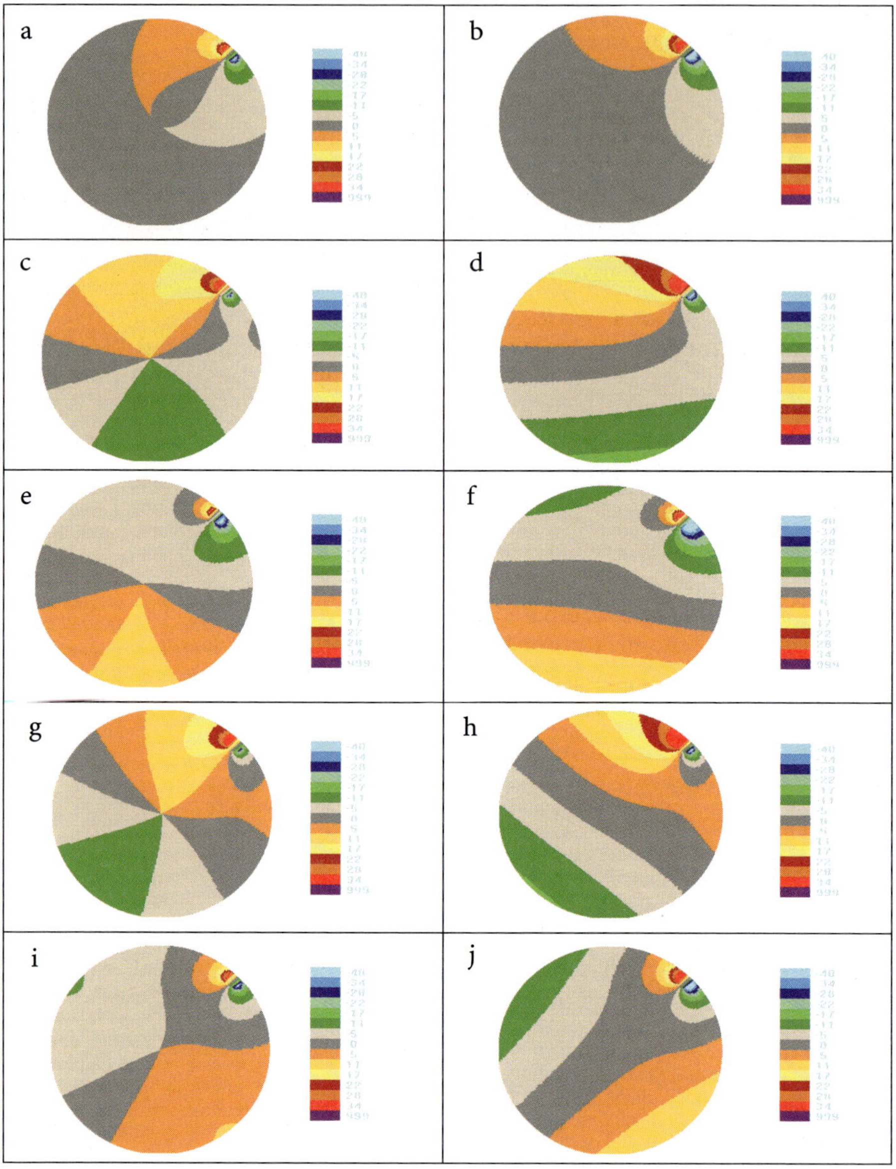

Fig. 18.6. The simulated Doppler radial velocity V_r (**a,c,e,g,i**) and $V_r \times D$ (**b,d,f,h,j**); **a** and **b** are under no environmental mean flow, **c** and **d** are under westward environmental mean flow ($V_M = 10$ m/s, $\theta_M = 180°$), **e** and **f** are under eastward environmental mean flow ($V_M = 10$ m/s, $\theta_M = 0°$), **g** and **h** are under southwestward environmental mean flow ($V_M = 10$ m/s, $\theta_M = 135°$), **i** and **j** are under southeastward environmental mean flow ($V_M = 10$ m/s, $\theta_M = 45°$)

18.5.2 The Result of Simulated Test

For a Rankine combined vortex, with vortex center located at (100, 100), maximum wind is 40 m/s and the radius maximum wind is 10 km, its Doppler radial velocities are simulated in five environmental horizontal mean flows (Fig. 18.6). Under no environmental mean flow, eastward and southeastward environmental mean flow, the distributions of the simulated radial velocities (V_r) and the product ($V_r \times D$) of the radar radial velocity value (V_r) and the distance from observation site to radar center (D) are characteristic of approximately axisymmetric feature. Meanwhile, under westward and southwestward environmental mean flow, their distributions are of very asymmetric feature.

Table 18.1. The position and the radius maximum wind (RMW) errors of Simulated Test with Geometric Axisymmetric Center Positioning Method (GACPM) (km)

Scheme	Real position of center and RMW			Estimated result					
	X_{C0}	Y_{C0}	RMW_0	X_C	Y_C	RMW	ΔX_C	ΔY_C	ΔRMW
$V_M = 0$ m/s	100.0	100.0	10.0	100.26	100.26	9.88	0.26	0.26	0.12
$V_M = 10$ m/s, $\theta_M = 180°$	100.0	100.0	10.0	100.01	100.01	9.88	0.01	0.01	0.12
$V_M = 10$ m/s, $\theta_M = 0°$	100.0	100.0	10.0	99.72	99.73	9.88	−0.28	−0.27	0.12
$V_M = 10$ m/s, $\theta_M = 45°$	100.0	100.0	10.0	99.88	99.53	9.88	−0.12	−0.47	0.12
$V_M = 10$ m/s, $\theta_M = 135°$	100.0	100.0	10.0	100.32	100.32	9.88	0.32	0.32	0.12

Table 18.2. The Position and the radius maximum wind (RMW) errors of Simulated Test with Velocity Distance Azimuth Display Positioning Method (VADAPM) (km)

Scheme	Real position of center and RMW			Estimated result					
	X_{C0}	Y_{C0}	RMW_0	X_C	Y_C	RMW	ΔX_C	ΔY_C	ΔRMW
$V_M = 0$ m/s	100.0	100.0	10.0	100.11	100.11	9.90	0.11	0.11	0.10
$V_M = 10$ m/s, $\theta_M = 180°$	100.0	100.0	10.0	100.13	100.15	9.90	0.13	0.15	0.10
$V_M = 10$ m/s, $\theta_M = 0°$	100.0	100.0	10.0	100.14	100.09	9.92	0.14	0.09	0.08
$V_M = 10$ m/s, $\theta_M = 45°$	100.0	100.0	10.0	100.11	100.11	9.90	0.11	0.11	0.10
$V_M = 10$ m/s, $\theta_M = 135°$	100.0	100.0	10.0	100.12	100.20	9.94	0.12	0.20	0.06

But it also is evident that their distributions are of a particular dipole pattern feature. After using the Geometric Axisymmetric Center Positioning Method (GACPM) and Velocity Distance Azimuth Display Positioning Method (VADAPM) respectively with simulated data, the center and the radius maximum wind of vortex are estimated. The results of simulated tests show that the center position errors of GACPM are less than 0.5 km, the radius maximum wind errors of GACPM are less than 0.12 km; the center position errors of VDADPM are less than 0.2 km, and the radius maximum wind (RMW) errors of VADAPM are less than 0.10 km (Table 18.1 and Table 18.2).

18.6 Actual Test of Typhoon Case

The simulated test results show that the Geometric Axisymmetric Center Positioning Method and the Velocity Distance Azimuth Display Positioning Method are two better tools in detecting the center and the radius maximum wind of vortex, even for these vortexes with asymmetric distribution under westward and southwestward environmental mean flow. The following tests will be made with actual radial velocity data to certificate the capability of these two methods.

18.6.1 The Data of a Typhoon Case

1. Typhoon Case: Typhoon Nari (No. 0116).
2. Radar location: WSR-88D in Wufengshan Mountain in Taiwan.
3. Scan Time: From 0000UTC to 1300UTC 16 September 2001.
4. Scan Time Interval: One hour.
5. Angle of Elevation: 0.5°.
6. Grid Length in Radial Direction: 250 m.

18.6.2 The Result of Actual Test for a Typhoon Case

Figure 18.7 gives the distribution charts of the observed Doppler radial velocities (V_r) and ($V_r \times D$) when Typhoon Nari (0116) was approaching the northeast of Taiwan province in September 2001. It is evident that their distributions of (V_r) and ($V_r \times D$) are characteristic of approximately axisymmetric feature and show a particular dipole pattern feature.

Using the Geometric Axisymmetric Center Positioning Method and Velocity Distance Azimuth Display Positioning Method respectively with real observed Doppler radar radial velocity data of Typhoon Nari (0116), its center and the radius maximum wind are estimated. The results show that Typhoon Nari had the trend to move west-south-westward. It is identical with the real motion direction. In the course of estimating the center position, the best track is

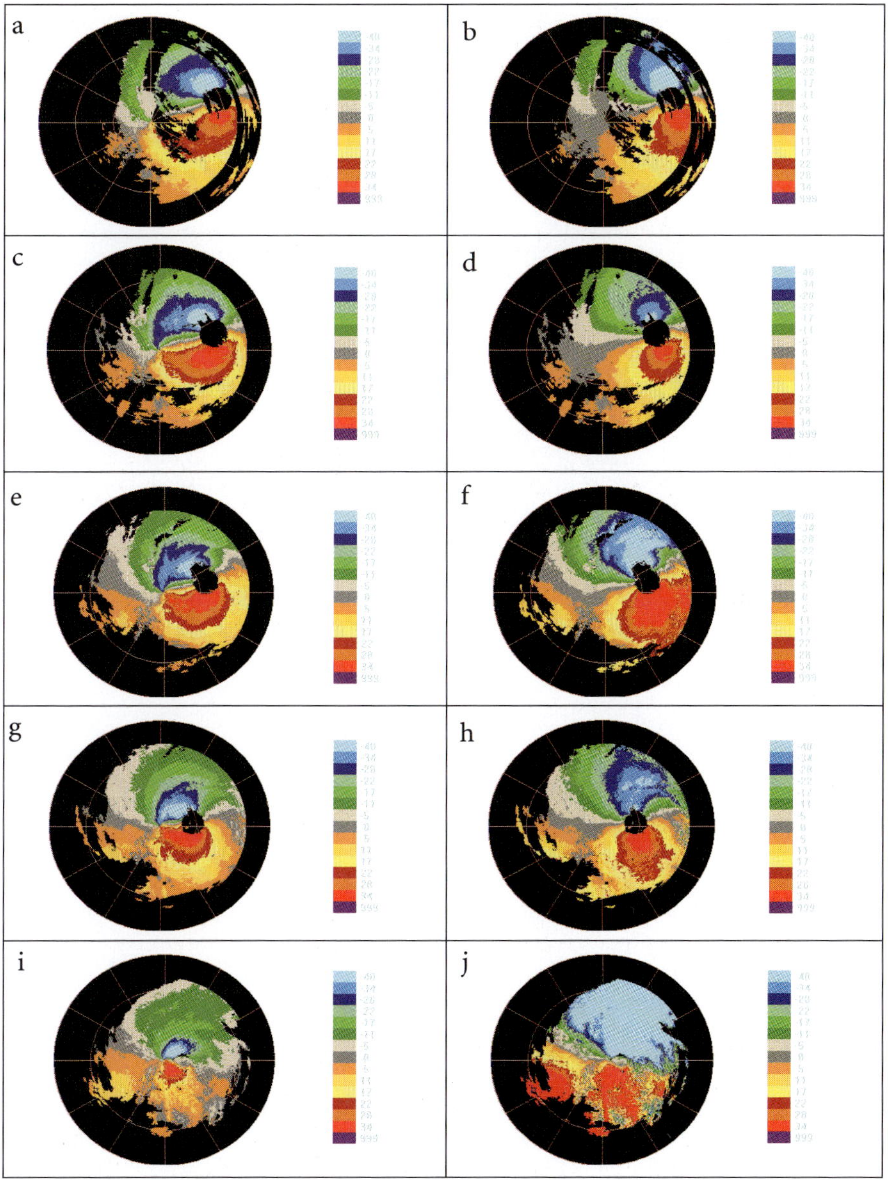

Fig. 18.7. The observed Doppler radial velocity V_r (**a**,**c**,**e**,**g**,**i**) and $V_r \times D$ (**b**,**d**,**f**,**h**,**j**), **a** and **b** are at 0000UTC, **c** and **d** are at 0300UTC, **e** and **f** are at 0600UTC, **g** and **h** are at 0900UTC, **i** and **j** are at 1200UTC

served as the real-time center position. Test results also show that the center position errors of GACPM and VADAPM are from 3 to 13 km and sometime the errors are less than 3 km (Table 18.3, Table 18.4).

Table 18.3. The position and the Radius Maximum Wind (RMW) errors of Actual Test with Geometric Axisymmetric Center Positioning Method (GACPM)

Real-time Center			The Geometric Axisymmetric Center Positioning Method (GACPM)				
			Result of Test		Error		RMW (km)
Time (UTC)	Lat. (°N)	Long. (°E)	Lat. (°N)	Long. (°E)	Lat. (°N)	Long. (°E)	
00	25.5	123.1	25.52	123.10	0.02	0.00	32.74
01	–	–	25.48	123.00	–	–	33.99
02	–	–	25.42	122.91	–	–	30.62
03	25.5	122.8	25.38	122.87	0.12	0.07	36.20
04	–	–	25.38	122.76	–	–	31.41
05	–	–	25.37	122.68	–	–	31.96
06	25.4	122.6	25.35	122.57	0.05	0.03	33.15
07	–	–	25.25	122.50	–	–	30.79
08	–	–	25.24	122.47	–	–	34.02
09	25.2	122.4	25.13	122.44	0.07	0.04	28.43
10	–	–	25.06	122.33	–	–	24.17
11	–	–	25.04	122.20	–	–	19.31
12	25.0	122.2	24.99	122.17	0.01	0.03	14.18
13	–	–	24.99	121.93	–	–	8.17

Table 18.4. The position and the Radius Maximum Wind (RMW) errors of Actual Test with Velocity Distance Azimuth Display Positioning Method (VADAPM)

Real-time Center			The Velocity Distance Azimuth Display Positioning Method (VADAPM)				
			Result of Test		Error		RMW (km)
Time (UTC)	Lat. (°N)	Long. (°E)	Lat. (°N)	Long. (°E)	Lat. (°N)	Long. (°E)	
00	25.5	123.1	25.53	123.13	0.03	0.03	33.69
01	–	–	25.52	123.03	–	–	32.45
02	–	–	25.47	122.98	–	–	30.76
03	25.5	122.8	25.38	122.84	0.12	0.04	36.48
04	–	–	25.38	122.75	–	–	38.82
05	–	–	25.33	122.67	–	–	31.22
06	25.4	122.6	25.32	122.59	0.08	0.01	33.37
07	–	–	25.31	122.53	–	–	34.83
08	–	–	25.23	122.47	–	–	39.83
09	25.2	122.4	25.12	122.47	0.08	0.08	29.58
10	–	–	25.09	122.34	–	–	28.94
11	–	–	25.06	122.27	–	–	22.63
12	25.0	122.2	25.04	122.12	0.04	0.08	24.31
13	–	–	24.91	121.98	–	–	22.32

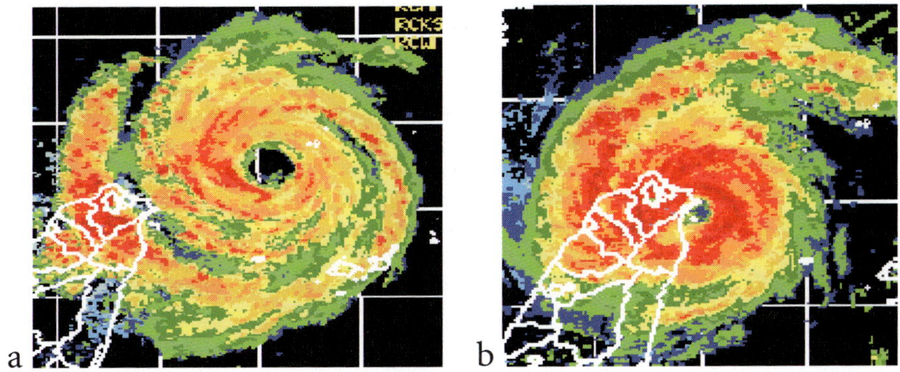

Fig. 18.8. The Radar Echo Images of Typhoon Nari on 16 September 2001. (**a**) 0000UTC, (**b**) 1300UTC (Source: Taiwan Province Meteorological Bureau)

From the result of actual test, it is evident that the radiuses maximum wind gradually reduced when Typhoon Nari was approaching toward and was making landfall over Taiwan. It is consistent with the lessening of the radiuses of eye wall echo (Fig. 18.8). The reason may be that its low-level circulation converges toward the typhoon center with orographic forcing effect when Nari was approaching the land. And the radiuses maximum winds estimated by the two methods are about the same from 00 to 09 UTC. But after 09 UTC the difference between the radiuses maximum winds estimated by GACPM and the radiuses by VADAPM are increasing. These reasons maybe that asymmetric component of typhoon circulation was increasing by orographic effect or radial velocity observed by radar would become greatly deformed when Typhoon Nari was very close to the radar.

18.7 Conclusions and Discussion

1. The Geometric Axisymmetric Center Positioning Method (*GACPM*) and the Velocity Distance Azimuth Display Positioning Method (*VDADPM*) give a good position of a typhoon for operational typhoon forecasting. In this case, their accuracy of positioning is less than 13 km, sometimes less than 3 km.
2. The results also show that the radius maximum wind gradually reduced when Typhoon Nari approached towards and was making landfall over Taiwan. It is consistent with the lessening of the radius of eye wall echo.
3. Because *GACPM* and *VDADPM* are based on the hypotheses in which the typhoon circulation is axisymmetric and there is the radius maximum wind (RMW) in the typhoon circulation, a big error will occur when the asymmetric component or environmental mean flow is evident.
4. Because of the velocity ambiguity in observed Doppler radial velocity field, the technique of velocity unfolding (dealiasing) directly influences the accuracy of the center position.

Acknowledgements

Many thanks to Mr. Paoliang Chang at Taiwan Province Meteorological Bureau for providing a lot of Doppler radar data and for many beneficial discussions about some technical details.

References

Anthes RA (1982) Tropical Cyclones: Their Evolution, Structure and Effects. Amer. Meteor. Soc. Boston, 208.

Chang P-L (2000) Varied Analysis of Landfalling Typhoon Circulation: A Case Study on Ttyphoon Herb (1996), The Ph. D. Thesis of Taiwan University Atmospheric Sciences, 155.

Desrochers PR and Donaldson RJ Jr. (1992) Automatic tornado prediction with an improved mesocyclone detection algorithm. Weather and Forecasting, 7, 373–388.

Donaldson RJ Jr. (1970) Vortex signature recognition by a Doppler radar. J. Appl. Meteor. 9, 661–670.

Griffin IS, Burpee RW, Marks FD Jr. and Franklin JL (1992) Real-time airborne analysis of aircraft data supporting operational hurricane forcasting. Weather and Forecasting, 7, 480–490.

Jou BJ-D, Deng SM and Chang P-L (1996) The Application of Doppler Radar in the Decision of typhoon center and radius maximum wind, Atmospheric Science (Taiwan Province), Vol. 24, 1–24.

Jou BJ-D and Lee WC etc (2001) Application System of Doppler Radar Data on Severe Weather Forecasting in Taiwan Province.

Lee WC, Jou BJ-D, Chang P-L and Deng SM (1999) Tropical cyclone structure retrieved from single Doppler radar observations. Part I: Interpretation of Doppler velocity patterns and the GBVTD technique on analytic tropical cyclone. Mon. Wea. Rev. 127, 2419–2439.

Lee WC and Marks FD (2000) Tropical cyclone structure retrieved from single Doppler radar observations. Part II: The GBVTD- Simplex center finding algorithm. Mon. Wea. Rev. 128, 1925–1936.

Li Wei, Tang Dazhang and Su Weidong (1999) Automatic Detection of Typhoon by Using a Single Doppler Radar, Journal of Nanjing Institute of Meteorology, Vol. 22, No. 2, 233–237.

Marks FD, Houze RA and Gamache JF (1992) Dual-aircraft investigation of the inner core of hurricane Norbert. Part: Kinematic structure. J. Atmos. Sci. 49, 919–942.

Senn HV and Hiser HW (1959) On the origin of hurricane spiral bands. J. Meteor. 16, 419–426.

Wilson JW and Wilk KE (1982) Nowcasting applications of Doppler radar. Nowcasting, K.A. Browning, Ed. Academic Press, 87–105.

Wood VT (1994) A technique for detecting tropical cyclone center using a Doppler radar. J. Atmos. and Oceanic Tech. 11, 1207–1216.

Wood VT and Brown RA (1986) Single Doppler velocity signatures interpretation of non-divergent environmental winds. J. Atmos. Oceanic Technol. 6, 114–128.

Wood VT and Brown RA (1983) Single Doppler velocity signatures: An atlas of patterns on clear air/widespread precipitation and convective storms. NOAA Tech. Memo. ERL NSSL-95, Nation Severe Storms Laboratory, Norman, OK.

CHAPTER 19

Application of Remote Sensing and GIS for Analysis of Forest Fire Risk and Assessment of Forest Degradation

Allen R. Riebau, John Jianhe Qu

Abstract Forest, shrub-steppe, and grassland fires have come under the inclusive terminology of wildland fire. In the past decade, fire occurrence and area of surface burned has increased dramatically. This increase in wildland fire frequency and intensity has been driven by climate variability and in some countries, worsened by fuel management policies. Many areas that have been burned are in remote areas of the world where collection of information on fires is either difficult or impossible on the ground. Satellite remote sensing has been applied to this problem with some remarkable successes in locating large fires, mapping the areas burned, and tracking smoke plumes. For smaller fires, pixel size limitations have led to use of aircraft based sensor development. To understand fire ecological consequences, parameters such as fire intensity, fuel consumed, and heat release rate to atmosphere must be understood. In parts of the Southwest Pacific Region, the potential for fire occurrence is exacerbated by lack of infrastructure common to more developed areas. The use of remote sensing, geographic information systems, and computer models will allow fires to manage to lower risks.

19.1 Introduction

During the past decade wildland fire occurrence in the United States has increased dramatically. The consequences of these increasing wildland fires have been devastating and include loss of life and structures. Increasingly, it has become recognized that fires are becoming more intense and more difficult to control. This has resulted from a number of issues including fire management policies that have resulted in even forest canopies and extreme fuel build-up. Another exacerbating factor has been droughts and overly warm winter seasons increasing the spread of forest pests. Rangeland systems are also suffering from drought events, with increasing fire severity. Although the United States is a developed and wealthy country, the sheer volume of fires has stretched its resources to detect fires by traditional means beyond the breaking point. Additionally, fire smoke is causing increasing concern in the United States with the size of plumes reaching beyond regional scales (Fig. 19.1; Fox and Riebau, 2000). During the period from 1991 to 2000 there were an average

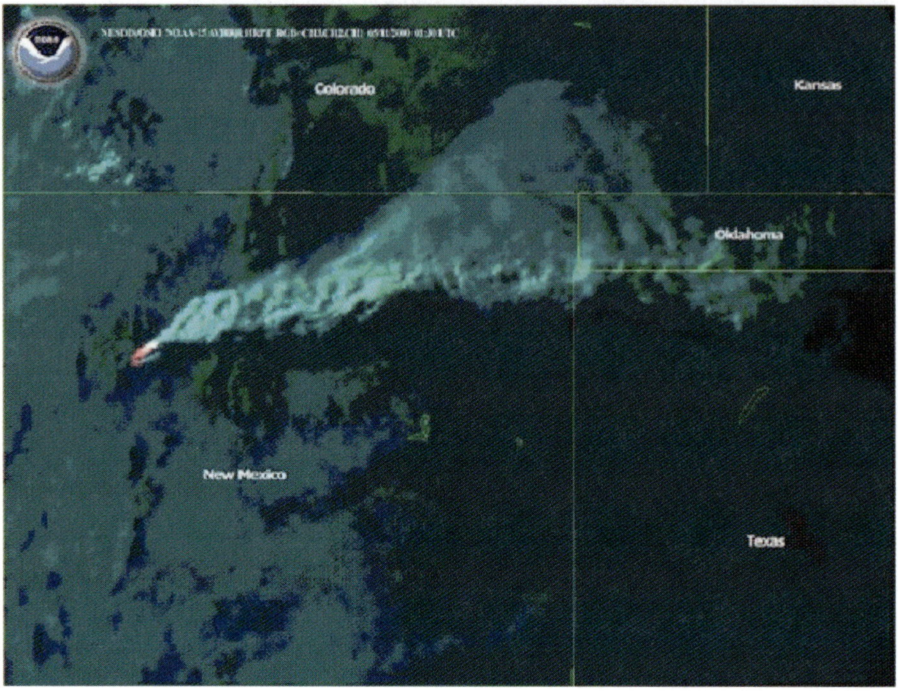

Fig. 19.1. Smoke Plume from the New Mexico El Centro Wildfire as imaged by GOES Satellite covers a large region (Source: NOAA NESDIS)

of 80,303 wildfires in the U.S each year burning an average of 3.7 million acres (Fig. 19.2). In 2002 there were 88,458 fires in the United States burning 6,937,584 acres costing an estimated $ 1.6 billion for fire suppression (source: US National Interagency Fire Center). In addition, more than 1.4 million acres of forests and rangelands were deliberately burned on average in each of the last six years in prescribed burns to reduce the danger of wildfires (in 1999 2,240,104 acres and in 2000 1,077,314 acres, for example; Source: U.S. National Interagency Fire Center). To address this growing problem, the United States has increased its investment to manage fire beginning in 2001 by approximately $ 1.8 billion per year under an initiative named the National Fire Plan (USDA, 2001).

To understand wildland fire one must understand fire fuels. Fuels have a number of characteristics that influence fire behavior and emissions to the atmosphere. The first is fuel loading, or the amount of fuel that is available to burn. Fuel loading is also closely related to fuel type, which in natural vegetation incorporates such issues as the amount of oils and volatile compounds in fuels that affect the combustion process. Another factor that must be considered is the structure of the fuel bed itself, which includes issues such as the occurrence of "ladder fuels" or those forest fuels that allow fire to spread upwards into the

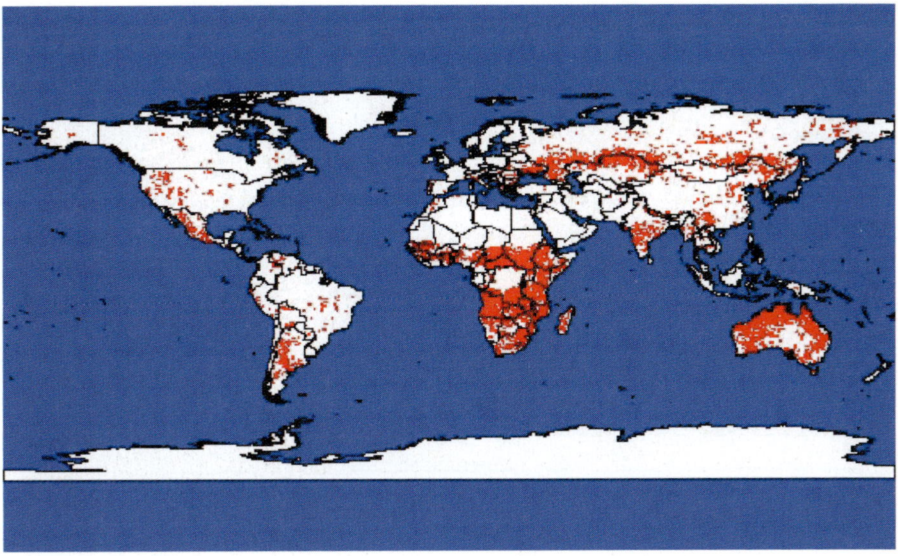

Fig. 19.2. Planetary area of wildland fire (in *red*) for the severe fire occurrence year 2000 (Source: FireGlobe)

forest canopy. Finally, there are very important ephemeral physical properties of fuels such as fuel temperature and moisture, which change with weather.

For mapping all fuels parameters, geographic information system (GIS) technology has been applied, often with the GIS capabilities being joined with descriptive and predictive models. Newer satellite sensors are also allowing direct detection of fuels characteristics on meaningful scales that can be displayed using GIS technologies.

Fire itself is controlled by the atmosphere in a complex reciprocal series of events. First, the risk of fire is a function of both current weather and weather over the past several days or weeks. Although most wildland fires are human ignited, the occurrence of dry lightening is also of paramount importance for fire ignition in areas that are remote. Fire behavior, which includes such measures as fire line intensity, heat release rates, and rate of spread, is also a function of the state of the atmosphere.

Finally, smoke emissions are controlled by the atmosphere's influence on fuel conditions and the combustion process itself. Fire also influences the atmosphere in return, with large intense fire often creating weather anomalies at the fire front. From a practical standpoint, wildland fire managers often combine weather, fuels, and fire information in simple indices to help them estimate how severe or dangerous fires are likely to be (Tables 19.1 and 19.2).

Fire emissions to the atmosphere are becoming of increasing importance. At several sets of temporal and spatial scales, fires are a major concern in the global carbon cycle. Inventories of fires and fire emissions in large areas of the world

Table 19.1. Fire Behavior, Controllability, and Fireline Intensity (Source: USDOI Bureau of Land Management)

Burning Index	Fireline Intensity BTU/S/Ft	Flame Length FT	Narrative
0–28	0–50	2.8	Most prescribed burns are conducted in this range.
38	100	3.8	Generally represents the limit of control for manual attack methods.
78	500	7.8	The prospects of control by any means are poor above this intensity.
96	700	9.6	The heat load on people within 30 feet of the fire is dangerous.
108	1,000	10.8	Ave this intensity, spotting, fire whirls, and crowning should be expected.

Table 19.2. Severe Fire Behavior Potential Related to Relative Humidity and Fuel Moisture Content (Source USDOI Bureau of Land Management)

RH%	1-HR F.M. %	10 HR F.M. %	Relative ease of change ignition and spotting, general burning conditions
> 60	> 20	> 15	Very little ignition; some spotting may occur with winds above 9 mph.
60	15–19	12–15	Low ignition hazard - campfires become dangerous; glowing brands cause ignition when RH is < 50%.
30–45	11–14	10–12	Medium ignitability - matches become dangerous; "easy".
26–40	8–10	8–9	High ignition hazard - matches always dangerous; occasional crowning, spotting caused by gusty winds; "moderate" burning conditions.
15–30	5–7	5–7	Quick ignition, rapid buildup, extensive crowning; any increase in wind causes increased spotting and crowning, loss of control; fire moves up the bark of trees igniting aerial fuels; long distance spotting in pine stands; dangerous burning conditions.
< 15	< 5	< 5	All sources of ignition dangerous; aggressive burning, spot fires occur often and spread rapidly, extreme fire behavior probable; critical burning conditions.

were almost completely unreliable until the application of remote sensing in the 1970s (Qu and Omi, 1993). On a local scale, cities have had extreme smoke events that have endangered public health and welfare (Chubarova et al., 2003). In the United States, fire smoke has become highly regulated under air

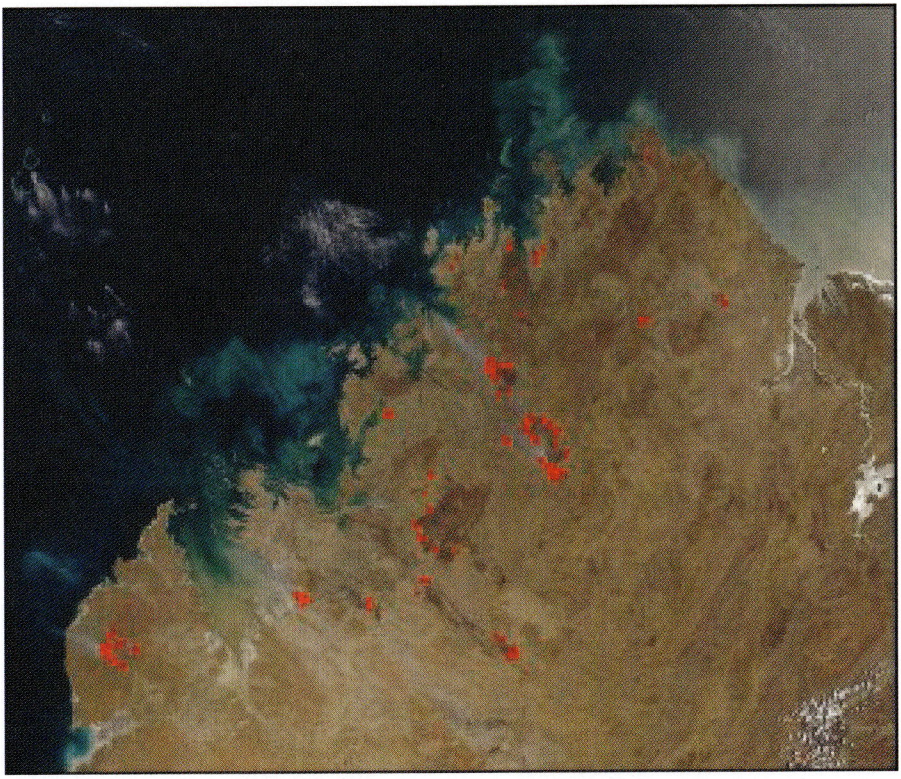

Fig. 19.3. MODIS image of wildfires in Western Australia's Kimberly Range (Source: NASA)

quality law (Fox and Riebau, 2000). Remote sensing of smoke has become common practice for many of these issues and the relatively new Moderate Resolution Imaging Spectroradiometer (MODIS) (Fig. 19.3) instrument has proven very useful both for location of fires (Fig. 19.4) and smoke plume trajectories (Fig. 19.5).

The use of simulation models is also increasing in fire management. First several systems of models for fire risk have been developed and applied. Additionally, new modeling tools for fire behavior are being developed and applied. Perhaps one of the newest approaches to fire management in the United States has been the application of mesoscale meteorological models to fire management, allowing predictive capability of important fire weather parameters with high terrain definition, temporal resolution and projection, and linkage with more traditional fire models. Finally, these new models have been applied to predict the impacts of fire smoke (Riebau et al., 2003).

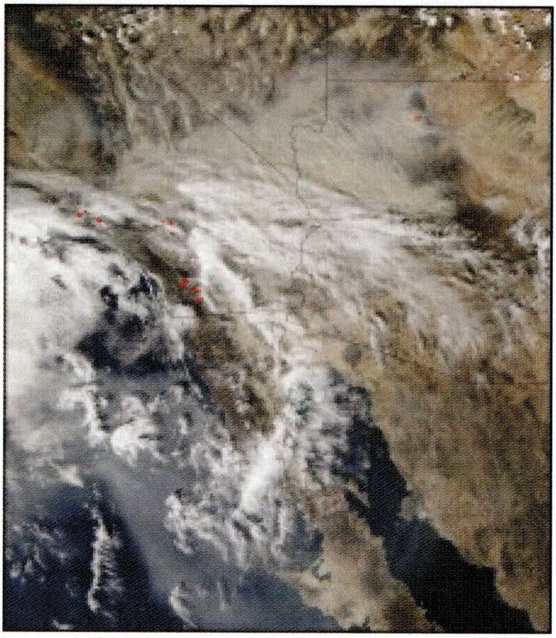

Fig. 19.4. MODIS images of Southern California smoke plumes from fires in the fall of 2003, showing how fires in mountains areas can be transported into urban centers (Source: NASA)

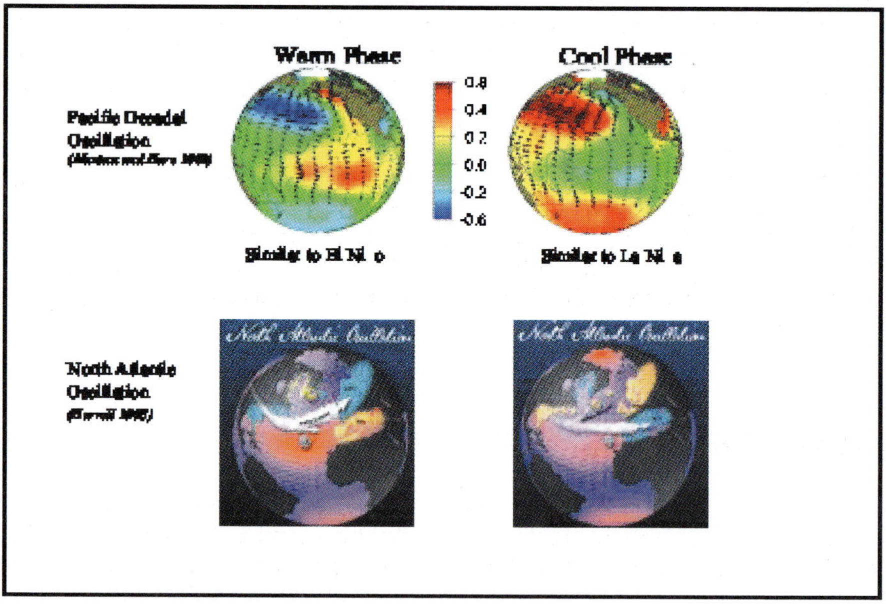

Fig. 19.5. Ocean surface temperature phenomena influence the temperature regimes of the atmosphere and are strong predictors of fire occurrence and severity. In these graphics red is warmer and blue corresponds to colder regions (Source: Ron Nielson, USDA Forest Service)

19.2 Climate Change, Climate Variability, and Fire Danger

Atmospheric conditions and forest fuel properties (especially, fuel moisture) are two major factors in potential wild fire severity. Because of the significant inter-annual variability in atmospheric conditions, fire severity may change dramatically from one year to another. The El Niño and Southern Oscillation (ENSO), for example, can result in increased chances of droughts in some regions of the U.S. and floods in other regions and, as a result, affect wild fires (Qu et al., 1994, and Qu and Wolf, 2003). The USDA Forest Service reported the lowest rangeland fire occurrence and area burned since record-keeping began in 1906 during the 1982–83 strong ENSO event. Fire activity is statistically weak in the southern hemisphere during ENSO events. The influence of climate on fire regimes was apparent in the regional synchrony of large fires with reduced fire activity with low ENSO indexes (Aber et al., 2001). ENSO can also contribute to the inter-annual variability of vegetation in the North America (Bachelet et al., 2001). It is critical to these events when assessing the potential severity of the forthcoming fire season to enable fire management to respond under the expected conditions (Fig. 19.5).

A large number of efforts have been made toward developing such capacity using traditional techniques mainly based on calculation and analysis of meteorological data. Monthly fire potential and burning index (e.g., Fosberg Burning Index, Fosberg 1978) were estimated in monthly fire weather prediction. Global and regional climate models have been used to obtain high-resolution distribution of seasonal trend of precipitation and temperature, which then are used to predict fire risk using statistical techniques (Fox and Riebau, 2000). Some comprehensive fire danger tools, for example, the U.S. National Fire-Danger Rating System (NFDRS, Deeming et al. 1978, Burgan et al., 1998) and the Canadian Forest Fire Danger Rating System, issue a fire occurrence index or fire weather index for estimating the possibility of potential wildland fires. These systems include calculations of indices (e.g., Keetch-Byram Drought Index (KBDI)) and the Canadian Drought Index, which are useful for long-term fire potential assessment.

19.3 Remote Sensing Applications of Fire Fuels Monitoring

Satellite remote sensing is a new technique for wild fire monitoring and fire danger assessment. Satellite instruments like the Advanced Very High Resolution Radiometer (AVHRR) and the Moderate Resolution Imaging Spectroradiometer (MODIS) provide global high-resolution products like the Normalized Difference Vegetation Index (NDVI) and Surface Temperature (ST), which have shown close relationships with fuel moisture status. Thus, remote sensing can overcome some problems facing the traditional techniques, including low spatial resolution and unavailability of meteorological data over part of forest regions.

Remote sensing mapping and characterization of vegetation and fire fuels parameters and datasets were clearly defined as technology goals in strategic documents such as the US National Fire Plan and have been required by international fire management and science communities. Applications of satellite remote sensing have been made during many large wild fire areas, including monitoring fires lines and total fire smoke (Hao et al., 2003). The Wildland Fire Assessment System (WFWAS) developed by the USDA Forest Service, became operational in middle 1990s, and has been providing useful information such as a "greenness" map using AVHRR-NDVI since (Burgan et al. 1998).

Satellite remote sensing also has great potential to provide information for calculation of seasonal fire danger. The fuel moisture detected by satellite remote sensing is mostly the moisture of live fuel, which predominately represents long-lag moisture (e.g., 1,000 hour fuel moisture), a determining factor of seasonal fire danger. Thus, it is possible to develop a capacity to assess the seasonal fire danger using remote sensing of fuel moisture and other products. These products are limited by spatial resolution of 1 km or more and thus are generally only representative at meso or synoptic scales.

The estimation of forest fire danger from satellite remote sensing data is an important research area, with potential for great practical application. Fuel moisture is an important index for fire potential. Over the past decade, research on fire danger estimation from remote sensing data has been concentrating on determining fuel moisture. The accurate estimation of fuel moisture using remote sensing data is very difficult, most of the approaches use proxy variables as indices of fuel moisture, and most available techniques can be placed into two categories: 1) methods based on the relationship of Land Surface Temperature (LSTs) and Normalized Difference Vegetation Index (NDVI); and, 2) methods based on regression analysis of vegetation indices directly.

As a key research instrument of the NASA Earth Observing System (EOS) missions, the Moderate Resolution Imaging Spectroradiometer (MODIS) instrument was launched onboard the NASA Terra and Aqua satellites. Because the MODIS instrument senses the earth's entire surface in 36 spectral bands, it spans the visible (0.415 μm) to infrared (14.235 μm) spectrum at nadir, with spatial resolutions of 1 km, 500 m and 250 m, respectively. One significant strength of MODIS, especially important to fuel property estimation, is that strong absorbance of water in the middle infrared region (1.3 to 2.5 μm) makes this band most suitable for the estimation of forest fuel moisture content. This band is not included in most satellite systems, suggesting that MODIS should be useful for improving the capacity of applying satellite remote sensing data to more accurately estimating fuel moisture content and fire danger indices.

Qu et al. (2003) computed fuel moisture leveraged on Burgan's algorithm using both MODIS vegetation index measurements. MODIS provides two VI products, NDVI ad EVI (enhanced vegetation index) globally over land with 250 m and 500 m spatial resolution. EVI was developed to improve vegetation signal sensitivity in high biomass regions and reduce the influences of canopy background signals and atmosphere.

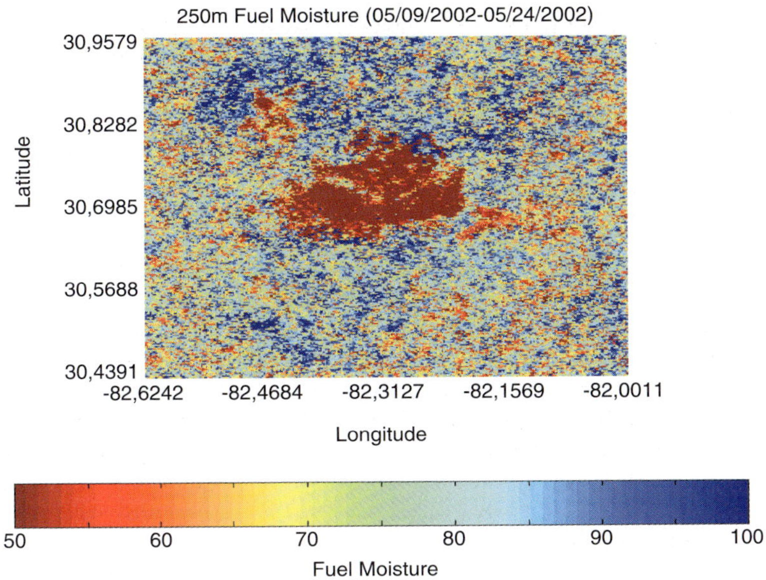

Fig. 19.6. Fuel moisture measurement in Okefenokee National Wildlife Refuge using the MODIS instrument depicts areas where fire is likely to consume vegetation (Source: Qu and Lui)

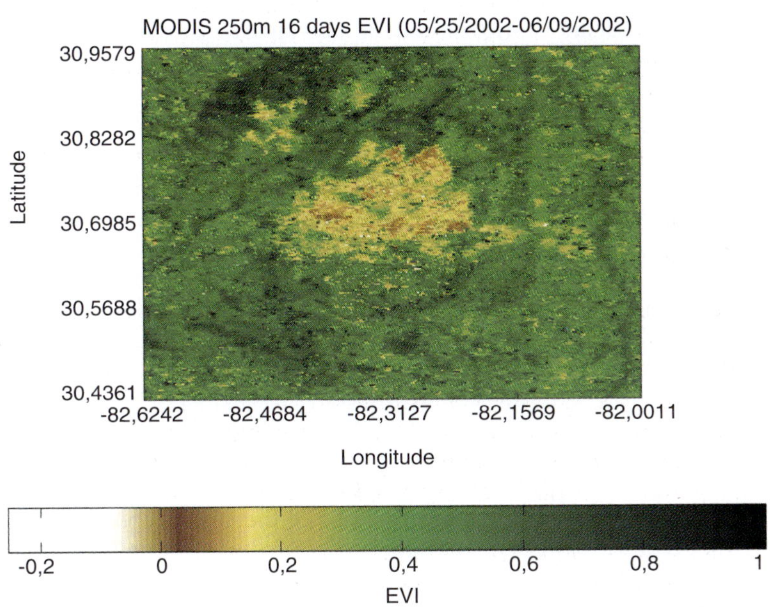

Fig. 19.7. NVDI/EVI measurement in Okefenokee National Wildlife Refuge using the MODIS instrument shows where fire burned, corresponding well with MODIS derived fuel moisture (Source: Qu and Lui)

A swamp wildfire in the Okefenokee National Wildlife Refuge (Georgia, USA) spread over 11,000 acres during May 2002. Figure 19.6 shows fuel moisture before this large fire during May 2002. The 16-day-composited 250 m NDVI image shows the vegetation change after the fire (Fig. 19.7). A good agreement between the fuel moisture and burn areas can was demonstrated in this study.

19.4 GIS Applications and Fire Information Mapping

The GIS as a tool can provide the functions of spatial and temporal geographic classification, analysis, and result mapping to support wildland fire management. GIS is an important tool to integrate the information of risk conditions, fuel conditions, forest value, NDVI, land quality, weather patterns, topographic data and other fire risk related factors. GIS can thus compute and generate fire danger maps based on the knowledge base of traditional and state-of-the-art remote-sensing fire-risk model.

A new major program, LANDFIRE, will use GIS technology to map all wildland fire fuels across the USA at 30-meter spatial resolution. LANDFIRE will develop a comprehensive package of spatial data layers, models, and tools in support of analyses for prioritization and planning to initiate the implementation of the National Fire Plan, both at the national and local level. LANDFIRE is a mid-scale project targeting map accuracies of 60 to 80 percent for the sub-watershed level (10,000 to 40,000 acres). The spatial datasets for LANDFIRE will be maintained at a 30-meter pixel size. LANDFIRE is intended to be the safety net for land management agencies that do not have local-scale information, and the project is not a substitute for finer scale, local mapping efforts. It is intended to be scalable from sub-watersheds to a national level. While LANDFIRE will develop many layers that will be key for mapping the wildland/urban interface, LANDFIRE will not create a wildland/urban interface map (US DOI, 2002).

19.5 Fire Smoke and Air Quality

Emissions from fires are lofted high into the atmosphere and can significantly impact air quality on local, regional, and even global scales. These emissions can cause problems with air quality regulatory programs for particulates, ozone, and visibility. The smoke produced is the most obvious air quality impact, but certainly not the only one. Forest fire plumes Forest fire plumes are known to contain a highly variable mix of gaseous and particulate pollution and are difficult to inventory (Sestak et al., 2002). This mix is farther modified by atmospheric reactions and surface deposition as the plume drifts downwind.

In September 2000, the USDA Forest Service's Deputy Chief for Research and Development issued a request for NFP research proposals to the Research branch of the Forest Service. In response to this request and these research needs, scientists within the atmospheric research program of the USDA Forest Service developed a research and product development strategy built around the concept of regional modeling consortia. Scientists at five USDA Forest Service research stations are developing five regional Fire Consortia for the Advanced Modeling of Meteorology and Smoke (FCAMMS), as shown in Fig. 19.8. Each regional Consortium has been established as a multi-agency coalition of researchers, fire managers, air-quality managers, and natural resource managers at the federal, state, and local levels.

The missions of the consortia are to:

- Increase understanding of fire-weather, fire danger, fire behavior, and the transport, diffusion and impact of smoke from fire;
- Develop and implement new technologies related to regional fire-weather and air-quality;
- Enhance our ability to use fire for land management purposes; and,
- Improve tools for fire fighters by better predicting the dangers of wildfire.

In order to deliver this mission, the FCAMMS are developing and maintaining contemporary computational and data processing resources for US fire fighting and smoke management use. The computer requirements needed to simulate high-resolution meteorology include very large data processing and

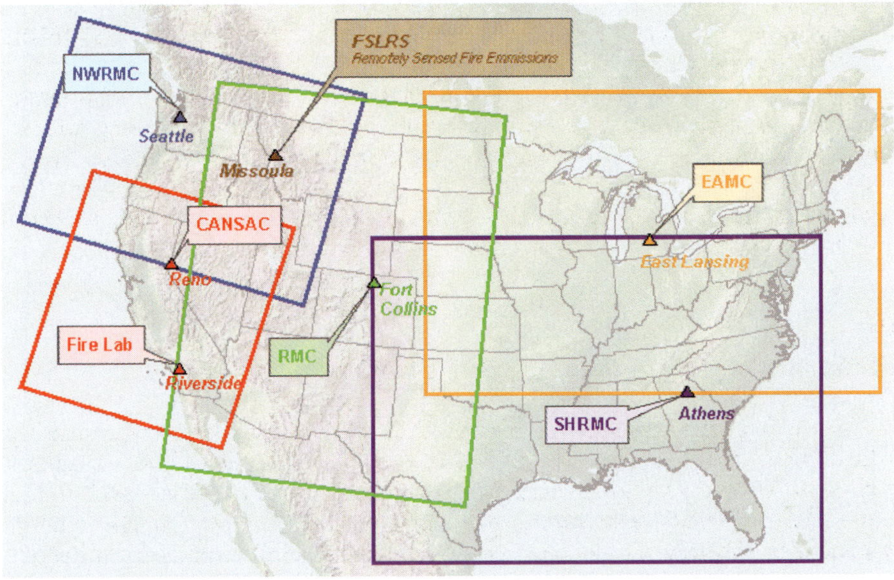

Fig. 19.8. Locations of the five Fire Consortia for Advanced Modeling of Meteorology and Smoke (FCAMMS) established under the United States of America National Fire Plan and the 12 km spatial grid weather modeling domains covered by the individual consortia (Source: USDA Forest Service)

storage capacities, high-speed calculation capabilities, broadband communications facilities and web-based, map-based graphical displays (Fig. 19.9). In general, these capabilities are being provided through the use of clusters of multiple PC processors operating under LINUX systems, utilizing large RAID disk array data stores and relying on the World Wide Web for communication. To a large extent these represent new and different ways for Forest Service to do business that have increased the challenge to consortium members. FCAMMS have also led to significant improvements in the internal capacity of the USDA Forest Service Research and Development through hiring scientists and technicians with modern computer and atmospheric sciences skills.

A major justification for the FCAMMS is to provide intelligence for managing smoke from forest burning. Smoke management has strategic planning, tactical planning and operational aspects. For the purpose of strategic

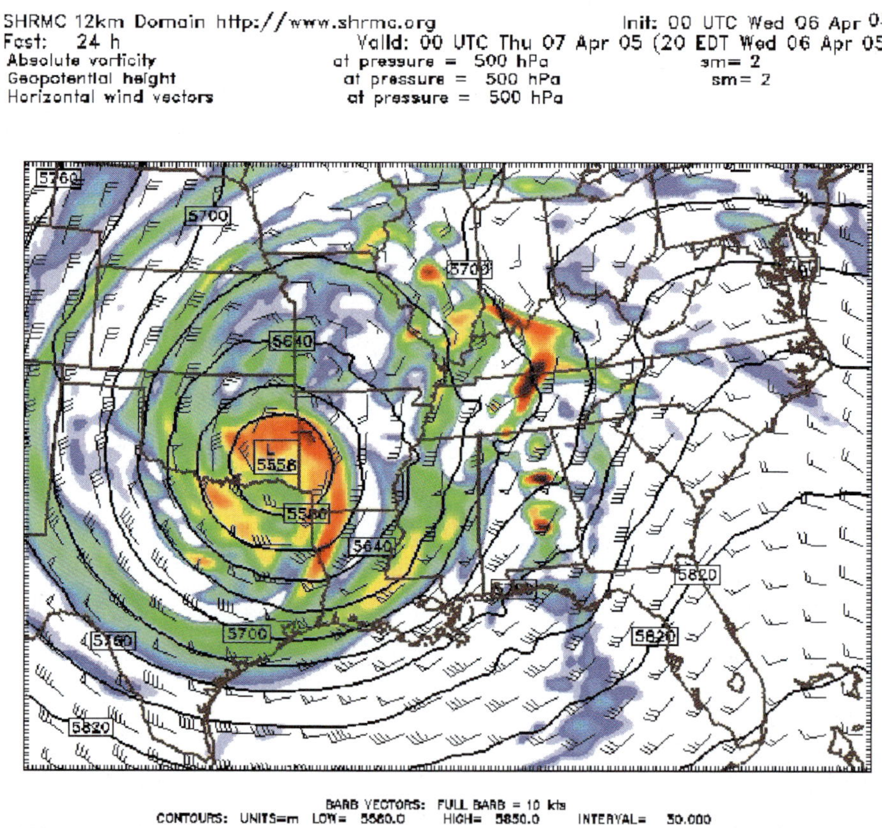

Fig. 19.9. The Southern High Resolution Modeling Center (SHRMC) produces fire weather and fire indices forecast up to 48 hours into the future (Source: USDA Forest Service)

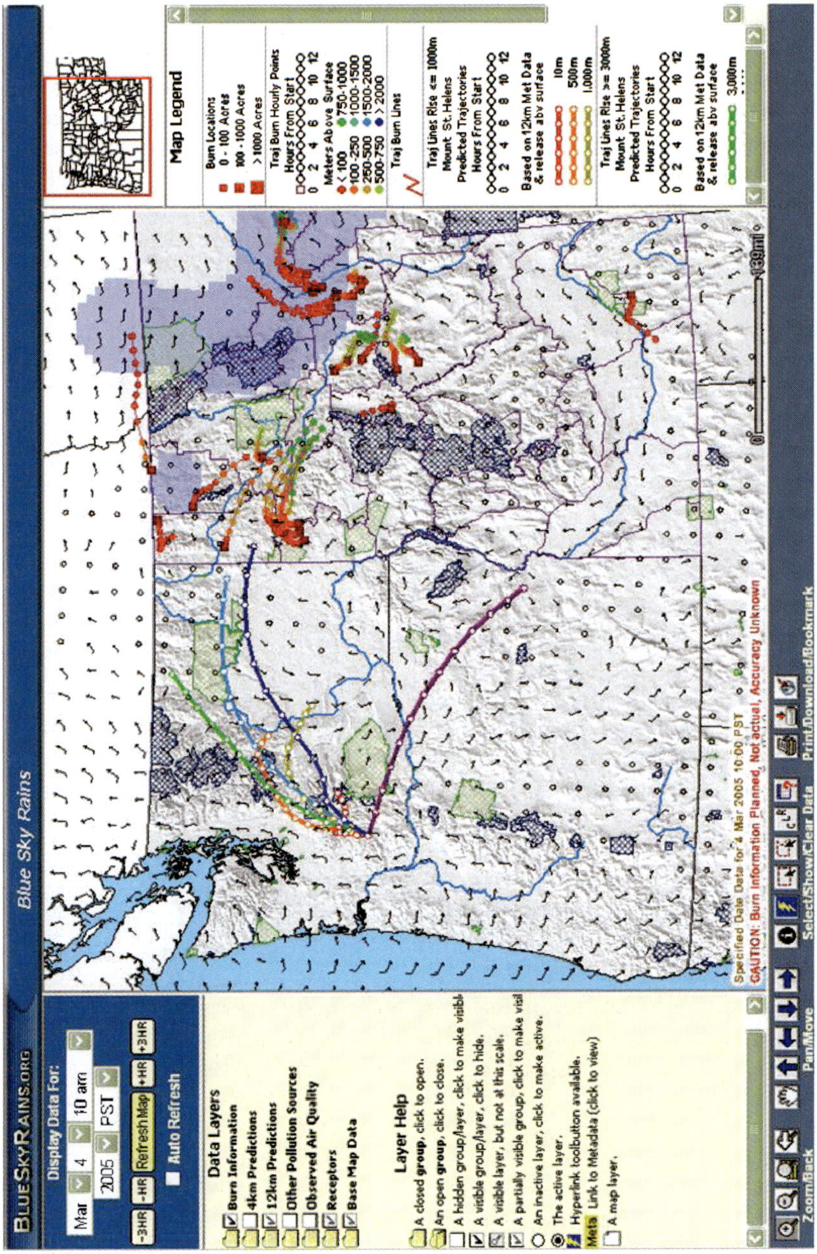

Fig. 19.10. The BlueSkyRAINS combines fire modeling, mesoscale meteorology modeling, and smoke dispersion modeling within a web-based GIS framework. The new system is planned to be implemented across the USA by 2006 (Source: USDA Forest Service)

planning applications are needed that can consider regional scale impacts on issues such as the contribution to regional haze and ozone. For tactical planning, there is a need for a permitting tool; one that allows some indication of where smoke is likely to go once a fire is ignited. Finally, for the operational application, there is a need for real-time forecasts of smoke impacts from individual and groups of potential and actual fires. Although the FCAMMS will address all of these applications the latter, operational application is currently well developed through the BlueSky modeling framework. The BlueSky smoke-modeling framework is a flexible framework for obtaining smoke impacts by combining fire location information, forest fuels, outputs from MM5 (Greel et al., 1988) meteorological simulations and air quality dispersion models (e.g., the US EPA CALPUFF model) to illustrate where smoke will go and how much of it will get there. BlueSkyRAINS links the BlueSky smoke modeling framework with the US EPA Rapid Access INformation System (RAINS) based web serving technology. RAINS utilizes the ArcIMS / geographic information system to allow for data overlays from a variety of geographical data. The capability of the BlueSkyRAINS to provide information about the likely trajectories and ground level concentration impacts from prescribed fires is a major advance of FCAMMS (Fig. 19.10).

19.6 Summary and Conclusions

The use of remote sensing for wildland fire is rapidly growing. Notable successes have been made in detecting fires and mapping fire areas, especially for large fires. To understand the ecological consequences of fire and fire emissions effects to the global carbon cycle, remote sensing tools will need further development for measuring fire intensity and actual fuel consumed. Also needed are tools that can give fuel loading and condition parameters such as fuel moisture content and temperature in near real-time (every five to 10 hours). Higher spatial resolution is also needed, with high-resolution detectors such as the MODIS and its 250-meter resolution perhaps still not fine-scale enough for agricultural and other small fires. Fire smoke is also an issue of high concern. Remote sensing and GIS can be used for fire smoke, with remote sensing of fire location and emissions parameters being tied to weather and dispersion models for display of ground-level concentrations on the earth's surface.

References

Aber J, Neilson RP, McNulty S, Lenihan JM, Bachelet D, Drapek RJ (2001) Forest Processes and Global Change: Predicting the Effects of Individual and Multipule Stressors. Bioscience. Vol. 51 No. 9. 2001. 735–751.

Andrews PL, and Bevins CD (1992) BEHAVE Fire Modeling System Redesign and Exopnasion. Fire Management Notes, 59(2): 16–19.

Bachelet, Dominique, Neilson RP, Leniham JM, and Drapek RJ (2001) Climate Change Effects on Vegetation Distribution and Carbon Budget in the United States. Ecosystems. Vol .4. 164–185.

Battye W and Battye R (2002) Development of Emissions Inventory Methods for Wildland Fire. Final Report. Prepared for: Thompson G. Pace, U.S. Environmental Protection Agency EPA Contract No. 68-D-98-046. Prepared by EC/R Incorporated.

Burgan, Klaver, Klaver (1998) Fuel Models and Fire Potential from Satellite and Surface Observations, International Journal of Wildland Fire, 8(3): 159–170.

Burgan RE and Hartford R A (1993) Monitoring Vegetation Greenness with Satellite Data. United States Department of Agriculture, Forest Service, General Technical Report INT-297, Intermountain Forest and Range Experiment Station, Ogden, Utah. 13 pages.

Chubarova NYe, Prilepsky NG, Riebau AR, Shoettle AW, Musselman R, Uliumdzhieva NN, Yurova AYu, Potter B and Rublev AN (2003) A Natural Fire Experiment in Central Russia: Meteorology, Radiative Transfer and Optical Properties of the Atmosphere and Resulting Effects on Boreal Forest Plants. AMS Orlando Florida Specialty Conference on Wildfire. 5 pgs.

Deeming JE, Burgan RE and Cohen JD (1977) The National Fire-Danger RatingSystem – 1978. Gen. Tech. Rep. INT-39. Ogden, UT: U.S. Department of Agriculture, Forest Service, Intermountain Forest and Range Experiment Station. 63 pp.

Fosberg MA (1978) Weather in wildland fire management: the fire weather index. Proc. of the Conference on Sierra Nevada Meteorology, South Lake Tahoe, 1–4.

Fox DA and Riebau AR (2000). Technically Advanced Smoke Estimation Tools (TASET). Project Final Report. Colorado State University, Cooperative Institute for Research of the Atmosphere. Fort Collins, Colorado. 99 pgs.

Grell GA, Duhia J and Stauffer DR (1994) A description of the Fifth Generation Penn-State./NCAR Mesoscale Meteorological Model (MM5). NCAR Technical Note NCAR/TN – 398 + STR, 138 pages.

Haines DA (1988) A lower atmospheric severity index for wildland fire. National Weather Digest. Vol 13. No. 2:23–27.

Hao, WeiMin, Salmon JM, Nordgren BM, Kaufman YJ and Rong-Rong L (2003) Daily Emissions of CO and Pm2.5 from a Large Fire in the US Using MODIS Data. IGBP Workshop on the Integrated Land Ecosystem – Atmosphere Process Study. 4 pgs.

Pittock A (1990) The enhanced greenhouse effect and its agricultural impact. In the conference proceedings for *Climatic variations and change: Implications for agriculture in the Pacific Rim*, 46–55. Davis, CA: University of California, Public Service Research and Dissemination Program. 12 pgs.

Qu J, Kafatos M, Hao X, Riebau A and Liu Y (2003) Estimating of Fire Danger Potential Using MODIS Measurements, Processing of ISPRS (CD-ROM Publication).

Qu J and Wolf H (2003) Wildfire behavior science & modeling strategic planning, presentation for USDA/FS R&D, *National Fire Plan*, January 2003, 16–17, New Orleans, LA.

Qu J et al. (2002) Responding to the Challenge of Producing and Distributing MODIS Data at the NASA/GES DISC via the Remote Sensing Information Partners (RSIP) Program, *Proceeding of IGARSS' 2002* (CD-ROM publication).

Qu J, Serafino G, Kafatos M, Lathrop R and Trimble J (2001) New application of Earth science remote sensing data at NASA/GDISC: Remote Sensing Information Partner (RSIP) with Rutgers University. In *Proceedings. International Geoscience and Remote Sensing Symposium (IGARSS) 2001*, July 9–13, at University New South Wales, Sydney (CD-ROM, publication).

Qu J, Sestak ML, Riebau AR and Smith LR (1994) Study of El Nino and Southern Oscillation (ENSO) Impact on Drought and Wetness in the Western United States, CONFERENCE ON CLIMATE VARIATIONS ; 6th : Jan 1994 ; pp. 101–104.

Qu J and Phil Omi (1993) Potential Impact of Global Climate Changes on Wildfire Activity in USA. The proceeding of the 12th International Fire and Forest Meteorology, October 26–29, 1993, at Jekyll Island, George, USA.

Riebau AR, Fox DG, Hirami P, McClelland T, and Fisher R (2003) Fire Consortia for Advanced Modeling of Meteorology and Smoke – FCAMMS: A National Paradigm for Wildland Fire and Smoke Management. AMS Specialty Conference on Wildland Fire. 12 pgs.

Sestak ML, O'Neill S, Ferguson S, Ching J and Fox D (2002) Integration of Wildfire Emissions Into Models-3/Cmaq With The Prototypes: Community Smoke Emissions Modeling System (CSEM) and BlueSky. CMAS Conference extended abstracts.

United States Department of Agriculture (USDA) Forest Service (2001) "National Fire Plan Research and Development 2001 Business Summary". 36 pgs.

United States Department of the Interior (US DOI) (2002) LANDFIRE Prototype Information Bulletin # 1. 3 pgs.

Conclusions and Recommendations

Conclusions

Participants in the meeting concluded that:

1. With increasing incidence of natural disasters around the world, a comprehensive assessment of their impacts on agriculture, forestry, and fisheries and strategies for mitigation of natural disasters is critical for sustainable development, especially in the developing countries.
2. Progress in impact assessment has been made. The impact assessment of natural disasters is extended from agriculture to social and economy, local to regional, and national and global level.
3. Natural disasters are affected by various modes of climate. Strategies for adaptation and preparedness are required for these modes. This will have extremely profound implications for natural resources managers and agriculturalists as they attempt to adapt to and mitigate the effects of natural disasters.
   A. Climate Change. We have much evidence that the climate is changing. The IPCC TAR has concluded that most of the warming in the last half of the 20th Century is due to human activities.
   B. Climate Variability. Variability is most important in the near-term from a natural disaster viewpoint. These include both short term [ENSO] and longer-term variability forced by the oceans, including the PDO/IPO, NAO, and other such oscillations.
   C. A New State of Climate. The climate may be moving, and perhaps rather rapidly, to a new systematic state. A new systematic state may be inherently more variable (having more extremes) due to a warmer and thus more energetic atmosphere.
4. Variability of climate, if encompassing more intense and frequent extremes, will result in the occurrence of natural disasters that are beyond our socio-economic planning levels. This will stretch national response capabilities beyond their capacity and will require new adaptation and preparedness strategies.

5. Ecosystems, at least parts of them may be relics. This idea has not caught on with natural resource managers and perhaps not agriculturalist, but it certainly is difficult for policymakers.
6. We have no excuse for non-cooperation. Consortia approaches are a key feature in changing and adapting along the potential path to a new systematic climate state: International – Regional – Local – Community.
7. Each type of extreme event has its own particular climate, cultural, and environmental setting, and mitigation activities must use these settings as a foundation for proactive management.

Recommendations

1. Current definitions of natural disasters are primarily meant to address socio-economic management issues related to natural disasters. It is important to develop a more comprehensive definition of natural disasters for applications in agriculture, rangelands, forestry, and fishery sectors.
2. Assessment of the impact of natural disasters on agriculture, rangelands, forestry, and fisheries requires the design of a comprehensive data base in accordance with the users needs. There is a need for an integrated data management system, from adequate collection to quality control, analysis, presentation and also metadata not just meteorological data, but data specific to certain kinds of disaster, e.g., lightning incidence, soil moisture, and fire danger indices. Presentation should make use of best available technology, e.g., GIS and Internet.
3. Effective management of, and preparedness for, natural disasters requires free and unlimited access to relevant databases that will allow monitoring, assessment, and prediction. It is recommended that all agencies responsible for these databases develop good collaborating links for the exchange of information included in these databases.
4. There is an urgent need to assess the forecasting skills for natural disasters to determine those where greater research is needed. Lack of good forecast skill in drought, for example, is a constraint to improved adaptation, management, and mitigation.
5. There is a need to recognize the importance of understanding glacial lake outburst floods (GLOFS) as a new natural disaster and develop appropriate assessment and preparedness strategies.
6. Given the growing incidence of dust and sand storms around the world, it is essential to include measurements of aeolian sedimentation loads in the standard agrometeorological stations of NMHSs. It is also essential to include a routine and comprehensive analysis of wind speed and direction data and disseminate this information to the users. These data should be applied to analyze the impact of sand storms on agriculture. Use of air

quality networks to aid in data collection on dust and sand storms may also be examined.

7. Current natural disaster management is largely crisis driven. There is an urgent need for a more risk-based management approach to natural disaster planning in agriculture, rangelands, forestry, and fisheries. An effective risk management approach would include a timely and user-oriented early warning system with rapid dissemination of information to users.

8. Recognizing that early warning is an important component of preparedness, it is recommended that:
   A. The concept of the drought monitor map product be promoted as a tool for all drought-prone countries to better understand drought severity using multiple indicators.

   B. A white paper on the methodology for preparation of a drought monitor map be developed with recommendations for minimum, maximum, and optimum data layers needed to successfully accomplish this task. Recommendations on the importance of inter-agency collaboration should be included to accomplish this task.

   C. The feasibility of organizing joint training workshops on national and regional drought monitor products under the auspices of WMO and the NDMC should be examined. Indices used in China in their agrometeorological bulletin could be effective training tools. The first of these workshops should be organized in China.

9. Given the importance of storm surges to coastal lowlands, it is essential that WMO, in collaboration with other international and regional agencies, develop an integrated coastal management approach in reducing the impacts of natural disaster on agriculture, rangelands, forestry, and fisheries.

10. The issue of distinguishing long-term climate variability (e.g., IPO) and long-term climate change is important, as is the need to consider the impacts of both on agriculture, water resource management, and disasters such as bushfires. This is important because there are implications for long-term sustainability of certain types of activities, especially agriculture. There is modeling work at some overseas institutions (e.g., the Hadley Centre) that would be of relevance here. These issues need to be drawn to the attention of national policymakers.

11. Agricultural risk zoning is an essential component of natural disaster mitigation and preparedness strategies. Given the complex nature of databases, GIS and remote sensing should be employed to facilitate strategic and tactical applications at the farm and policy levels.

12. A number of modern tools and methodologies for the monitoring and prediction of natural disasters, such as storm surges, tropical cyclones, drought, floods, etc., are now available. It is recommended that an integration of GIS, remote sensing, simulation models, and other computational

techniques be used to develop more effective early warning alerts of natural disasters.

13. It is recommended that countries develop policies aimed at effective natural disaster management. Such policies should emphasis preparedness and incentives over insurance, insurance over relief, and relief over regulation.
14. Education and training is an important component of natural disaster management in agriculture, forestry and fisheries. It is recommended that strategies for education and training address the needs at national, regional, and international levels in order to exploit the synergies and share experiences.
15. Best practice strategies for developing and implementing education and training programs in support of drought preparedness policies should be documented from the experiences of countries such as Australia, China, India, and the United States. Elements of this include consulting the users; developing skills in an influential "target group," user-producer workshops; and adequate feedback channels.
16. Community involvement and education is essential in preparedness and mitigation. The "Community Fireguard" example for bushfire-prone areas in Australia is a good example.
17. The growing frequency of natural disasters requires effective use of the media to better inform and educate the general public and policymakers about the potential impacts of natural disaster and the need to adopt preparedness strategies.
18. Given the need for guidance in natural disaster management, case studies of China and other countries in agriculture, rangelands, forestry, and fisheries and the application of storm surge forecasting at the local level must be documented.
19. Given the regional and global nature of natural disasters, it is essential to promote and foster collaboration between agencies and between international and regional programs and build partnerships.
20. Given the complexity of issues involved in natural disaster management, it will be necessary to develop virtual networks with distributed functions to enhance cooperation on issues related to improved management of and preparedness for natural disasters.
21. More attention should be given to the impacts of potentially increasing frequency and severity of extreme events associated with global change and appropriate mitigation strategies.
22. Attention needs to be given to minimizing damage due to extreme events of infrastructure underpinning agriculture, rangelands, fishery and forestry.
23. The WMO Technical note on drought and agriculture published in 1975 should be updated to include the major advancements in understanding

these complex interrelationships. This revision should emphasize early warning, monitoring, and prediction techniques; vulnerability and impact assessment tools; preparedness and mitigation strategies. Case studies of China and India should be included in the revision. The publication could be published in hard copy, CD-ROM, and made internet accessible.

24. The IPCC Assessment Report 4 should examine whether or not climate extremes are becoming more frequent and, if so, for which natural hazards and climate extremes.
25. There should be more research into the physical behavior of crop growth and moisture regimes to develop better agricultural mitigation strategies.
26. There is a need and opportunity for agrometeorologists to supply design requirements for new satellite sensors. This applies in particular to drought and rangeland and forest fires from a disaster mitigation viewpoint.
27. Effective feedback mechanisms between affected parties (farmers, pastoralists, foresters, and fishermen) and those responsible agencies for mitigation and relief should be developed.

Index

C

D

E

F

G

H

O

P

Q

R

S

T

Library of Congress Catalog Card Number: 72-600370

First Printed 1973—CMH Pub 90-18

For sale by the Superintendent of Documents, U.S. Government Printing Office
Washington, D.C. 20402

Foreword

The United States Army has met an unusually complex challenge in Southeast Asia. In conjunction with the other services, the Army has fought in support of a national policy of assisting an emerging nation to develop governmental processes of its own choosing, free of outside coercion. In addition to the usual problems of waging armed conflict, the assignment in Southeast Asia has required superimposing the immensely sophisticated tasks of a modern army upon an underdeveloped environment and adapting them to demands covering a wide spectrum. These involved helping to fulfill the basic needs of an agrarian population, dealing with the frustrations of antiguerrilla operations, and conducting conventional campaigns against well-trained and determined regular units.

As this assignment nears an end, the U.S. Army must prepare for other challenges that may lie ahead. While cognizant that history never repeats itself exactly and that no army ever profited from trying to meet a new challenge in terms of the old one, the Army nevertheless stands to benefit immensely from a study of its experience, its shortcomings no less than its achievements.

Aware that some years must elapse before the official histories will provide a detailed and objective analysis of the experience in Southeast Asia, we have sought a forum whereby some of the more salient aspects of that experience can be made available now. At the request of the Chief of Staff, a representative group of senior officers who served in important posts in Vietnam and who still carry a heavy burden of day-to-day responsibilities has prepared a series of monographs. These studies should be of great value in helping the Army develop future operational concepts while at the same time contributing to the historical record and providing the American public with an interim report on the performance of men and officers who have responded, as others have through our history, to exacting and trying demands.

All monographs in the series are based primarily on official records, with additional material from published and unpublished

secondary works, from debriefing reports and interviews with key participants, and from the personal experience of the author. To facilitate security clearance, annotation and detailed bibliography have been omitted from the published version; a fully documented account with bibliography is filed with the Office of the Chief of Military History.

The author of this monograph, Major General William B. Fulton, was intimately involved in the early development of the riverine warfare concept as commander of the 2d Brigade, 9th Infantry Division, which arrived in Vietnam in January 1967 and immediately began combat operations in the Mekong Delta. In March the brigade moved into Dong Tam, the base created by dredging sand from the bottom of an arm of the Mekong River. In early June the brigade teamed with Navy Task Force 117 to form the Mobile Riverine Force. During World War II General Fulton fought in Italy and during the Korean War served on the staff of Army Forces Far East Advance. He is presently the Director of Doctrine Evaluation and Command Systems in the Office of the Assistant Chief of Staff for Force Development and is also Systems Manager for Surveillance, Target Acquisition, and Night Observation Systems.

Washington, D.C.
15 August 1972

VERNE L. BOWERS
Major General, USA
The Adjutant General

Preface

This monograph describes U.S. Army Riverine planning and operations in the Republic of Vietnam during the years 1966 through 1969. Since the personal experience of the author was with preparations for riverine operations and the initial operations themselves, emphasis has been placed on these activities through early 1968. In summarizing operations conducted in the balance of the three-year period, particular attention has been called to significant trends or changes in riverine operations in Vietnam, a co-operative enterprise of the U.S. Army and the U.S. Navy.

Looking back from the vantage point of early 1972, this study attempts to reconstruct the events and describe the situation as it was from 1966 through 1969, using official records, reports, and personal interviews.

The author is indebted especially to the officers of the 9th Infantry Division who helped in the research and writing of this monograph. Major Johnnie H. Corns was initially the intelligence officer and later the operations officer of the 2d Brigade during 1966 and 1967. His research, writing, and continuous editing have been indispensable in preparation of this study. Colonel Lucien E. Bolduc, Jr., commanded the 3d Battalion, 47th Infantry, and was later operations officer of the 9th Infantry Division during preparations for and initial conduct of riverine operations. His contributions in preparing the 2d Brigade of the Old Reliables for these operations and in preparing this manuscript are fully appreciated. Colonel Thomas C. Loper, 9th Infantry Division Engineer in 1967 and 1968, contributed both in the enactment and the recording of the riverine story. The author also wishes to acknowledge the typing assistance of Miss Ann M. Faherty and Miss Judith A. Secondo, who patiently saw the manuscript through several drafts.

Washington, D.C.
15 August 1972

WILLIAM B. FULTON
Major General, U.S. Army

Contents

Charts

Diagrams

Tables

Maps

Illustrations

All illustrations are from Department of Defense files.

RIVERINE OPERATIONS, 1966-1969

CHAPTER I

Background

During 1965 and 1966 when the possibility of creating a U.S. riverine force for operations in the Mekong Delta was being discussed, there were three basic considerations that weighed heavily in favor of the force: a tradition of past American success in riverine operations, particularly Union operations in the Mississippi basin during the Civil War; the success more recently achieved by the French in riverine operations during the Indochina War under conditions that appeared to have changed little during the years that had intervened; and, most important, a situation in the Mekong Delta that seemed ripe for exploitation by a riverine force.

Previous American Experience

American forces conducted riverine operations as early as the Revolutionary War and made use of them frequently during the century that followed. In each case, the operations were part of a larger land campaign. While they were not always tactically successful, they usually made a substantial contribution to the success of the campaign.

In 1775, on the heels of the battles of Lexington and Concord, both the Americans and British sought to gain control of the strategic Hudson River-Lake Champlain-St. Lawrence River waterway system connecting New York and the British centers in Canada. Forces under Colonels Benedict Arnold and Ethan Allan seized the British forts at Ticonderoga and Crown Point; Arnold then armed a captured schooner and pressed north along Lake Champlain to attack the British base at St. Johns on the Richelieu River above the lake. In a daring surprise attack, he captured or destroyed a large number of boats and considerable amounts of supplies and equipment before returning to Lake Champlain. By means of this lake and river operation, Arnold effectually blunted the British ability to counterattack.

Pressing their advantage, the Americans then undertook an invasion of Canada, with one line of advance up the lake and river

chain. Their attack on Quebec was repulsed, and by the spring of 1776 the British, heavily reinforced from the home country, regained the offensive. The British began to assemble a fleet at St. Johns. Some of the larger vessels were actually built in England, knocked down for shipment across the ocean, and reassembled at St. Johns. Spurred on by Benedict Arnold, the Americans sought to keep pace with the British at their own Skenesborough shipyard, near the southern end of Lake Champlain. On 11 October 1776 the Americans, with a fleet of fifteen boats hastily constructed of green timber and manned by a scratch force of backwoodsmen and recruits from the seacoast—"the flotsam of the waterfront and the jetsam from the taverns"—engaged the formidable British force of twenty-nine vessels at Valcour Island, just north of Ticonderoga. The poorly organized and ill-equipped American force lost its ships one by one, and the Americans were forced to retire to Crown Point. The British tenaciously pursued and drove Arnold's force out of Crown Point into Ticonderoga at the southern tip of the lake. Although Arnold had lost the engagement, his small flotilla delayed the British advance until the following year, when Major General John Burgoyne attempted a thrust southward to cut the colonies in two and was defeated at Saratoga.

Some thirty-five years later in the War of 1812, U.S. forces again fought the British on American waterways. In 1813 Commodore Oliver H. Perry assembled a fleet on Lake Erie to counter the British threat from Canada, and by defeating the British fleet in a series of engagements gained control of the Great Lakes, permitting a subsequent American advance into Canada. During the next year the British made three major attempts to exploit inland waterways: one from Chesapeake Bay against Washington, a second overland from Canada toward New York, and a third from the Gulf of Mexico against New Orleans. Each British effort was met by an opposing American riverine force. In August 1814 British forces pushing up the Chesapeake were met by a force under Commodore Joshua Barney. After several engagements in which Barney's initiative and dogged persistence frustrated the British advance, the British disembarked at Benedict, Maryland, and then marched the thirty-one miles to Washington, where they burned the public buildings. The British drive down Lake Champlain was blocked in September 1814 by American Commodore Thomas Macdonough in the river battle of Plattsburg.

A few months later, in December 1814, the British moved against New Orleans, connected to the Gulf of Mexico by the lakes

and bayous of the Mississippi Delta. While the British assembled forces off the coast, American Commodore Daniel T. Patterson prepared his river forces to oppose the attack. He placed obstructions in the waterways leading directly to the city, and with five gunboats and two tenders met the advancing British forces on Lake Borgne. Outgunned and outmaneuvered, Patterson was thoroughly beaten. Despite his defeat he continued with the remnants of his shallow-water navy to oppose the British advance with skill and energy, thus gaining enough time to permit General Andrew Jackson to ready the defenses of the city. Then Patterson, with the *Carolina* and the *Louisiana*, his two largest ships, deployed on the river, joined forces with Jackson in a defense that resulted in ultimate defeat of the British.

From 1835 to 1842 Army and Navy forces fought in the Florida Everglades against the Creeks and the Seminoles in a particularly frustrating and indecisive campaign. To navigate the inlets along the coast as well as the waterway mazes of the swamps, a mosquito fleet of some 150 craft—schooners, flat-bottomed boats, bateaux, and canoes—was assembled and manned by soldiers, sailors, and marines. Operating in conjunction with Army Colonel William J. Worth, Navy Lieutenant John T. McLaughlin transported men, equipment, and supplies, patrolled waterways, and conducted raids. Because of its mobility and striking power, the riverine force, well adapted to conditions in the Everglades, was able to reduce the Indian threat.

A few years later during the War with Mexico, 1846-1848, the American Navy conducted two river raids which, although not particularly decisive in themselves, made a strong impression on one of the younger participants, Lieutenant David Dixon Porter, who was to use the experience in another war. The raids were conducted some seventy-five miles up the Tabasco River against San Juan Bautista, an enemy crossing point and operating base. Commodore Matthew C. Perry utilized steamers to tow two sailing schooners and smaller craft up the winding river against a strong current. On his first raid he bombarded the city without debarking any troops. On his second, he landed a force of 2,500 sailors and marines, which had been formed into infantry and artillery divisions, and assaulted the city from the shore, supported by fire from the force afloat that included four steam warships, six schooners, some "bomb rigs" that mounted mortars, and a number of ships' boats. While Perry was ashore leading his men, the ships and boats were left under the command of Lieutenant Porter.

Extensive riverine operations were conducted during the Civil War. Along the coasts of the Confederacy, the Union launched joint Army-Navy expeditions that gradually gained footholds in the vicinity of Mobile, Alabama; Savannah, Georgia; Charleston, South Carolina; and Wilmington, North Carolina. At the same time there was a concerted Army-Navy effort to control the inland waterways of the Mississippi basin and thus cut the Confederacy in two. At the outset, Union Commodore John D. Rodgers assembled a fleet of steam-powered, shallow-draft vessels at Cairo, Illinois, where the Ohio River flows into the Mississippi. The gunboats were well armed and well armored, could carry a dozen large guns—8-inch, 32- and 42-pounders—and ranged up to a 50-foot beam and a length of 175 feet, with ironclad, sloping superstructures. Rodgers used his craft to patrol and gather intelligence of the Confederate forces that lay within striking distance. Relying in part on the intelligence Rodgers obtained, Brigadier General Ulysses S. Grant decided to move against Confederate forces at Belmont, Missouri. Rodgers transported 3,000 of Grant's men and provided fire support for the operation. When substantial Confederate reinforcements arrived unexpectedly to bolster the garrison at Belmont, Grant prudently re-embarked and withdrew, demonstrating that river craft may be as useful in a withdrawal as they are in an advance.

Soon afterward, Commodore Andrew H. Foote, who succeeded Rodgers, and General Grant requested permission to undertake offensive river operations. In January 1862 their plan for a joint Army-Navy operation was approved. In February the combined forces moved up the Ohio River and then the Tennessee River to attack Confederate Fort Henry. The force used regular civilian transports as well as the specialized shallow-water gunboats. Grant landed his men five miles above the fort and advanced overland. The gunboats began the attack alone, firing as they approached the fort. They closed to point-blank range, silenced the Confederate shore batteries, and unexpectedly forced the surrender of the garrison. It was a riverine victory for Union forces, and the news was welcome in Washington, where it contrasted dramatically with disappointing reports from other areas.

The victory at Fort Henry was followed by Navy gunboat raids up the Tennessee River. The raids ranged across the entire state of Tennessee, probed into Mississippi, and even pushed into Alabama as far as Muscle Shoals. At the same time, General Grant and Commodore Foote moved against nearby Confederate Fort Donel-

son, which lay on the Cumberland River and covered the land routes to Nashville. Grant moved most of his troops directly overland, while Foote descended the Cumberland to the Ohio, then sailed up the Tennessee. The battle began when Foote moved his gunboats against the Confederate shore batteries. Heavy fire from the shore soon forced the gunboats to withdraw; each of the six ships took twenty or more hits and two ships were put out of action. Casualties were high and included Commodore Foote, who was seriously wounded. Later, under renewed Union attack from river and shore, the Confederate garrison surrendered. While Fort Henry was essentially a Navy victory, it was a combination of Army and Navy forces that proved decisive at Fort Donelson.

Shortly thereafter the Union Navy in the Gulf of Mexico moved inland against New Orleans. Forces under Admiral David Glasgow Farragut first crossed the sandbar at the mouth of the Mississippi, then rerigged ships for river operations. Farragut directed a five-day bombardment by a specially prepared mortar flotilla to soften up the Confederate defenses along the river. He assembled his force and in a daring night maneuver ran the gauntlet of forts and Confederate warships that blocked the way. Once he had penetrated the Confederate defenses, he moved on unopposed to his objective. With a force of less than 3,500 men, he succeeded in capturing the South's largest city.

Although he had achieved success, Farragut was nonetheless impressed by some of the difficulties which his deepwater force had encountered on the river. He wrote, "Fighting is nothing to the evils of the river—getting on shore, running afoul of one another, losing anchors, etc."

A few weeks later, in June 1862, Farragut moved up the Mississippi and tried to seize Vicksburg but was repulsed. He succeeded, however, in running by the Confederate fortifications to link up with Navy forces under Captain Charles Henry Davis, who had relieved Commodore Foote. The combined flotillas then operated in support of Grant. Confederate resistance at Vicksburg prevented further passage along the river and delayed Union attempts to split the Confederacy. It was nearly a year before Grant could force the surrender of the city. During Farragut's June operations, Union Navy forces in the vicinity of Vicksburg were under the command of Rear Admiral David Dixon Porter, who had earlier taken part in river warfare in Mexico. Throughout the campaign of the lower Mississippi, the Confederates skillfully used water mines—torpedoes, as they were called—to counter the riverine

threat. Some forty Union craft fell victim to these torpedoes. Confederate snipers and shore batteries as well as such navigational difficulties as shallow water and tree stumps also helped to frustrate Union attempts on Vicksburg.

In the end, Navy forces from the north succeeded in running the gauntlet at Vicksburg. Grant's brilliant maneuvering first to the south and then to the east of the fortress defeated the Confederate forces that might have come to the relief of Vicksburg, permitting Grant to bring about its surrender. The fall of Vicksburg split the Confederacy in two: Union control of the Mississippi, the Ohio, the Cumberland, and the Tennessee proved decisive in the final Union victory.

For many years after the Civil War the United States had little occasion to engage in riverine warfare. Perhaps the most notable U.S. river operation in the early twentieth century was the American Navy's patrol of the Yangtze River after the Boxer Rebellion in China. During the twenties and thirties a flotilla of U.S. Navy gunboats known as the Yangtze Patrol maneuvered along some 1,500 miles of meandering river, faced with occasionally swift currents, and with water levels that were subject to sudden and violent change—up to twenty-four feet in twenty-four hours. In addition to the difficulties of navigating the Yangtze, the patrol had to cope with snipers, ambushes rigged by warlords, and the necessity to conduct rescue missions that sometimes required the use of landing parties. During World War II American forces were used in some operations that might be termed riverine warfare, but it was not until the Vietnam War that the United States organized a complicated river flotilla to fight on inland waterways.

French River Warfare in Indochina, 1945–1954

During the nineteenth century the French had used the waterways of the Mekong and the Red River Deltas in their conquest of Indochina. When French forces returned to Vietnam in 1945 after World War II in an attempt to regain control of the country, they used conventional military tactics at first and such riverine operations as were launched were characterized by improvisation, but in the next nine years the French developed a remarkable riverine force.

In the north, in Tonkin, where most of the fighting occurred, the delta formed by the combined Red, Black, and Clear Rivers, and in the south, in Cochinchina, the delta formed by the Mekong, the Bassac, the Dong Nai, the Saigon, and the Vam Co rivers

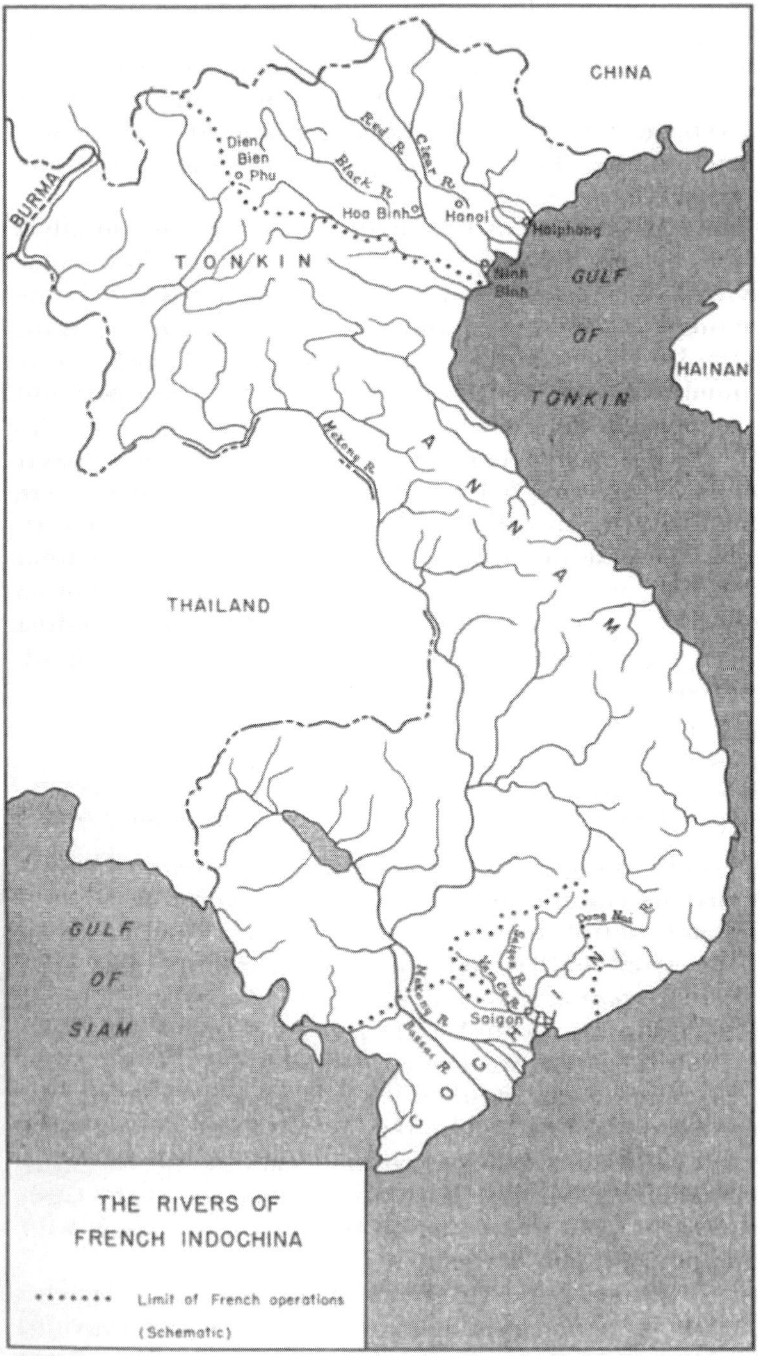
CHINA
BURMA
Dien Bien Phu
Red R.
Clear R.
Black R.
Hoa Binh
Hanoi
Haiphong
Ninh Binh
TONKIN
GULF OF TONKIN
HAINAN
Mekong R.
ANNAM
THAILAND
Dong Nai R.
Saigon R.
Saigon
Bassac R.
COCHIN
GULF OF SIAM
THE RIVERS OF FRENCH INDOCHINA
Limit of French operations (Schematic)

MAP 1

offered mazes of meandering streams connected here and there by straight canals. (*Map 1*) Even in peacetime most of the traffic in the north of Vietnam was by waterway. In wartime, with roads and railroads frequently cut by enemy action, the French estimated that 90 percent of traffic was by inland waterway. It was not surprising, therefore, that the French should turn to these waterways for military operations.

The French started out with craft that were available locally—native or left by the Japanese—which they modified with armor and armament. In addition, they received from the British some LCI's (landing craft, infantry) and a few LCA's (landing craft, assault) and LCT's (landing craft, tank). The French Army created a number of river flotillas that were used for transport and river patrols. By 1946 the French Navy had organized river flotillas and in 1947 these were designated *divisions navales d'assaut* (naval assault divisions), abbreviated to Dinassauts. The divisions were organized initially to provide transport with fire support escort. In time their composition changed, and one flotilla differed from another according to the area in which it operated. Each flotilla had from twelve to eighteen craft, ranging from LCVP's (landing craft, vehicle or personnel) to LSSL's (landing ships, support, large), and contained at least these elements:

Command and fire support.	1 LCI or LSSL
Transport.	1 LCT
Landing and support.	2 LCM's and 4 LCVP's
Patrol and Liaison.	1 harbor patrol boat

The flotilla could transport and land a force of approximately battalion size and its equipment. It could also support the landing force by fire and control the waterways in the vicinity of the landing area. The river flotilla could do some patrolling, and even some raiding in support of the main operating force ashore. Toward the end of the campaign as the Viet Minh massed their forces, the single river flotilla proved inadequate and the French organized river task forces consisting of several flotillas, reinforced by additional troop transports. In addition to these naval units, the French had other specialized units, organized and manned by the Army for river operations; some were built around vehicles of the Crab and Alligator type and some supply units were equipped with LCM's (landing craft, mechanized).

Command relations between French Army and Navy elements varied, but the Navy generally operated the vessels and provided support to Army operations. In Tonkin the Commander, River

Forces, maintained close liaison with Headquarters, Ground Forces, North Vietnam, and to all intents and purposes the four flotillas in the area operated in support of the Army. In Cochinchina the River Forces Command provided support to the Commanding General, Ground Forces, South Vietnam, as agreed with the Commander, French Naval Forces, Indochina.

These arrangements appeared to the French commanders, even in retrospect, to have been satisfactory, although there seems to have been some Army complaint that there were not enough landing craft, and some Navy complaint that the local Navy commanders had no organic infantry capable of defending Navy bases, either fixed or afloat. The naval command was responsible for all river forces regardless of their parent service. Thus, French Army troops transported by a Dinassaut were under operational control of the French Navy. Co-ordination with the Army was achieved through river posts, which maintained direct liaison with ground force commanders in their areas. A continuing difficulty was the problem of Dinassaut base defense. Generally, each Dinassaut had a small Navy Commando detachment or an Army light support company attached to it. But the Dinassaut did not have regular battalion-strength landing forces permanently assigned; these would have permitted greater autonomy and better control of an area between major operations. The infantry landing forces, when made available by the Army for a particular operation, usually lacked training or experience in river warfare.

A river assault flotilla generally deployed in column. Leading the column was an opening group of minesweepers and a guide for navigation, followed by a fire support ship. Next came the troop transports and, bringing up the rear, the command and support vessels. (*Diagram 1*) When a flotilla was attacked it usually attempted to force passage, using what the French called "the ball of fire," or heaviest possible volume of immediate fire directed at enemy positions on shore. One theory held that when there was reason to suspect an ambush, a direct assault against the enemy was most effective; such improvised assaults, however, were rare. In an assault the opening group consisted of a guide ship and minesweepers. Close behind was the shock group, consisting of a command vessel, one or more fire support ships, and several landing craft carrying the assault troops. The remainder of the force trailed at a distance of 1,000 to 1,500 meters. Upon arrival at the assault site, the shock group would fire a preparation bombardment and the assault troops would then land. Once the landing site was se-

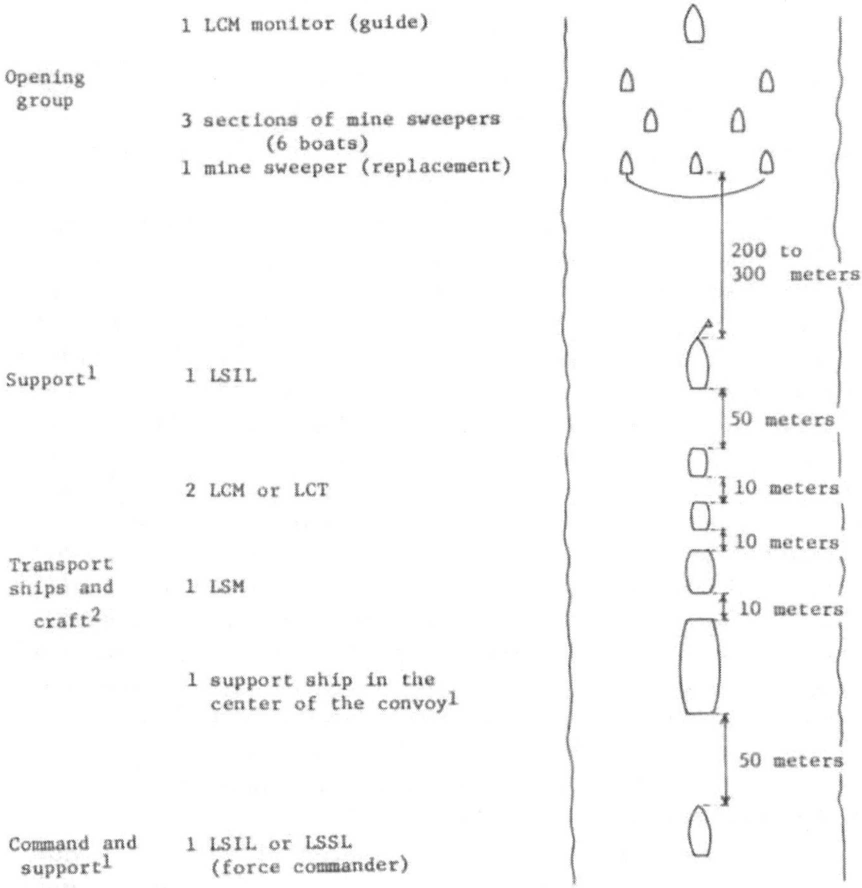

[1] When there was but one support ship, it took the lead and was the command ship. When there were two support ships, the command ship was in the rear; when there were three, the command ship was in the center. The lead support ship might also be an LSSL, an armed LCT, or a section of LCM monitors (2 boats).

[2] Sometimes the transport craft were in two columns, lashed in pairs.

Diagram 1. A typical French river convoy in North Vietnam.

cured, the remainder of the landing force would beach and unload. During the landing and until the river craft were withdrawn, the force afloat provided fire support, protected the flanks of the land-

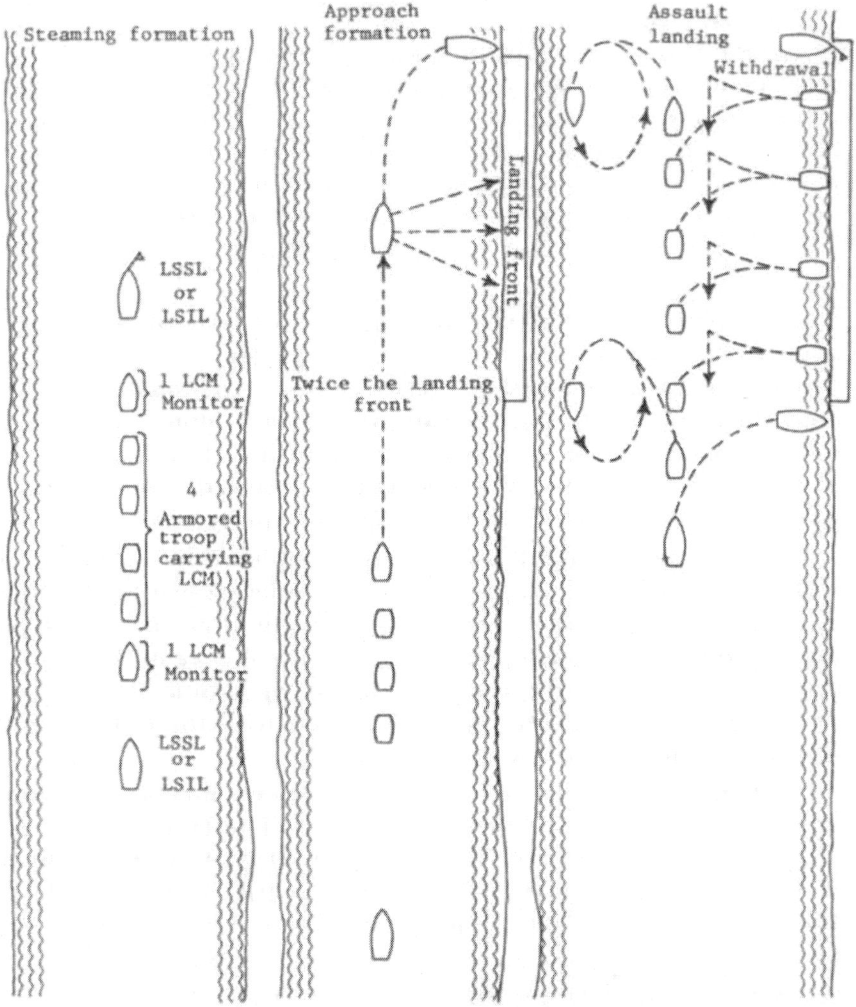

Diagram 2. French assault landing.

ing site, and patrolled in the vicinity to provide additional security. Some craft were used for logistic support, and the larger ships served temporarily as command posts for the ground unit commanders. (*Diagram 2*)

French reports show that during the entire campaign from 1945 through 1954 losses from attack while vessels were at anchor exceeded those suffered while a convoy was under way. Most Viet Minh attacks against French river forces were carefully planned

and made use of some combination of swimmers, drifting mines, artillery and mortar fire, and occasionally overland assaults. French countermeasures had spotty success. Nets were used with ebb and flow of the tide to trap mines; security watches were set up; all floating objects were systematically fired upon; patrol craft were sent out; grenade attacks were made at irregular intervals and on unidentified objects; and lights were installed at entrances to anchorages.

Harassing enemy mortar and artillery fire proved particularly effective, causing the French to withdraw some forces that otherwise would have remained on station. French commanders discovered they could not rely on passive measures alone. They were sometimes obliged to shift anchorages during darkness to avoid planned Viet Minh fire. When boats were beached it proved necessary to cover the beach front with a ground unit.

The river assault flotillas were rarely employed in coastal operations, since craft modified for river use were less seaworthy, had less cargo capacity, and had armored superstructures that hindered troop transfers. Troops experienced in riverine assaults required extensive additional training for coastal operations before they were able to master going down cargo nets with their equipment, even when the sea was calm.

The tempo of French riverine operations increased steadily from 1946 until the end of the war in 1954. The Dinassauts played a key role in the battles for the Red River Delta from the summer of 1951 through early 1952. During the battle of Ninh Binh, 29 May–18 June 1951, 100 kilometers south of Hanoi on the Day River, a Dinassaut reinforced the French garrison, provided fire support, cut enemy communications, and helped to force the Viet Minh to withdraw.

During the battle of Hoa Binh, 14 November 1951–24 February 1952, sixty kilometers west of Hanoi on the Black River, two Dinassauts escorted convoys through a sector in which one bank of the river was controlled by the enemy for a distance of about fifty kilometers. As the Viet Minh increased pressure on the lines of communication connecting the garrison at Hoa Binh with the main French forces to the east, the overland routes were severed. The Dinassauts, however, continued to run the gauntlet until 12 January 1952, when an entire river convoy was ambushed and took heavy losses. Finally, when the decision was made to evacuate Hoa Binh—Operation Armaranth—the Dinassauts supported the operation by reopening the Black River corridor, ferrying retreating

troops and refugees, then fighting their way out along the river toward Hanoi. Later, two Dinassauts played a key role in the offensive phase of Operation Lorraine, 29 October–8 November 1952, northwest of Hanoi on the Clear and Red Rivers.

By the end of the war, French riverine forces were fully committed and were taking heavy casualties. In over a dozen ambushes fourteen naval craft were sunk or damaged, with nearly 100 men killed or wounded. (*Table 1*)

TABLE 1—VIET MINH RIVER AMBUSHES, 5 JANUARY–16 FEBRUARY 1954

Date	Waterway	Group or Type of Craft	Enemy Weapons	Comments
5 January...	Song Thai Binh	*Dinassaut* 4	Mine	LSIL sunk; Navy, 10 killed or missing, including 2 officers; Army, 40 wounded
13 January..	Red River	*Dinassaut* 12	Automatic weapons	....
14 January..	Bamboo Canal	*Dinassaut* 4, *Javeline*	Mine	....
15 January..	Nam Dinh Giang	LCT 9.063, barge convoy, and LCM	Bazooka, automatic weapons	1 barge sunk; Navy, 4 wounded; Army, 1 killed
20 January..	Song Thai Binh	Platoon river patrol boat	Bazooka, automatic weapons	1 launch sunk; 2 killed, 3 Army wounded
22 January..	Song Thai Dinh	*Dinassaut* 1	Bazooka, automatic weapons	Detected by Morane observation plane, neutralized by artillery fire
1 February..	Red River	LCI with LCM	Mines	LCI damaged
1 February..	Van Uc Canal	Platoon river patrol boat	Mine	....
2 February..	Day River	*Pertuisane*, LCM, river patrol boat	Bazooka, automatic weapons	River patrol boat burned, LSSL *Pertuisane* damaged; 4 Navy killed or missing, 2 wounded
4 February..	Upper Red River	LCM and *Mytho Dinassaut* 12	Bazooka, automatic weapons	1 *Mytho* sunk, 1 Navy wounded
5 February..	Song Thai Binh	*Dinassaut* 1	Mine, bazooka, automatic weapons	....

TABLE 1—VIET MINH RIVER AMBUSHES, 5 JANUARY–16 FEBRUARY 1954—Continued

Date	Waterway	Group or Type of Craft	Enemy Weapons	Comments
9 February . .	Nam Dinh Giane	*Dinassaut* 3	Bazooka, automatic weapons	LSSL *Arquebuse* and 2 LCM's damaged; 3 Navy killed, including 2 officers; 15 Navy wounded
14 February . .	Middle Red River (Hung Yen)	*Dinassaut* 12	Bazooka, automatic weapons	LCT 9047 damaged; Army, 1 killed, 1 wounded; Navy, 4 killed
16 February . .	Nam Dinh Giang	*Dinassaut* 13	Bazooka, automatic weapons	LCT 9033 damaged, 1 boat sunk, 1 LCM damaged; 1 killed, 7 seriously wounded

After the war the French analyzed their experience with river warfare in Vietnam. French matériel, they concluded, was generally adequate, although increased armament might have improved performance. Operational techniques were generally adequate, but French riverine forces proved highly vulnerable in base defense and susceptible to water mines. Command relationships were satisfactory but effectiveness of the forces would have been greatly improved by permanently marrying Army assault forces to Navy lift capacity. In short, there was a need to organize an amphibious force made up of permanently assigned Army and Navy elements.

It was the observation of the French that the enemy, the Viet Minh, were remarkable infantrymen, but they made no attempt to launch armed water craft against French river forces. Rather, Viet Minh action was in the main restricted to the banks of the waterways, from which the enemy tried to intercept French movements by means of ambush, fixed obstacles, or use of remotely controlled floating mines. Enjoying excellent intelligence and the support of the population, the Viet Minh proved meticulous in their planning and masterful in the use of terrain and camouflage. Because of their excellent fire discipline, they usually enjoyed the advantage of surprise. The French thought that the Viet Minh tended to disperse their weapons too much along the banks rather

than mass them, thus reducing the effectiveness of the surprise and firepower they could bring to bear against individual elements of the French floating force.

The French operated in both the Red River and Mekong Deltas, and their experience was to prove applicable in some part for U.S. river forces operating more than a decade later in the Mekong Delta.

The Mekong Delta

The Mekong Delta extends from Saigon south and west to the Gulf of Thailand and the border with Cambodia. With an area of about 40,000 square kilometers and an estimated eight million inhabitants, it constitutes about one-fourth of the total land area of South Vietnam, and contains about one-half of the country's population. (*Map 2*) The delta is generally a flat alluvial plain created by the Mekong River and its distributaries. Much of the land surface is covered by rice paddies, making the area one of the world's most productive in rice growing. It is by far the most important region in South Vietnam.

MAP 2

SOLDIER STRUGGLES THROUGH DELTA MUD

The delta had poor overland communications. Crisscrossed by a complex network of rivers, canals, streams, and ditches, the low, poorly drained surface was subject to extensive and prolonged inundation. There was only one major hard surface road, Route 4, which extended from Saigon south to Ca Mau, traversing the delta and linking many of the larger towns. Secondary roads were poorly surfaced, and by the mid-1960's had deteriorated because of lack

of maintenance and enemy action. The waterway system made frequent bridging necessary; bridge capacities were mainly in the 10- to 20-ton range. In short, the roadnet was of limited use for military operations.

Any movement off the roads was limited. It was best during the dry season—November to March—when paddies were dry and would support light tracked vehicles and artillery pieces; it was poorest during the wet season—May to October—when paddies were inundated. It was restricted all year round by the network of rivers, canals, streams and ditches. There were many swamps, marshes, and forests, generally bordering the seacoast.

In comparison with most of the other areas of Vietnam, the Mekong Delta was good for air operations, since landing zones were abundant and weather seldom prevented helicopter operations. Once troops debarked from helicopters, however, their movement was sharply restricted by ditches and waterways.

In sharp contrast to the limited overland transportation, the delta had a highly developed inland waterway system. There is evidence that the inhabitants began to improve natural drainage as early as 800 A.D., and succeeding generations have continued the work. As a result, the 2,400 kilometers of navigable natural waterways were supplemented by about 4,000 kilometers of land-cut canals of varying width and depth, and in good to poor condition.

The wet season, the time of the southwest monsoon, permits deliberate flooding of rice paddies, but also causes some unavoidable flooding as rivers overflow their banks. Waterways grow wider, sharply curtailing cross-country traffic and limiting the amount of firm ground suitable for placing artillery. Off-road movement of troops and vehicles can cause damage to crops and drainage systems, and was therefore unpopular with the farmers. Further, the high humidity and extensive flooding aggravated the foot problems of the soldier, who got his feet wet from frequent water crossings. One-third of the delta is marsh, forest, or swamp forest. In the north lies the Plain of Reeds, a flat, grassy basin with almost no trees. During the wet season it is generally inundated to a depth of two to three meters; during the dry season, much of the plain dries out to the extent that large grass fires are frequent. There are relatively few points in the delta that lie more than a few hundred meters from a navigable waterway; even the marshes, swamps, and forests can be reached by water. In short, the delta is well suited to riverine operations.

Military operations on the waterways of the Mekong Delta had to be governed by the life of the delta. By any standard, the delta is densely populated; it has an average of about 200 inhabitants per square kilometer, about the same density as Massachusetts. The people are concentrated along the waterways that constitute their principal transportation routes. Transportation, essential to the people and to the national economy, felt the impact of military operations, and the U.S. commander was obliged to maneuver his forces so as to interfere as little as possible with the normal flow of civilian transport.

Where people lived there were trees and bushes cultivated for their fruit, shade, or decoration. Since these offered protection to an enemy seeking concealment, military engagements often took place in populated areas. Thus a commander was faced with the task of inflicting damage on an armed enemy, and at the same time trying to avoid damage to the unarmed local inhabitants. High population density makes it difficult to conceal military activity from observation; weaknesses may be noted and word passed to the enemy, moves may be telegraphed before they can achieve their purpose.

Seacoasts of the delta have extensive mangrove swamps, including the Rung Sat southeast of Saigon, and the U Minh Forest along the west coast; vegetation on the tidal mudflats is dense, root structure is high and tangled, and covering is thick, making access difficult and cross-country movement arduous. The paddy land is interspersed with thickets of trees and bushes in patterns that vary: in one area vegetation is clustered polka-dot style about scattered dwellings; a few kilometers away, it is in continuous strips along waterways. There are large cultivated plantations—rows of palm trees grown on embankments which alternate with wide drainage ditches. In the wide-ranging operations characteristic of riverine warfare, it was necessary to adapt tactics to the lay of the land.

Riverine operations in the delta also had to deal with the strong influence of sea tides along the inland waterways throughout the region. Tidal fluctuations range up to about four meters. The twice daily tidal flow influences the velocity of currents, and has an important bearing on the feasibility of navigation in many waterways. Variations in sea tides, together with the complex nature of the interconnecting waterways, make tidal effects exceptionally difficult to predict with confidence. As a consequence, planning for water operations involves a good deal of guesswork, for the time required to travel from one point to another cannot be predicted.

The condition of the inland waterways added obstacles to military movement. By 1960 many canals had deteriorated from neglect; of some 1,500 surveyed that year, nearly half required "restoration," and could carry traffic only at full tide. Some waterways had been deliberately obstructed by the enemy. In the smaller streams there were many fishtraps that restricted passage. Bridges are always to be dealt with in waterborne movement, since frequently there is not enough clearance to permit passage of vessels. This problem is further complicated by variation in water levels due to tidal fluctuations, which may permit passage for short periods twice a day only.

Finally, it was difficult to base troops in the Mekong Delta since firm ground was limited. Military needs for land bases tended to conflict with the needs of the population, so that a floating base offered a clear advantage.

Enemy Forces

As elsewhere in Vietnam, the enemy had in 1966 and early 1967 both a political and a military organization in the delta. In mid-1966 estimated strength of the Viet Cong in the IV Corps Tactical Zone, which corresponds roughly to the delta, was 82,545 men. Of these, 19,270 were combat troops; 1,290 were support troops; 50,765 were local part-time guerrillas; and 11,220 were working as political cadre. At the time, no North Vietnam Army forces were reported in the IV Corps Tactical Zone. In the summer of 1966 the office of the assistant chief of staff for intelligence estimated that organized military forces in the IV Corps Tactical Zone consisted of three regimental headquarters, 28 battalions (8 of which were in the regiments), 69 separate companies, and 11 separate platoons—an estimate that agreed closely with that of U.S. Military Assistance Command, Vietnam. *(Map 3)*

It was believed that Viet Cong logistics depended on support from the population, on captures from Republic of Vietnam units, and on supplies furnished by the Democratic Republic of Vietnam—primarily weapons and ammunition—infiltrated by sea, or by land from Cambodia. The base area was believed to be fundamental to the Viet Cong logistical system. Base areas such as the Plain of Reeds, the U Minh Forest, and the Cam Son Secret Zone west of My Tho served as sites for political, military, logistical, and training installations, and were supplemented by "combat villages" on South Vietnamese territory, organized and controlled by the Viet Cong. Cambodia provided a rear service area for 10,000 or

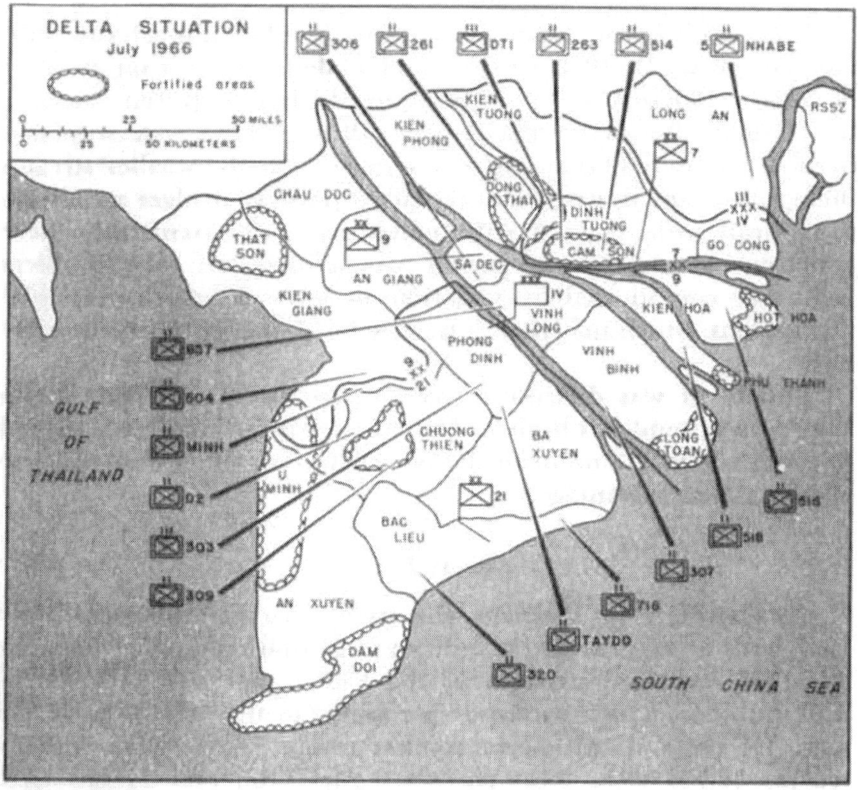

MAP 3

more main and local Viet Cong forces in Vietnamese provinces that offered free access to Cambodian territory.

Viet Cong forces in the delta were as well armed as those elsewhere in Vietnam. The new family of 7.62-mm. weapons manufactured by the Chinese Communists were first captured in the delta on 11 December 1964 by Republic of Vietnam armed forces. The capture included copies of the AK47, the SKS carbine, the RPD light machine gun, and a quantity of M43 intermediate 7.62-mm. cartridges of Soviet design. During the same month, the Vietnam Army also captured the first RPG2 antitank grenade launcher, capable of penetrating six to seven inches of armor at ranges of 100 to 200 meters—a potent weapon for use against riverine craft.

Viet Cong communications equipment, while not abundant, appeared adequate. The enemy was able to interfere with and intrude on voice circuits of U.S. and South Vietnam forces, blanking out transmissions and employing tactical deception.

Viet Cong methods of operation in the delta were not sub-

stantially different from those employed by the enemy in other regions. In 1966 it was reported that Viet Cong activities had been primarily small unit operations—harassment, terrorism, and sabotage, with the delta accounting for approximately one-third of all Viet Cong initiated incidents in South Vietnam. The enemy, however, was also able to mount battalion-size attacks. Several times the Viet Cong demonstrated a willingness and an ability to slug it out with government forces. In January 1963 at Ap Bac a Viet Cong force engaged a superior Army of Vietnam force that was attempting to surround the Viet Cong by using heliborne assault in conjunction with conventional ground movement. Five helicopters were destroyed and nine damaged as the Viet Cong inflicted heavy casualties and withdrew. In December 1964 the Viet Cong 9th Division with two regiments seized the Catholic village of Binh Gia. During the next four days the enemy ambushed and virtually destroyed the Vietnamese 33d Ranger Battalion and 4th Marine Battalion and inflicted heavy casualties on armored and mechanized relief forces.

Republic of Vietnam Forces

In 1966 most of the Mekong Delta was included in the IV Corps Tactical Zone, although Gia Dinh Province, Long An Province, and the Rung Sat Special Zone in the north were part of the III Corps Tactical Zone. The IV Corps Tactical Zone was in turn subdivided into three division tactical areas. In the north the 7th Division had its headquarters at My Tho; in the center was the 9th Division with headquarters at Sa Dec; and in the south was the 21st Division with headquarters at Bac Lieu. In 1966 Army of Vietnam assigned strength in IV Corps Zone averaged 40,000. In addition to the three divisions, there were five Ranger battalions and three armored cavalry squadrons.

South Vietnam paramilitary forces included Regional Forces, Popular Forces, Civilian Irregular Defense Group (CIDG) troops, and National Police. In 1966 Regional and Popular Forces manned outposts and watchtowers scattered throughout the delta. Poorly supported and highly vulnerable to Viet Cong attack, both these forces had high desertion rates. The Civilian Irregular Defense Group troops were employed generally along the Cambodian border as part of an effort to seal the frontier against Viet Cong and North Vietnamese movements of men, equipment, and supplies. The National Police were organized after the manner of the French *gendarmerie,* and exercised similar functions.

Vietnam naval forces attached to the Vietnamese 4th Naval Zone in the delta included six river assault groups, which had evolved directly from the French Dinassauts, and eleven coastal groups known collectively as the Junk Fleet. The assault groups were under the operational control of the IV Corps commander with the primary mission of supporting Vietnam Army riverine operations. Each group could lift a Vietnamese infantry battalion, giving the IV Corps Zone the capability of a six-battalion lift. The river assault groups were in 1966 being used in their primary role only 10 percent of the time. Vietnam Army division commanders apparently preferred airmobile operations to riverine operations, hence the river assault groups were employed in support of small unit operations by Regional or Popular Forces under the control of province chiefs. Often they were used simply as escort for commercial craft.

U.S. Forces

The American military first entered the Mekong Delta in 1957 when U.S. Navy advisers replaced French advisers. By early 1966 the American advisory effort had permeated the Republic of Vietnam military structure. Advisers included nearly 700 officers and over 2,000 enlisted men from the U.S. Army, Navy, and Air Force. American Army advisers were present at corps, division, and sector (province) level, and with the IV Corps Tactical Zone Area Logistics Command. The U.S. Navy Advisory Group, Vietnam, provided advisers to the Vietnamese Navy's six river assault groups and eleven coastal groups.

In 1965 the U.S. Army 13th Combat Aviation Battalion had been sent to the delta to support Republic of Vietnam operations, and by August of that year the battalion had four assault helicopter companies and one reconnaissance airplane company. By mid-1966, U.S. naval forces included Task Force 115, with the code name MARKET TIME, and Task Force 116, known as GAME WARDEN. MARKET TIME had the mission of patrolling coastal areas to prevent resupply of Viet Cong and North Vietnam forces by sea. GAME WARDEN had the mission of interdicting enemy lines of communications and assisting government forces in repelling enemy attacks on river outposts of the Regional and Popular Forces.

The Situation in 1966

In April 1964 the American Ambassador, Henry Cabot Lodge, Jr., had remarked: " . . . I would not be surprised to see the

Mekong Delta totally cleared of Communist forces by the end of 1965." In the summer of 1966, however, the Military Assistance Command, Vietnam, reported that about one-third of all Viet Cong actions against the government in South Vietnam occurred in the IV Corps Tactical Zone, and estimated that the Viet Cong controlled 24.6 percent of the population of IV Corps.

Government influence varied widely from province to province. An Giang Province was considered "pacified," while in An Xuyen Province it was estimated that government forces controlled as little as 4 percent of the land. In many areas government forces claimed control by day, but acknowledged that the "night belongs to the Viet Cong." With few exceptions, the large towns served as either province or district capitals and in 1966 were largely under government control. All, however, were occasionally subject to terrorist incidents, mortar or rocket attacks, or assaults upon outlying guard posts. The Viet Cong was, furthermore, choking off the flow of rice to market: in 1963 rice arriving at the market in Saigon reached a high of about four million metric tons; in 1966 amounts declined to about three million tons, making importation of rice necessary. Far from being "totally cleared of Communist forces," in 1966 the delta was more than ever under Viet Cong control and something different was going to have to be done to change the situation. The introduction of additional forces might help, but such forces would require bases, which were difficult to provide. It was against this background that the basic decisions for the creation of an American riverine force were made in late 1965 and early 1966.

CHAPTER II

The Concept of a Riverine Force

In July 1965 the staff of the United States Military Assistance Command, Vietnam (MACV), was drafting the campaign plan for 1966, the purpose of which was to help the government of South Vietnam establish control over the people. The staff concluded that it was possible to secure government control over the I, II, and III Corps Tactical Zones, a conclusion largely based on the presence of American and other Free World Military Assistance Forces in these corps areas. In the IV Corps Tactical Zone, however, where there were no U.S. ground forces to bolster the efforts of the Vietnam armed forces, the staff questioned whether the objective of the campaign plan could be attained. The armed forces of Vietnam were considered capable of maintaining their position in the IV Corps, but not of reducing Communist control over significant portions of the population and terrain in the Mekong Delta. While additional American and other Free World Military Assistance Forces were planned for the other corps areas, none were planned for the IV Corps area. In the opinion of the staff, there could be no substantial progress in the IV Corps Tactical Zone unless U.S. ground forces were introduced.

As a preliminary step in studying the possibility of sending U.S. forces into the delta, Brigadier General William E. DePuy, J-3 (assistant chief of staff for military operations) of the MACV staff, directed his planners to survey the delta for land suitable for basing ground troops. A team dispatched to examine My Tho, Vinh Long, Sa Dec, and Can Tho concluded that all land suitable for large tactical units was either heavily populated or occupied by Republic of Vietnam armed forces. If U.S. units were based in the delta, they would have to share already crowded areas or displace a portion of the population. Since neither of these courses was acceptable, the planning staff then searched for other means of basing troops. Obtaining river sand as fill material and building up an area to accommodate a division base was considered. U.S. Military Assistance Command engineers estimated that with the dredge equipment at hand seventeen and a half dredge-years

would be necessary to fill an area large enough to base a division. Since adequate dredge equipment was not available in Southeast Asia, immediate steps were taken to procure additional dredges from the United States, but these were not expected to arrive until late 1966. Although not entirely satisfied with this solution, the planning staff believed the building of a base area by dredging operations to be a sound choice.

In addition to a land base, other means of operating in the delta were needed, and the planning staff turned to the experience of the French in Indochina for answers. The French had used small landing craft extensively to exploit the net of waterways from land bases. The American staff planned to use not only small craft but also a group of larger landing craft that would house and support a riverine force. These ships, the planners believed, might also be able to move along the coastline and major rivers and serve as mobile bases. It was known that during World War II the Navy had converted LST's (landing ships, tank) into barracks ships, and the planners believed that such ships could provide a base for a brigade force afloat. As the concept of an American river flotilla took form, the planners concluded that the LST barracks ships could be altered to furnish a helicopter flight deck and that barges could be provided on which helicopters would be able to land for maintainance. The ships could also be equipped with weapons for defense.

In their initial survey to find anchorage sites in the delta, the planners of the Military Assistance Command suggested seven that might serve. An LST could resupply the river force by traveling from Vung Tau across a stretch of the South China Sea into the selected anchorage sites on the Mekong and Bassac Rivers.

Captain David F. Welch of the U.S. Navy, who headed the Plans and Requirements Division of J-3, the Operations Directorate. U.S. Military Assistance Command, Vietnam, believed that the afloat force concept merited full study and suggested that a task force be set up under a U.S. Navy commander. He discussed the matter with Rear Admiral Norvell G. Ward, Commander, U.S. Naval Forces, Vietnam, who agreed that the idea had possibilities and that some naval ships could be made available.

In early December of 1965, during the monthly Commander's Meeting of the Military Assistance Command, Vietnam, General DePuy briefed General William C. Westmoreland and his commanders on the concept of a Mekong Delta riverine force that would employ an Army brigade with a comparable Navy organiza-

LST WITH ARMORED TROOP CARRIERS AND MONITORS

tion, and that would operate from various anchorages within the delta. The plan called for the development of two land bases, one for an infantry division headquarters and one brigade, and another for one brigade to be located in the northern part of the delta, probably in Long An Province. A third brigade would be based on the water. With the use of a mobile floating base, the mingling of U.S. troops with the Vietnamese population could be reduced—a prime consideration in view of the reluctance of the Vietnamese to accept U.S. ground forces within the delta. General Westmoreland declared the idea of a floating base "most imaginative," and directed that a team be sent to brief Headquarters, Pacific Command, and solicit its support.

When General Westmoreland had accepted the idea, reconnaissance was conducted to find a land base for a division headquarters and one brigade. General DePuy and Colonel Sidney B. Berry, Jr., Senior Advisor, 7th Division, Army of Vietnam, selected a base site approximately eight kilometers west of the town of My Tho, where the Kinh Xang Canal enters the My Tho branch of the Mekong River. According to French hydrographic maps, adequate sand deposits for fill material existed nearby. A request was then sent to the Joint Chiefs of Staff in Washington for an

additional infantry division to employ in the IV Corps area. For planning purposes the unit was designated Z Division.

During the latter part of December a MACV team headed by General DePuy met with the Pacific Command staff and outlined a general plan for the riverine force. After a period of discussion, the Pacific Command staff accepted the feasibility of the proposal. In the discussion the concept that the river force operating within the Mekong Delta would be a joint Army-Navy task force was emphasized. It was felt that barracks ships, then in storage, could be utilized to house part of the force, and that these barracks ships could be supported by small landing craft, patrol boats, and helicopters. Such support would enable the river force to conduct operations within a defined radius of the floating base. In addition, the mobility of the small craft would help to protect the floating base.

Upon acceptance by the Commander in Chief, Pacific, of the feasibility of a riverine force, the Joint Chiefs of Staff tentatively approved the employment of an Army division in the Mekong Delta to put the plan in operation. The force was to be provided barracks ships and LCM-8's. The matter of land bases was left unanswered in the joint staff action of the Commander in Chief, Pacific.

In Washington the Vice Chief of Naval Operations, Admiral Horacio Rivero, Jr., supported the concept of a riverine force and approved a proposal to send a planning group to Saigon to work with the MACV staff. The group was asked to develop a complete plan and to specify the means to support it. Headed by Captain David Bill, U.S. Navy, Office of the Chief of Naval Operations, and consisting of representatives of the Bureau of Ships, the Marine Corps, and the Amphibious Command, Pacific, the planning group arrived in Vietnam in January of 1966. Together the MACV staff and the Navy group studied in detail the experience of the French and Vietnamese with river assault forces in order to establish a similar American force, but one with greater capabilities. Under the leadership of Captain Welsh and Captain Bill, requirements were drawn up for self-propelled barracks ships (APB's), LST's, large covered lighters (YFNB's), large harbor tugs (YTB's), landing craft repair ships (ARL's), and a mine countermeasures support ship (MCS); all were to carry appropriate armament for the area of operations. The LCM-6 would be used instead of the LCM-8, which was in limited supply.

In a message of 19 February, the Commander in Chief, Pacific,

SELF-PROPELLED BARRACKS SHIP WITH AMMI BARGE MOORED ALONGSIDE

requested that the Commander, U.S. Military Assistance Command, Vietnam, develop and submit a detailed plan of operations and logistical support for the employment of a Mekong Delta Mobile Afloat Force. The plan was to cover force composition and phasing, tasks to be executed, methods of execution, navigable areas, operating areas, afloat force locations, afloat force base, provisions for U.S. Air Force, Navy gunfire, and helicopter support, and command relationships. The Commander in Chief, Pacific, Admiral Ulysses S. Grant Sharp, Jr., also asked for amplification of the logistics involved: depots and ports from which support would be provided, additional facilities required at the logistic support base, requirements for additional lighterage, and specification of other resources needed to support operations. Representatives of Admiral Sharp would meet with the MACV staff to "assist in the development of the foregoing and to facilitate subsequent evaluation and review." The MACV-Navy group continued to plan, following the guidance of the Commander in Chief, Pacific.

The planning culminated in a MACV study, Mekong Delta Mobile Afloat Force Concept and Requirements, dated 7 March 1966, which was forwarded on 15 March 1966 to the Commander in Chief, Pacific, for approval. The study articulated fully for the

first time the concept for the Army-Navy force afloat. It later proved to be a far-sighted and comprehensive blueprint for the preparation of both Army and Navy components and the conduct of operations. Most of the planned features of the force as stated in this document later materialized in the operational force.

The study recommended that the force contain a U.S. Army reinforced brigade consisting of three infantry battalions, an artillery battalion, and other combat and combat service support. The force would be based aboard U.S. Navy ships that would include 5 self-propelled barracks ships, 2 LST's, 2 large harbor tugs, and 2 landing craft repair ships. In addition, two U.S. Navy river assault groups would provide tactical water mobility. Each assault group would be capable of lifting the combat elements of one reinforced infantry battalion. A small salvage craft would be necessary to recover damaged ships or craft. The reinforced brigade would be organized under the current standard ROAD (Reorganization Objective Army Divisions) tables of organization and equipment, with limited augmentation. Certain equipment specified in the tables, such as tents, mess facilities, 106-mm. recoilless rifles, antitank wire-guided missiles, and all wheeled vehicles except artillery prime movers were to be deleted from the force requirements. The number of 90-mm. recoilless rifles in each rifle company was to be reduced from six to three to improve the mobility of the weapons squads. The 4.2-inch mortars would accompany the force and be moved by water or air to field positions as necessary. Radios would be either ship-mounted or man-portable. Coxswains of plastic assault boats were to be designated in the proposed tables and trained upon arrival in Vietnam. Enough troops from each of the units afloat would be left at a land base to maintain equipment left in storage. The plan provided for an augmentation of three countermortar radar sections, each manned by nine men, to operate and maintain ship-mounted countermortar radars. A mobile Army surgical hospital team, U.S. Air Force tactical air control parties which included forward air controllers, Vietnam Army liaison troops, and additional ANPRC-25 radios were to be furnished from sources outside the parent division of the brigade.

Each river assault group, later designated river assault squadron, was to consist of the following: 52 LCM-6's to serve as armored troop carriers, 5 LCM-6's to serve as command and communication boats, 10 LCM-6's to serve as monitors, 32 assault support patrol boats, and 2 LCM-6's to serve as refuelers. A salvage force would include: 2 2,000-ton heavy lift craft, 2 YTB's for salvage, 2 LCU's

Armored Troop Carrier

(landing craft, utility), and 3 100-ton floating dry docks. The Mobile Afloat Force concept stated the specific tasks the force was to accomplish: secure U.S. base areas and lines of communication required for U.S. operations; conduct offensive operations against Viet Cong forces and base areas that posed a threat both to the national and to the IV Corps Tactical Zone priority areas for rural construction in co-ordination with Republic of Vietnam armed forces and other U.S. forces; isolate the most heavily populated and key food-producing areas from Viet Cong base areas; interdict Viet Cong supply routes; and in co-ordination with the Vietnam armed forces provide reserve and reaction forces in the IV Corps Tactical Zone.

Because the Viet Cong forces were dispersed throughout the Mekong Delta, the river force would be required to operate in many locations. Initial operations would be conducted in the provinces to the north of the Mekong River—Go Cong, Dinh Tuong, Kien Phong, Kien Tuong, and as far north as the Plain of Reeds. (*See Map 5.*) Once the Viet Cong main force units were neutralized in these areas, with the U.S. Navy Game Warden forces con-

COMMAND AND COMMUNICATIONS BOAT

trolling Viet Cong cross-river movements, the center of gravity of U.S. operations would shift south into the provinces between the Mekong and the Bassac and eventually south of the Bassac River.

It was decided to use a brigade for initial operations because the experience of U.S. forces in Vietnam indicated that the brigade was the smallest U.S. unit that could be safely and economically kept in the delta. It was large enough to defend its base against heavy enemy attack and at the same time to have forces available for offensive operations.

The formal plan of the Military Assistance Command required the construction of a land base by dredging. Since land reclamation would be expensive and time-consuming, however, the plan also recommended the creation of a water base as well in order to use American forces most effectively. The plan indicated that the land base near My Tho could be made available at the end of 1966 or in early 1967.

Criteria for the selection of anchorages for the floating base were established. The base was to be reasonably near an airfield;

be within or contiguous to a relatively secure area; have access to a land line of communications; have enough room to anchor the entire force; and have a river bank that would permit debarking of artillery to provide fire support to the force. Sites that met most of the above criteria were found in the vicinity of My Tho, Vinh Long, Cao Lanh, Tra Vinh, Can Tho, Long Xuyen, and Dai Ngai, which was in the vicinity of Soc Trang.

Recognizing that the My Tho land base would not be large enough to accommodate a division headquarters until early in 1968, the plan called for the Z Division to go initially to Ba Ria, which lies north of Vung Tau. The division base was to include a division headquarters, a support command, one brigade, and a storage area for the heavy equipment and rolling stock of the two brigades sent into the delta—one to be land-based near My Tho rather than in Long An Province, and a second to be the Army component of the Mobile Afloat Force. Vung Tau was to be the port for staging and resupplying the Mobile Afloat Force. Minimum security measures would be needed for the force when it was at Vung Tau and when it was traveling from there to the Mekong River. During navigation of the major rivers, however, continuous security measures would be necessary. Such movements were to be considered ventures through hostile territory and were to be handled much the same way as running a tactical land convoy.

The plan provided for continuous air cover, both by fixed-wing tactical aircraft and armed helicopters. Advance, flank, and rear security on the rivers was to be provided by U.S. Navy river assault groups. GAME WARDEN forces would provide intelligence to assist in reconnoitering the route. The mine countermeasures were to be chain drags pulled along either side of the channel to cut command detonation wires of river mines, and mine-hunting sonar. Canals terminating in the main river might harbor hostile boats and were therefore to be reconnoitered by both boat and air. Armed helicopters, in conjunction with the Navy assault groups, were to cover the banks of the rivers and search out possible ambush sites. The major ships were to be at a condition of highest watertight integrity during transit, and fixed weapons were to be manned to counter enemy attack. Major ships of the force were to be moved in daylight hours.

At the anchorage of the afloat base, an area security plan would be in effect, with Army security forces on shore and reinforced by artillery. Outposts including foot and boat patrols would be used. The barracks ships would be moored near the center of the river

and their countermortar radars could provide coverage of the surrounding area. All weapons would be manned, with an Army-Navy reaction force standing by while the ships were at anchor.

The plan recognized three principal enemy threats. The first was infiltration by small mortar or recoilless rifle teams which would attempt to penetrate the infantry perimeter ashore. These teams would probably fire several rounds quickly and then withdraw rapidly. The second threat was that of a large-scale, overt attack. The third and possibly the most critical threat to the floating base would come from the water—floating mines, swimmer saboteurs, and suicide boats. Measures to counter the threats were left to the commanders immediately concerned.

The Mobile Afloat Force concept provided for a variety of tactical operations. The force would remain in an enemy base area as long as operations could be profitably conducted—about four to six weeks. Ground operations would last four to five days, after which troops would be allowed a rest of two or three days to dry out and to repair equipment. It was believed that the force could conduct four operations a month. A brigade would be deployed in the delta for as long as six months, with the possibility of rotation with either the brigade at My Tho or that at Ba Ria. The force would obtain intelligence and develop plans prior to its arrival in a base area, and operations would commence as soon as the floating base arrived at its anchorage. The brigade would rely heavily on local Vietnamese intelligence organizations.

To acquire knowledge of the area in which the force was operating, elements of the brigade, embarked in Navy assault group boats, helicopters, and plastic assault boats, would be employed to gather data on stream and canal depths, height of banks, areas for beaching, possible artillery positions, helicopter landing zones, the local population, and the enemy.

Full-scale operations would be conducted by deploying the force into the combat area by water, land, and air. Security for the floating base would require from a company to as much as a battalion in the immediate base area. Operations could range up to fifty kilometers from the floating base and, in exceptional cases, beyond. Security measures similar to those for relocation of the major ships would be taken for the movement of the assault craft formations during operations. Tactical air and armed helicopters were to fly cover, and artillery would displace by echelon as required to provide continuous fire support. The assault support patrol boats were to furnish forward, flank, and rear security, and to sweep for shore-

ASSAULT SUPPORT PATROL BOAT. *A high-speed armored boat used for waterway interdiction, surveillance, escort, mine-sweeping, and fire support.*

commanded mines. Preparatory fire could be delivered by the armament of the Navy assault groups, by artillery, and by air. Landing of the force would be in accordance with the tactical plan and the armored troop carriers would then withdraw to rendezvous points or return to the afloat base for resupply. At least one armored troop carrier would be designated as a floating aid station located at a prearranged point in the operational area of the battalion that it was supporting. Some elements of the force could deploy by water or air beyond the immediate landing areas to cut off withdrawing enemy troops and some could be held afloat as a reserve. The river force would prevent the enemy from withdrawing across the major rivers. Assault support patrol boats and monitors not required to protect the troop carriers would be used to provide flank and rear security, close-in fire support, and forward command post protection.

Command and control would be exercised through the use of command and communications boats, which were to provide space and communications for battalion and brigade command groups. Command helicopters could also be used to control operations.

MONITOR

Helicopters would be used to deploy a portion of the ground force, to position artillery, and to deliver reserves for the purpose of blocking the enemy or exploiting success. Armed helicopters would be employed for escort and fire support. Large Viet Cong forces encountered in these operations were to be neutralized or destroyed by fire and land maneuver. Should no large enemy force be encountered, the Mobile Afloat Force would be redeployed to another area of operations or the area would be systematically searched and enemy resources captured or destroyed. It was estimated that a battalion could conduct an operation throughout an area of forty square kilometers in four to six days. After troops had closed in on the final objective they would be met and withdrawn by the Navy assault group boats or possibly by helicopters. The withdrawal phase was considered critical because of the possibility of ambush and mines. Alternate routes and decoys would be used, but the basic concept for withdrawal would be the same as that of the movement to contact. Mine-clearing devices and air cover would be used. *(Diagram 3)*

All available means of fire support would be employed by the force, using the air and ground operations system common throughout Vietnam. Sorties would be allocated upon planned requests; emergency requests would be honored in accordance with priorities. The Direct Air Support Center at IV Corps Tactical Zone headquarters in Can Tho would be responsible for the allocation of

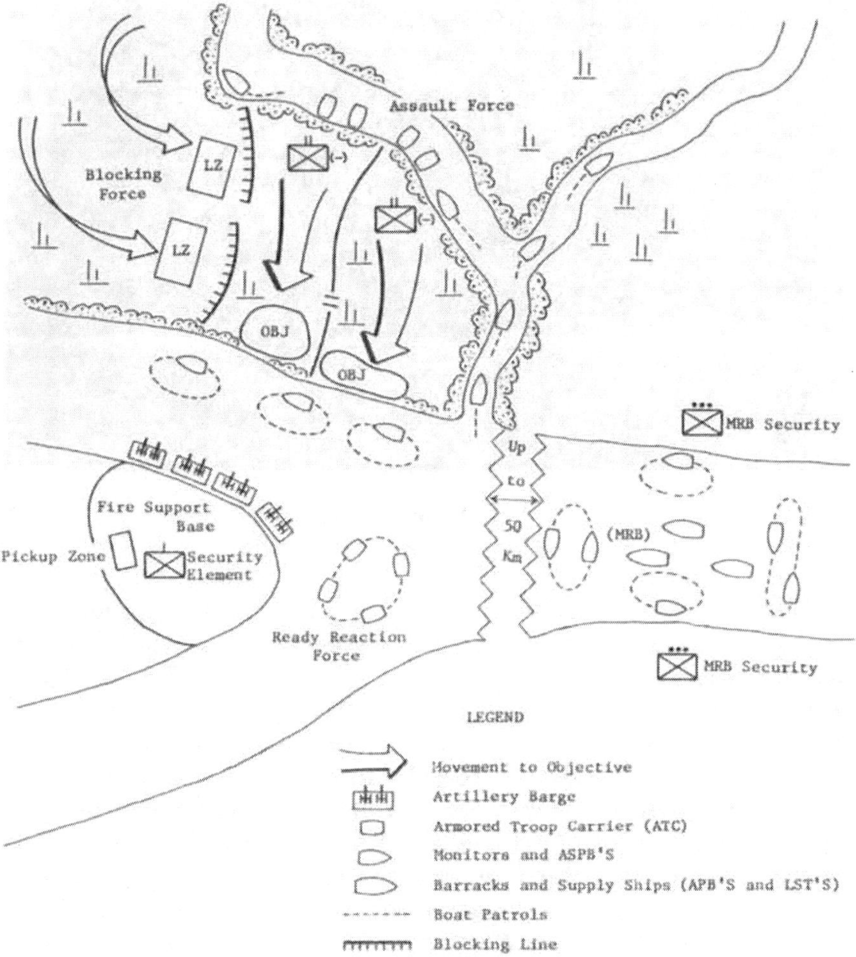

Diagram 3. Riverine operation and base defense.

tactical air support with airborne forward air controllers utilized to the maximum. Artillery support would be furnished in the traditional manner, providing fire support to all echelons of the force. Gunfire support would also be supplied by assault support patrol boats and monitors. Army helicopter gunships would be requested and allocated by the division in the same manner as for other divisional brigades.

The Vung Tau Area Support Command would provide common items of supply and logistic support peculiar to the Army; the Navy would furnish items peculiar to the Navy. A small Navy

support facility at Vung Tau would probably be necessary. Two LST's from Vung Tau would take care of all resupply except Class III. Each would be capable of carrying ten days' dry cargo for all elements of the force. One LST would remain on station with the force for seven days as a floating supply point. At the end of the seven days, the LST on station would be relieved by the second ship carrying ten days of supply, and would return to Vung Tau for replenishment.

Self-propelled barracks ships would be stocked with thirty days of frozen and chilled components of Class I and thirty days of dry components prior to deployment. Resupply of Class I would be scheduled to draw on the initial provisioning to a fifteen-day level by the time an APB returned to Vung Tau. The Navy would be responsible for operating all messes aboard ships, utilizing both Army and Navy mess personnel.

Resupply of petroleum products would be accomplished by having the craft of the Navy assault groups refuel from the APB's and the supply LST. The supply LST would refuel the APB's and other ships of the flotilla as required. Gasoline for the assault boats and vehicles would be supplied from two LCM-6 tankers, each with a capacity of 10,000 gallons. The LST would be refilled by commercial barge at ten-day intervals. Fuel for helicopter companies would be provided to the base by commercial contractor. U.S. Army, Vietnam, would supply what the commercial contractors could not.

The LST supply ships would carry that portion of the Class V basic load not carried on the APB's and ATC's. Army Class V carried aboard an APB would be limited to that portion of the basic load carried by the individual soldier and ten days' supply for Army weapons mounted on the ship. A normal allowance would be carried for the Navy weapons. The Army element of the force would prepare loading plans for that portion of the basic load to be carried on the ATC's. An operating allowance of Class V would be carried on the Navy assault group craft for all mounted weapons, with a ten-day resupply for these weapons on the supply ship.

Resupply of forces operating away from the Mobile Afloat Force anchorage would be tailored to the specific operation and would vary with the size of the force, duration of the operation, and distance from the anchorage. Resupply would be accomplished primarily by ATC and helicopter. Airlift and other means of transportation could be called upon to augment the capacity of the organic force for resupply as warranted by the area and nature of

operations. Matériel to be evacuated to Vung Tau or to the division base at Ba Ria would be carried by the resupply ships on their return to Vung Tau. Aircraft would be used for evacuation and for the replacement of urgently needed matériel between supply ship turnarounds. The Army and Navy components would retain maintenance responsibility for organic matériel, including parts supply for service-peculiar items. Each component would provide repair stores and load lists accordingly.

Army responsibility for maintenance would be limited to organizational maintenance of all matériel and direct support maintenance of weapons, vehicles, signal equipment, and assault boats and motors. Shops on the APB's would be used by unit armorers and maintenance men to perform organizational maintenance. Billeting space for a maintenance detachment of one officer, three warrant officers, and twenty-seven enlisted men, and shop space of approximately 1,500 square feet would be required on the ARL's to meet the Army direct support requirements.

All Navy ships and smaller craft were to arrive in Vietnam in operational condition. Maximum operational time would be available prior to regular overhaul. Two ARL's constituted the major repair and maintenance assets for the force and would normally remain with the force to function as advanced tenders. ARL shop spaces and equipment would be designed for a broad range of repair and maintenance of all craft and ships of the force and for overhaul of all assault craft. In addition, shop spaces were to be manned by the Army to provide for repair and maintenance of Army equipment.

Medical care would remain a service responsibility aboard ship. The Army would provide medical service for all elements of the force when they were away from the flotilla on an operation. Two medical evacuation helicopters would be stationed at the airfield closest to the area of operations to evacuate patients from battalion aid stations located on ATC's or from the battlefield to a mobile Army surgical hospital or an APB. Further evacuation to hospitals in the Saigon or Vung Tau area would be by helicopter or fixed-wing aircraft. Hospitalization would be in the Saigon, Bien Hoa, and Vung Tau areas. A mobile Army surgical hospital located at the land base near My Tho, in conjunction with the brigade to be based there, and a surgical team located on an APB, would provide surgery and medical treatment necessary to prepare critically injured or ill patients for evacuation to hospitals. A helicopter pad, sixteen-bed dispensary, and surgical suites with equipment neces-

sary for one surgical team would be provided on each APB.

Communications would be provided by the permanently installed Army and Navy radio equipment aboard each ship and assault craft. This equipment would provide command, tactical, and logistical communication links to higher and subordinate headquarters. Tactical communications ashore would be by man-packed radios and by airborne relay when required.

General command arrangements had been set forth previously by General Westmoreland. The plan indicated that they would be further developed by study. The command relationships of the force were to be a major topic of discussion for many months after the publication of the Mobile Afloat Force concept.

CHAPTER III

Riverine Preparations in the United States and in Vietnam

During the final preparation of the Mobile Afloat Force plan in South Vietnam, the 9th Infantry Division was activated at Fort Riley, Kansas, on 1 February 1966 under the command of Major General George S. Eckhardt. This was the one infantry division to be organized in the United States during the fighting in Vietnam —the so-called Z Division that had been scheduled for operations in the Mekong Delta. It was probably no coincidence that the division had been designated the 9th U.S. Infantry; General Westmoreland had seen extensive service with the 9th in World War II, having commanded the 60th Infantry and having served as chief of staff of the division during operations in both France and Germany.

Because of a shortage of men and equipment the activation order provided for incremental formation of the division. Division base elements such as the headquarters and headquarters company, division support command, and brigade headquarters and headquarters companies were activated first. Activation of the battalions of each brigade was phased, commencing in April for the 1st Brigade, May for the 2d Brigade, and June for the 3d Brigade. The artillery and separate units were scheduled for activation during April and May. Some of the division's officers who had been previously assigned to the Department of the Army staff had learned that the 9th Infantry Division was scheduled to operate in the southern portion of the III Corps Tactical Zone and the northern portion of the IV Corps Tactical Zone and that it was to provide a floating brigade. This information was not discussed officially but was known to the brigade commanders and the division artillery commander.

The division was organized as a standard infantry division composed of nine infantry battalions of which one was initially mechanized. It had a cavalry squadron and the normal artillery and supporting units. The division training program was limited to

eight weeks for basic combat training, eight weeks for advanced individual training, and eight weeks for both basic and advanced unit training—a total of twenty-four weeks. This compression of training time eliminated four weeks from each of the unit training periods and the four weeks usually allowed for field training exercises and division maneuvers—a total of twelve weeks from the normal Army training time for a division. Although it was not generally known in the division, the division training period as established in Army training programs had been reduced in order to conclude at the time of the beginning of the Vietnam dry season in December 1966 when the MACV plan called for the introduction of U.S. ground forces into the Mekong Delta.

The normal Army training programs were followed for the basic combat and advanced individual training. General Eckhardt, perceiving that the existing training programs had limitations for combat in Vietnam, by means of a personal letter gave his brigade commanders the latitude to make innovations and to modify training in order to prepare their men for the physical conditions and the tactics of the enemy in Vietnam. The training given by the brigade commanders was based on lessons learned and standing operating procedures of United States units then fighting in Vietnam. Although aware of possible employment of his unit in the Mekong Delta, each of the brigade commanders required the training he deemed advisable to prepare his unit for operations in any part of Vietnam. Although the brigade afloat had not been designated, Colonel William B. Fulton, Commanding Officer, 2d Brigade, felt that the mission could ultimately be assigned to his brigade. Because the training period was short, however, he elected to adhere to normal basic training in counterinsurgency for his units.

Colonel Fulton established a training course for the brigade and battalion commanders and their staffs that was designed to develop proficiency in command and staff actions for land operations in Vietnam. The class was held every ten days in a map exercise room with a sand table and map boards depicting selected areas of Vietnam. The sessions lasted approximately five hours. Three days before the class a brigade operations order was issued to the battalion commanders, requesting each battalion to prepare plans and orders in accordance with the brigade tactical concept. At the start of the session the brigade staff outlined the situation. Each battalion commander was then required to furnish a copy of his orders and to explain why he deployed his units as he did. There was a general critique, after which a new situation was

assigned for study. Each commander and his staff, which included representatives of artillery, engineer, aviation, and other supporting elements, then prepared the next set of orders. These in turn were presented to the group for analysis and critique.

The commanders and staff studied various forms of land movement by wheeled and tracked vehicles on roads and cross-country. Next, air movement was considered, including troop lift and logistics computations, various formations, and the selection of landing and pickup zones. This in turn was followed by a study of water movement by small craft.

In conjunction with command and staff training, the brigade was developing a standing operating procedure, which was reviewed at sessions of the command and staff course. The course began in May when the brigade was activated and was separate from the Army training programs being undertaken by the units. Officer and noncommissioned officer classes in the subject matter covered in the Army training programs were conducted at the battalion level.

Following the map exercises on movement, a series of exercises was conducted involving the organization and security of the base area and patrolling outside of the base in the brigade tactical area of responsibility. Subsequently, exercises were conducted combining both air and ground movement in search and destroy operations.

The brigade patterned its standing operating procedure and its methods of tactical operations primarily after those of the 1st Infantry Division. At the time that the command and staff training course was in full swing, several unit commanders from the 1st Infantry Division were returning from Vietnam to Fort Riley and Junction City, Kansas, where they had left their families when they departed with the 1st Division in 1965. Extensive interviews were conducted by the brigade commander and staff, who incorporated the information obtained into the standing operating procedures and the command and staff course.

The command and staff course did not specifically deal with riverine operations. Some of the map exercises were plotted in the northern delta immediately adjacent to the Saigon area and Di An where the 1st Division was operating, and one problem explored water movement. Too little was known about riverine operations to incorporate them fully into training. Furthermore, Colonel Fulton felt that the basic operational essentials should be mastered first; riverine operations could be studied after unit training was completed.

Review of the Mobile Afloat Force Concept

On receipt of the MACV Mobile Afloat Force concept in mid-March 1966, the Commander in Chief, Pacific, Admiral Sharp, requested that it be reviewed by the Commander in Chief, Pacific Fleet, Admiral Roy L. Johnson. The latter generally concurred in the plan but pointed out that while the Mobile Afloat Force concept provided for maintaining a brigade in the delta for up to six months, it might be necessary to rotate the APB's for maintenance and upkeep every two to three months.

Admiral Sharp had questioned the command arrangements. Under the Mobile Afloat Force plan it had been recommended that the Navy commander be charged with the security of the mobile base, while the Army brigade commander would provide support. Admiral Johnson, on the other hand, believed that the Army commander should be responsible for base security with the Navy commander providing supporting fire and protection against waterborne threat. He also questioned whether the Mobile Afloat Force could search junks effectively and protect naval craft against water mines and ambushes. He expressed concern that hydrographic charts of the delta waterways were incomplete, and that river assault craft were not properly designed and were, furthermore, too noisy.

Except for operational control, the Navy units were under the control of the Commander in Chief, Pacific Fleet. The Commander, Amphibious Force, Pacific, and Commander, Service Force, Pacific, had specific responsibilities and part of the subsequent success of the force stemmed from the professional manner in which the Navy fulfilled its obligations. In the case of logistics, support was given not only by units in Vietnam such as the harbor clearance units of Service Force, Pacific, but also by other units of the logistic support system, afloat and ashore, as set forth in the support plan of the Service Force. Each of the commanders concerned felt personal responsibility for the performance of those of his units that would be operating under Military Assistance Command, Vietnam.

As staffing of the Mobile Afloat Force proceeded, General Westmoreland continued to call attention to the need for beginning immediately U.S. operations in the Mekong Delta. In a message to Admiral Sharp on 11 May, with respect to intensification of the efforts of Vietnam armed forces and early initiation of U.S. operations in the Mekong Delta, he stated that "enemy access to Delta resources must be terminated without delay."

The planned deployment into the delta had appeared in MACV Planning Directive 3-66 published 21 April 1966. The directive had emphasized widening the range of operations in the northern coastal areas of South Vietnam and close-in clearing and securing operations around Saigon. It did not provide for any major American effort in the delta during the rainy season of May-November 1966, but the possibility of short operations by units such as the Special Landing Force was cited. These operations were to have as their target the mangrove swamps along the coastal areas in the southern III Corps and northern IV Corps Tactical Zones. Plans for the operations were based on the success of operation JACKSTAY in the Rung Sat Special Zone in 1966.

General Westmoreland expected to send forces in late 1966 and early 1967 from the III Corps Tactical Zone to the Plain of Reeds and other northern delta areas. The planning provided for the Commanding General, II Field Force, Vietnam, Lieutenant General Jonathan O. Seaman, to assume command of U.S. tactical operations in the IV Corps Tactical Zone, co-ordinating operations with the Commanding General, IV Corps, through the American senior adviser who was to be a brigadier general. (Colonel William D. Desobry, Senior Advisor, IV Corps Tactical Zone, was promoted to brigadier general in August of 1966.)

On 29 May 1966 General Westmoreland was briefed on the deployment of the 9th Infantry Division to the IV Corps Tactical Zone. He approved the plan and ordered his staff to discuss with General Seaman an alternate location for the 9th Division base. General Westmoreland directed that the Mobile Afloat Force plan to locate a division headquarters and one brigade at Ba Ria be reconsidered. He called attention to his previous decision that the 9th Division would be placed under General Seaman to facilitate tactical operations along the III and IV Corps border and that the Commanding General, 9th Division, would not become the senior adviser to the IV Corps Tactical Zone. General Westmoreland pointed out further that the introduction of a division force into IV Corps would require discussion with General Cao Van Vien, chairman of the Joint General Staff of the Republic of Vietnam.

On 9 June General Westmoreland suggested that the delta might well be a source of stabilization of the Vietnamese economy. The delta could produce enough rice for the entire country if it were kept under government control; other areas of the country would then be free to industrialize. The delta was also the source

of nearly 50 percent of the country's manpower. It therefore followed that development of the region had to be accelerated; sending in a U.S. division would aid in this acceleration.

On 10 June General Westmoreland discussed with General Vien and Lieutenant General Dang Van Quang, Commanding General, IV Corps Tactical Zone, the possible introduction of U.S. forces into the IV Corps Tactical Zone. On 13 June the matter came up for discussion in the Mission Council meeting. General Quang, who had made a statement to the press some weeks earlier against the stationing of U.S. troops in IV Corps, but in the meantime had apparently had a change of heart, now expressed in the meeting a desire that a U.S. brigade be stationed in IV Corps. General Westmoreland told the council that a final decision on the matter of basing American troops in IV Corps would be made in October. Dredges had already been ordered, and the proposed site would be ready by December. He further stated that the troops would be located about eight kilometers from My Tho, which would be off limits to U.S. troops, and that travel through My Tho would be sharply restricted. Since the base would be completely self-sufficient, it would be no drain on local resources. When Ambassador Henry Cabot Lodge, Jr., and the political counselor expressed reservations, General Westmoreland agreed with them that it would be preferable to use Vietnamese troops, but pointed out that up to this time Vietnamese troops had not been completely successful in the delta, and important Viet Cong units were still operating there.

The land base had been selected by General Westmoreland himself from four sites submitted by the engineers as suitable for building by dredging. The sites were designated W, X, Y, and Z, and the one near My Tho chosen by General Westmoreland was W. The general's staff immediately referred phonetically to Site W as Base Whisky, the word used in the military phonetic alphabet for the letter W. General Westmoreland felt that the site should be given a significant name in keeping with its role as the first American base camp in the Mekong Delta. He asked the official MACV translator to give him several possible Vietnamese names for the base, such as the translation of "friendship" or "co-operation." The translator's list included the Vietnamese term *Dong Tam*, literally meaning "united hearts and minds." General Westmoreland selected this name for three reasons: first, it signified the bond between the American and Vietnamese peoples with respect to the objectives to be achieved in the delta. Second, it connoted an appro-

LANDING CRAFT REPAIR SHIP WITH ARMORED TROOP CARRIERS

priate objective compatible with the introduction of U.S. forces into the populous delta where their prospective presence had evoked some official concern. Third, Dong Tam was a name which Americans would find easy to pronounce and remember. Having chosen the name of Dong Tam, General Westmoreland asked General Vien, the chairman of the Joint General Staff, his English translation of the name. General Vien confirmed that "united hearts and minds" was the literal translation. Thus the name Base Whisky was changed to Dong Tam, which became a well-known landmark during the subsequent co-operative efforts of American and Vietnamese troops in the delta.

Later in the month, at the Honolulu Requirements Planning Conference, the Mobile Afloat Force was included in requirements for the calendar years 1967 and 1968.

> The Mekong Delta Mobile Afloat Force will provide a means to introduce, employ and sustain substantial U.S. combat power in that vital area. Introduction of the MDMAF [Mekong Delta Mobile Afloat Force] at the earliest practicable date, whether it be an increment of that force or all of it, will provide a capability for more rapid achievement of U.S. objectives in that area.

These requirements as set forth in the conference provided for the arrival of the first component of that force by April 1967 and the second, final component by March 1968.

On 5 July Robert S. McNamara, Secretary of Defense, approved activation and deployment of a Mobile Afloat Force consisting of two river assault groups. At the time of the approval, he reduced the number of self-propelled barracks ships from five to two and eliminated one landing craft repair ship from the force. He was not willing to provide for the total package requested for the Mobile Afloat Force because he felt that the force could be fully tested with the equipment he had approved. Included in the cut was the salvage force, which required two heavy lift craft of 2,000 tons, two YTB's altered for salvage, two LCU's, and three 100-ton floating dry docks. Only one YTB was authorized. Secretary McNamara's decision was to have an appreciable impact on the preparation for and the operations of the Mobile Riverine Force as it was constituted in June of the following year. Subsequently General Westmoreland, through Admiral Sharp, requested reconsideration of the decision to field only two self-propelled barracks ships and two river assault groups; again four river assault groups and at least four barracks ships were requested.

In view of the request, the Joint Chiefs of Staff asked for an evaluation of the planned employment and of the additional effectiveness which these ships and craft would contribute to the force. Such an evaluation already had been completed by the MACV staff in May. The evaluation pointed out the lack of firm ground for stationing major troop elements and noted that "the time-consuming process of dredging" required to base additional units on land justified the additional two barracks ships. Admiral Johnson called attention to the fact that the Mobile Afloat Force concept also provided for a mobile brigade independent of a land base. One brigade of the 9th Division was to be stationed at Dong Tam in early 1967. When barracks ships became available, a reinforced battalion from a brigade in III Corps Tactical Zone would be put afloat.

The two river assault groups approved by Secretary McNamara for fiscal year 1967 would be stationed at Dong Tam. Omission of the repair boat to provide mobile maintenance would preclude the permanent basing of assault groups with the two barracks ships in the first increment of the Mobile Afloat Force. Either assault group would be available on call from a land base to provide lift for a

battalion afloat. Some of the boats of an assault group would remain to protect the barracks ships.

Units of the 9th Division not stationed at Dong Tam or aboard APB's would be based in III Corps Tactical Zone north and east of the Rung Sat Special Zone, permitting extensive operations into the special zone and IV Corps by assault group craft. With the arrival of a third assault group, which would include an ARL, one group could be permanently assigned to the forces afloat. This would leave two river assault groups assigned to Dong Tam for lifting battalions of that brigade or the other brigade from the Vung Tau area for operations in the upper delta. When the second increment arrived, at least two river assault groups would be needed for the floating base and one each to support the other two brigades of the 9th Division.

General Westmoreland strongly recommended to Admiral Sharp that the two additional APB's and assault groups be included in the calendar year 1967 force requirements and that they be activated and deployed at the earliest practicable date. Admiral Sharp supported General Westmoreland's position and forwarded it to the Joint Chiefs of Staff on 16 July, with a further justification of the two river assault groups on the grounds that "projection of U.S. combat operations into the Delta is an objective of major importance." Admiral Sharp stated that with three thousand kilometers of navigable waterways, an absence of adequate roads, and a lack of helicopters, the 9th Division, which "will be the principal riverine ground combat force," would require river assault group support. Four groups (two organic to the Mobile Afloat Force and two additional) could lift about half of the riverine ground force at any one time. In addition, river assault group craft would be used in reconnaissance and patrolling missions and resupply operations, would reinforce GAME WARDEN and MARKET TIME operations when necessary, and would support operations to open and secure important water routes.

In August while the question of whether to increase the number of Navy boats was being decided at higher headquarters, in Vietnam Mobile Afloat Force preparations were nearing completion. On 1 August, MACV published Planning Directive 4-66—Operations in the Delta. The directive called for employment of riverine forces "regardless of whether based on land (Dong Tam or elsewhere) or on MDMAF [Mekong Delta Mobile Afloat Force]." Composition of a river assault group was established as 26 armored troop carriers, 16 assault support patrol boats, 5 monitors, 2 com-

mand communications boats, and 2 LCM-6 refuelers. This composition was to vary little throughout the entire period of river assault group operations.

The MACV plan would operate in three phases. During the first, the Construction Phase, 1 July 1966-31 January 1967, all actions required to prepare the ground and facilities for occupation of the base would be completed. In the second, the Preparation and Occupation Phase, also to run from 1 July 1966 through 31 January 1967, all actions required to prepare the Army and Navy units to occupy bases, and the actual occupation would be completed. Preparation of the forces would proceed concurrently with base construction. The Improvement and Operations Phase would begin when the Army and Navy units had occupied the bases and were ready to begin combat operations. All actions necessary to conduct and sustain combat operations from the Dong Tam base would be undertaken during this phase and base facilities would be improved and expanded as necessary.

9th Infantry Division Studies the Mobile Afloat Force

In July 1966 the 9th Division at Fort Riley had been furnished copies of the plan and requirements of the Mobile Afloat Force. These were studied by the division staff and, after approximately two weeks, the chief of staff, Colonel Crosby P. Miller, assembled the brigade commanders. In very broad terms, Colonel Miller outlined the intended area of operations for the division and referred to the provision for a brigade afloat. This briefing aroused the curiosity of Colonel Fulton, commander of the 2d Brigade. He requested copies of the complete plan for study by himself and his staff, and the division commander approved. An intensive analysis was then made by the appropriate staff officers and the study was returned to division headquarters.

Exhibiting the foresight that had characterized its planning effort, MACV sent one of the principal planners for the Mobile Afloat Force to the United States on normal rotation and placed him on temporary duty with the 9th Division for a week. This officer, Lieutenant Colonel John E. Murray, Field Artillery, was extremely enthusiastic about the project and was familiar with all aspects of the plan and with the area of intended operations in the delta. Colonel Fulton arranged for Colonel Murray to spend some time with his brigade staff and battalion commanders to discuss all facets of the project. Through questions and answers, a clear

understanding of what was intended was conveyed to the brigade officers.

Later in the week, Colonel Murray addressed the division staff and subordinate commanders. He outlined the Mobile Afloat Force plan, discussed the environment, and sketched the nature of intended activity. He also explained that the Marine Corps had developed a basic riverine manual entitled Small Unit Operations in the Riverine Environment. This document was obtained by the 2d Brigade for study from the division G-3 (assistant chief of staff for operations), Lieutenant Colonel Richard E. Zastrow.

In early May the Department of the Army had informed the 9th Division that it would be sent to the Republic of Vietnam, beginning in December. The assistant division commander, Brigadier General Morgan E. Roseborough, headed an advance element that left for South Vietnam in late August to plan deployment of the division.

The Coronado Conference and Doctrine

On the 12th of September, Colonel Fulton was informed by General Eckhardt that a conference would be held on the 17th of September at Coronado Naval Base, California, and was designated the division representative at the conference. The conference had been co-ordinated by Headquarters, U.S. Continental Army Command, to examine the joint training implications which would be imposed on the 9th Division and the Army U.S. training base by participation in the Mobile Afloat Force. General Eckhardt directed Colonel Fulton to prepare a brief of the Army views on the Mobile Afloat Force that would include a plan for logistic support. General Eckhardt further asked that he receive a briefing on the presentation to be made at the conference. It was also decided that the division support commander, Colonel John H. Barner, should accompany Colonel Fulton to Coronado to handle the logistical aspects.

On 19 September, accompanied by Captain Johnnie H. Corns of his staff, Colonel Barner, division G-2 and G-3 representatives, and a representative of the division signal office, Colonel Fulton proceeded to Coronado. The conference was held under the auspices of the Commander, Amphibious Command, Pacific, Vice Admiral Francis J. Blouin. Navy attendees included representatives from the Chief of Naval Operations, Commander in Chief, Pacific Fleet, the Amphibious Training Center, the U.S. Marine Corps, and Commander, Service Force, U.S. Pacific Fleet. The Army dele-

gation also included representatives from Continental Army Command headquarters, U.S. Army Combat Developments Command, Fifth Army headquarters, and Sixth Army headquarters. The commander of River Assault Flotilla One, Captain Wade C. Wells, who was to be the U.S. Navy component commander of the Mobile Afloat Force, his chief of staff, Captain Paul B. Smith, and the rest of his staff attended. The conference was chaired by Rear Admiral Julian T. Burke, who had the additional responsibility of preparing a U.S. Navy doctrinal manual for riverine operations. Presentations were made of the organization and operations of the two components as well as the broad problem of command relationships of the joint force. Afterward several working groups were established to deal with command and control, joint staff arrangements, training, logistics, communications, and medical support.

During the conference Captain Wells informed Colonel Fulton that the River Assault Flotilla One chief of staff and representatives of N-1, N-2, N-3, and N-4 staff sections, as well as a communications officer, were going to Vietnam as an advance party in early October. Upon learning that the 9th Division had an advanced planning element at Headquarters, U.S. Army, Vietnam, Captain Wells agreed that the two advance elements should be the basis for co-ordination in Vietnam. Captain Wells and Colonel Fulton also made informal arrangements to co-ordinate mutual problems that might arise between the time of the conference and the departure of the 2d Brigade for Vietnam, which was tentatively scheduled for early January 1967. While there were no official provisions for direct communications and co-ordination between the two component commanders of the Mobile Afloat Force, this early meeting proved extremely beneficial in resolving matters that could have impaired the entire undertaking.

Returning to Fort Riley, Kansas, Colonel Fulton and Captain Corns briefed General Eckhardt and his staff on the results of the Coronado conference. General Eckhardt then designated Colonel Fulton as the executive agent for the division on riverine matters, and specified that this responsibility carried with it the designation of commander of the floating brigade and the U.S. Army component for the Mobile Afloat Force. Colonel Fulton still chose not to incorporate riverine operations in his training program because of the great amount of normal training to be accomplished. It was implied, however, at the conference that there would be a training period in Vietnam of approximately two to three months during which the Army and Navy components would be able to train

their forces for the joint riverine operations. Nevertheless when Colonel Fulton informed Lieutenant Colonel William B. Cronin, Commanding Officer, 2d Battalion, 47th Infantry, and Lieutenant Colonel Guy I. Tutwiler, Commanding Officer, 4th Battalion, 47th Infantry, of the riverine mission, they decided as part of their normal training programs to incorporate techniques and equipment for crossing small rivers.

While at Coronado, Colonel Fulton explored with the commander of the Amphibious Training School the possibility of conducting a riverine course for his brigade staff, the battalion commanders and their staffs, and the brigade supporting unit commanders. This school was a repository of amphibious doctrine and concepts as well as lessons learned in the MARKET TIME and GAME WARDEN operations, which were, respectively, the U.S. Navy offshore and river operations in South Vietnam. All the U.S. Navy advisers for the Vietnamese river assault groups were trained under Amphibious Training School auspices. The Navy SEAL (sea-air-land) teams, which were rotated in and out of South Vietnam, were also trained there. Many of the school faculty had already completed tours of duty in both the U.S. and Vietnamese navies. The idea of conducting a riverine course was acceptable to the commander of the Amphibious Training School, and plans for instruction were developed through subsequent correspondence between the brigade and the Amphibious School during the period October through December. Also it was agreed that the Amphibious Training School would provide a team at Fort Riley during December 1966 for the purpose of training selected men from brigade and battalion in techniques of waterproofing, small boat loading and handling, and combat offloading from transports. Also included was instruction in water safety techniques that was to prove immensely valuable once operations commenced.

In early October Colonel Fulton was informed that he would accompany the division commander to Vietnam on an orientation visit to reconnoiter the riverine environment and get a preview of the requirements for riverine operations. The division commander and a small staff left on 9 October for a three-week visit to Vietnam. Colonel Fulton's arrival coincided with that of the advance party of the River Assault Flotilla One staff. Colonel Fulton and Captain Smith, chief of staff of the flotilla, were able to visit and analyze the proposed training site in the vicinity of Ap Go Dau, adjacent to the Rung Sat Special Zone, approximately fifteen kilometers south of Bearcat. Bearcat, ten miles south of Long Binh, was to be

the 9th Division base instead of Ba Ria, which had been specified in the original Mobile Afloat Force plan. Bearcat was to be expanded into a base capable of accommodating the entire division until the 2d Brigade base at Ap Go Dau could be built. Colonel Fulton and Captain Smith agreed on the site near Ap Go Dau, and plans were developed by Company B, 15th Engineer Battalion, to construct the joint training base. A phased training schedule based on actual combat operations from the new location was also tentatively agreed upon. The shift of the division base from Ba Ria to Bearcat was to have no significant effect on the Mobile Afloat Force plan.

Colonel Fulton also visited the Dong Tam construction site, the 7th Division headquarters of the Army of Vietnam, and General Desobry, senior adviser of IV Corps Tactical Zone at Can Tho. Discussions centered on projected operations of U.S. forces in the Mekong Delta. It was especially fortuitous that Colonel Fulton, General Desobry, and the Senior Advisor, 7th Division, Colonel John E. Lance, Jr., had been on the faculty of the Army War College during the period 1962 to 1965. This professional association proved to be very valuable during the ensuing months as the 2d Brigade planned for and conducted operations in the Mekong Delta.

While at Headquarters, U.S. Army, Vietnam, Colonel Fulton learned that the Army had not completed the preparation of the riverine doctrine, a task assigned to it by MACV Directive 3-66. In discussing the task, Colonel Fulton found that Major John R. Witherell, the U.S. Army Combat Developments Command liaison officer, had developed an active interest in the Mobile Afloat Force plan and had started preparation of a rather detailed manuscript which dealt with organization and tactics of units as envisaged by the Mobile Afloat Force planners. At Colonel Fulton's suggestion Major Witherell agreed to propose to Headquarters, Combat Developments Command, that it undertake the drafting of a test field manual on riverine operations. In making this recommendation, Major Witherell planned to furnish his draft manuscript. The meeting proved to be beneficial since during the Coronado conference the doctrine matter had been explored with both Navy representatives and Major Donald R. Morelli, the Combat Developments Command representative at the conference. It was agreed that if the manual was undertaken, the writing should be done at the Amphibious Training School at Coronado, the best source of information on riverine warfare.

Upon his return to Fort Riley, Colonel Fulton was visited by a representative from Combat Developments Command who had outlined a manual based on Major Witherell's manuscript. Colonel Fulton suggested that this be accepted as the basis for the manual which was to be prepared at Coronado with representatives from the various Army service schools and Combat Developments agencies under the leadership of the Institute of Combined Arms Group, Fort Leavenworth. Colonel Fulton stressed that the manual should be available when the brigade and battalion staffs departed for South Vietnam on 14 January 1967. The need for the manual was quite apparent since the U.S. Marine Corps Fleet Marine Force Manual 8–4, the only doctrinal manual on riverine operations, dealt with only small boat tactics and did not cover joint riverine operations.

Final Decisions on Deployment

Secretary of Defense McNamara during his October visit to South Vietnam was briefed on the Mobile Afloat Force, and the need for two additional river assault squadrons was stressed at that time. The designation river assault group had been changed to river assault squadron in order to avoid confusion with the Vietnamese river assault groups. During this briefing it was emphasized to Mr. McNamara that if the objectives of the MACV campaign plan were to be achieved, U.S. ground operations in the IV Corps area were needed to assist the Republic of Vietnam armed forces. It was further pointed out to Mr. McNamara that roughly 50 percent of the population and 68 percent of the rice-producing area in the Republic of Vietnam were in the Mekong Delta. In this briefing it was explained that the 9th Division, due in Vietnam in December 1966 and January 1967, was to be the principal river ground force, and that the river assault boats needed to provide tactical mobility would conform to standard U.S. Navy organization structure, with two assault squadrons of about fifty boats each under command of River Assault Flotilla One. The Mobile Afloat Force with the two approved river assault squadrons would provide a good start, but there was a need for at least two additional assault squadrons by the end of calendar year 1967 in order to sustain the momentum of the riverine operations. The 9th Division, now planned with seven infantry battalions and two, rather than one, mechanized battalions, would require the support of two river assault squadrons. Two additional battalions from the U.S. 25th Division and the Australian Task Force would bring the total to

eleven infantry battalions for riverine operations. No decision was made by Mr. McNamara on the two river assault squadrons during his visit.

On 21 November General Westmoreland suggested to the Mission Council that it was feasible to deploy a battalion to Dong Tam in January 1967, and requested the council's endorsement of this action. Anticipating the council's concurrence, General Westmoreland directed the planning and preparation for deployment of a brigade in February 1967 if it was deemed feasible by II Field Force, Vietnam. By 29 November, Ambassador Lodge had approved this deployment.

On 1 December, Headquarters, Military Assistance Command, Vietnam, published a plan for logistical support for the Mobile Afloat Force. This plan provided for two kinds of support—one for the units based at Dong Tam and one for the units afloat. The land base commander was assigned responsibility for the logistical support for Dong Tam, and the mobile riverine base commander would have responsibility for the mobile riverine base, while service-peculiar supply would be the responsibility of the component commander concerned. Saigon was designated as the primary supply source for the Dong Tam base, with Vung Tau as the alternate, and land lines of communication were to be used wherever possible. The mobile riverine base would be supported by Vung Tau, with Saigon as the alternate.

After evaluating the progress of the base construction at Dong Tam, II Field Force reported on 4 December that it was feasible to support a battalion at Dong Tam in late February 1967. The planning and liaison machinery went into high gear early in December. Elements of the 9th Division would be available for training at the Vung Tau base in early January. The commander of Naval Forces, Vietnam, had shifted all efforts to prepare the base for riverine training when the proposed training site at Ap Go Dau was abandoned in favor of Vung Tau. He recommended liaison between 9th Division and Vietnamese and U.S. agencies. He also asked Admiral Johnson to provide a suitable support ship at Vung Tau about 7 January and at the same time to place aboard it one river assault squadron staff and one river division staff. The commander of River Assault Flotilla One and his staff, less the advance element, were to leave Coronado in mid-February and a second river division was to leave in late February.

General Westmoreland directed the commander of II Field Force to prepare to send an infantry battalion task force from the

9th Division to Dong Tam, to add forces later to increase it to a brigade, and to advise him of the arrival dates. The Commanding General, U.S. Army, Vietnam, was to support the task forces and Senior Advisor, IV Corps Tactical Zone, was directed to plan for the provision of Vietnam Army security forces, co-ordinate Vietnamese and U.S. security arrangements, and prepare the Vietnamese people for the presence of U.S. troops at Dong Tam.

The first elements of the 9th Division landed at Vung Tau on 19 December, and on 20 December General Westmoreland estimated that the battalion task force would move to Dong Tam on 25 January following two engineer companies that were to arrive there on 7 January. He calculated that strength would increase to brigade level in late February or early March.

The Mobile Afloat Force, conceived and approved during 1966, was one of the most important MACV accomplishments of the year. This force was expected to play a major role in the control of the delta, not only in a military sense, but also economically and politically. Further, the entire delta campaign could well be a key to the success of the combined operations of the United States and the Republic of Vietnam.

2d Brigade, 9th Infantry Division, Arrives in Vietnam

Upon completion of training at Fort Riley at the end of November 1966, the 2d brigade began preparation for overseas movement. During this time, selected men from the brigade and battalions were given training by the Marine Training Team from the Naval Amphibious School. On 3 January, the brigade commander and staff, the commander and staff of the 2d Battalion, 4th artillery, the commanders and staffs of the 3d and 4th Battalions, 47th Infantry, and the S-2, 2d Battalion, 47th Infantry, went to Coronado to attend the ten-day riverine course that had been established at the request of the brigade commander. The course provided a great deal of useful information on operations of the Vietnamese river assault groups, U.S. Navy SEAL teams, Viet Cong intelligence operations in the delta, and the riverine environment. The ten days gave the commanders and staffs of the brigade's attached and supporting units the opportunity to concentrate on purely riverine problems for the first time.

When the course ended on 12 January, the Combat Developments Command writing group had completed the first draft of Training Text 31-75, Riverine Operations. Colonel Fulton now learned that although the Navy had provided advice and consulta-

tion, it would not formally accept the manual, nor would the commander of River Assault Flotilla One acknowledge the text as a source of doctrine to which he would subscribe. When agreement had been reached in September that a manual would be written, Navy acceptance was tacitly understood, but since it was not forthcoming the question of agreement on joint procedures remained. The brigade commander and his officers, however, considered the training text a sound new source of riverine doctrine and concepts on which subsequent training in Vietnam could be based. Improvement could be made in the text after the experience of actual operations.

The 2d Brigade officers who had attended the Coronado riverine course arrived at Bien Hoa on 15 January and proceeded to Bearcat. Shortly afterward the remainder of the advance party arrived by air from Fort Riley; included were all the squad leaders from the three infantry battalions, two platoon sergeants and two platoon leaders from each of the companies, and all company commanders. The main body was en route by water from San Francisco under the supervision of Lieutenant Colonel Thomas F. O'Connor, the brigade executive officer. The other officer cadre consisted of the company executive officers, two platoon leaders from each company, and the battalion executive officers with small staffs. The purpose of the large brigade advance party was to give most of the tactical unit combat leaders battle experience before the main body arrived at the end of January. To this end the unit advance parties were sent to operate with the U.S. 1st and 25th Divisions for approximately two and one-half weeks.

The main body debarked at Vung Tau on 31 January and 1 February 1967 and reached Bearcat on 1 February. On 7 February the 2d Brigade, with the 2d, 3d, and 4th Battalions, 47th Infantry, and 3d Squadron, 5th Cavalry, commenced a one-week operation in the Nhon Trach District of Bien Hoa Province just north of the Rung Sat Special Zone. The 2d Brigade had an excellent opportunity to shake itself down operationally in a combat environment and to compensate for the lack of a brigade field training exercise which it had been unable to conduct at Fort Riley because of the short training period.

Rung Sat Special Zone Operational Training

During the latter part of January the 3d Battalion, 60th Infantry, an element of the 3d Brigade, 9th U.S. Division, had begun riverine training with the advance River Assault Flotilla One ele-

ments that were aboard USS *Whitfield County* (LST 1169), anchored in Vung Tau harbor. The 3d Battalion, 60th Infantry, was to be the first infantry unit sent to Dong Tam and was to arrive there by the end of January. It would be followed in training by the 3d Battalion, 47th Infantry, which was then participating in the 2d Brigade operation in Nhon Trach.

On 10 February, two companies and the staff of the 3d Battalion, 47th Infantry, under Lieutenant Colonel Lucien E. Bolduc, Jr., left for riverine training in the Rung Sat Special Zone. The Navy crews with which the battalion would work had received on-the-job training with the Vietnamese Navy river assault groups in the delta.

On 15 February, as the brigade was returning from the Nhon Trach operation to its base camp at Bearcat, an order was received from II Field Force directing that an entire battalion conduct operations in the Rung Sat Special Zone beginning 16 February. The order was prompted by a Viet Cong attack on a freighter navigating the Long Tau, the main shipping channel connecting Saigon and Vung Tau, on 15 February. It brought to an abrupt halt the organized training for Colonel Bolduc's battalion, which had accomplished only three of the scheduled ten days of training; full-scale combat operations would have to begin. The battalion commander was aboard USS *Whitfield County* with two companies when Colonel Fulton talked to him. Colonel Fulton had already ordered the remaining companies of the 3d Battalion, 47th Infantry, as well as a direct support battery of the 2d Battalion, 4th Artillery, into the northern portion of the Rung Sat Special Zone. Colonel Bolduc's mission was to disrupt enemy activities in the major base areas of the Viet Cong.

The resulting operation initiated on 16 February 1967 and terminated on 20 March was designated RIVER RAIDER I. It was the first joint operation by U.S. Army and U.S. Navy units that were later to constitute the Mobile Riverine Force. The 3d Battalion, 47th Infantry, was supported by River Assault Division 91 of River Assault Squadron 9. (*Chart 1*)

Joint operations centers were maintained twenty-four hours a day both at land bases and aboard the APA *Henrico*, and joint plans were made for each project. River Assault Flotilla One provided rear area support and planning assistance for river squadron operations. Squadron 9's assault division was commanded by Lieutenant Charles H. Sibley, who operated LCM-6's and a command vessel borrowed from the Vietnam Navy since the squadron's boats

CHART 1—RIVER ASSAULT SQUADRON ORGANIZATION

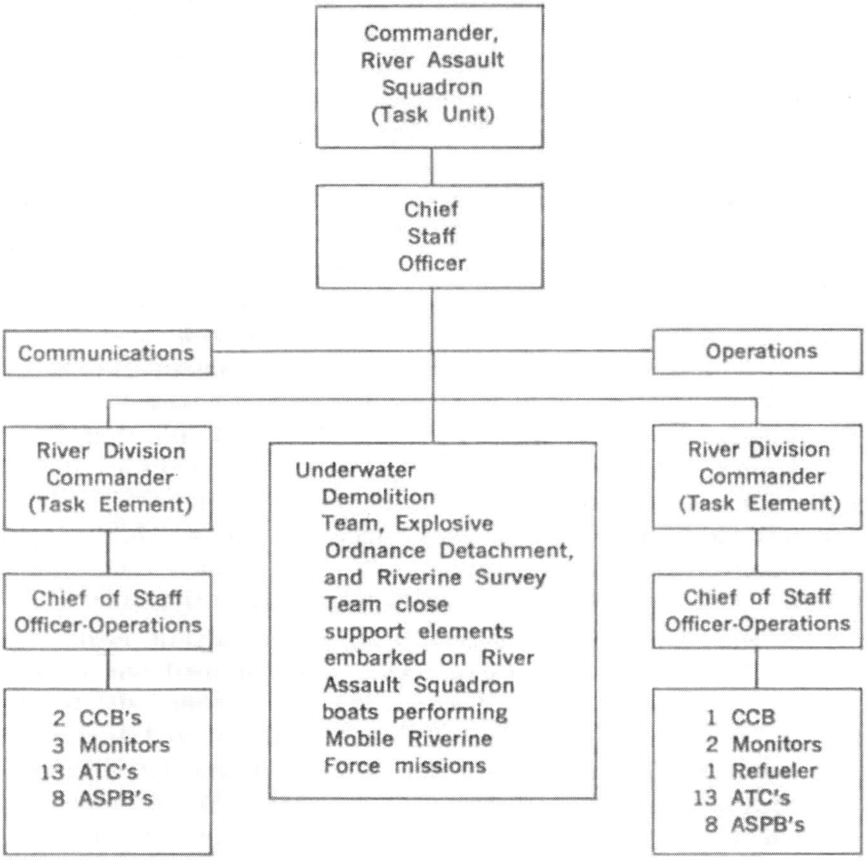

had not yet been delivered. Vietnam River Assault Group 26 provided mine-sweeping support and escorts for movement on narrow and dangerous waterways.

Particularly important was the support provided by the Infantry Advisor, Rung Sat Special Zone, Major McLendon G. Morris, USMC, who joined the battalion command group at the outset of the operation and remained till the battalion's operations were completed. Major Morris furnished liaison to the Senior Advisor, Rung Sat Special Zone, at Nha Be and invaluable advice and assistance. His extensive knowledge of and experience in the zone, his familiarity with complex fire support procedures, and his ability to get support on short notice made a significant contribution to RIVER RAIDER I's success.

During the operation boats of River Assault Division 91 moved

troops to and from the barracks ship by ATC at all hours of the day and night, delivering them primarily to friendly ambush sites, sometimes to the battalion land base. They remained in ambush sites at night, and one boat was normally kept within a fifteen-minute standby range of the land base as transportation for the platoon-size ground force. Division 91 was called on a number of times and was twice instrumental in the capture of sampans and documents. Its boats patrolled the waterways and gave great flexibility to the river force.

Use of deception in the conduct of night operations was found very effective in the Rung Sat Special Zone. Usually a rifle company would be positioned around the battalion command post during daylight hours. Under cover of darkness this company would be withdrawn and transported to a new area where it would establish a perimeter, then move out to place ambushes. It would then be in a position to begin search and destroy or strike operations at first light. On one occasion a complete riverine assault, with artillery fire and radio transmissions, was staged as a feint while troops remained quietly aboard landing craft that resumed their patrol stations from which small landings were subsequently made.

For operations on small waterways plastic assault boats and water safety devices were useful. The battalion used one 27-foot engineer boat and several 13-man inflatable rubber rafts to advantage, but they, like the plastic assault boats, offered no protection from small arms fire and their slow speed and the inevitable bunching of troops made them highly vulnerable. Water movement was essential, however, since the Viet Cong moved primarily by sampan; no amount of trudging through mangrove swamps would outmaneuver an enemy who sought to avoid contact.

While salt water damaged clothing, the need for exchange was not much higher than in normal field use. The jungle boot stood up well under protracted use; jungle fatigues were washed overnight aboard the support ship, and a small direct exchange stock was maintained. The principal damage was to weapons and ammunition, which salt water corroded. It was necessary to break down weapons and scrub them with a mixture of cleaning solvent and oil on each return to the ship. The 7.62 metal link belt ammunition was frequently so badly corroded that it had to be discarded.

Only essential equipment was carried, the normal load consisting of seven magazines for the M16 rifle, 200 rounds for the machine gun, and twelve rounds of 40-mm. grenades. Each squad

carried 100 feet of nylon rope, a 10-foot rope with snap link per man, and a grappling hook with 50 feet of line—items that were invaluable in water crossing operations as well as in detonating booby traps.

Ambushes were found to be most successful on well-traveled waterways. While airborne infrared devices for detecting people or things by their difference in temperature from the surrounding area proved valuable in the sparsely populated Rung Sat Special Zone, they were later less successful in the heavily populated delta.

The common rule that an ambush should be moved after being tripped did not always apply to water ambush in the Rung Sat Special Zone. On one occasion, an ambush was tripped three times in one night, with the result that seven of the enemy were killed and three sampans and two weapons captured.

Lack of positions suitable for placing artillery and the great distances involved produced a large zone in which the enemy was not subject to friendly fires. The zone was sparsely populated, with friendly civilians concentrated in a few widely dispersed villages separated by areas which the government warned citizens not to enter. Here the U.S. effort was to keep the enemy out of areas where troops had already discovered and destroyed large Viet Cong base camps, factories, munitions, stocks of rice, and other matériel.

As a result of the experiences in RIVER RAIDER I, Colonel Bolduc submitted several suggestions that assisted later operations of the Mobile Riverine Force. He advised that any unit operating in the Rung Sat Special Zone for other than a limited objective should either be a riverine unit or receive riverine training at the outset of the operation. Techniques were relatively simple and easily learned; troops that were well conditioned and adequately commanded could operate in the zone for a long time without suffering adverse effects. The longer the operation, the longer the troops needed to rest and dry out.

Since the enemy scrupulously avoided contact, current tactical intelligence was of paramount importance. Units operating in the Rung Sat therefore had to seek intelligence aggressively from all sources and agencies. Quick translation of enemy documents was also important. In one case the battalion captured at night from a group of five sampans documents and maps showing the location of a Viet Cong regional headquarters and various stops made by the sampan owners along the route they had traveled. One document showed delivery of arms to specified Viet Cong units and compromised an entire Viet Cong signal system. Although the general nature of

the contents and their importance was immediately apparent to the battalion commander, he was obliged to send all the documents to the senior adviser of the Rung Sat Special Zone. Although he dispatched them immediately, no translation was received for more than a week—too long a delay to permit timely exploitation. During the last stages of the operation, all documents were quickly translated by the S-2 and an interpreter before being sent to higher headquarters so that information could be acted upon at once.

Airmobility was essential to effective riverine operations; it was necessary to have a command helicopter capable of carrying a commander, a fire support co-ordinator, an S-2 intelligence officer, an S-3, and necessary radio communications. When naval helicopters were used a Navy representative had to be aboard. When not required for direct control of operations, the helicopter was fully utilized for reconnaissance and liaison.

One UH-ID helicopter was placed in direct support of the battalion to permit resupply and emergency lift of widely scattered troops. The battalion could fully utilize five transport helicopters and two armed helicopters on a daily basis for about ten hours. They were needed for airmobile assaults, positioning of ambushes, troop extractions and transfers, ground reconnaissance of beaches and helicopter landing zones, checking sampans, and return to areas previously worked in order to keep the enemy off-balance. In addition, there was a need for night missions using one helicopter with two starlight scopes and two armed helicopters in order to interrupt Viet Cong sampan traffic along the myriad waterways. Finally, airlift was necessary to make use of any substantial finds in the area of operations. Without helicopter transport, it was frequently necessary to destroy captured matériel—munitions, cement, large rice stores—because it could not be moved.

Fire support was diverse and highly effective. Artillery fire support bases were established with as many as three separate artillery batteries (105-mm., towed) employed at the same time from different locations. Naval gunfire was used continuously throughout the operation, including indirect fire of several destroyers and the direct and indirect fire of weapons organic to boats of Division 91, U.S. Navy, and River Assault Group 26, Vietnam Navy. Other fire support was provided by tactical air, and by U.S. Air Force, Army, and Navy fixed-wing and helicopter gunships.

Fire support presented a number of co-ordination and clearance problems. The main shipping channel was patrolled constantly by U.S. Navy river patrol boats and aircraft which, because of daylight

A Wet But Peaceful Landing

traffic in shipping, required close control to prevent damage to U.S. or South Vietnamese forces and equipment. An additional problem was that in some areas clearance from the government of Vietnam was required for each mission. U.S. ground clearance for indirect fire was co-ordinated by the artillery liaison officer. Getting government clearance at first proved to be time-consuming. Requests had to be submitted to the senior adviser of the zone at Nha Be. Any aircraft in the area of operation would cause a cease fire in the entire Rung Sat. A zonal clearance system was worked out by all parties concerned and proved highly successful. It consisted of a circle with a radius of 11.5 kilometers drawn on the fire support map using the fire support base (battery center) as the center. The circle was then subdivided like a pie into eight equal parts or zones and each was numbered from one to eight. Thereafter, to obtain clearance all that was required was to ask permission to fire into the zone in question. Artillery and mortar fire could thus be applied to the zone or zones where it was needed and withheld from the rest of the area.

Troops on combat operations in the Rung Sat Special Zone

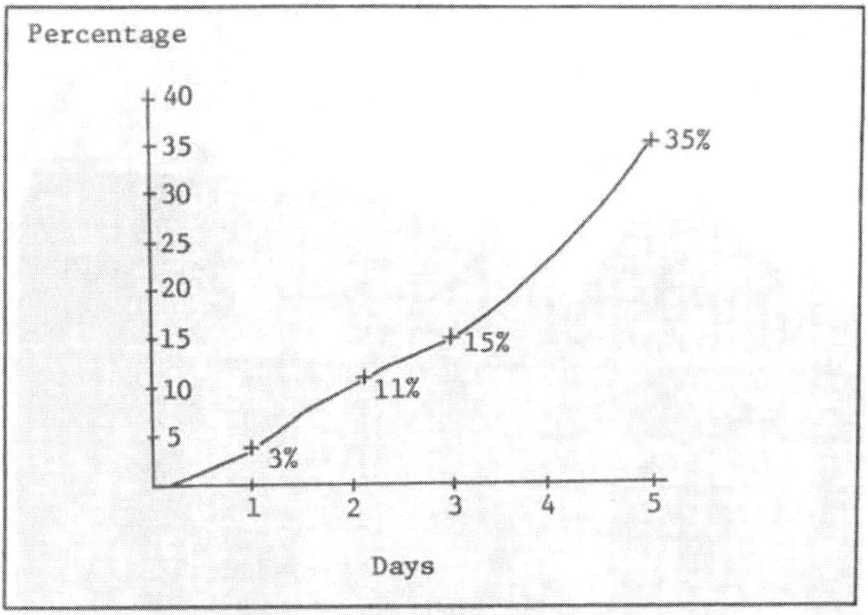

Diagram 4. Foot disease incident rate.

were continually in mud and the salty, dirty water could not be used for bathing. Certain measures were therefore taken to safeguard the health of the troops. Men stayed on combat operations for forty-eight hours at a time and were then sent to the troopship in the Vung Tau area where adequate shower facilities were available and every man had a bunk for the night. All companies received instructions on care of the feet, which included thorough washing, drying, and daily inspection of the feet by medics. The battalion surgeons carried out frequent inspections and all cases of dermatophytosis were treated with fungicidal ointments and powders. *(Diagram 4)*

An experiment to determine the efficiency of a silicone ointment was carried out with seventy-six soldiers who used the ointment daily. During some twenty days, six cases of immersion foot occurred, four of them on men who used the silicone. It was found that the system of rotating troops for a forty-eight-hour drying out period and conducting inspections of feet to insure necessary treatment of conditions as soon as they arose was of far greater benefit than use of the preventive ointment.

Positioning mortars ashore in the Rung Sat was difficult and time-consuming because of lack of firm ground, and was done only

in the area of the battalion forward command post, which was seldom moved. For support of wide-ranging operations more mobility than this semipermanent location of the mortars was required. During RIVER RAIDER I, two 81-mm. mortars were installed in the forward portion of an LCM. The mortar boat was nosed into the bank, engines kept running to advance or back in accordance with the tide, and steadied against the current by quartering lines running from the stern forward and outward at an angle of about 30 degrees to the bank on either side of the bow ramp. Most of the mortar crews were used to establish local security on the bank. This arrangement permitted a high degree of mobility, rapid positioning for firing, minimum wasted effort by the gun crews, and an ample supply of ammunition close at hand. The mortar boat was used both day and night throughout the operation, and provided flexible, mobile fire support for all types of maneuver.

The major operational success of RIVER RAIDER I was the capture of substantial stores of water mines and the destruction of facilities for constructing water mines. It is highly probable that these losses suffered by the Viet Cong account in part for the very limited use of water mines against riverine forces during later operations in the northern Mekong Delta as well as in the Rung Sat Special Zone.

After RIVER RAIDER I, the 4th Battalion, 47th Infantry, took up operations in the Rung Sat Special Zone, making use during April and early May of the experiences of the 3d Battalion, 47th Infantry, and improvising other techniques. During this time so-called Ammi barges were obtained from the Navy for use alongside the LST of River Flotilla One. With this type of ponton, rope ladders were not needed and the training time for riverine troops was drastically reduced. The Ammi barges provided as well some much needed space for cleaning and storage of ammunition, crew-served weapons, and individual equipment.

CHAPTER IV

Initial Delta Operations

Movement to Dong Tam

When the 3d Battalion, 60th Infantry, was sent to Dong Tam in January 1967, the Headquarters, 3d Brigade, U.S. 9th Infantry Division, also went to Dong Tam to direct construction of the base and to conduct operations in Dinh Tuong Province. The 5th Battalion, 60th Infantry (Mechanized), was stationed at Tan Hiep, an airstrip about eight kilometers northeast of My Tho along Highway 4. The 3d Brigade conducted the first of a series of operations in late February 1967; these operations were usually limited to one battalion and one or two companies of another battalion because so many troops were needed to protect the construction work at Dong Tam Base.

Implementing earlier planning, General Westmoreland decided to move Headquarters, 2d Brigade, to Dong Tam so that it would become operational at that location on 10 March. The 3d Brigade was moved north to Tan An, the capital of Long An Province, with the mission of conducting a consolidation operation to assist the pacification program in the southern portion of the III Corps Tactical Zone. With the arrival of Headquarters, 2d Brigade, at Dong Tam, the brigade began a demanding 90-day period of performing four separate but related tasks: the defense and construction of the Dong Tam Base; limited offensive operations in Dinh Tuong and Kien Hoa Provinces; operations in the Rung Sat Special Zone to protect the shipping channel; and planning with Task Force 117 to move aboard the Navy ships.

Conducting operations in Dinh Tuong Province was to prove as important in the seasoning of the battalions of 2d Brigade as was the riverine training in the Rung Sat Special Zone. The threat posed to Dong Tam by the Viet Cong 514th Provincial Battalion, the 263d Main Force Battalion, and local force sapper and infantry companies and guerrillas demanded a varied offensive campaign to protect the base as well as to reduce the Viet Cong influence.

Although Dinh Tuong Province was bordered on the south by

the My Tho River, the limited navigability of smaller streams greatly limited boat movement within the province. Most canals were blocked by the debris of destroyed bridges or by earthen dams erected by the Viet Cong to prevent their use by civilians or military forces. Therefore, although the brigade was scheduled for a riverine role, it was to carry on between March and June a variety of operations common to all U.S. infantry units in Vietnam. These included throwing a cordon around hamlets and searching them in the hope of capturing members of the local Viet Cong political organization and guerrilla squads and platoons; brigade and battalion reconnaissance in force operations; and extensive patrolling. Troops moved by helicopter, boat, wheeled and tracked vehicles, or on foot. During the wet season the brigade operated in the Dong Tam area by boat and helicopter chiefly. When aircraft were not available, wheeled vehicles could carry troops on Highway 4 to find the enemy near suspected Viet Cong bases. The troops then moved on foot into the operational areas to the north and south of the road.

During March and April the brigade commander and staff, assisted by the senior U.S. adviser of the 7th Vietnam Army Division, Colonel Lance, assembled good intelligence covering Dinh Tuong and Kien Hoa Provinces. In addition, through the U.S. senior adviser to IV Corps, General Desobry, arrangements were made for periodic briefings of the brigade staff by the G–2, IV Corps advisory staff. This information was used in a series of co-operative operations undertaken by the 2d Brigade with the 7th Vietnam Army Division. These were to provide beneficial combat experience for the brigade before it embarked as the Army component of the Mobile Riverine Force.

The diverse tasks of the brigade seldom permitted more than two infantry battalions to be used in offensive operations. Thus in covering large base areas without exact intelligence of enemy locations, participation of additional forces was co-ordinated with the 7th Vietnam Army Division to search thoroughly and to block escape routes. The Commanding General, 7th Vietnam Army Division, approved of such operations because they permitted him to deploy small forces into Viet Cong base areas, for which he would otherwise have been obliged to use more Vietnamese troops than he could spare from their security missions. Throughout the experience of the Mobile Riverine Force, it was to prove necessary to obtain additional troops, whether they were other U.S. units, Vietnam Army units, or Vietnamese marines, to augment the two

battalions routinely operating within the Mobile Riverine Force.

During the period 15 February–10 May, the 2d Brigade, although physically based at Dong Tam, conducted infantry operations with the 3d and 4th Battalions, 47th Infantry, in the Rung Sat Special Zone and southern Bien Hoa Province. While the 3d Battalion, 47th Infantry, was located in the Rung Sat during February and March, the 4th Battalion, 47th Infantry, was used in the southern portion of Bien Hoa Province, south of Bearcat and bordering the Rung Sat. The boats of the Navy's River Assault Squadron 9 supported Army operations in the Rung Sat during this period.

Final Mobile Riverine Force Preparations

At the same time active co-ordination and planning for the Mobile Afloat Force was begun by the 2d Brigade staff with the advance staff of River Assault Flotilla One. General Eckhardt assigned Lieutenant Colonel John R. Witherell, who had been the Combat Developments Command liaison officer to the G-3 Section, U.S. Army, Vietnam, to 2d Brigade. As a major, Colonel Witherell had initiated the first outline draft of the riverine doctrine manual. He was made the deputy for riverine planning for the 2d Brigade, augmenting the brigade staff. At this time, Lieutenant Colonel James S. G. Turner, U.S. Marine Corps, was placed on duty with Headquarters, 9th Division, by the Commandant of the Marine Corps as an observer and liaison officer for riverine operations. At the request of the brigade commander, Colonel Turner was placed on special duty with the brigade staff. Colonels Witherell and Turner proved to be valuable additions because the brigade staff was involved in planning for and conducting daily operations of the two maneuver battalions, including Dong Tam Base defense operations, as well as in formulating plans to establish the brigade at Dong Tam. In the early draft plans prepared by Colonel Witherell, the term Mekong Delta Mobile Afloat Force was dropped and the title Mobile Riverine Force was formally adopted.

As of mid-January, two APB's, *Colleton* and *Benewah*, were scheduled to arrive in Vietnam during May or June 1967. The two ships would provide only 1,600 berthing spaces for Army troops. When troop space for artillery support, base security forces, and the brigade staff was subtracted, there would remain spaces for a combat force of approximately one reinforced battalion. It was considered essential to increase the combat strength of the brigade to at least two battalions. Since only two APB's were available, it

NON-SELF-PROPELLED BARRACKS SHIP

was recommended by the 2d Brigade that a towed barracks ship, an APL, with a berthing capacity for 660 troops be made available. This recommendation was forwarded through division to the commander of II Field Force, who, in turn, requested the commander of the naval force in Vietnam to make the APL available. He proposed that if the APL was diverted, Navy men be provided to operate and maintain only the equipment peculiar to the Navy and that Army troops be required to perform all other support functions such as cleaning, preventive maintenance, messing, laundry service, and security. This proposal was satisfactory to the Army and APL-26 was diverted to Dong Tam in May, along with two sea-going tugs to move the barracks ship. An additional 175 berths were provided by building bunks on the top side of the APL under a canvas canopy.

Frequent co-ordination meetings were held with Captain Wells aboard his flagship at Vung Tau. During the last two weeks of March Colonels Witherell and Turner moved aboard the flagship to participate with the Navy staff in the development of a Mobile Riverine Force standing order. The brigade operations officer, Major Clyde J. Tate, continued to concentrate on the current

ARTILLERY BARGE

brigade operations. Colonel Fulton felt that the Mobile Afloat Force plan had not provided effective artillery support for the Mobile Riverine Force operations. The armored troop carriers were to lift the 105-mm. howitzers and their prime movers. The artillery battalion, after being moved by water, would debark at a suitable place along the river and establish firing positions. Terrain reconnaissance conducted during his October 1966 visit had convinced Colonel Fulton that the off-loading of the prime movers from an ATC as provided for in the plan would greatly restrict operations because of the varying tides of four to thirteen feet and the steepness of the river banks. Consequently, the Commanding Officer, 3d Battalion, 34th Artillery, Lieutenant Colonel Carroll S. Meek, began to experiment with barge-mounted artillery. He placed a 105-mm. howitzer on an Ammi barge, using cleats and segments of telephone poles against which the trails of the howitzer rested. Successful firing demonstrated the feasibility of this method. By use of aiming stakes placed ashore routine fire support could be provided from the barges anchored securely against the river bank.

During these experiments, the Deputy Commanding General,

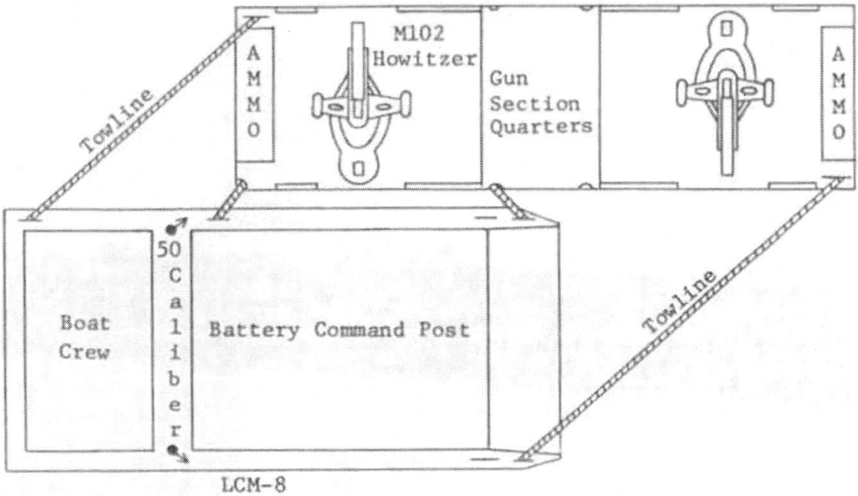

Diagram 5. Artillery barge towing position.

U.S. Army, Vietnam, Lieutenant General Jean E. Engler, placed a liaison officer from the G-4 Section on temporary duty with the brigade. This officer's sole mission was to procure special equipment requested by the brigade, thus eliminating the procedure of requesting items through normal channels and greatly facilitating the acquisition of equipment. After Colonel Meek's initial firing experience with the barge-mounted artillery, the liaison officer was asked to arrange fabrication of an experimental artillery barge. The work was done in a few weeks at Cam Ranh Bay, where small pontons were welded together into an artillery barge. The barge was then towed down the coast and brought up the Mekong River to Dong Tam by a commercial ocean-going tug. Further firing experiments proved the barge unsatisfactory in that the high bow of the ponton was nearly perpendicular to the surface of the water, making it hard to maneuver and tow, especially against the current, tide, or prevailing wind. The barge was redesigned so that it floated lower in the water and had a sloped bow, and the design was forwarded by the G-4 liaison officer to Cam Ranh Bay. As a result of successful experiments with the new design, six barges were fabricated, each barge to accommodate two tubes of 105-mm. howitzer, M102. (*Diagram 5*) Each barge was to be towed by an LCM-8 throughout the areas in which the Navy assault craft supported Army forces. The decision to develop the artillery barges was the most important equipment decision made by the Army for the

ARTILLERY FIRES FROM BARGES ANCHORED ON RIVER BANK

Mobile Riverine Force because the barges provided effective artillery support at all times. Had General Engler not had the foresight to furnish liaison and to make special arrangements to obtain unique equipment, the initial effectiveness of the Mobile Riverine Force would have been greatly impaired.

Also during the March-April period, Colonel Fulton decided that there was not enough helicopter landing space aboard the ships of the river force. Each APB had one helicopter landing pad, and the LST could accommodate the landing of one helicopter. Through the same expedited development process used for the artillery barges, a helicopter landing barge was provided that could accommodate three UH-1 helicopters and would have a refueling capacity for the helicopters of 1,500 gallons of JP-4. This barge would be towed by an LCM-8 of the Army transportation boat company.

As more U.S. Navy craft arrived in Vietnam and the initial training was completed, River Assault Flotilla One elements were sent from the Vung Tau-Rung Sat Special Zone training area into Dinh Tuong Province. On 10 April the first river assault division

HELICOPTER BARGE

moved to Dong Tam with 18 ATC's, 1 CCB, and 2 monitors and immediately began waterborne operations with the 3d Battalion, 47th Infantry, which had also left Rung Sat Special Zone for Dong Tam. Although these operations on the Mekong River were small in scale during the month of April, the men soon learned that unlike the rivers of the Rung Sat Special Zone, where obstacles were few, the delta waterways had many man-made obstructions such as low bridges, fish traps, and earthen blocks on canals.

During April a small staff group of River Assault Flotilla One went to Dong Tam to work closely with the 2d Brigade staff in developing the final plans for Mobile Riverine Force operations, which were to begin on 2 June. The first embarkation was to include the brigade headquarters and staff; 3d Battalion, 34th Artillery; Company C, 15th Engineers; and the division support elements. In the final planning phase for this embarkation it was decided that a larger supply LST would be needed to provide an additional 375 U.S. Army berthing spaces. Originally two class 542 LST's were scheduled as the logistics support ships to rotate between Vung Tau and the riverine base. Since the 542 class is

much smaller and provides less billeting space, a request was made to River Assault Flotilla One for one of the larger 1152 class LST's. Again, the Navy accommodated the Army request. Acquisition of the larger LST was to prove very advantageous in that the entire brigade aviation section, which included four OH-23 helicopters, and the maintenance section were placed aboard the LST and were operated from the flight deck. Limited air resources were now available for combat operations regardless of the remoteness of the location of the mobile riverine base from 9th Division support.

Both the brigade headquarters and the infantry battalions were internally reorganized so that all equipment not necessary for riverine operations could be left at the Dong Tam base under the control of unit rear detachments. Spaces for jobs not needed by the Riverine Force, such as for drivers and mechanics, were converted to spaces with shipboard and combat duties. The combat support company of each battalion was reconstituted into a reconnaissance and surveillance company. Men assigned to spaces whose function was not required in the Mobile Riverine Force were placed in these two companies and thus two additional rifle platoons were created in each of the two battalions. When these two platoons were added to the reconnaissance platoon normally found in the combat support company, a fourth rifle company for each battalion was organized. This arrangement was to prove most satisfactory because it provided for base defense and other security missions and at the same time left three companies for battalion tactical operations.

2d Brigade Operations

While planning of the riverine organization of the infantry battalions was taking place, in April and May the battalions were actually conducting land operations. The 3d Battalion, 47th Infantry, and 3d Battalion, 60th Infantry, were increasing their proficiency in airmobile operations and learning to work with the officials of Dinh Tuong Province and Army of Vietnam division forces.

On 1 April 2d Brigade forces made a raid into northern Kien Hoa Province on the My Tho River south of Dong Tam, using barge-mounted artillery. During darkness, one battery was moved west on the river from the land base, followed closely by infantry mounted in ATC's. Once opposite a manned station of a Viet Cong communications liaison route, the artillery poured surprise, direct fire into the target area. To protect a Ham Long District company, which had occupied a blocking position south of the target, time

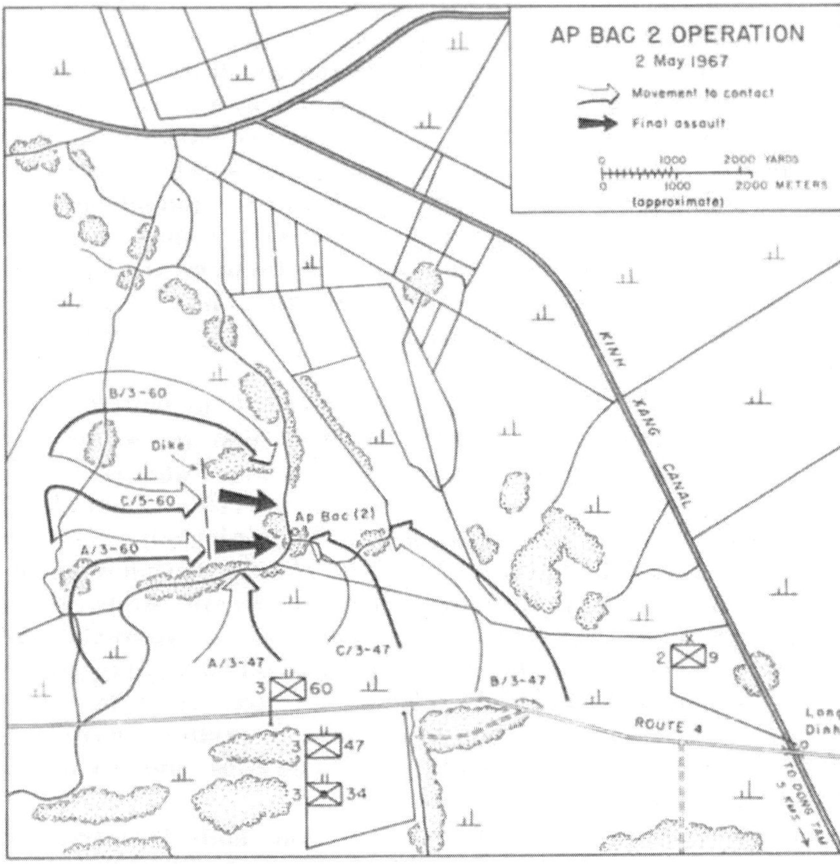

MAP 4

fuzes were used. Immediately after the artillery fire, infantry troops landed and swept through the station. The artillery then anchored against the north bank of the river and prepared to provide indirect fire support. The raiders captured enemy ammunition and a portion of the labor force of the enemy communication-liaison platoon.

On 2 May the 3d Battalion, 47th Infantry, and 3d Battalion, 60th Infantry, participated in the 2d Brigade's heaviest fighting to date. The target of the brigade assault was the 514th Local Force Battalion in the Ap Bac 2 area, which had been identified by the commanding general of the 7th Vietnam Army Division as a normal operating area for the Viet Cong. (*Map 4*) The ground over which the battle would take place was alternately rice paddy and thick patches of vegetation with heavy foliage along the streams and canals. The brigade plan was to conduct an airmobile operation in

co-operation with the 7th Vietnam Army Division. On the morning of the operation, however, shortly after the scheduled time for the airlift of the first company, the brigade was notified that no airmobile company was available. The brigade commander decided to send the unit by truck, a move requiring approximately two hours. At the same time the brigade forward command post elements moved north by ATC's on the Kinh Xang Canal to set up a post at Long Binh. The 3d Battalion, 60th Infantry, truck convoy was followed by the 3d Battalion, 47th Infantry, on the same route. The companies of both battalions were off-loaded facing north from Highway 4.

By 0830 both battalions were moving north to their first objectives. Elements of the 3d Battalion, 60th Infantry, on the west met no resistance. Units of the 3d Battalion, 47th Infantry, moved forward cautiously during the morning, covering about 1,500 meters and receiving light sniper fire. At approximately 1250 both Companies A and B, 3d Battalion, 47th Infantry, encountered the enemy. Company A came under heavy automatic weapons fire immediately after emerging from a wood line; it deployed in an attempt to move against the enemy position, but was hampered by heavy undergrowth that made it difficult to see the enemy firing positions. The company sought to improve its position, and called in fire support, including artillery fire, gunships, and eventually three air strikes on the enemy position.

About 1300, Company B, 3d Battalion, 60th Infantry, was placed under operational control of the commander of 3d Battalion, 47th Infantry, and directed to establish a blocking position north of the enemy location. At the same time Company A, 3d Battalion, 60th Infantry, and Company C, 5th Battalion, 60th Infantry, were ordered to close enemy escape routes on the west. Companies B and C, 3d Battalion, 47th Infantry, were ordered to block northeast of the position of Company A, 3d Battalion, 47th Infantry.

Company C, 5th Battalion, 60th Infantry, a mechanized company, using full-tracked armored personnel carriers, moved through inundated areas which had been thought impassable to occupy a position west of the enemy. Company C, 5th Battalion, 60th Infantry, and Company A, 3d Battalion, 60th Infantry, then made the final assault from the west into the enemy position. Company C, 5th Battalion, 60th Infantry, greatly enhanced the firepower of the assault by employing the .50-caliber fire of its full-tracked armored carriers in the assault line. The organic infantry attacked on foot alongside the carriers. The enemy resisted with heavy rifle

and automatic weapons fire. Company A, 3d Battalion, 60th Infantry, on the right flank of the assault, charged into the southern line of bunkers, and finally overran it. Seeing that their line was broken, many Viet Cong tried to escape to the east but were killed or wounded as both attacking companies made their final assault into the bunker line. By 1820 enemy resistance was overcome.

Although not a riverine operation, the action at Ap Bac contributed to the brigade tactical experience which was later carried into riverine operations. In developing operational plans, the brigade staff recognized that tactical control measures have to provide for maximum flexibility in controlling the maneuver of units. If the original plan cannot be executed, there must be sufficient latitude to redirect the effort of the participating units according to the control measures previously issued. Check points, a series of intermediate objectives, and an area grid system provide flexibility for redirection of units and shifting of boundaries and fire co-ordination lines.

The Ap Bac operation included in the words of an infantryman "finding, fixing, fighting, and finishing" the enemy. Through the use of simple control measures, units of the two battalions were moved into blocking positions to cut off enemy routes of withdrawal, and other units were maneuvered into assault positions. When this maneuver was combined with use of both air and artillery supporting force, the enemy had to remain in his position or expose himself in an attempt to withdraw.

The operation made it clear that units must be prepared to bring in artillery and air support as close as possible to the front lines. Each battalion commander used an observation helicopter to co-ordinate fire in close support of his unit. The infantry units, once having found the enemy, had to be prepared through fire and maneuver to conduct a co-ordinated assault against the enemy positions. In this engagement, by the use of close supporting artillery fire, the infantry units were able to launch a co-ordinated attack which overcame the enemy.

Mechanized infantry was used most successfully at Ap Bac. The unit commander had to search for routes through very swampy and marshy areas. Once his armored personnel carriers (APC's) were deployed, he dismounted his infantry and moved APC's and infantry in a forward rapid assault on the enemy. The volume of fire delivered by the APC's into the bunker line, along with high explosive and white phosphorous rounds of artillery, enabled the infantry to overcome the enemy in bunkers and foxholes. Once

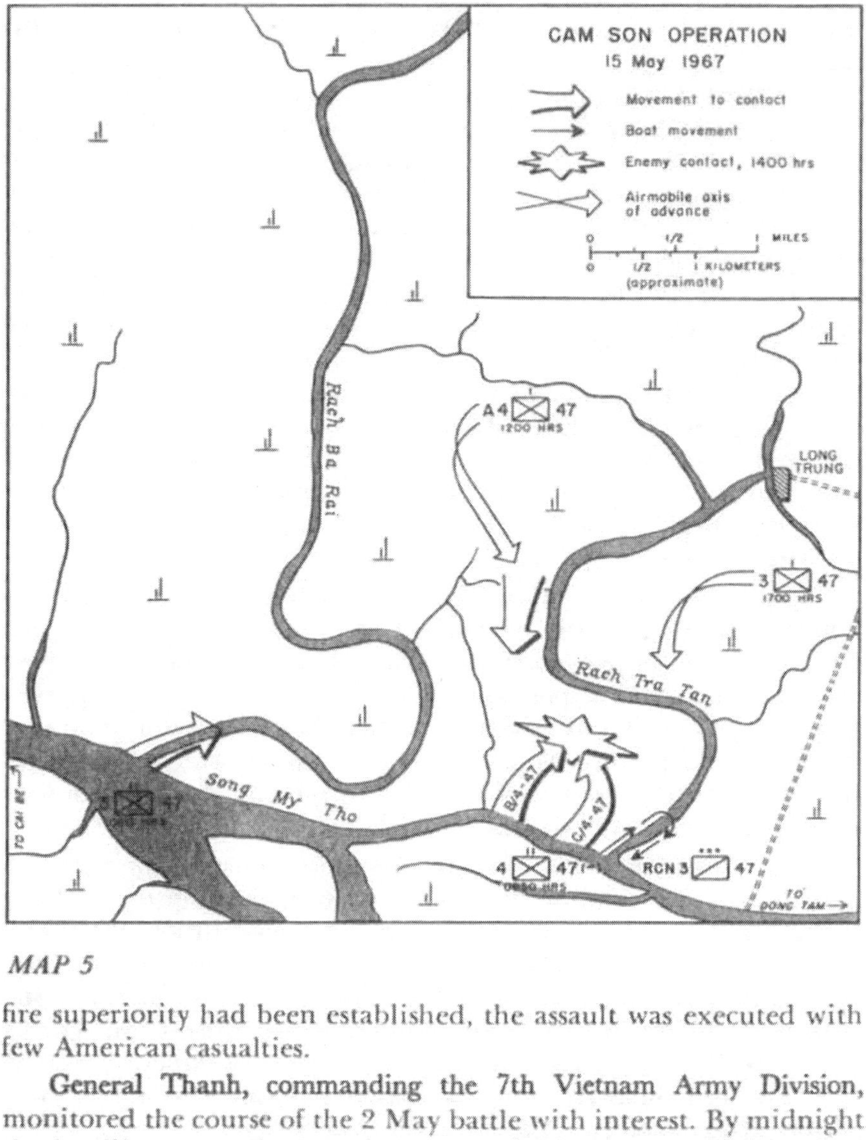

MAP 5

fire superiority had been established, the assault was executed with few American casualties.

General Thanh, commanding the 7th Vietnam Army Division, monitored the course of the 2 May battle with interest. By midnight the intelligence staffs of the brigade and 7th Vietnam Division had estimated that the enemy consisted of the 514th Local Force Battalion with two companies and its heavy weapons. An enemy attack launched at dusk of 2 May on the command post of the 3d Battalion, 60th Infantry, just south of Highway 4, appeared to be the rest of the 514th Local Force Battalion trying to relieve pressure on its units north of Highway 4. On the strength of this estimate, General Thanh sent a Ranger battalion on 3 May into the area which he

Armored Troop Carriers in Convoy Battle Line

considered the most probable location to which the enemy would flee. His Rangers encountered and attacked an estimated company of the 514th Local Force Battalion during the afternoon. The casualties inflicted upon the 514th by the U.S. 2d Brigade and the 7th Vietnam Army Division severely reduced its combat effectiveness. Reports received later suggested that the 263d Main Force Battalion was forced to undertake training of new recruits and operational tasks of the 514th Battalion because of the losses in men and weapons.

On 13 May the 4th Battalion, 47th Infantry, completed operations in the Rung Sat Special Zone and joined the brigade at Dong Tam. Colonel Tutwiler's battalion participated in its first major riverine operation in the Dong Tam area on 15 May as the 2d Brigade conducted the first of several operations in the Cam Son Secret Zone. (*Map 5*) The Cam Son area was considered one of four major Viet Cong bases in Dinh Tuong Province by the intelligence staffs of both the province and 7th Vietnam Army Division. The operation of 15 May relied entirely on intelligence provided by these Vietnamese organizations.

The brigade plan was to search the southern area of Cam Son along the Rach Ba Rai and Rach Tra Tan streams and to capture or destroy the Viet Cong, their supplies, and their equipment. A forward brigade command post was established approximately two kilometers north of Cai Be, and a barge-mounted artillery base was established on the southern bank of the My Tho River, five kilometers southeast of the mouth of the Rach Ba Rai. The two infantry battalions were supported by 22 ATC's, 2 monitors, and 2 CCB's of River Assault Flotilla One. Commander Charles H. Black, the operations officer of River Assault Flotilla One, joined the brigade command group and co-ordinated the support of the Navy assault craft.

The operation began at 0815 when the 3d Battalion, 47th Infantry, landed from assault craft at the mouth of the Rach Ba Rai and began to search northeast along the stream. At 0830 the 4th Battalion, 47th Infantry, landed two companies from assault craft on the north bank of the My Tho River approximately halfway between the mouths of the Rach Ba Rai and the Rach Tra Tan. At 1200 Company A, 4th Battalion, 47th Infantry, was airlifted from Dong Tam, where it had been in readiness as a reserve, and was landed three kilometers north of the river on the west side of the Rach Tra Tan to act as a blocking force against any enemy encountered. After landing the Army troops, the boats of the Navy task force proceeded to blocking stations along the waterways.

At 1400 Companies B and C of the 4th Battalion, 47th Infantry, met a Viet Cong force equipped with small arms, light machine guns, and rocket launchers. Company A maneuvered south and met the enemy within two kilometers of the My Tho river. Although Companies B and C made little progress in moving against the enemy fire, artillery and close air support maintained pressure on the Viet Cong.

In an effort to move more troops to the northeast and rear of the Viet Cong force, the Reconnaissance Platoon, 3d Battalion, 47th Infantry, was moved by ATC into the Rach Tra Tan. When the low tide and enemy antitank rockets prevented the assault craft from penetrating upriver beyond enemy positions on the banks, the platoon was withdrawn.

By 1630 it became apparent that some enemy forces were escaping by moving to the northeast, away from both Companies B and C, which were moving from the south, and from Company A, which was moving from the northwest. At this point Colonel Fulton directed Colonel Bolduc to move one of his companies by helicopter

to establish a blocking position northeast of the battle. This was accomplished by 1700, but the company failed to find the enemy, and by 2000 all contact with the Viet Cong was lost.

This fight called attention to the limitations imposed upon maneuver or assault craft during low tide and the importance of artillery and close air support against an enemy in well-prepared bunkers and firing positions. Although most American casualties did not require evacuation because their wounds were superficial, both Army and Navy were concerned by the vulnerability of men aboard the assault craft to rocket fragments. The Mobile Riverine Force was to find vulnerability of troops aboard assault craft one of its continuing problems.

On 18 and 19 May the 2d Brigade moved a forward command post by motor convoy north of Cai Be in western Dinh Tuong Province to control the operations of 3d Battalion, 47th Infantry, and 3d Battalion, 60th Infantry. To reduce the distance the supporting helicopters would have to fly, River Assault Flotilla One assault craft moved both battalions by water thirty-six kilometers west of Dong Tam to landing sites on the north bank of the My Tho River. With the infantry battalions positioned within seven kilometers of the center of the target area, one airmobile company was able to insert each battalion rapidly into the area. In the event of substantial fighting the helicopters could turn around quickly in building up forces.

During this operation the battalions of the 2d Brigade met few of the Viet Cong, but the brigade command post was attacked by mortar, recoilless rifle, and machine gun fire during the early hours of 19 May. A ground attack was thwarted when the Reconnaissance Platoon, 3d Battalion, 47th Infantry, led by Second Lieutenant Howard C. Kirk III intercepted enemy movement toward the western perimeter of the position; assisted by artillery, gunships, and an armed illumination aircraft, the reconnaissance platoon broke up the attack.

Throughout the month of May, a maximum effort was made to establish procedures for close co-operation between Army and Navy elements. Colonel Fulton and Commander Black agreed that the helicopter was invaluable for command and control of riverine operations. Finding important terrain features is a very difficult task for the commander on the surface, but a simpler task for the airborne commander. Foot troop and assault craft maneuver were facilitated by information furnished by the airborne command group. During darkness and in marginal flying weather, the com-

mand and communications boats were valuable to brigade and battalion commanders in their forward positions. The combined use of command helicopters and command boats by brigade and battalion commanders permitted close supervision and control of the Mobile Riverine Force in combat.

The operations of the 2d Brigade in Dinh Tuong Province from March through May had already provided invaluable operational experience. All of the brigade's units had met and withstood the enemy; thus a feeling of confidence had been engendered in commanders and men. Of even greater importance, however, was the displacement of Viet Cong units from the Dong Tam area to the western portion of Dinh Tuong Province, from which the enemy could not readily launch a large ground attack on the base. The 2d Brigade had become operational at Dong Tam on 10 March, only seventy-two hours after an engagement involving a platoon of the 3d Brigade and a larger Viet Cong unit less than eight kilometers from the base, and forty-eight hours after the enemy launched an attack on Dong Tam with recoilless rifles and mortars.

In the next ninety days the enemy made one successful sapper attack that damaged a construction vehicle outside the camp and an American patrol ambushed an enemy force of platoon size three kilometers west of the base. To prevent enemy fire and ground attack on the base, a patrolling system was executed by the base defense battalion, using division artillery radar. Patrols were made on foot, by tracked and wheeled vehicles, by boat, and by helicopter, and were sent out by day and by night.

Offensive operations came to be guided by knowledge of the habits of the Viet Cong, who made use of certain areas and routes for bases and for communication, liaison, and supply. Although information provided by the Vietnam Army division and province intelligence agencies seldom arrived in time to serve as the basis for an attack, it was useful in tracking the enemy's routes, bases, and length of stay in bases. It permitted a rough guess as to when a Viet Cong battalion would be in any one of the four major Dinh Tuong base areas at a given time. Operations could be conducted against one or more bases on a priority basis. By co-operation with the Vietnam Army and using economy of force, the brigade could enter the first priority base area with most of two battalions, the Vietnamese forces could enter the second priority area, and the brigade would use a small force to reconnoiter a third base area. Helicopter transport was essential and air cavalry support highly desirable. It was not, however, until May that a portion of Troop

D, 3d Squadron, 5th Cavalry, was used by the 2d Brigade.

By the end of May, the brigade had found the enemy in three of the four Dinh Tuong base areas on many occasions and had contributed to a period of relative security for the men attempting to complete construction of the Dong Tam Base as well as for the province itself.

Mobile Riverine Force Campaign Plan

During early April it had become apparent that the Navy's concept of the employment of the Mobile Riverine Force differed slightly from the original Mobile Afloat Force concept. Captain Wells contended that the force should act as a separate force, divorced from 9th U.S. Division operations and control, with deployments deep into the delta. A period of intensive discussions between the component commanders ensued. Differences were not resolved until the preparation of a wet and dry season campaign plan for the Mobile River Force as well as a draft letter of instruction from the 9th U.S. Division commander to the brigade commander. This letter was prepared by the brigade staff and carried to the commander of the 9th Infantry Division, General Eckhardt, with the recommendation that it be issued by Headquarters, 9th U.S. Division, to the Mobile Riverine Force Army component commander.

General Eckhardt accepted the draft letter and draft campaign plan and personally took them to Captain Wells, who, as the commander of River Assault Flotilla One and Task Force 117, the operational designation of the Navy component of the riverine force, accepted both. These two documents adhered to the original Mobile Afloat Force plan for operation of the river force in conjunction with the other 9th U.S. Division elements in the southern III Corps Tactical Zone and in the northern portion of the IV Corps Tactical Zone north of the Mekong River.

The opinion of the Navy component commander that operations should be conducted exclusively in IV Corps was not supported by Headquarters, MACV, Planning Directive Number 12-66, dated 10 December 1966, entitled Command Relations for Riverine Operations in South Vietnam. Essentially, this document placed Army forces conducting riverine operations in III and IV Corps Tactical Zones under the operational control of the Commanding General, II Field Force, with a stipulation that he might exercise operational control through designated subordinate U.S. Army headquarters. The commander of U.S. Army, Vietnam,

would exercise command, less operational control, of all U.S. Army units engaged in riverine operations. Similarly, Navy riverine forces were placed under the operational control of the commander of Naval Forces, Vietnam, and under the command, less operational control, of Commander in Chief, Pacific Fleet.

The planning directive recognized the unique command arrangements peculiar to the Mobile Afloat Force concept. This was not to be a joint task force and the specific command relationships had not been planned under the original concept. The planning directive took note of the fact that "the conduct of riverine operations by Army and Navy forces is a new concept which will require the utmost coordination and cooperation by all concerned. As operations are conducted and lessons learned, command relations will be revised as necessary to insure effective and workable arrangements." There was to be no single commander of the Mobile Riverine Force.

The MACV document stipulated that the base commander for all bases, whether ashore or afloat, would be the senior Army commander assigned. It further stipulated that the relation between the Army and Navy units stationed on both Army and Navy bases would be one of co-ordination and mutual support, with the Navy providing its appropriate share of forces for local base defense, including naval gunfire support and protection against waterborne threats. Operational control of Navy units in these circumstances was to be exercised by the base commander through the Navy chain of command.

Command relationships for riverine operations were further defined as commencing when troops began embarking on assault craft to leave the base and ending when troops had debarked from the boats upon return to the base. During riverine operations, the Army commander would control all participating Army forces and the Navy commander would control all participating Navy forces. The Navy would provide close support to the Army during these operations according to the definition by the Joint Chiefs of Staff:

> When, either by direction of higher authority or by agreement between the commanders concerned, a force is assigned the primary mission of close support of a designated force, the commander of the supported force will exercise general direction of the supporting effort within the limits permitted by accepted tactical practices of the service of the supporting force. Such direction includes designation of targets or objectives, timing, duration of the supporting action, and other instructions necessary for coordination and for efficiency.

Command relationships between the Army and Navy elements

or components during the operational planning phase would be one of co-ordination. This particular aspect of the planning directive posed a basic dilemma in that co-ordination meant agreement at the lowest possible level as to where the force would deploy and how it would fight. If the agreement could not be obtained during the planning phase, the planning directive made provision for the following:

> It is the responsibility of commanders at all levels through liaison, cooperation, coordination, and good judgement to make every effort possible to solve all command relationship problems in the most expeditious and workable manner consistent with the problem. Problems which cannot be solved satisfactorily will not be forwarded to the next higher headquarters until the commanders concerned have made every effort to reach agreement.

In the event that agreement could not be reached, the division of component authority meant that each of the component commanders would proceed up his respective chain of operational control to obtain a solution. A solution would require co-ordination between the headquarters of the parallel Army and Navy echelons. For example, in the event that there was a disagreement between the Navy component commander and the Army component commander of the Mobile Riverine Force, the Army commander would have to pass the problem to the commander of the 9th U.S. Division. It is notable that there was no Navy echelon equivalent to the 9th Infantry Division under the Navy chain of operational control. The commander of the 9th Infantry Division would then have to pass the matter to the commander of II Field Force, who, in turn, would co-ordinate with the parallel Navy commander. (*Chart 2*)

The command relationships had complicated planning of the location of initial operations, and prompted Captain Wells to challenge the role of the 2d Brigade as an integral part of the 9th U.S. Division's over-all operations. This initial disagreement was resolved by Captain Wells' acceptance of the 9th Division letter of instruction. Determination of the mission and area of operation of the Mobile River Force in relation to the 9th Infantry Division, however, was a continuous source of friction between the brigade and flotilla commanders. Generally the commander of River Assault Flotilla One agreed to the Joint Chiefs of Staff provision that the supported force would direct operations: "Such direction includes designation of targets or objectives, timing, duration of the supporting action, and other instructions necessary for the co-ordination and for efficiency."

CHAPTER V

Plans and Operational Procedures for the Mobile Riverine Force

While the Army elements were being moved aboard the ships of River Assault Flotilla One, the staffs of the flotilla and brigade completed the planning for initial tactical operations. Lacking the unity of organization for planning that is present in a joint staff, the brigade and flotilla commanders informally agreed upon the planning roles of each staff.

Intelligence

Colonel Fulton and Captain Wells decided that Mobile Riverine Force operations would be planned and conducted on the basis of intelligence collected from areas in which major enemy forces were most likely to be met and in which the abilities of a river force could be best used. Once an enemy force and area had been agreed upon as a target by the two component commanders, the brigade S-2 and flotilla N-2, working together closely, would collect intelligence to support further planning. The N-2 would concentrate on the navigability of waterways and threats to the Navy ships and assault craft. The S-2 would gather intelligence information necessary to conduct land operations. Both the S-2 and the N-2 would be highly dependent upon the intelligence disseminated by other U.S. and Vietnamese headquarters because the Mobile Riverine Force had a large area of interest and would need to move considerable distances from one location to another.

Since the S-2 and N-2 could not supply all the information needed by the riverine force, intelligence was obtained from several agencies. A seven-man Mobile Intelligence Civil Affairs Team was formed by the brigade to co-ordinate with these agencies. This team was formed from the brigade S-2 and S-5 sections, the prisoner of war interrogation team, the psychological operations team, and military police platoons. It moved into the operational areas of the force to collect intelligence from Vietnamese political officials and from intelligence officers of military units. Working close

CHART 2—MOBILE RIVERINE FORCE COMMAND STRUCTURE

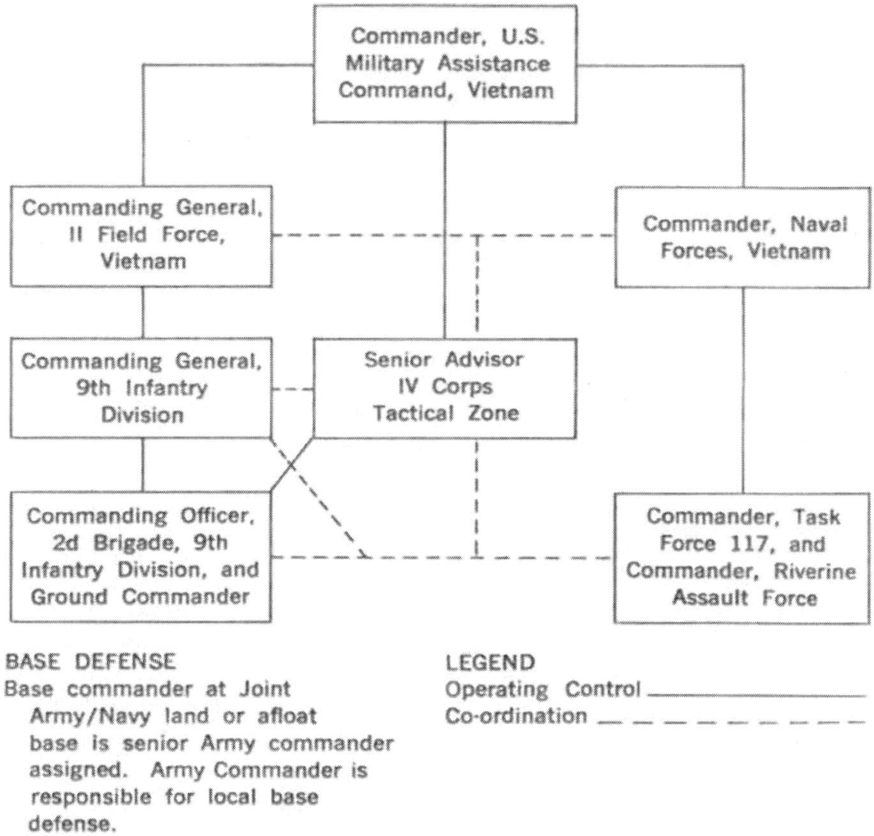

Brigadier General George G. O'Connor, Assistant Commander, 9th U.S. Infantry Division, was responsible to the division commander for the division operations in IV Corps. He provided broad guidance and considerable latitude to the brigade commander in dealing with the command relationships and operational procedures of the Mobile Riverine Force. Without this degree of flexibility, it is unlikely that the brigade and flotilla commanders could have resolved as they did many fundamental issues without recourse to their respective superiors.

Based on the fundamental issues agreed upon and the campaign plan outline of where and how the river force was going to fight, planning continued as the 2d Brigade and its subordinate units began to board the Navy ships of the Mobile Riverine Force at Dong Tam on 31 May.

to the intelligence centers of Vietnamese organizations, the team aided in the exchange of tactical intelligence as well as in the co-ordination of a limited number of civil affairs activities. The Mobile Intelligence Civil Affairs Team was not usually moved into an area during the planning phase of an operation for fear of detection by enemy counterintelligence, but was moved in at the beginning of an operation.

Another means of intelligence collection for the riverine force was the riverine survey team. Controlled by Task Force 117, the team accompanied the task force on missions and recorded depth soundings and clearance measurements, thus providing reliable navigation data for subsequent operations. Under the direction of the 9th Infantry Division engineer, Vietnamese Regional and Popular Forces also collected information on bridges so that the riverine force could be sure of vertical and horizontal bridge clearance on a waterway.

While they were collecting information, the S-2 and N-2 briefed the brigade and flotilla commanders and the planners on the area of the impending operation, and this intelligence along with other information was incorporated in planning guidance provided by the commanders. It was at the point of the formulation of planning guidance that the command relationships of the Mobile Riverine Force received one of its greatest tests. By close co-operation, Colonel Fulton and Captain Wells decided upon the guidance to be issued to their staffs for each operation, and these agreements in time came to be accepted as an informal set of standing operating procedures. They were, however, subject to change at any time by either commander and were not reduced to a written, formally approved document. Task Force 117 published a standing order that contained the results of considerable Army and Navy planning but was not jointly approved.

Planning

The brigade commander or a higher echelon Army commander usually selected the enemy target and area of operations. The two commanders then agreed upon the general task organization, the tentative duration and timing of the operation, and the location of the mobile riverine base to support the operation. The planning often included great detail in order to insure that major issues were resolved early, permitting the two staffs to plan efficiently.

From the outset, Colonel Fulton and Captain Wells recognized

the need to plan for future efforts while operations were in progress. Colonel Turner, who had been designated riverine adviser, acted as the planning co-ordinator, with the assistant N-3 (plans), the assistant S-3 (plans), the N-2, and the S-2 forming the planning staff. The planning for operations to take place ten days hence, the planning for the next operation, three to five days hence, and the execution of the current operation were accomplished at the same time, with the S-3 and N-3 dealing with the current operation.

The planning staff, using the commanders' guidance, began by outlining the scheme of maneuver in the objective area. From there the planners worked backwards, covering in turn the landing or assault, water movement, and loading phases. In preparing operations orders they took into consideration factors peculiar to riverine operations such as tides, water depth, water obstructions, bridge clearance, distance of the mobile riverine base from beaches, availability of waterway routes into the area of operations, suitability of river banks for landing sites, and mooring for barge-mounted artillery. Operation orders followed the format of the standard, five-paragraph field order, and were authenticated by the S-3 and the N-3. The bulk of the information peculiar to riverine operations was in the annexes, which included intelligence, operations overlay, fire support plan, naval support plan, signal arrangements, logistics, civil affairs, and psychological operations information. The intelligence annex supplied information on terrain, weather, and the enemy situation. Waterway intelligence was usually provided in an appendix that covered hydrography and the enemy threat to assault craft. A map of the waterways to be used was furnished, showing tides, widths and depths of streams, obstacles of various types, bridges, shoals, mud banks, and other navigation data. The enemy threat to assault craft was covered in descriptions of recent enemy action along the waterways, location of enemy bunkers, kinds of enemy weapons likely to be used in harassing action and ambushes, and water mine and swimmer danger to the small boats. The operations overlay annex showed schematically the area of operations, plan of maneuver, and control measures such as checkpoints. The naval support plan annex contained the naval task organization and referred to the operations order for information concerning the situation, mission, and concept of operations. It assigned specific support tasks to subordinate naval elements together with the necessary co-ordinating instructions. An appendix to the naval support plan provided waterway data.

Battalion orders for operations in a riverine area followed the format used in the brigade orders and were authenticated by the battalion S-3. The supporting river assault squadron water movement plan was prepared by the river assault squadron operations officer in co-ordination with the battalion S-3. It was, in effect, the river assault squadron operations order. This plan did not become a formal part of the battalion operations order but was attached to it. The water movement plan contained the task organization, mission, schedule of events, boat utilization plan, command and communications instruction, and co-ordinating instructions. The schedule of events paragraph listed events critical to the operation and the time they were to take place, for example, where and when each unit was to load, and its scheduled time of arrival at critical water checkpoints. The boat utilization paragraph included numerical designations of boats and the companies to embark on these boats. The co-ordinating instructions paragraph included information on submission of reports, radio silence, recognition lights, reconnaissance by fire, and the use of protective masks.

Once approved, both the brigade and battalion operations orders became the authority for fulfilling the intent of the scheme of maneuver and for providing combat and combat service support. Prior to each operation, a briefing was conducted on the flagship of the Mobile Riverine Force; it included presentation of the plan of each battalion and squadron, as well as the latest S-2 and N-2 intelligence. Briefings seldom required decisions by the flotilla or brigade commander because problems had been resolved between the Army and Navy elements at each subordinate echelon and between echelons during planning. Beginning in June of 1967, plans and orders were in far greater detail than the Army brigade and battalion commanders believed was needed or even optimum for routine land warfare. The detailed nature of the plan, however, included agreements at the Mobile Riverine Force command level; the Navy commander desired to define clearly the tasks to be performed by each service. For the Navy commanders and staffs, the detailed plan was less objectionable, being similar to the planning associated with amphibious, ship-to-shore operations. After June of 1967 as time passed both Army and Navy commanders realized that matters such as the loading of a battalion from a barracks ship onto the assault craft of a river assault squadron and the configuration of a battalion river assault squadron convoy were standing operating procedure. The Mobile Riverine Force was nevertheless to rely heavily on detailed operations orders

Mobile Riverine Force Briefing Aboard USS Benewah

rather than a formal, written standing operating procedure for many months of its operations.

Operational Concept and Procedures

The initial operations of the Mobile Riverine Force during the period 1 June to 26 July 1967 were designated Operation Coronado I. From 3 June through 10 June operations were conducted in the vicinity of Dong Tam, and were designed to make the land base more secure and to test the operational procedures of the force. A tactical operations center manned jointly by the Army and Navy provided a focal point for communications and for monitoring operations and could fulfill the needs of the riverine brigade and flotilla commanders separately.

At the levels of maneuver battalion and river assault squadron, the staffs operated independently but maintained close, harmonious relationships. The maneuver battalion staff outranked and outnumbered the river assault squadron staff. Because of a limited staff, the naval river assault squadron was unable to perform staff functions which corresponded to those of the infantry battalion it

supported. This limitation contributed to the centralization of planning at the flotilla level. The two boat divisions organic to the river assault squadrons also were limited in their planning. The rifle company commanders transacted their business with naval enlisted men, who did not have the authority to make decisions. Time could be lost at the company and boat division level in situations requiring quick decision.

Operations consisted of co-ordinated airmobile, ground, and waterborne attacks, supported by air and naval forces. The ground and naval combat elements traveled to the area of operations by river assault squadron craft, helicopter, or a combination of these means. While the boats would serve as a block and could transport troops in the early stages of an operation, once the troops had engaged the enemy more speed was necessary and air transportation of forces was generally more satisfactory.

The basic offensive maneuvers used by the Mobile Riverine Force were to drive the enemy against a blocking force, with the open flanks covered by Army helicopter gunships, or to encircle him. These maneuvers were based on the estimate that the enemy would choose to fight only when he thought he could inflict heavy casualties or when he was surprised and forced to fight. If he chose not to fight, he might attempt to take advantage of the concealment afforded by the numerous tree-lined waterways by dispersing into small groups and leaving the area of operations under the protection of small delaying and harassing fire teams. It was necessary for U.S. forces to establish blocking positions quickly and to be able to deploy troops rapidly. Several maneuver battalions were needed, but the Mobile Riverine Force had usually only two battalions, a situation that provided further incentive for co-ordinating operations with Vietnamese or other U.S. units.

Control of the riverine forces was aided by reliable and versatile parallel Army and Navy communications systems, by operations conducted over flat terrain by units within mutually supporting distances, and by command helicopters used at the brigade and battalion levels. Troop movements were controlled and co-ordinated from the flagship. The focal point for control was the joint tactical operations center, the Army element of which was under the supervision of the brigade executive officer. Staff representation came from the S-3, S-2, N-3, air liaison, Army aviation, and artillery fire support sections, and the center was operational at all times.

During operations the brigade employed a forward command group, usually located at the fire support base, and composed of the

Troops Prepare to Embark From Ammi Barges

brigade commander, the brigade S–3, the brigade assistant fire support co-ordinator, and the air liaison officer. The brigade command group was generally aloft in a command helicopter in daylight hours. The battalion command post was fragmented into a forward and a rear tactical operations center. The forward post was located in a command boat that accompanied the troops during their waterborne movement and normally consisted of the battalion commander, S–3, assistant S–3, artillery liaison officer, an operations sergeant, an intelligence sergeant, and two radio telephone operators. The river assault squadron commander and his operations officer were aboard the same command boat. The rear command post was located aboard ship at the mobile riverine base and usually consisted of the battalion executive officer, S–1, S–2, S–3, S–4, an operations sergeant, and two radio telephone operators. When a command and control helicopter, usually an observation type, was available, the battalion commander and his artillery liaison officer controlled the operation from the air.

After the briefing that took place before each operation, battalion commanders and supporting river assault squadron com-

MONITORS AND ASSAULT PATROL BOATS HEAD IN TO SHORE

manders issued final instructions to their staffs and subordinate commanders. These confirmed or modified previously issued orders. Each rifle company was allocated three armored troop carriers, and arrangements were made to load bulk cargo and ammunition prior to troop embarkation. At the designated loading time for a particular company, three armored troop carriers were tied alongside the Ammi barges moored to the ship and three platoons of a company embarked simultaneously. In most instances, the loading of a complete company was restricted to twenty minutes. The Navy stationed expert swimmers on the barges prepared to rescue men who might fall overboard. Once a company was loaded, the ATC's proceeded to a rendezvous where they waited for the remainder of the battalion to load. Upon completion of the battalion loading, the ATC's moved at a specified time across a starting point and thence to the area of operation. Army elements observed radio silence; the naval radio was used only for essential traffic control.

The boats moved across the starting point in column formation, the leading assault support boats providing minesweeping and fire support. Each minesweeper was equipped with a drag hook that

ASSAULT CRAFT GOING IN TO LAND TROOPS

was tethered with a steel cable to a high-speed winch. Two of the leading minesweepers preceding on the flanks of the column dragged these hooks along the bottom of the waterway in an effort to cut electrical wires rigged to underwater mines. The river assault squadron commander traveled aboard the leading command and control boat. The battalion commander was either in the control boat or in a helicopter aloft. Each supporting river division was

divided into three sections, each section consisting of a monitor and three ATC's supporting the one rifle company. These sections were in such order in the column as to maintain infantry company integrity and to place units on the beach in the order specified by the battalion commander. While underway, the river assault squadron commander exercised command and control of the boats. The rifle company commanders monitored the Navy command and control and the battalion command net in order to stay abreast of developing situations. Radar was the principal navigational aid during the hours of darkness.

Water checkpoints were designated along the routes as a means of control, and as march units passed these a report was made to the joint tactical operations center. Whether on large rivers or small streams, river assault units were faced with the threat of hostile attack. However, when the boats left the large rivers for the smaller waterways the danger from recoilless rifles, rocket fire, and mines increased. The waterways negotiated by the boats varied from a width of 100 meters where the smaller waterways joined the river to widths of 15 to 25 meters. The formation that the boats assumed was similar to that used on the main river except that the leading minesweepers moved to positions nearer the column. While the column proceeded along the small waterways, suspected areas of ambush (if in an area cleared by the Vietnamese government) were often reconnoitered by fire from the small boats and by artillery.

When the lead boats of the formation were approximately 500 meters from the beach landing sites, artillery beach preparations were lifted or shifted. The leading minesweepers and monitors then moved into position to fire on the beaches on both sides of the waterway. This fire continued until masked by the beaching ATC. If no preparatory fire was employed, the boats landed at assigned beaches in sections of three, thus maintaining rifle company integrity. Boats within a section were about 5 to 10 meters apart, and a distance of 150 to 300 meters (depending on the beach landing sites selected for each company) separated each boat section or assault rifle company. After the units had disembarked, the boats remained at the beaching site until released by the battalion commander. Upon release, the river assault squadron craft either moved to provide fire support to the infantry platoons or moved to interdiction stations and rendezvous points. Subsequent missions for the boats included independent interdiction, resupply, and extraction of troops.

TROOPS GO ASHORE FROM ARMORED TROOP CARRIER

After landing the troops secured initial platoon and company objectives and moved forward in whatever formation the terrain and mission required. (*Diagram 6*) When contact with the enemy was made, commanders immediately acted to cut off possible enemy escape routes. The most frequently used tactic was to move units to blocking positions to the flanks and rear of the enemy and direct extensive artillery fire, helicopter gunship fire, and air strikes into the enemy positions. After this softening up process, troops swept the area.

The Mobile Riverine Force had little trouble finding suitable helicopter landing and pickup zones. Pickup zones used to mount airmobile operations were normally located in the vicinity of the fire support base or on terrain adjacent to the waterways. When there were not enough aircraft to move an entire company at one time, platoons were shuttled into the area of operations. The brigade standing operating procedure called for helicopter gunships to escort each serial, and for an airborne forward air controller to operate along the route of aerial movement.

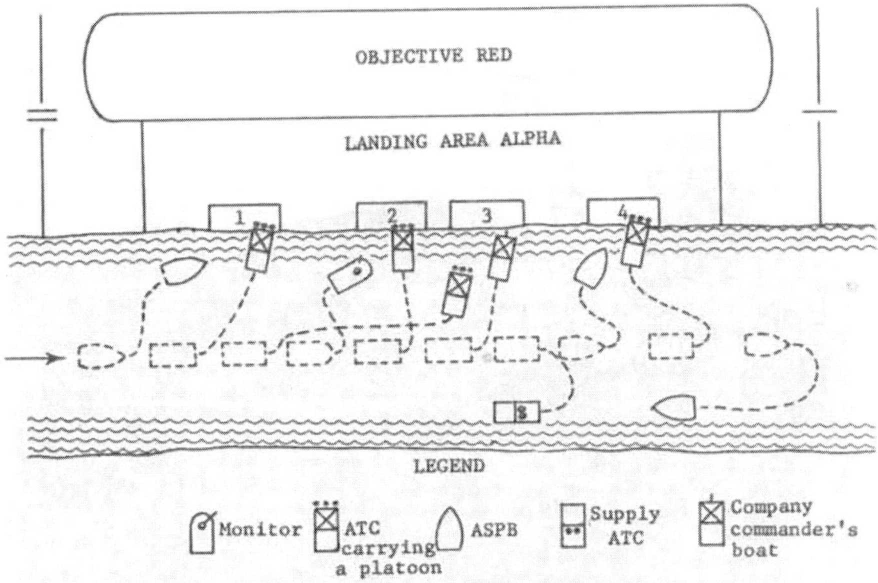

Diagram 6. Typical company landing formation.

During each day plans were made to arrive at a night defensive position before dark unless troops were engaged with or pursuing the enemy. After resupply, the night defensive position was organized against possible enemy attack.

The fourth rifle company of each battalion was usually employed at the fire support base under operational control of the base commander; for mobile riverine base security it was under direct control of the Mobile Riverine Base commander.

When the river-assault craft remained overnight on the small waterways in the operational area, the brigade and battalion commanders planned for security of the boats, which were usually kept near infantry elements ashore. The boat fire was integrated with the fire of other units.

Withdrawal of riverine forces from the area of operations was accomplished chiefly by boat. When units operated far inland from navigable waterways, however, they were extracted by helicopter and taken to a landing zone near a navigable waterway. They were then transported by boat to the Mobile Riverine Base or to another area of operations. Sites designated for picking up troops were often on streams lined with dense banana groves, coconut groves, or nipa palm that made it difficult for both the ground unit

commander and the river assault squadron commander to spot. Ground and water movement to the site was therefore controlled and co-ordinated from the battalion commander's helicopter. The commander marked specific beach and unit locations with colored smoke and directed the Army and Navy elements to them by radio.

The same boat that was used to put an Army element ashore was also used to bring it out. Each boat carried an identifying number and flew a distinctive colored pennant from the mast that could be recognized by troops who were behind dikes and vegetation. Each boat also carried three running lights tiered near the top of the mast. Through the use of varying color combinations in the top, center, and bottom running lights, the boat designated for each boarding element could be readily identified.

Tides, draft, and maneuver room were primary considerations when boats were scheduled to rendezvous with Army forces for pickup because tides not only affected the water depth in canals, but caused significant changes in current velocity and direction. Traveling a distance of thirty kilometers upstream could take up to six hours against an ebbing tide, whereas the same route might be covered in four and a half hours on an incoming tide.

The threat of ambush during troop removal was considered as great as during movement to landings. When possible, an alternate to the route used for entering an area was selected for picking up troops. Beaching of the boats was usually accomplished with little difficulty because the infantrymen on shore had reconnoitered the banks for obstacles. Army security elements were placed 100 to 150 meters inland from the extraction site. After the troops were aboard, the boats resumed tactical formation and moved along the canal or stream and into the major waterway where the Mobile Riverine Base was anchored.

During offensive operations, the commander of the Army element of the Mobile Riverine Force in his position as base commander for all joint Army and Navy bases was also responsible for the defense of the Mobile Riverine Base. The Navy provided its appropriate share of forces, which included those for gunfire support and protection against waterborne threats. Defense of the base also involved arrangements for curfew and other restrictions for water craft with appropriate Vietnamese civilian and military officials in the immediate vicinity.

River assault divisions were assigned the responsibility for waterborne defense on a rotating basis. They conducted patrols to control river traffic, enforce curfews, and deter attack by water;

swept the vicinity of the Mobile Riverine Base for water mines; sent demolition teams to inspect underwater anchor chains, hulls of ships, and pontons for limpet mines; and dropped random explosive charges in the anchorage area to deter enemy swimmers.

One rifle company under the base commander was responsible for shore defense of the Mobile Riverine Base. Normally the company employed one platoon on each bank of the river and held the remainder of the company in reserve. Since anchorage space was 2,000 to 2,500 meters in length, it was impossible to conduct a closely knit defense with the relatively small number of troops committed. Platoon ambush sites were established on the banks opposite the flotilla flagship and during daylight hours security patrols were dispatched to provide early warning of enemy attack. The security troops were reinforced as needed by elements of the company defending the fire support base. Artillery and 4.2-inch and 81-mm. mortar fire as well as fire from the boats was planned in support of the defense of the base. The flat trajectory of naval weapons fire required careful planning for both shore and waterborne defense.

The planning and operational procedures initially used by the Mobile Riverine Force were refined as the staff and the force gained experience. The first significant test of these procedures came during operations in Long An Province in the III Corps Tactical Zone.

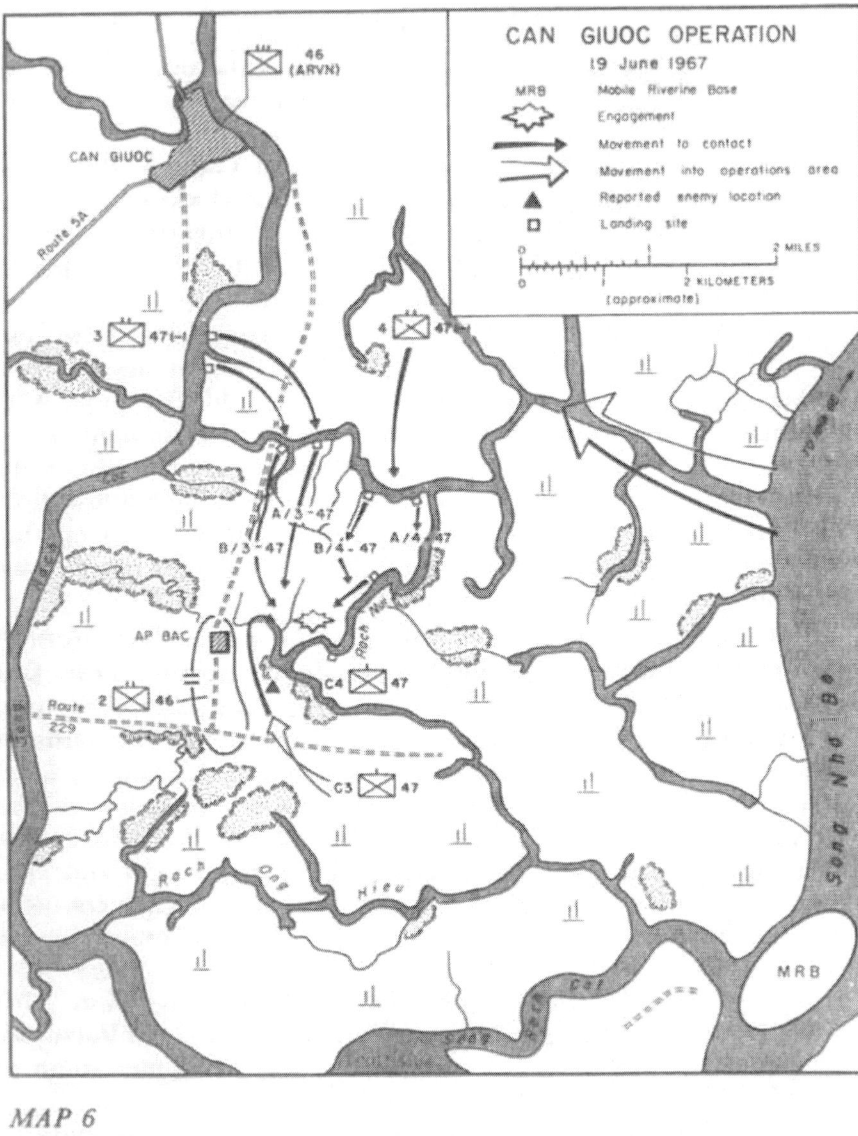

MAP 6

kilometers of a location that had been a remote Viet Cong base area enjoying considerable security. To attack this area, Colonel Fulton and Captain Wells had agreed to anchor the riverine ships as close as practicable to the area of operations. The time needed by assault craft to enter the area and the turnaround time of both helicopters and water craft conducting resupply and medical evacuation would thus be reduced. The base was moved into its anchorage on the evening before operations. The risk of disclosing the intended area of operations by this move was accepted because the

Before entering Long An Province, the brigade commander briefed the battalion and company commanders on their new area of operations. The battalions of the 2d Brigade had fought in Dinh Tuong and northern Kien Hoa Provinces where trees lined the banks of most streams and major canals. Although the soldiers were familiar with moving in the rice paddies near Dong Tam, they had seldom encountered such wide expanses of open paddy as existed in eastern Long An Province. These open areas provided observation and fields of fire that dictated greater dispersion of men and greater reliance on the "scouts out" technique. A long proven infantry tactic in open terrain, scouts out provided for point and flank security men to move out on foot as much as 500 meters from the remainder of their squad to provide early warning.

In further preparation for entering eastern Long An Province the brigade commander, S-2, and S-3 visited the 3d Brigade, 9th Infantry Division, and the 2d Battalion, 3d Infantry, 199th Infantry Brigade (Separate). Both of these units had operated in Long An and furnished helpful information. The group also visited the headquarters of the 46th Regiment, 25th Division, Army of Vietnam, and the Can Giuoc District headquarters. Although such visits risked exposing planned operations to compromise, the risk was accepted in order to exploit local intelligence and take advantage of operating in conjunction with other U.S. and Vietnamese units. For the Can Giuoc activities, the Mobile Intelligence Civil Affairs Team set up operations with the 2d Battalion, 46th Regiment, Army of Vietnam, and Can Giuoc District headquarters on 19 June, the first day of Mobile Riverine Force operations in Long An.

Can Giuoc Operation

Can Giuoc District contained a good network of navigable waterways, permitting the assault craft to enter an area that intelligence reports indicated was used extensively by Viet Cong regional forces for rest and training. Civilian travel in the locality was chiefly by water, since most bridges and the one ferry were no longer serviceable. Previous military actions here had depended for movement on boats and aircraft. The U.S. 9th Infantry Division's 3d Brigade was disposed in Long An Province west of Can Giuoc District to lend security by its operations to Highway 4; however, Can Giuoc District was outside the brigade's routine area of operations.

The fact that the Mobile Riverine Force could move its afloat base permitted the establishment of a brigade base within three

CHAPTER VI

III Corps Operations and the Threat to Dong Tam

On 11 and 12 June 1967 the fully constituted Mobile Riverine Force made its first major move. Leaving Dong Tam, it sailed down the Mekong River and across a stretch of the South China Sea to a temporary anchorage at Vung Tau, and thence to an anchorage southeast of Saigon at Nha Be—a total distance of sixty miles. At Vung Tau it received additional Navy assault craft that had arrived from the United States. Since the assault support patrol boats that had been scheduled to join the force at this time were delayed, armored troop carriers continued to perform minesweeping and security operations until arrival of the assault support patrol boats in September and October. Upon reaching the Nha Be anchorage, the force returned to the Vietnamese Navy the borrowed craft it had used since February. With the new assault craft, the flotilla had 52 armored troop carriers, 10 monitors, 4 command boats, and 2 refuelers.

From the Nha Be anchorage operations were conducted from 13 through 17 June in the Rung Sat Special Zone as part of the U.S. 9th Division operation, GREAT BEND, which brought greater security to the Long Tau shipping channel. Although the force found a base camp believed to have been the recent site of the headquarters controlling Viet Cong actions in the Rung Sat Special Zone, no contact was established with the enemy.

On 18 June the Mobile Riverine Force moved eight miles to a mobile riverine base anchorage at the junction of the Soi Rap and Vam Co Rivers in preparation for operations in the Can Giuoc District of eastern Long An Province. One battery of supporting 155-mm. self-propelled artillery was moved by LCU to the west bank of the Soi Rap River adjacent to the mobile riverine base at the confluence of the Soi Rap and Vam Co Rivers. Ammunition supply for the battery was transported from Vung Tau by Army LCM-8's while the barge-mounted artillery continued to receive ammunition support from the LST supply ship in the base.

force had not been in the area before and the Viet Cong were not thought to be familiar with the capabilities of the ships and assault craft. The brigade and flotilla commanders agreed that assault craft would go up only those streams wide enough to permit them to make 180-degree turns. Colonel Fulton and Captain Wells also agreed to use a portion of the assault craft as a blocking force once Army troops had landed. The boats were to enter the waterways as the tide was rising, permitting greater speed and a long period of high tide level immediately following entry.

The Mobile Riverine Force sent five companies of the two U.S. infantry battalions into the operations area by assault craft. (*Map 6*) The sixth company, Company C, 3d Battalion, 47th Infantry, moved by water to an air pickup zone to stand by as reserve. The troops entered the area and searched south, using assault craft to cross water barriers and to provide communications and fire support. The 2d Battalion, 46th Infantry, Army of Vietnam, moved during darkness and established a blocking position near the town of Ap Bac, oriented to the east.

At approximately 1000 Colonel Fulton was notified through the Mobile Intelligence Civil Affairs Team, located near Can Giuoc District headquarters, that the district chief had information of a Viet Cong battalion-size force due east of the blocking position of the 2d Battalion, 46th Infantry. The brigade commander issued a fragmentary order at 1010 informing the commanders of the 3d and 4th Battalions, 47th Infantry, and the U.S. adviser to the Vietnam Army battalion of the reported enemy strength and location. The order directed the two U.S. battalions to continue their maneuver south toward the newly designated objective—the reported location of the enemy battalion. Colonel Fulton also directed Colonel Tutwiler, commanding the 4th Battalion, 47th Infantry, to prepare to assume control of Company C, 3d Battalion, 47th Infantry, and to deploy the company by helicopter south of the reported enemy location.

Colonel Tutwiler moved Company C at 1105 by air south of the reported enemy location with the mission of moving northeast to reconnoiter the enemy position. He sent Company C of his battalion by assault craft south on the stream and landed the company at 1135 northeast of the reported enemy position.

By 1150 Company C of the 3d Battalion had swept the reported enemy location without finding the Viet Cong. Company C of the 4th Battalion was moving west from its landing area when it was fired upon from the north. At the same time Company A of the 4th

Battalion, moving south but at a distance of some 800 meters north of Company C, 4th Battalion, encountered heavy automatic weapons and small arms fire from its front and right flank. The enemy position had turned out to be north of the reported location. Company A, lacking cover, suffered heavy casualties. The Viet Cong had occupied well-fortified firing positions in an L-shape along the north bank of the stream that separated Company A from both Company C of the 3d Battalion and Company C of the 4th Battalion, 47th Infantry.

Colonel Tutwiler shifted Company B of his battalion, which had been moving on the battalion right flank, into the area behind Company A to assist by fire and with medical evacuation. He held his Company C on the south side of the stream to maintain contact and manéuvered Company C, 3d Battalion, north across the stream. Once on the north bank, Company C, 3d Battalion, fought slowly to the east, with its right flank on the stream. Major H. Glenn Penny, executive officer and acting commander of the 3d Battalion, 47th Infantry, meanwhile moved his Companies A and B southeast to link up with Company C of the 3d Battalion on that company's north flank. At 1545 Major Penny resumed operational control of his Company C. Company B attacked east on line with Company C while Company A moved into a blocking position 600 meters north of the left flank of Company B. By 2000 darkness and enemy fire prompted a halt to maneuvers and placed Companies B and C, 3d Battalion, some 600 meters west of Company C of the 4th Battalion where the units remained throughout the night of 19-20 June.

In the movement of the two U.S. battalions toward the first reported enemy location, the naval assault craft were limited to movement on the stream paralleling the 4th Battalion's left flank. When Company C of the 4th Battalion was fired upon at 1150, the naval assault craft that had brought in the company provided fire support for it and, later, for Company C of the 3d Battalion as it crossed the stream west of the actual enemy positions.

Although the rifle companies and the naval assault craft quickly returned the enemy fire, artillery and helicopter gunship fire could not be immediately brought to bear because the heavy casualties initially suffered by Company A of the 4th Battalion prevented the company commander from determining the dispositions of his platoons. Artillery fire began at approximately 1200; it was delivered at the request of the forward observers and co-ordinated by both ground and airborne artillery observers. Support was provided

during the daylight hours and well into the night by helicopter gunships, medical evacuation helicopters, and assault craft.

The movement of Companies A, B, and C, 3d Battalion, 47th Infantry, required the entire afternoon, but not all companies were able to link up before dark. The 2d Battalion, 46th Vietnam Army Infantry, remained in its original position during the engagement because Colonel Fulton decided that maneuver of the battalion to the east would be complicated by the presence of U.S. companies between the Vietnam battalion and the enemy.

Viet Cong and U.S. troops exchanged fire throughout the daylight hours and into the night. U.S. troops found that the enemy's reinforced bunkers could be destroyed only by 90-mm. recoilless rifle fire and helicopter rockets and bombs; the 20-mm. and 80-mm. rounds fired by the assault craft penetrated a bunker only after multiple hits.

Casualties of Company C of the 3d Battalion were evacuated during the afternoon and evening of the battle by plastic assault boat and helicopter. Company C of the 4th Battalion was unable to use the assault craft for medical evacuation and relied on helicopters. Whether casualties were taken directly by helicopter from a ground pickup zone or from an armored troop carrier aid station after treatment, they were sent either to the barracks ship USS *Colleton* or to a field hospital.

During the afternoon and evening of 19 June most of the enemy losses came from U.S. supporting fire concentrated on a small area. A portion of the enemy battalion eluded the U.S. blocking positions during the night, probably by crossing the stream to the west of Company C, 4th Battalion, 47th Infantry, and then moving south. On 20 June the 4th Battalion, 47th Infantry, searched south to regain contact with the enemy. Moving by helicopter and on foot the battalion found the enemy north of the Rach Gion Ong stream at Ap Nam and, aided by a company of the 2d Battalion, 60th Infantry, encircled and eliminated an enemy platoon. The company from the 2d Battalion, 60th Infantry, was provided from the 3d Brigade by the Commanding General, 9th Infantry Division.

Following the battles of 19–20 June, the Mobile Riverine Force remained in the Can Giuoc area, capitalizing on the knowledge obtained of the area and attacking small elements of the Viet Cong 5th Nha Be Battalion and local guerrillas with whom battle had been joined on 19 June. On 23 June Company A, 4th Battalion, 47th Infantry, which had taken heavy losses, was moved by air to Dong Tam to begin a two-week period of refitting and retraining. Operations were conducted in the Rung Sat Special Zone, the Can

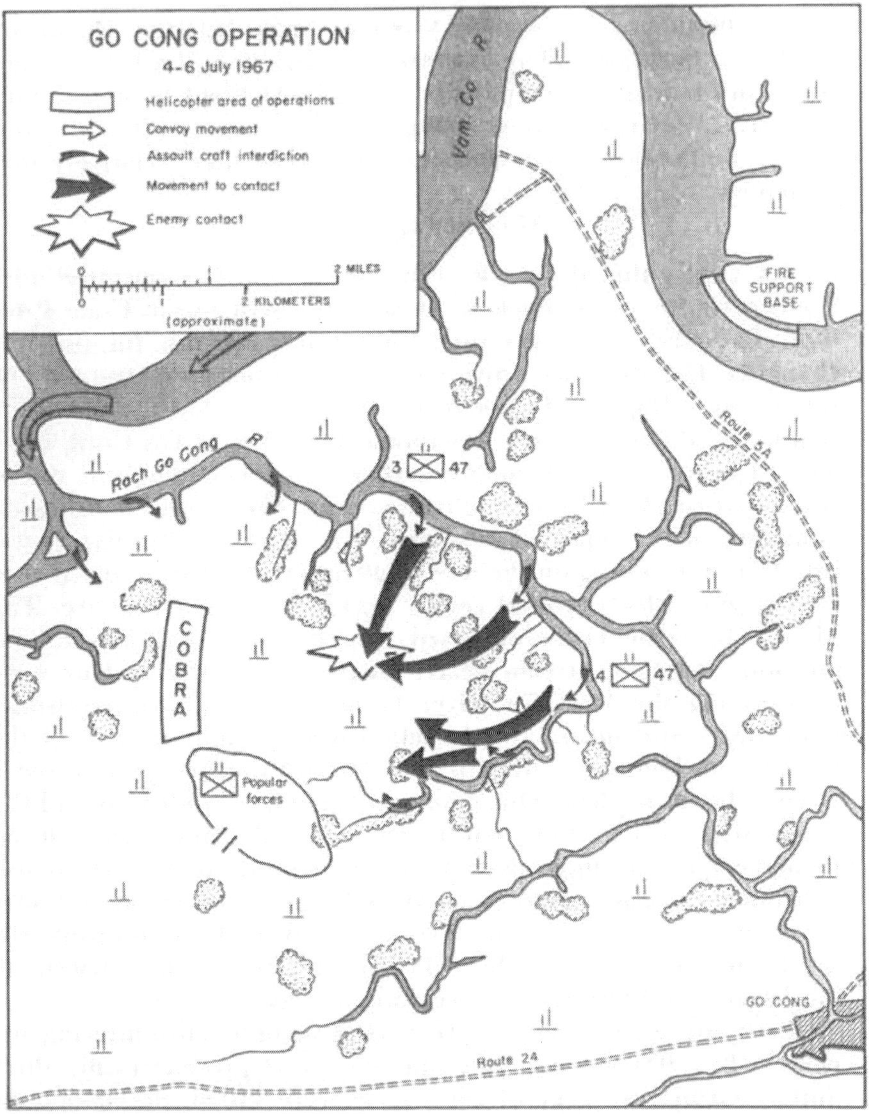

MAP 7

Duoc District of Long An Province, and Go Cong Province of IV Corps. Companies on independent missions entered the smaller waterways by ATC at night to establish widely dispersed ambushes. Infantry platoons set up ambushes ashore in co-ordination with ATC's and monitors to interdict land and water routes. Co-operating with the 199th Infantry Brigade (Separate) and the 2d

and 3d Battalions of the 46th Vietnam Army Infantry Regiment. the force ranged in widely dispersed formations to find the enemy. An assault helicopter company provided by II Field Force was used extensively to move troops when the riverine assault craft were limited by the lack of navigable streams or by their relatively slow movement.

Go Cong Operation

On 4 July the Mobile Riverine Force again co-operated with Vietnamese forces to attack an enemy base area in Go Cong Province. The base, reportedly used for training recruits for the Viet Cong My Tho Province units, lay in an area nearly surrounded by waterways. (*Map 7*) The scheme of maneuver for the operation required a Popular Forces provisional battalion of Go Cong Province to move overland during darkness to block the land escape route that the Viet Cong might use—if attacked—on the southwest boundary of the reported base area. To assist the Popular Forces battalion in blocking on the west, helicopter gunships watched over an area into which ground troops were not allowed to move. The Mobile Riverine Force deployed into the area and moved into positions along the stream nearly encircling the enemy base area.

Entering the Vam Co River before daylight, five companies under the command of two battalion headquarters moved off the Vam Co and into the streams that formed a three-quarter circle around the enemy base. Once the companies had landed, the ATC's, augmented by all other assault craft available, moved to stations to prevent enemy movement across the small streams. An assault helicopter company stood ready at first light to move one infantry company as a reserve under brigade control. This company was set down before daylight by ATC's at a pickup zone adjacent to the barge-mounted artillery fire support base.

Early on 4 July troops were landed without encountering the enemy. By 1800 the sweeping operation had produced only three minor skirmishes; three of the enemy were killed, ninety-one detained, and two weapons were captured. There had been no U.S. or Popular Forces casualties. The U.S. battalions established dispersed ambush sites and shortly after dark began to encounter small groups of Viet Cong in a number of places. These encounters continued sporadically throughout the early evening; several of the enemy were killed and more were detained.

On 5 July, discovering cleverly disguised bunkers and so-called spider holes in which the enemy had escaped detection the previous day, U.S. troops made a more methodical search of the area.

On the same day the Viet Cong who had been detained were sent to a combined interrogation center established by the Mobile Intelligence Civil Affairs Team and Vietnamese provincial officials at the province capital. Later, the interrogation reports provided information that the enemy elements encountered had been a platoon-size cadre and a newly recruited Viet Cong company. The cadre was reportedly equipped only with small arms, and the company had few weapons. The search for the Viet Cong continued through 5 July and ambushes were again established during the night.

The Go Cong operation ended on 6 July without casualties to U.S. or Popular Forces troops. Throughout the operation, the assault craft guarded the waterways, stopping over seven hundred native watercraft and screening the people, some of whom were held for further interrogation. On 4 July the reserve company had been inserted by the assault helicopter company into the area near. the COBRA zone. This company, operating under control of its parent battalion, had subsequently searched and maintained patrols and ambushes in the area. The impact of the operation on the enemy was not fully known until several days later when the interrogation center had completed its work and a Popular Forces company had re-entered the area to check information provided by the prisoners. Enemy losses included sixty-six killed, sixty-two prisoners of war, and five weapons. Under the Vietnamese government program of granting amnesty to Viet Cong who returned to the government, seven returnees, or Hoi Chanhs, were received.

The operation emphasized the importance of using a combination of land, air, and water transportation to take advantage of terrain, and the necessity for close co-operation with South Vietnamese officials and intelligence units. The combined transportation means enabled the Mobile Riverine Force to gain control over an area quickly. By seizing sampans along the waterways the Mobile Riverine Force prevented the enemy from using them to escape.

During this operation the Army initiated air resupply directly from the support LST. Previously supplies were moved from the LST by boat to a landing zone ashore, whence they were carried by helicopter to the field, an operation that required four separate transfers and considerable time. With the new method, loads were broken to unit size on the LST, each helicopter was loaded directly from the LST deck, and the material was delivered to the troop landing zone. To resupply two battalions using 10 UH–ID's (one assault helicopter company) required one hour and ten minutes.

This time was bettered as the crew teamwork and the landing cycle on the LST's were smoothed out.

During the period of the Go Cong operation, another innovation was fielded by the Mobile Riverine Force that significantly increased flexibility. The first of several armored troop carriers was modified by adding a helicopter landing pad sixteen feet square over the troop compartment. An H-23 successfully operated from the new pad on 4 July and a UH-ID on 5 July.

On 15 July Captain Wells requested that the Commander, Naval Forces, Vietnam, obtain twelve Vietnamese national policemen of the river police branch for assignment to Task Force 117. The experience of searching over seven hundred native boats during the Go Cong operation highlighted the need for more men to screen waterway traffic. Although no Vietnamese river policemen were subsequently assigned, the force had the co-operation of the river police on most operations.

Further Can Giuoc Operations

On 21 July 1967 the Mobile Riverine Force concluded a two-day operation employing two U.S. infantry battalions in the southern part of Can Giuoc District. During the preceding thirty days, the force had conducted three operations in Can Giuoc District, killing 316 Viet Cong, taking 15 prisoners, capturing 68 weapons, and detaining 337 people. This fourth operation was part of a program to strike at different sections of Can Giuoc until the whole district had been fully covered. Reports that company-size groups of Viet Cong were in the area had been received by the combined intelligence center, the 46th Vietnam Army Regiment, and Can Giuoc District headquarters at Can Giuoc.

At 0545 on 20 July the 3d Battalion, 47th Infantry, and the 4th Battalion, 47th Infantry, boarded ATC's that joined in column with the heavy-gunned monitors and steamed west from the Mobile Riverine Base along the Vam Co River. From the fire support patrol base adjacent to the mobile riverine base the artillery supported the operation with two barge-mounted batteries of 105-mm. howitzers and one land-based battery of 155-mm. self-propelled howitzers from the 2d Battalion, 35th Artillery. At dawn Company C of the 4th Battalion, 47th Infantry, landed at the fire base as an airmobile or waterborne reserve force. Three ATC's stood by as water transport until a company of UH-ID helicopters arrived to support the operation and to provide transportation of the reserves to any part of the battle area. When the convoy left the base

ARMORED TROOP CARRIER WITH HELIPAD

it had a unique combat element. Both reconnaissance platoons of the two battalions had been designated Team Recon and were under the control of the 4th Battalion.

At 0700 the ATC's landed along a four-kilometer stretch of the east bank of the Song Rach Cat River, and 800 troops moved ashore under the watchful support of the monitors. The 3d Battallion, 47th Infantry, landed on the north beaches, the 4th Battalion on the south beaches, with Team Recon going in on the right flank. The landing was unopposed and all units quickly secured initial objectives beyond the beach, moving steadily to the east.

The operational area was bounded by the Song Rach Cat on the south and west and by the Song Nha Be on the east. On the north, a company of the 3d Battalion guided its left flank along Highway 229 running generally west to east. From the Song Rach Cat units moved slowly across the rice paddies, picking up all male Vietnamese, who were then evacuated and checked for identification at the combined intelligence center at Can Giuoc.

Overhead the brigade and battalion commanders directed movement from helicopters. Artillery observers and Air Force forward air controllers surveyed the battlefield, waiting to bring in supporting fire. By 1000 the companies had moved east approximately 2,000 meters without meeting the enemy while the Navy support craft moved up close behind them on the narrow streams with guns ready. At 1030, Company A in the center of the 3d Battalion formation reported fire from well-fortified bunkers. It was reported that 10 to 15 Viet Cong were sighted, but later a prisoner gave the number as 50 to 60. A point scout from Company A, operating well forward, was the only friendly combatant killed by enemy fire in the operation. As troops maneuvered to bring small arms fire upon the enemy, close supporting fire by 105-mm. and 155-mm. artillery was adjusted forward of the advance. Shortly after 1100 the first air strike began to lay 750 pound bombs and napalm on the Viet Cong bunkers. Of the five strikes called in on the target, three were directed while the artillery continued to attack.

Units began moving rapidly to block enemy escape routes. Companies B and C of the 3d Battalion to the north and south of the enemy deployed on a line facing the enemy. Company B, 4th Battalion, 47th Infantry, maneuvered northeast across the rice paddy to block to the east. Company A of the 4th Battalion moved to join the block of the 3d Battalion's Company C on the south.

During the whole maneuver, the Navy ATC's and monitors were used to help find the enemy. Moving forward, the assault craft took up positions alongside the ground troops to add their fires to the battle and assist the infantry to cross the Rach Ong Hieu, a small stream. When Company A, 4th Battalion, 47th Infantry, arrived from the south at the Rach Ong Hieu the Navy ATC's provided lift across the stream, accomplishing in minutes what might have taken hours, and furnishing fire support as needed. By 1300 all blocks were in position with the exception of Company C of the 4th Battalion. At 1415 this unit was lifted by helicopters from the artillery fire support patrol base near the Mobile Riverine Base and landed to complete the block on the east next to Company B of the 4th Battalion. The encirclement of the Viet Cong was complete by 1515.

This encirclement was accomplished with a smoothly operating team of infantry, artillery, Navy and Air Force components. The tricky tides were known, and the boat commanders knew when and how far up a stream they could go to provide support. The infantry commanders maneuvered their troops across the rice pad-

dies carefully, alert for booby traps and ambushes. Just in advance of lead elements, artillery delivered white phosphorus rounds which exploded 200 feet above the ground. Constantly re-adjusted, this marking fire insured that rapid and accurate artillery fire could be placed on the ground in front of the U.S. troops when they were engaged with the enemy. The Navy moved up close to the infantry positions to add the fire of the monitor's 40-mm. guns and 81-mm. mortars to reduce Viet Cong bunkers and machine gun emplacements. The Air Force and the artillery worked in close co-ordination, simultaneously laying on air strikes and artillery fire, while the Navy monitors maintained direct fire. When the helicopter company arrived in the battle area with Company C, 4th Battalion, 47th Infantry, it was able to land in a zone already secured by a U.S. ground unit, thus speeding the encirclement.

Action continued during the afternoon of 20 July, with units tightening their ring on the enemy as helicopters shuttled back and forth bringing a resupply of munitions and food. Seventeen Viet Cong dead had been accounted for by nightfall of the 20th. The pressure of artillery and small arms fire was maintained. The infantry positioned 90-mm. recoilless rifles and began to punch holes in the enemy bunkers. With darkness the Viet Cong fire became sporadic. U.S. troops maintained vigilance over the area during the night, using artillery illumination and allowing the Viet Cong no rest.

On 21 July a methodical sweep of the battle area revealed fifteen enemy bodies, one pistol, and seven rifles. More than eighteen bunkers had been reduced to rubble. Ten prisoners of war were taken, one of whom stated that on meeting U.S. troops the Viet Cong company had split into small groups and attempted to get out.

On 22 July the first helicopter barge was delivered to the Mobile Riverine Base. It was prepared for movement with the 3d Battalion, 34th Artillery, to the brigade forward command post at the barge-mounted artillery fire support base on 24 July.

On 24 July, in co-operation with the 3d Battalion, Vietnam Marine Corps, the 3d Battalion, 46th Vietnam Infantry Regiment, and GAME WARDEN patrol boats, the Mobile Riverine Force began operating in the northern part of Can Giuoc District in an effort to reduce the enemy threat to the district capital. (*Map 8*) At 0320 elements of the 3d Battalion, 34th Artillery, left the Mobile Riverine Base and went into position to support the operation at Fire Support Base TANGO. Elements of the 3d Battalion, 46th Vietnam

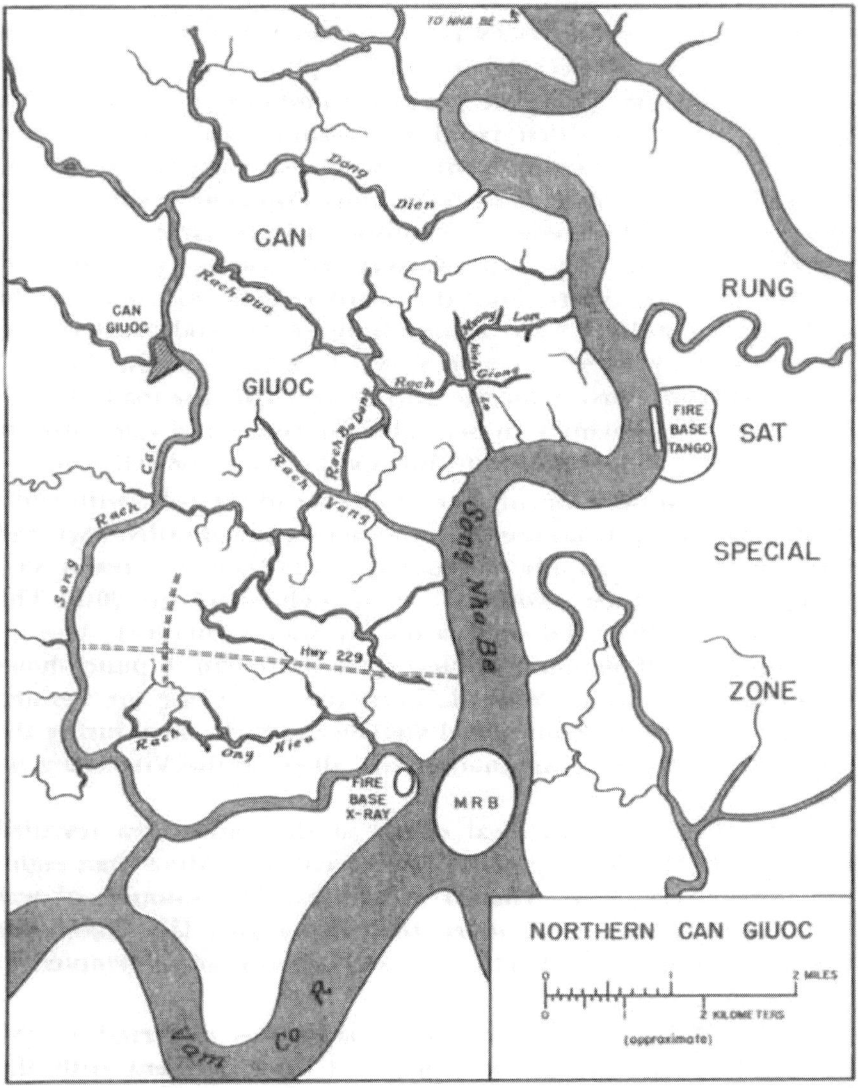

MAP 8

Infantry, moved from their cantonment area at Can Giuoc directly east across the Song Rach Cat to land unopposed. They moved northeast through selected control points to the Rach Dua River without meeting the enemy.

The 2d Brigade troops approached the area from the south, moving up the waterways in ATC's protected by monitors. The 3d Battalion, 47th Infantry, reinforced by the reconnaissance platoon of the 4th Battalion, 47th Infantry, went ashore on the

north bank of the Rach Vang River, moved north, and secured its first objectives without opposition. The 4th Battalion remained afloat near the mouth of the Rach Vang as the brigade reserve. Company C of the 4th Battalion was landed at Fire Support Base X-RAY, from which it would later be deployed as needed by helicopter. The waterborne units of the 3d Battalion patrolled the waters of the Kinh Lo, Rach Giong, Rach Ba Dang, and Rach Vang to detect the enemy. The 2d Battalion, 3d Infantry, 199th Infantry Brigade, also participated in the operation, landing two companies by air south of the Dong Dien stream. The Vietnamese Marine battalion was next to enter the operation, having been lifted by ATC's from loading points at Nha Be to beaches along the north bank of a stream called Muong Lon. The landings were unopposed, and all units quickly secured their initial objectives.

At 0915 one company of the 4th Battalion, 47th Infantry, was inserted by ATC's at a beach on the east bank of the Rach Ba Dang and at 1110, the battalion's remaining two companies were brought into the operation, one landing by helicopter and the other by assault craft west of the Rach Ba Dang. A few bunkers and booby traps were found and destroyed. Team Recon discovered freshly broken ground guarded by booby traps; further search disclosed an arms cache containing two 57-mm. recoilless rifles, one 75-mm. recoilless rifle, and one 81-mm. mortar with bipod. Other elements of the 3d Battalion, 47th Infantry, also discovered arms. The caches contained a total of sixteen individual and four crew-served weapons.

Throughout the day of 24 July no enemy was sighted. On the morning of 25 July the Vietnam Marine Corps battalion and the 3d Battalion, 47th Infantry, cleared the remainder of the assigned areas of operation. The marines consolidated in place along the banks of Muong Lon facing north, while the 4th Battalion, 47th Infantry, was taken aboard ATC's from beaches on the south bank of the Rach Dua and Rach Giong Rivers to begin movement south. In the northern operational area the 2d Battalion, 3d Infantry, had moved to the east, while the Vietnam marines held their position. The 3d Battalion, 46th Vietnam Infantry, had swung about to search again to the southwest, arriving at the original beaches about 1540. By 1715 all units were on their way to their home stations by water or air.

A Hoi Chanh who had rallied to the republic during the operation said that his unit had been directed to hide its weapons and equipment, break up into small groups, and temporarily cease operations in the Can Giuoc area. He said that his unit had oper-

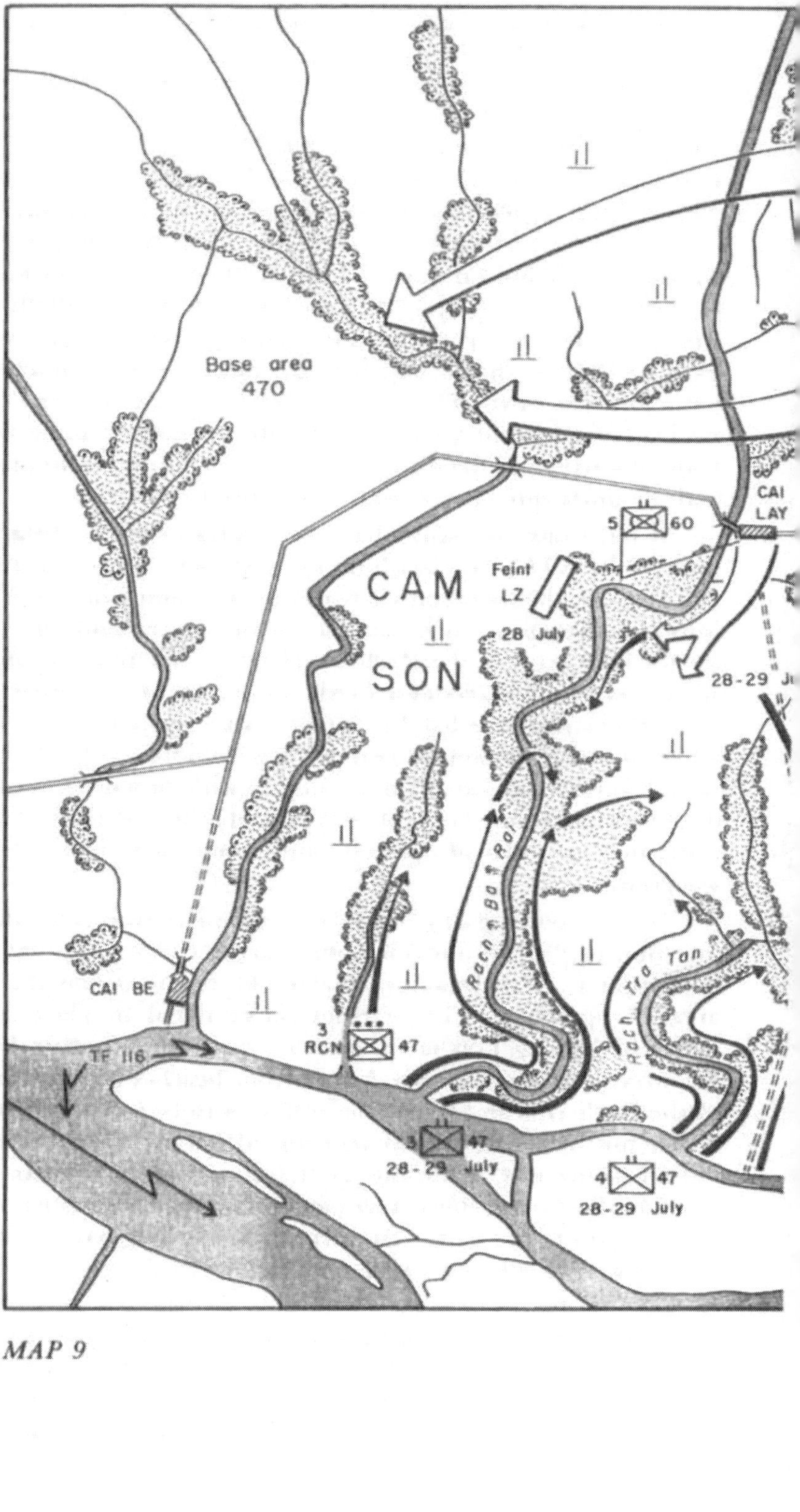
Base area
470
CAI
LAY
5 60
CAM
SON
Feint
LZ
28 July
28-29 J
Rach Ba Rai
Rach Tra Tan
CAI BE
TF 116
3
RCN
47
3 47
28-29 July
4 47
28-29 July

MAP 9

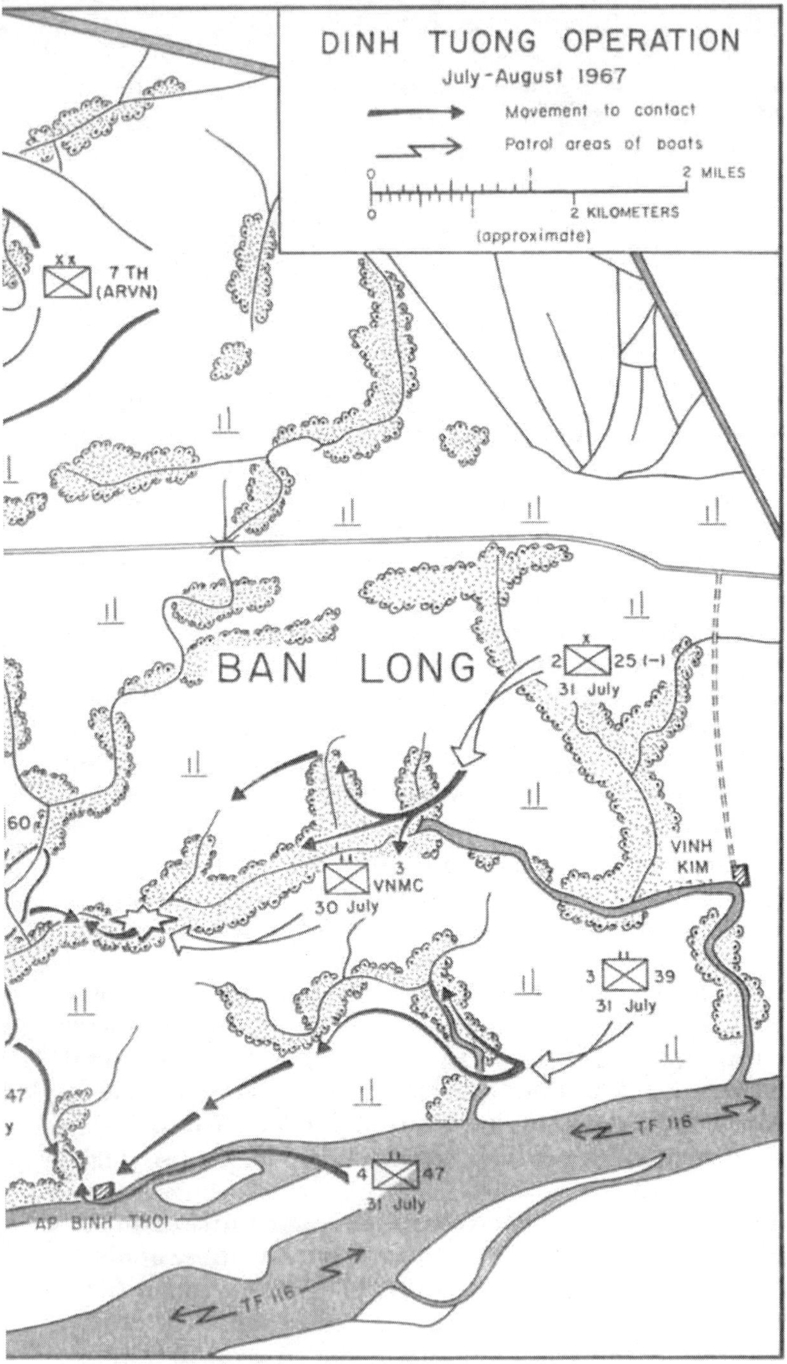

DINH TUONG OPERATION
July-August 1967
Movement to contact
Patrol areas of boats
0
1
2 MILES
0
1
2 KILOMETERS
(approximate)
7 TH
(ARVN)
BAN LONG
2
25 (-)
31 July
3
VNMC
30 July
VINH
KIM
3
39
31 July
4
47
31 July
AP BINH THOI
TF 116
TF 116

ated in Can Giuoc and Can Duoc Districts of Long An Province prior to battles with the Mobile Riverine Force.

At the end of Operation CORONADO I on 25 July the Mobile Riverine Force, operating in eastern Long An Province, the Rung Sat Special Zone, and Go Cong Province, had killed 478 of the enemy since 17 June. The force had been operating in areas seldom frequented by the South Vietnam Army and other U.S. units.

Dinh Tuong Operation

The Mobile Riverine Force ended operations in the Long An area because of reliable intelligence indicating an enemy buildup west of My Tho in Dinh Tuong Province. The force received word on 25 July that the Mobile Riverine Base would move on 27 July from the confluence of the Soi Rap and Vam Co Rivers to the vicinity of Dong Tam. During the afternoon of 25 July ground forces were picked up by the assault craft and returned to the Mobile Riverine Base. At 0200 on 27 July riverine assault craft began leaving the base for minesweeping and patrol stations along the route to Dong Tam. At 0550 the last ship of the force was proceeding south on the Soi Rap River. Because of the slow speed of the towed APL moving against the tide the journey took eleven and a half hours. This did not, however, delay the commencement of the operation in Dinh Tuong Province on 28 July. In just over forty-eight hours the Mobile Riverine Force was able to relocate a base supporting 3,900 men over a distance of sixty miles and to shift its area of operations a total of eighty-five miles to the area west of Dong Tam.

The Mobile Riverine Force was about to join the largest force with which it would co-operate in a single operation in its Vietnam experience. Intelligence indicated that a Viet Cong force of several battalions threatened My Tho and Dong Tam; the plan was to attack simultaneously three of the four Viet Cong base areas in Dinh Tuong Province from which an attack by the enemy might be staged. *(Map 9)*

On 27 July the 7th Army of Vietnam Division initiated operations designed to search for the Viet Cong from east to west, north of Highway 4. The search would cover the territory around the Ap Bac Viet Cong base and end in the eastern portion of enemy base area 470. On 28 July the Mobile Riverine Force would move into the Cam Son base area; on 29 July one or more battalions of the Vietnam Marine Corps would move into the Ban Long base area. The U.S. 9th Division was to place the 5th Battalion, 60th

Infantry (Mechanized), under the operational control of the 2d Brigade for the Cam Son operation, and was to hold the 3d Battalion, 39th Infantry, on call at that battalion's Long An base camp. The 1st Brigade, U.S. 25th Division, was to arrive at Dong Tam on 28 July for commitment under the operational control of the 9th U.S. Division. The 7th Army of Vietnam Division was given the 44th Ranger Battalion by the commander of IV Corps for insertion by helicopter into the northwestern portion of Cam Son on 28 July. The U.S. Navy Task Force 116 (GAME WARDEN) was to patrol the My Tho River from My Tho to Sa Dec with thirty river patrol boats. To facilitate command and control, the U.S. 9th Division moved a forward command post to Dong Tam for the operation.

The Mobile Riverine Force operation in Cam Son began during the night of 28 July with the movement of the 5th Battalion, 60th Infantry (Mechanized), from its Long An base camp along Highway 4 to the town of Cai Lay. A battalion command post was established there as the maneuver companies continued south into the Cam Son area. As the mechanized battalion moved into the area of operations from the northeast, the 3d and 4th Battalions of the 47th Infantry moved by assault craft into the waterways in the southern portion of Cam Son.

Troop D, 3d Squadron, 5th Cavalry, minus the aerial rifle platoon, was operating with the Mobile Riverine Force for the first time. The troop had a reconnaissance mission covering the eastern Cam Son and western Ban Long areas on the brigade's flank. One rifle company of the 3d Battalion, 60th Infantry, was on call to the 2d Brigade for airmobile employment from Dong Tam.

Two assault helicopter companies were available to the 2d Brigade from II Field Force. At 0800 on 28 July they conducted feint landings at two landing zones just north of known enemy fortifications in northern Cam Son for the purpose of delaying enemy movement north until both the 5th Battalion, 60th Infantry, and the 44th Ranger Battalion could move into blocking positions near the feint landing zones.

On 28 July troops of the 7th Army of Vietnam Division north of Highway 4 received a little enemy fire and the Mobile Riverine Force encountered a few snipers in Cam Son. Troop D detected a squad of the enemy in the Ban Long area and killed five men with gunship fire.

Throughout the late morning on 29 July both the 3d and 4th Battalions, 47th Infantry, discovered widely dispersed small enemy groups. From their movements it was deduced that these groups were trying to fall back to the north into a fortified area. As the 3d Battalion moved companies into this area from the north and south on the east bank of the Rach Ba Rai, the enemy resisted. Between the hours of 1700 and 1900, Company C of the 3d Battalion pressed north; small enemy groups moving northeast delayed the company's advance. Sometime during these two hours an ATC was hit by B-40 rockets and 57-mm. recoilless rifle fire; twenty-five men—Army and Navy—were wounded by fragments. The 5th Battalion, 60th Infantry, fired upon individuals and small groups of the enemy during the late afternoon and evening. None of the Mobile Riverine Force units were able to fix the enemy position and by 1930 contact with the enemy was lost. Most of the Viet Cong dead were found when the infantry moved into areas which had been attacked by artillery and helicopter gunships.

By the night of 29 July, the fact that the 7th Vietnam Army Division north of Highway 4 had found few Viet Cong and that the Viet Cong in central Cam Son had evaded major battle prompted General O'Connor, 9th Division, and Colonel Lan, commanding the Vietnam marines, to consider the probability that the enemy was in the Ban Long area. Although what the Mobile Riverine Force had learned about the enemy in central Cam Son was far from conclusive, the highly evasive tactics of the enemy encountered by the force was considered an indication that the Viet Cong units that the force had met might be protecting the movement of larger units into the Ban Long area. This was in keeping with the pattern of movement in southern Dinh Tuong Province, and while no enemy was detected in Ban Long by D Troop on 29 July, an enemy squad had been attacked there by gunships on 28 July.

At approximately 2000 on 29 July Colonel Lan selected a landing zone for his 3d Battalion of Vietnam marines to begin landing by helicopter in Ban Long on 30 July. The area selected by Colonel Lan as a likely enemy position was near the place where D Troop had discovered the enemy on 28 July.

When it landed on 30 July the 3d Vietnam Marine Corps Battalion met heavy resistance. The troops immediately attacked the enemy, who were in a wooded area north of the landing zone. More marines were airlifted into the landing zone, still under enemy fire. For five hours the marines attacked prepared Viet Cong de-

fensive positions. The enemy armament consisted of light and heavy machine guns and mortars. The marines relied on helicopter gunships and artillery. The 3d Marine Battalion fought the enemy throughout the afternoon, taking a few casualties and reporting more and more of the enemy killed and more equipment captured.

General O'Connor, noting the long east-west belt of trees that could provide concealment for the enemy, employed the 1st Brigade, 25th U.S. Division, east of the marines' location to block movement of the enemy to the east. In midafternoon General O'Connor directed Colonel Fulton, commanding officer of the 2d Brigade, to assist the 3d Marine Battalion in evacuating casualties and in preparing for an attack against the Viet Cong. Colonel Fulton directed Lieutenant Colonel Bruce E. Wallace, commander of the 3d Battalion, 47th Infantry, to establish a blocking position west of the battle area, facing east, a maneuver accomplished by 2000. The 4th Battalion, 47th Infantry, in the field for three days, was returned to the Mobile Riverine Base with instructions to be prepared to deploy early on 31 July. The 5th Battalion, 60th Infantry, was detached to 9th Division at 1230.

During the action on 29 July several assault boats were hit with small arms, rocket, and recoilless rifle fire. The monitor most seriously hit suffered no major structural damage. Hits and minor damage were received by five other craft, but the boats were able to absorb punishment and remain in operation.

At approximately 2030 on 30 July, Colonel Lan requested illumination for a night attack against the enemy positions. The subsequent attack by the 3d Vietnam Marine Corps Battalion silenced several 12.7-mm. machine guns and inflicted losses on the enemy. The attack was stopped by the 3d Marine Battalion commander because his own losses were heavy. Illumination was maintained during the night: at approximately 0430, 31 July, the enemy counterattacked to the east, with heavy losses to both sides.

Later in the morning of 31 July, the 5th Battalion, 60th Infantry, was again placed under the operational control of Colonel Fulton. At 0835, the battalion went to the assistance of the 1st Squadron, 6th Vietnamese Army Cavalry, and the 44th Ranger Battalion, which had met the enemy while moving to assist the 3d Marine Battalion. By 0825 the 5th Battalion, 60th Infantry, occupied a blocking position northeast of the 3d Battalion, 47th Infantry, where it remained during the day. During the morning of 31 July, the 3d Battalion, 39th Infantry, also was placed under the operational control of Colonel Fulton, bringing the Mobile Riv-

erine Force to four battalions and Troop D, 3d Squadron, 5th Cavalry. The 3d Battalion, 39th Infantry, landed south of Vinh Kim and searched west.

Reports from the 44th Ranger Battalion and 3d Marine Battalion indicated that the Viet Cong, tentatively identified as the 263d Main Force Battalion, had dispersed during the predawn counterattack and were making their way south. Airborne reconnaissance by the 3d Battalion, 39th Infantry, revealed movement south in the direction of the village of Ap Binh Thoi.

Colonel Fulton directed the 3d Battalion, 39th Infantry, to conduct reconnaissance using helicopters southeast of the battle area and working west to Ap Binh Thoi. He directed Colonel Wallace to move his battalion southeast and to search the area as he closed on Ap Binh Thoi. This movement was initiated in midmorning and by 1700 both battalions were on the outskirts of Ap Binh Thoi. During that time Viet Cong were observed moving into Ap Binh Thoi in groups of twenty-five to thirty.

The 4th Battalion, 47th Infantry, augmented by elements of a Vietnamese field police force company, landed by ATC and entered Ap Binh Thoi. One company moved out to search northwest of the town while the rest of the battalion assisted the Vietnamese police in rounding up and interrogating people suspected of being Viet Cong. Several of the prisoners reported that the Viet Cong units encountered by the Vietnamese marines were the 263d Main Force Battalion and elements of the 514th Local Force Battalion. The prisoners were classified as Viet Cong when they confessed to having hidden their arms as they left the battle area during the early morning darkness. Company C, 4th Battalion, 47th Infantry, captured four men northwest of Ap Binh Thoi, all of them members of the 263d Main Force Battalion and one of them the battalion deputy commander. Of the more than 400 suspects detained by the National Police, 83 were from the 263d Main Force Battalion. The cordon around Ap Binh Thoi was completed by dark on 31 July, and no more of the Viet Cong were sighted.

The four-day operation, primarily because of the outstanding performance of the 3d Vietnam Marine Battalion, caused severe losses to the Viet Cong 263d Main Force Battalion, destroyed major fortifications in the Cam Son base area, probably thwarted planned enemy operations against Dong Tam, and eased the pressure on Highway 4. It also demonstrated the ability of U.S. and Vietnamese forces to work together. After the operation intelligence sources reported that the enemy had attempted to organize boats for a crossing of the My Tho River into Kien Hoa Province but were

stopped by the cordon at Ap Binh Thoi and the river patrols. During the five-day operation on the My Tho River more than fifty patrol craft were employed: 283 Vietnamese watercraft were stopped and searched by naval craft from GAME WARDEN and the Mobile Riverine Force in the most ambitious attempt to control river traffic during the force's operations in 1967.

During operations from 28 July to 1 August, a large number of soldiers in the 3d Battalion, 47th Infantry, developed foot trouble as a result of spending five successive days on flooded land. The 4th Battalion, which had had one night of rest aboard ships of the Mobile Riverine Base, experienced a similar but lower rate of fungus infection and immersion foot.

The Mobile Riverine Force Returns to III Corps

CORONADO III was conducted in the Rung Sat Special Zone from 5 through 17 August after the Mobile Riverine Base had moved from Dong Tam to an anchorage at the junction of the Soi Rap and Vam Co Rivers. The purpose of the operation was to disrupt possible Viet Cong attacks on shipping in the Long Tau channel. Although no attacks were made on shipping during the operation, the Mobile Riverine Force fought sporadically with the Viet Cong and captured some munitions.

The Commander of Naval Forces, Vietnam, provided the following evaluation of the Mobile Riverine Force through CORONADO III:

> Perhaps the best evaluation of MRF achievements can be attained by examining the results of frequent riverine operations in Can Giouc District of eastern Long An Province. In its first operation in this area on 19 and 20 June, the MRF became engaged in the toughest fight it has experienced to date. Over 250 of the enemy were killed at a cost of 46 US KIA and 140 wounded. As the MRF returned to Can Giouc for further operations in late June, during July and again in late August, it never failed to make contact with the enemy. The size of the enemy units encountered has grown smaller and the percentage of prisoners taken versus enemy killed has risen steadily. Increasingly large caches of weapons have been uncovered. River Assault craft now move freely through areas where two months ago ambush with RPG–2's or recoilless rifles could be anticipated at any moment. In summary, it appears that a VC haven and stronghold, rarely ventured into by ARVN or FWMAF in the past, has been reduced to an area containing only scattered and poorly organized VC Guerillas.

Ben Luc Operation

CORONADO IV was conducted from 19 August through 9 September, with operations in Long An, Co Cong, and Kien Hoa

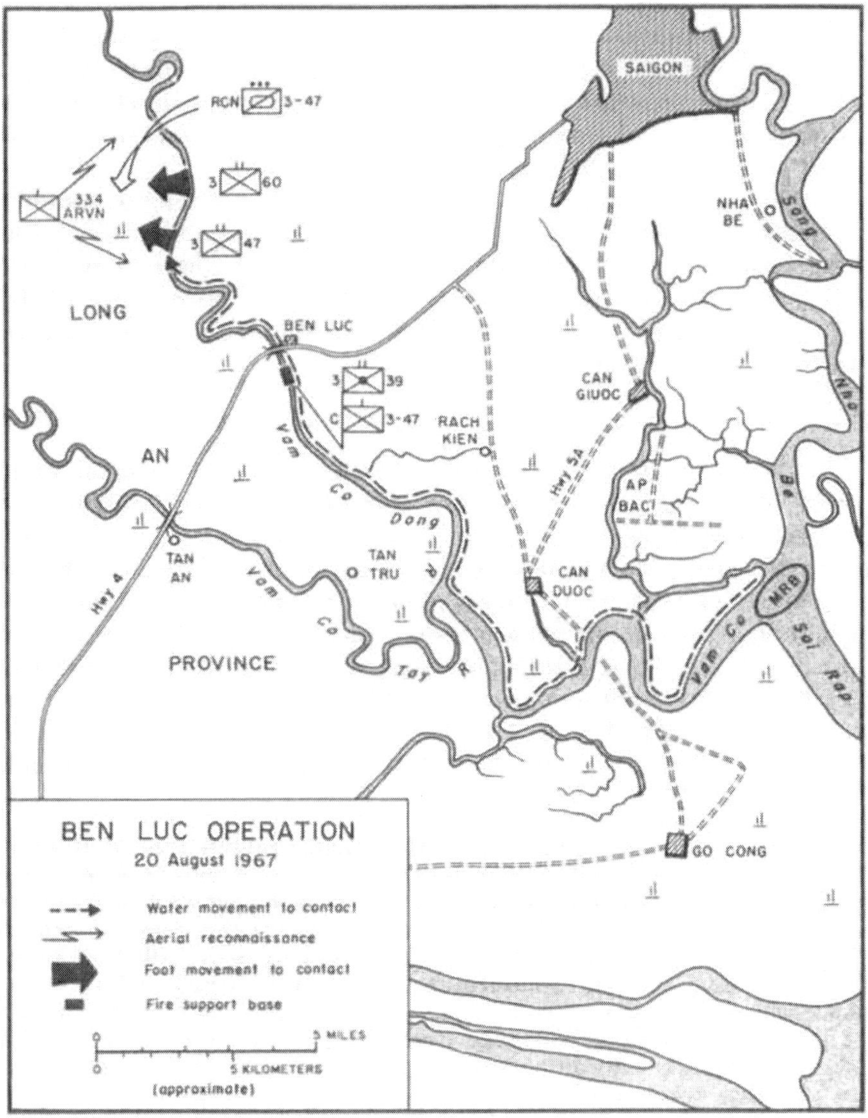

MAP 10

Provinces. On 20 August the assault craft left the Soi Rap and Vam Co River anchorage of the Mobile Riverine Base just after midnight and landed troops at 0904 in the target area north of Ben Luc on the Vam Co Dong River, a distance of over fifty kilometers. (*Map 10*) Because waterways off the Vam Co Dong in the target area were not navigable, the two infantry battalions were

landed to move west and search the area on foot. Company C of the 3d Battalion, 47th Infantry, remained at a pickup zone near the fire support base at Ben Luc as an airmobile reserve. To intercept enemy forces, the 334th Aviation Company's supporting gunships maintained surveillance of the areas forward of the infantry battalions.

At 1400 the movement of the 3d Battalion, 47th Infantry, forced an enemy platoon from a lightly fortified area. The gunships were placed under control of the battalion commander, and were directed against the enemy, who was attempting to evade contact with the ground troops. The gunships sighted the enemy troops and took them under fire, after which the reconnaissance platoon of the 3d Battalion, 47th Infantry, was brought in by helicopter to mop up. The platoon encountered some fire from snipers and co-ordinated fire of the supporting gunships on the enemy. A search of the area revealed that thirty-four of the enemy had been killed. From documents found in the area the enemy was identified as an element of the 506th Local Force Battalion.

The success of this operation was attributed to the surprise achieved by the ground troops landing from the river; the Viet Cong were forced into the open where gunships could take them under fire. During the operation, which ended on 22 August, the Mobile Riverine Force suffered six wounded and the enemy lost fifty-nine killed.

For the remainder of August and the first week of September, the Mobile Riverine Force operated in Can Giuoc District, encountering minor enemy resistance. On 7 and 8 September the force operated in conjunction with the 1st Brigade, 9th Division, in the southeastern part of Nhon Trach District of Bien Hoa Province and the northeastern part of the Rung Sat Special Zone. Few Viet Cong were encountered but a cache containing medical supplies, eight crew-served weapons, and ninety-seven small arms was discovered. This operation was one of the first in which Colonel Bert A. David commanded the 2d Brigade. Colonel David assumed command as Colonel Fulton, promoted to brigadier general, became assistant division commander of the 9th Infantry Division. General Fulton was responsible to General O'Connor, now commander of the 9th Division, for operations in the Mekong Delta. At this point in early September the Mobile Riverine Force had operated extensively in the III Corps provinces north of the Mekong River. As the second week of September began, the force prepared to return to Dinh Tuong, an area of great interest to the senior adviser of the IV Corps Tactical Zone.

CHAPTER VII

Cam Son to the Rach Ruong Canal

On 11 September 1967 the Mobile Riverine Force returned to Dong Tam to prepare for an operation that would start in the Ban Long area on 12 September, the first operation of CORONADO V. Brigadier General Nguyen Manh Thanh, commander of the Army of Vietnam 7th Division, furnished information that the Viet Cong 263d Main Force Battalion had been in the Cam Son and Ban Long area during the preceding ten to fourteen days. Although the 7th Division intelligence was known to depend largely on reports of agents and informers, the Mobile Riverine Force had found it highly reliable in identifying enemy base areas. In Dinh Tuong Province particularly the force repeatedly had found the Viet Cong in regions reported to be base areas.

Dinh Tuong Province and Coronado V

On 12 September the Mobile Riverine Force entered the Ban Long area with three battalions. The 3d and 5th Battalions of the 60th Infantry—the 5th again released from operational control of the 3d Brigade to operate under the 2d Brigade—relied on helicopter and overland movement to get into the major east-west forested portion of the Ban Long area. The assault craft of the force were unable to navigate the waterways of the area.

As the 3d Battalion, 60th Infantry, moved into the forest it found an enemy force in well-prepared positions. Assisted by artillery and close air support, the battalion advanced to the east. The Viet Cong, under the pressure of the infantry advance and supporting fire, attempted to evade to the north and northwest, exposing themselves along thinly vegetated rice paddy dikes. The 5th Battalion, 60th Infantry, northwest of the 3d Battalion, engaged a platoon of the enemy; fire from M113 armored personnel carriers and the battalion's mortar platoon killed or dispersed the enemy.

At approximately 1430 a provisional battalion of Dinh Tuong Province Regional Forces was sent in by helicopter northeast of

the 3d Battalion, 60th Infantry. The provisional battalion engaged widely dispersed enemy elements that had been forced from hiding by the 3d and 5th Battalions, 60th Infantry, attacks. Mobile Riverine Forces losses were nine soldiers killed and twenty-three wounded —all from the 3d Battalion, 60th Infantry. The three battalions accounted for 134 Viet Cong killed and 39 captured. Although the main forces of the enemy had escaped on 12 September, the Mobile Riverine Force continued to search the Ban Long area until it returned to the Mobile Riverine Base on 14 September. *(Map 11)*

Colonel David, after consulting with General Fulton, concluded that if an operation was launched in the Cam Son area on 15 September the enemy force, tentatively identified as the 514th Local Force Battalion, might be found. The plan was to strike the area in central Cam Son where the enemy's heaviest fortifications had been identified on previous operations. To reach this area before the major enemy force could escape, Colonel David decided to withhold preparatory and reconnaissance fire until the assault craft of the Mobile Riverine Force passed a wide curve in the Rach Ba Rai referred to as "Snoopy's Nose." Helicopter flights over the area were to be limited until the assault craft cleared Snoopy's Nose. The movement of the 5th Battalion, 60th Infantry, by ground vehicles into the area from Cai Lay was to be delayed until the 3d Battalion, 60th Infantry, and 3d Battalion, 47th Infantry, entered the Rach Ba Rai aboard ATC's. Finally, to provide a higher degree of flexibility if the enemy was found, the 2d Battalion, 60th Infantry, 3d Brigade, was designated a reserve force by the 9th Division. If required, the battalion would be employed by helicopter, staging from Dong Tam after movement from the battalion's Long An Province base.

Key elements of the maneuver were the landing of the 3d Battalion, 60th Infantry, north of an eastward bend in the Rach Ba Rai and the movement of the 5th Battalion, 60th Infantry, from the northeast. Both battalions would attack into a series of tree lines which the brigade planners believed had been used by enemy troops in the past to escape fighting.

As the 3d Battalion, 60th Infantry, moved up the Rach Ba Rai at approximately 0715 on 15 September, fire was withheld, and the assault craft moved steadily around Snoopy's Nose. By 0730 lead boats were nearing Beach White Two where a company of Lieutenant Colonel Mercer M. Doty's 3d Battalion, 60th Infantry, was to land when the boat formation came under heavy rocket, automatic, and small arms fire from both sides of the stream.

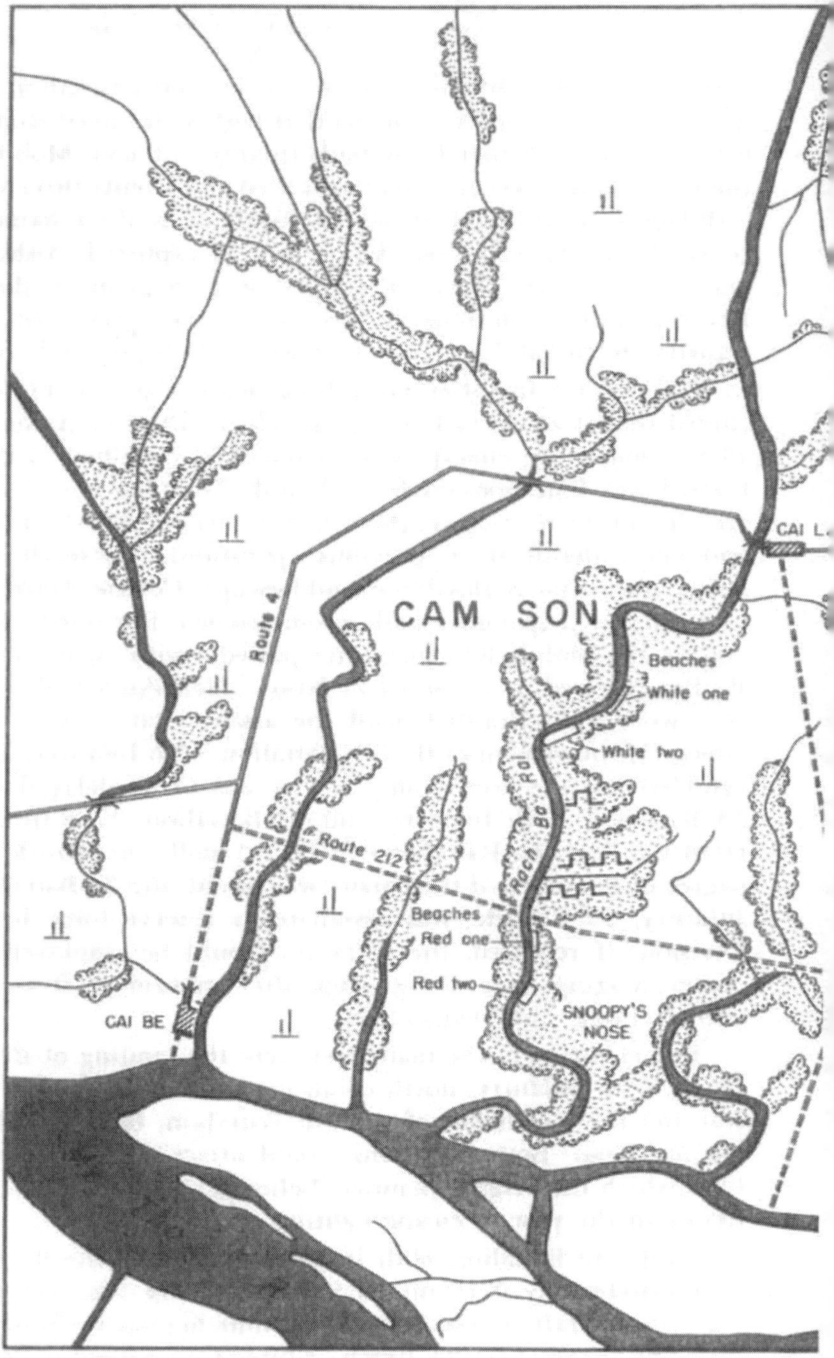
CAM SON
Route 4
Route 212
Beaches
White one
White two
Rach Ba Rai
Beaches
Red one
Red two
SNOOPY'S NOSE
CAI BE
CAI L

MAP 11

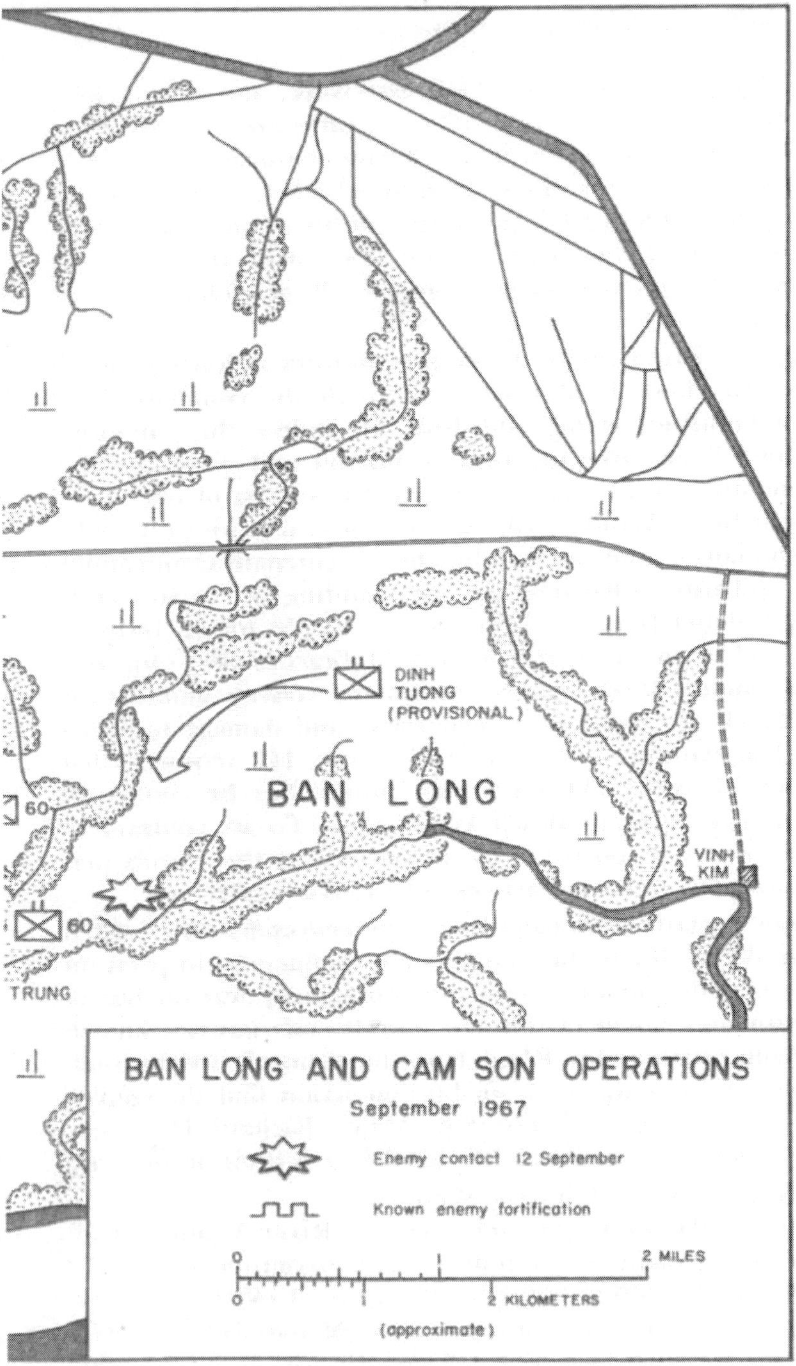
DINH
TUONG
(PROVISIONAL)
BAN LONG
60
60
TRUNG
VINH
KIM
BAN LONG AND CAM SON OPERATIONS
September 1967
Enemy contact 12 September
Known enemy fortification
0
1
2 MILES
0
1
2 KILOMETERS
(approximate)

Enemy fire was heavier from the east bank, and the organic firepower of the riverine assault craft was unleashed primarily to the right flank of the force. In the smoke and confusion the assault craft maneuvered to fire weapons or to avoid other craft temporarily out of control. One ATC proceeded north of the lead minesweepers and landed on Beach White Two with the company commander and one platoon of Company B of Colonel Doty's battalion.

During the critical fifteen to twenty minutes following initial enemy fire, the flow of information through the command and control communications net did little to reflect the situation. Colonel Doty, flying over the boat formation and observing the apparent mobility of all assault craft and the success of one ATC in arriving at Beach White Two, was convinced that his unit could continue and land at the assigned beaches. Lieutenant Commander Francis E. ("Dusty") Rhodes, Jr., commanding the assault craft supporting Colonel Doty, issued an order at 0758 for all boats to turn back and assemble in the vicinity of Beaches Red One and Two. Commander Rhodes' decision that the convoy should turn back was based on casualties to boat crews and damage to minesweepers. The standing orders of Task Force 117 required that minesweepers precede ATC's carrying troops, but he could not continue minesweeping to Beach White One. To act contrary to this procedure would constitute an action outside the "limits permitted by accepted tactical practices" of the Navy task force.

The boat captain who passed the minesweepers and landed his ATC at White Beach One was probably influenced to press on by the fact that the infantry company commander was on board. The successful movement of this one assault craft was not known to Lieutenant Commander Rhodes at the time of his decision. Colonel Doty did not waiver from his conviction that the convoy could and should continue. His S-3, Major Richard H. Sharp, aboard the command boat relayed Colonel Doty's decision to "send in the troops" to Commander Rhodes.

While the 3d Battalion, 60th Infantry, and River Assault Squadron 11 evacuated casualties, resupplied, and reorganized at the Red Beaches, the 5th Battalion, 60th Infantry, moved overland toward Beach White One from the northeast. The 3d Battalion, 47th Infantry, was commanded by Colonel Tutwiler, who had previously commanded the 4th Battalion, 47th Infantry. He held his battalion south of the congested area of Beach Red, prepared to resume movement on order.

At approximately 1000 Colonel Doty's battalion began to move upstream, supported by artillery gunships and helicopters. The volume of fire was as great as before but the convoy landed at Beaches White One and Two. Companies B and C had few men wounded in this second run, while Company A had eighteen men wounded in one platoon. Both the assault craft and the infantry, who had joined freely in returning the enemy fire, required resupply of ammunition. Once ashore, the 3d Battalion, 60th Infantry, attacked south against stubborn enemy resistance. The 3d Battalion, 47th Infantry, landed at Beaches Red One and Two and pushed north. The 5th Battalion, 60th Infantry, moved close enough to see the 3d Battalion, 60th Infantry, by early afternoon.

To encircle the enemy south of Colonel Doty's battalion, the 2d Battalion, 60th Infantry, was placed under the operational control of the 2d Brigade and landed by helicopter south of the 5th Battalion, 60th Infantry. By dark the 3d Battalion, 60th Infantry, unable to overcome the enemy resistance, was ordered back to improve its defensive position. One Vietnam Army battalion was landed by helicopter at approximately 1600, northwest of the 3d Battalion, 60th Infantry, and moved to set up a position along the west bank of the Rach Ba Rai. The four U.S. battalions were in an irregular arc on the east side of the Rach Ba Rai. The stream was a possible enemy escape route west, although the 3d Battalion, 47th Infantry, had seized a number of sampans just north of route 212 during the late afternoon. Assault craft supporting the 3d Battalion, 47th Infantry, and 3d Battalion, 60th Infantry, were positioned to observe and place fire along the stream; however, no U.S. boats were deployed into the stream unless there were American or Vietnamese troops ashore.

During darkness air and artillery illumination was maintained over the area and artillery fire was placed within the partially encircled area on likely enemy locations. Between 0200 and 0430, small groups of enemy soldiers were observed and fired upon forward of the 2d Battalion, 60th Infantry, and later the 3d Battalion, 47th Infantry. After 0430 on 16 September no Viet Cong were sighted. On 16 September the 5th Battalion, 60th Infantry, led the sweep into the area, followed by sweeps by the two southern battalions forward of their positions. Resistance was light; most of the enemy force encountered on 15 September had been killed or had slipped away during the night.

The Navy assault craft expended, during the period 0730-1600 on 15 September, 10,273 rounds of 40-mm. ammunition, sixteen

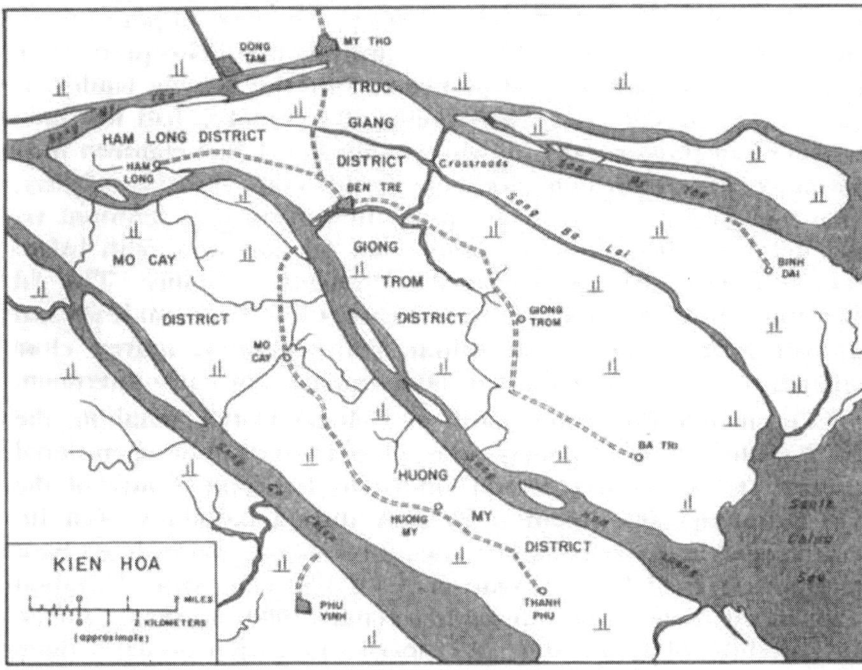

MAP 12

rounds of 81-mm., 7,445 rounds of 20-mm., 20,934 rounds of .50-caliber, and 40,216 rounds of .30-caliber. The operation ended on 16 September, after four days of heavy fighting in which the Americans and Vietnamese lost 16 killed and 146 wounded and the Viet Cong lost 213.

Following the Cam Son operation, the Mobile Riverine Force moved into Kien Hoa Province. (*Map 12*) Although operations during the remainder of September were widely separated in Ham Long, Giong Trom, and Huong My Districts, interrogation of local civilians revealed that they had prior knowledge of the operations. In Giong Trom District local inhabitants said that a Viet Cong unit, believed to be part of the 516th Local Force Battalion, had been in the area, but had left the night before the Riverine Force arrived. This experience became characteristic of Mobile Riverine Force operations conducted in Kien Hoa Province in late 1967. Operations usually involved brushes with enemy rocket launcher and recoilless rifle teams which delayed operations but inflicted few losses. Helicopters became invaluable during movement to detect and engage small enemy teams armed with antitank weapons. These operations in Kien Hoa Province saw the first use of the long-awaited assault support patrol boats.

During the period 5-7 October, the Mobile Riverine Force terminated major operations of CORONADO V with an operation in the Ban Long area. It was conducted in close co-operation with the 7th Army of Vietnam Division and resulted in a battle with the Viet Cong 263d Main Force Battalion. The Mobile Riverine Force had 1 killed and 26 wounded, while the Vietnam Army 7th Division suffered 6 killed and 36 wounded. The Viet Cong losses totaled 163 killed.

The Mobile Riverine Force moved to Vung Tau on 10 October and CORONADO VI commenced on 11 October in the Rung Sat Special Zone. Minor engagements took place and small lots of weapons and supplies were discovered. On the night of 16 October the Mobile Riverine Force sent out forty-eight ambush patrols and twenty-two boat patrols in order to insure the security of the Long Tau shipping channel. No ships were attacked during the Mobile Riverine Force operations.

On 20 October the Mobile Riverine Force moved to the juncture of the Soi Rap and Vam Co Rivers. From this base CORONADO VII was initiated, with operations in the northern part of Can Giuoc District to secure the waterways while elections were being held in the Republic of Vietnam. Both infantry and assault craft were widely dispersed to get as many American soldiers into the area as possible. The usual search of water craft was discontinued. After the elections the commander of the 46th Vietnam Army Regiment told Colonel David that the turnout for the election had been 83 percent of the registered voters.

CORONADO VIII was conducted in co-ordination with the Royal Thai Regiment and elements of the 3d Battalion, 5th Cavalry, on 27, 28, and 29 October in the southern Nhon Trach and the northern Rung Sat Special Zone. There were few encounters with the enemy.

Coronado IX

CORONADO IX began with the movement of the Mobile Riverine Base from its late October anchorage off Vung Tau to an anchorage in the My Tho River near Dong Tam. The series of operations that constituted CORONADO IX was conducted primarily north of the My Tho River and directed against the enemy bases of Dinh Tuong Province.

The Mobile Riverine Force arrived off Dong Tam 1 November 1967 and during the first days of the month concentrated on equipment maintenance and preparations for coming operations. The 3d Battalion, 60th Infantry, debarked and assumed the defense

mission for Dong Tam Base. The battalion's waterborne replacement was the 3d Battalion, 47th Infantry, which began tactical operations in the southeastern part of the Giao Duc District in western Dinh Tuong Province at 1500 on 2 November. During the nine-hour operation, the battalion destroyed 141 enemy bunkers and evacuated 1,100 pounds of rice and an assortment of enemy engineering and medical materials from a cache. The evacuated materials were delivered to the Dinh Tuong Province Headquarters for distribution within government programs.

On 5 and 6 November, a two-battalion operation was conducted in the Cam Son Secret Zone. Barge-mounted artillery gave support from a position on the north shore of the My Tho River. In this operation the Mobile Riverine Force destroyed 34 bunkers, captured 800 pounds of rice and 125 pounds of salt, and killed five of the enemy.

With the arrival of the 5th Battalion of the Vietnamese Marine Corps at a shore camp near My Tho on 6 November, the Mobile Riverine Force had a new and valuable asset. The 5th became the third maneuver battalion. Possessing four rifle companies and a heavy weapons company, it had a strength equal to that of the two embarked Army battalions. The Vietnamese marines brought a special esprit to the Mobile Riverine Force and fought extremely well throughout the CORONADO IX operation.

The battalion commander, Major Nam, participated fully with the advisers in the planning of Mobile Riverine Force operations. Major Nam, Colonel David, Captain Wells, and Captain Salzer, who succeeded Captain Wells on 2 December, agreed on plans and each commander approved and signed the original operation order prior to each operation. The Navy assault craft, Army artillery, and Air Force aircraft supported the Marine battalion as they did the Army battalions.

After a short time spent on training and maintenance, the Mobile Riverine Force resumed CORONADO IX on 9 November. Again the targets were the 263d and 514th Viet Cong Battalions in the Cam Son Base area. While the 3d Battalion, 47th Infantry, and the 5th Marine Battalion made beach assaults early on 9 November, the 4th Battalion, 47th Infantry, moved by ATC to the Dong Tam airstrip, and then in a series of helicopter lifts entered the area of operations. All landings were unopposed and few of the enemy were sighted.

On the second day the search for the enemy shifted to the east; troops were moved by boat and by helicopter into the Ban Long

PORTABLE FIRING PLATFORM WITH 105-MM. HOWITZER

and the Ap Bac base areas. Only the Marine battalion, which remained in its original area of operation, saw action, killing seven of the Viet Cong.

On 14 November, to increase security for Highway 4 and the Dong Tam Base area, a new operation began, with support from the artillery at Dong Tam. While the 3d Battalion, 47th Infantry, remained at the Mobile Riverine Base for maintenance, the other battalions were transported by boat and helicopter to the north along the Kinh Xang Canal and east of Dong Tam Base. The 5th Marine Battalion and the 4th Battalion, 47th Infantry, destroyed 63 bunkers and seized 2,000 pounds of rice and 200 pounds of salt. This operation continued through the night. Ambush patrols of platoon size were sent out along Highway 4 north of Dong Tam but saw few of the enemy.

On 16 November the Mobile Riverine Force began an operation with the 3d Brigade, 9th U.S. Division, and with elements of both the 7th and 9th Army of Vietnam Divisions along the border of Kien Phong and Dinh Tuong Provinces. The target was the

502d Local Force Battalion and 267th Main Force Battalion, which IV Corps and II Field Force intelligence estimated were in Base Area 470. Troops of the 3d Brigade were landed by helicopter in the northern part of the area along with the 3d Battalion, 47th Infantry, which was initially under the operational control of 3d Brigade. Since Base Area 470 lacked firm ground for artillery fire support bases, and artillery positioned along Highway 4 could not cover the area, experimental artillery firing platforms were used.

One platform accommodated an M102, 105-mm. artillery piece, ammunition, and enough space for the crews to operate. The legs of the platform were adjustable to various heights and a large metal "foot," mounted on the bottom of each leg, provided support in the mud of rice paddies. A CH-47 helicopter carried the platform, artillery piece, ammunition, and crew in four lifts. One battery of 105-mm. artillery was used throughout the operation. The second night of the 3d Brigade's operations, on 17 November, the Viet Cong attacked one of the brigade's fire support bases. The attack was repulsed several hours before the Mobile Riverine Force entered the operational area off the My Tho River on the Rach Ruong Canal. In order for the Riverine Force to enter the operational area, IV Corps Vietnamese engineers, who were accompanying the assault forces early on the morning of 18 November, had to remove the center span of the bridge on the Rach Ruong Canal.

Before the bridge span was removed, a barge-mounted artillery fire support base was set up so that the artillery could fire on any enemy troops encountered by the engineers. A 155-mm. fire support base was also established on Highway 4, six kilometers east of the bridge. For better control, the Mobile Riverine Base was moved upriver from Dong Tam to an anchorage near Sa Dec. While the 5th Marine Battalion conducted riverine assaults in the southern portions of the target area, the 4th Battalion, 47th Infantry, landed troops by helicopter in the north.

Early on 18 November both battalions discovered the enemy. In the Mobile Riverine Force fighting alone, forty-five of the enemy were killed, nearly half of them by helicopter gunships, and a large medical cache was uncovered. The Mobile Riverine Force had four wounded. Total losses for all American and Vietnamese units were 26 killed and 155 wounded, while the Viet Cong suffered 178 killed and 33 taken prisoner.

Following a period at the Dong Tam anchorage for rest and maintenance, the force returned to the Cam Son Base area on 23 November. After the first air attack—a B-52 strike in the heavily

populated Dinh Tuong Province—three battalions entered the area by riverine assault craft. Only a few of the enemy were found and eight of these were killed. Several caches were discovered.

From 27 to 30 November operations were conducted to clear the Kinh Xang Canal, which runs northwest by Dong Tam. Company D, 15th Engineer Battalion, removed all water blocks along the canal while the maneuver battalions searched the Ap Bac Base area. Five of the enemy were killed and two taken prisoner. Four major blocks were removed from the Kinh Xang Canal and sixty-two bunkers were destroyed. The opening of the upper reaches of this canal permitted the Mobile Riverine Force to use the assault craft of the Navy to attack the enemy's Ap Bac Base.

On 4 December the Mobile Riverine Base moved to Sa Dec and the 3d and 4th Battalions, 47th Infantry, with the 5th Marine Battalion, began operations to find and destroy elements of the 267th Main Force and 502d Local Force Battalions in western Dinh Tuong and eastern Kien Phong Provinces. The battle that ensued on 4-5 December proved to be one of the most severe the Mobile Riverine Force had yet experienced.

Early on 4 December the force encountered major elements of the 502d Local Force Battalion in a fortified base on the west bank of the Rach Ruong Canal. *(Map 13)* The enemy attacked the boats with rockets and automatic weapons and a decision was made to land the Marine battalion to the north of the enemy position. Shortly afterward the 3d Battalion, 47th Infantry, was landed south of the enemy position. Fighting was intense and Colonel David directed the 4th Battalion, 47th Infantry, which was standing by at a pickup zone, to land west of the enemy location. In midafternoon of 4 December, the Vietnamese Marine Battalion made a frontal assault that overran the enemy's major bunker complex. Supporting fire from armed helicopters and assault craft contributed largely to the success of the assault. To the south the 3d Battalion, 47th Infantry, encountered stubborn resistance from scattered enemy bunkers that prevented it from linking with the Vietnamese marines.

This action demonstrated the importance of a quick decision by a waterborne force commander when the force was under fire. Major Nam decided to land immediately on what he judged to be the enemy flank. The commander of the supporting river division immediately gave the order to his boat captains, and the landings were made. Total enemy casualties for 4 and 5 December were 266 Viet Cong killed, with the Vietnamese marines accounting for the

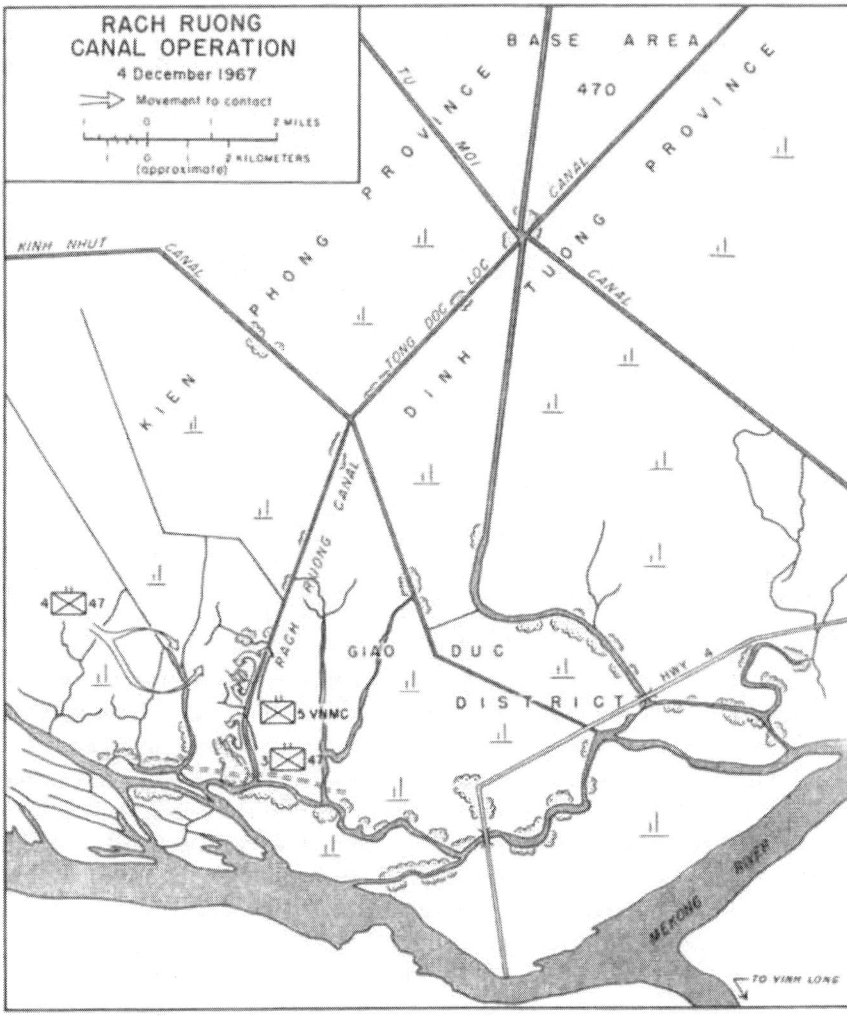

MAP 13

heaviest enemy losses. The Vietnamese marines had 40 killed and 107 wounded. Nine American soldiers were killed and 89 wounded.

The assault craft were able to land the marines with light casualties on 4 October chiefly because of a flame thrower aboard one of the armored troop carriers. A vehicle with a flame thrower had been driven aboard, and the ATC placed in the convoy to fire on targets as required. The flame thrower had not only a physical but also a psychological effect on the enemy manning bunkers south of the Marine landing site. The first successful test of a flame thrower aboard an ATC had been made on 4 October 1967.

On 6 December the Mobile Riverine Base moved to an anchorage on the Mekong River north of Vinh Long. After three days of local operations and maintenance work during which new fire support bases were established in support of forthcoming operations, the Mobile Riverine Force undertook operations in the southern part of Cai Be District and in the western part of Dinh Tuong Province against local guerrillas and their installations. Initial landings were made from ATC's but later in the first day of operations troops were landed by helicopter. Very few of the enemy were found.

On 14 December the Mobile Riverine Force again searched for the Viet Cong in Cam Son. The force made assaults by boat in the southern portion of the base, followed by helicopter landings by the 3d Battalion, 47th Infantry. It was not until the last day of the operation that the enemy was discovered and in a firefight nine of the Viet Cong were killed.

On 17 December the Mobile Riverine Base moved to Dong Tam to permit the Mobile Riverine Force to land troops from the two barracks ships. By the afternoon of 18 December the brigade and one battalion had moved ashore to Dong Tam. The USS *Colleton* left the area of operations to refit at Subic Bay in the Philippine Islands and the USS *Benewah* moved to Vung Tau in Vietnam to refit. At this time, the 3d Battalion, 60th Infantry, replaced the 4th Battalion, 47th Infantry, as part of the Mobile Riverine Force, and the 4th Battalion assumed the Dong Tam defense mission.

On 19 December a two-battalion operation was initiated in northern Cai Lay District while a third battalion operated near Dong Tam. On the night of 19 December, two companies of the 3d Battalion, 47th Infantry, were landed along the Kinh Xang Canal to set up ambushes. The other two battalions of the Mobile Riverine Force were landed by assault craft in the northern Ap Bac area. Effective artillery coverage was provided by the barge-mounted artillery and by Battery C, 2d Battalion, 35th Artillery (155-mm. self-propelled). This operation covered most of the north central part of Dinh Tuong Province and used both helicopters and boats to land troops. Although few of the enemy were found, the maneuver battalions killed eleven Viet Cong. Navigability of the Kinh Xang Canal was extended when engineers removed a major canal obstruction. The operation was concluded late on 22 December and the Mobile Riverine Force returned to the afloat base to conduct maintenance and prepare for Christmas truce operations.

The Mobile Riverine Force was instructed that operations should be defensive during the truce period; troops could fire on groups of enemy soldiers who "seemed to be trying to breed contact," or who were more than "platoonsize" in number. The truce ran from 1800 on 24 December through 1800 on 25 December. Prior to 1800 on the 24th, one Mobile Riverine Force battalion moved to each of the two areas in which saturation patrols were to be conducted. The 3d Battalion, 47th Infantry, moved to Long Dinh District, while to the southeast the 3d Battalion, 60th Infantry, moved into northern Kien Hoa Province. During the truce there was occasional enemy sniper and harassing fire.

The force's next operation was in Cai Lay District of Dinh Tuong Province. On 28 December the 3d Battalion, 60th Infantry, was landed by aircraft and waterborne craft in an area approximately eleven kilometers west of Dong Tam. It conducted operations directed westward toward Cai Be, with supporting fire from barge artillery batteries located to its south along the north shore of the My Tho River. Few of the enemy were seen during the operation; monitors and assault support patrol boats provided surveillance of inland waterways. On 29 December the 3d Battalion, 60th Infantry, continued operations to the west toward Cai Be, while the 4th Battalion, 47th Infantry, remained in the same area and destroyed eighty-five bunkers and captured two prisoners. During the day the 3d Battalion, 60th Infantry, found only a few of the enemy and concluded its operations in Cai Lay District.

On 30 December the 4th Battalion, 47th Infantry, was moved from Kien Hoa north and across the My Tho River to assist the 3d Battalion, 60th Infantry, in continuing operations in Cai Be and Cai Lay Districts. Although troops were shifted by boat from one place to another, the battalions could find no large groups of Viet Cong. On the following day both battalions went into positions where they stayed until the end of the New Year's truce at 0600 on 1 January 1968. In groups of platoon and company size, the battalions were dispersed to prevent the enemy from using important lines of communication during the truce period. Following the truce period the units arrived in Dong Tam on the afternoon of 2 January.

The next major mission of the Mobile Riverine Force was an operation in the eastern part of Vinh Long Province. On 7 January the battalions were landed by air and water but found no enemy until midafternoon, when a company of the 3d Battalion, 60th Infantry, came under heavy automatic weapons and small

arms fire from an estimated two platoons. Artillery fire and air strikes were called in on the enemy, who were firing from well-prepared, concealed positions. By the end of the fight, late on 7 January, two more companies of the 3d Battalion, 60th Infantry, were engaged. Twenty-eight of the enemy were killed and three weapons were captured. The Mobile Riverine Force concluded the operation the following day, killing two more of the enemy, and reached the riverine base late that afternoon.

The next target was the 261st Main Force Battalion in the Cai Be District, western Dinh Tuong Province. On 10 January assault craft and helicopters landed the 4th Battalion, 47th Infantry, and 3d Battalion, 60th Infantry, and by midafternoon Company A, 3d Battalion, 60th Infantry, was in heavy combat and had sustained moderate casualties. The battalion had been set down on a "hot" landing zone. Company E was flown in to support Company A. For both companies the fighting was heavy until after dark on 10 January. The following day fighting was sporadic. Over the two days, forty-seven of the enemy were killed. American forces suffered eighteen killed and fifty wounded; most of the casualties occurred at the landing zone where the enemy was in well-prepared, excellently camouflaged positions with good fields of fire.

On 12 January an operation was conducted in conjunction with the 3d Brigade in Binh Phouc District of Long An Province. The 3d Battalion, 60th Infantry, and 4th Battalion, 47th Infantry, moved by boat along the Cho Cao Canal to assigned beaches. In landing, troops met heavy enemy rocket and automatic weapons fire but killed seven of the enemy. The rest quickly escaped and there was no further fighting during the two-day operation.

On 14 January the 2d Brigade embarked on the USS *Benewah*, and was followed aboard on 21–22 January by the rest of the Mobile Riverine Force. A cordon was thrown around the village of An Quoi, near Dong Tam, by the 3d Battalion, 47th Infantry, which had the day before relieved the 4th Battalion, 47th Infantry, as a Mobile Riverine Force battalion. The object of the cordon operation was to capture or kill guerrillas, but it was unsuccessful; the operation brought CORONADO IX to a close.

Six Months in Retrospect

As the Mobile Riverine Force prepared for the *Tet* truce, it could look back over six months of unusual experiences. The force had entered the Mekong Delta in June as the first U.S. force free to conduct sustained operations against Viet Cong main force

units whose ability was respected. Republic of Vietnam armed forces were widely dispersed, with battalions of infantry and batteries of artillery fragmented to spread a degree of security throughout populated areas along the major land and waterway lines of communication. South Vietnamese offensive operations were limited in number and duration during the hours of darkness by the need to maintain security posts at political, military, and commercial centers and on highway bridges.

The 2d Brigade did not escape the duty of maintaining security at fixed installations; one of its battalions was required at Dong Tam. The Mobile Riverine Force as a whole—a force of nearly 4,000 men—was not tied to a static security mission, but was available for wide-ranging movement.

Whether operating in southern III Corps Tactical Zone or northern IV Corps Tactical Zone, the Mobile Riverine Force added to the capabilities of the corps commanders. It provided more troops with which to expand Vietnam armed forces and Free World Military Assistance Force influence. Also it possessed significant water mobility to complement air and ground movement. The contribution of the Mobile Riverine Force on the waterways is difficult to separate from its contribution as an additional combat force even if it had not relied extensively on water mobility. The Mekong Delta in June 1967 was a place where the operations of an airmobile brigade could also have had significant impact. The Mobile Riverine Force was, however, able to attack areas such as Cam Son, the Rung Sat Special Zone, eastern Long An Province, Go Cong, and western Dinh Tuong Province where the enemy relied heavily on waterways in bases seldom violated by operations of the South Vietnamese armed forces.

These areas were lucrative targets for a force free to operate within them. They were generally remote in terms of being outside the range of established artillery positions, accessible only with great difficulty by tracked vehicles, and requiring substantial use of air or water craft for troop movement. The mobility of the barge-mounted artillery and the Mobile Riverine Base enabled the Mobile Riverine Force to give effective artillery coverage to troops and to bring other substantial combat and combat service support to operations immediately adjacent to the base area. Further, the Navy assault craft, more heavily armed and armored than the craft of the Vietnamese river assault groups and enjoying far more artillery support, were able to enter base areas into which the craft of the river assault group dared not venture. Finally, the

Mobile Riverine Force was an unknown and, therefore, intimidating force to the Viet Cong. Intelligence estimates of June 1967 were that it would take the Viet Cong six months to obtain adequate knowledge of Mobile Riverine Force equipment, tactics, and techniques and to react with a program of mining and recoilless rifle and rocket attacks. Wide-ranging operations across the boundaries of responsibility of the major enemy organizations would gain more time before highly effective countermeasures were fielded by the enemy.

Operations were planned, therefore, so that the Mobile Riverine Force would move freely among the provinces north of the My Tho River; subsequently move south of the My Tho; eventually move into the southern Mekong Delta. In each area the force was to attack enemy main forces and local forces. By inflicting major losses on these units, the Mobile Riverine Force would take away the security provided to local guerrillas and the secret political organization by the presence of the Viet Cong main forces.

It was in the pursuance of this major mission that operations were conducted from June through December 1967. Within the widely separated base areas, the Mobile Riverine Force inflicted losses on the 5th Nha Be Battalion and 506th Local Force Battalion of the Long An and Gia Dinh areas, on the 514th Local Force Battalion, 263d Main Force Battalion, and the Viet Cong district companies of Dinh Tuong Province, on the 502d Local Force Battalion of Kien Phong Province, and on training cadres and companies of Long An and Go Cong Provinces. These units were most frequently met in areas where water mobility permitted U.S. forces to reach fortifications and caches at the heart of the enemy base. Most enemy bunkers were chosen for defense against likely helicopter landings, and covered trails entering the base from the limited land routes; few were designed to cover an approach by water.

The Mobile Riverine Force routinely requested and planned for the use of aircraft in the conduct of operations during the last half of 1967. Often an assault helicopter company was shared during the day with the 3d Brigade, 9th Infantry Division. This arrangement, while not ideal, provided aircraft which the commanders considered an inherent part of riverine operations.

During the six-month period, Mobile Riverine Force operations departed from the original planning in respect to duration of operations and the authority for and frequency of major relocations of the Mobile Riverine Base. Plans had called for a four- to five-day

operation, to be followed by a two- or three-day period for rest, skin care, and equipment maintenance. Beginning with Operation RIVER RAIDER I in the Rung Sat Special Zone, however, it was learned that foot problems caused by wetness and fungi increased rapidly after a few hours of exposure, and operations generally were limited to forty-eight hours, with a rest period of twenty-four to thirty-six hours. When it was necessary to keep troops on land for longer periods, a high percentage of foot problems occurred and rest periods of three to four days were required. Since the operations were shorter, eight instead of four operations took place each month.

A June 1967 planning directive of the Military Assistance Command, Vietnam, stated that the Mobile Riverine Base could be relocated within the assigned tactical area of responsibility by the base commander in co-ordination with the senior Navy commander embarked. For relocation to a new area of responsibility, the commanding general of II Field Force would co-ordinate the move with the commander of Naval Forces, Vietnam, and the senior adviser of IV Corps, as appropriate, keeping Headquarters, Military Assistance Command, Vietnam, informed. This procedure was followed; however, the base commander was able to initiate co-ordination for a move to a new area of operations and inform the commanding general of II Field Force, through the Commanding General, 9th Infantry Division. This practice was often followed; but the requirements of the planning directive issued by Headquarters, Military Assistance Command, Vietnam, were met on movements across the boundary between III and IV Corps.

The original plan to keep the Mobile Riverine Base in the same anchorage for four to six weeks was not adhered to; more than twenty major relocations of the anchorage were made during the six months of operations—an average time at one anchorage of less than two weeks.

During the months of October and November, the Mobile Riverine Force operations in Dinh Tuong and Long An Province, where substantial numbers of the enemy had been found during previous months, met with limited success. The Viet Cong had become familiar with riverine equipment and tactics and was using small teams armed with rockets and recoilless rifles. Stationed along the waterways entering enemy base areas, the teams harassed the Mobile Riverine Force, delaying operations and inflicting casualties.

It was clear that if the Mobile Riverine Force could find the large enemy units it could destroy them. In the 4–5 December engagement, the 502d Local Force Battalion had suffered heavy losses either because the Americans achieved surprise or because the enemy, aware of a pending attack, had decided to stand and fight. Despite this successful experience, both Colonel David and Captain Salzer noted that the enemy was becoming more difficult to find. On the eve of the *Tet* truce period, the Mobile Riverine Force published an order that would return the force to the scene of its last battle—western Dinh Tuong Province. The Viet Cong *Tet* offensive was to find the Mobile Riverine Force widely dispersed in that area and well removed from the scene of the major battles that erupted throughout the Mekong Delta on 30 January 1968. The Mobile Riverine Force would have to move quickly to the scene of the major actions during *Tet*.

CHAPTER VIII

Tet Offensive of 1968 and U.S. Reaction

Shortly before the beginning of the three-day cease-fire declared by the government of South Vietnam to celebrate *Tet,* the lunar new year, the Mobile Riverine Force was ordered to the western portion of Dinh Tuong Province and the eastern part of Kien Phong Province to prevent the enemy from using communication routes running east and southeast through the area. There had been continuous and credible intelligence reports of enemy activity within the area, which was readily accessible to assault craft. The Mobile Riverine Force planned to establish bases along the waterways to provide fire support, as it had done during operations in the preceding November and December.

The operation began with the movement north by stages of supporting artillery batteries; as each battery moved, an infantry unit was sent ahead to secure the next fire support base. Four bases were established, three of them north along the Rach Ruong Canal from the Song My Tho to an agroville, one of the many agricultural resettlement areas established by President Diem. Three CH-47 helicopters were used to set down the artillery on the agroville site and the battery was laid and ready to fire by 1800 on 29 January 1968, the beginning of the *Tet* truce. The fourth base was on Highway 4, about two kilometers north from the river. The Mobile Riverine Force elements met only sporadic sniper fire on the 29th although the CH-47's were fired upon with .50-caliber machine guns.

The brigade's two battalions, in addition to defending the fire support bases, actively patrolled in the area of operations. Surveillance was aided by the radars mounted on the assault support patrol boats. At 2100 on 29 January, fifteen rounds of 82-mm. mortar fire were received at one fire support base without causing casualties or damage. No further enemy activity occurred during the remainder of the evening of 29 January and the Mobile Riverine Force dispersed to ambush positions to prevent major enemy movement.

On 25 January 1968, the senior adviser of IV Corps Tactical Zone had warned all subordinate elements in the IV Corps that

during the impending *Tet* holiday cease-fire period the Viet Cong were expected to resupply and move into position for a post-*Tet* offensive. The warning was based on past experience and recent intelligence reports. On 29 January the senior adviser dispatched the following message:

> Desire immediate dissemination of following information to all provinces without delay. There are a number of positive intelligence indicators that the enemy will deliberately violate the truce by approaching friendly installations during night of 29 January or early morning of 30 January. All provincial senior advisors will take action to insure maximum alert posture through the *Tet* period. Be particularly alert for enemy deception involving use of friendly vehicles and uniforms.

About 1000 on 30 January, the Mobile Riverine Force was informed that the Military Assistance Command, Vietnam, had canceled the *Tet* truce because of Viet Cong attacks on cities in the three corps to the north. The force was directed to resume offensive operations "with particular attention to the defense of the Headquarters complexes, logistical installations, airfields, population centers and billets." The Mobile Riverine Force ordered increased and aggressive reconnaissance in force within the original area of operations, but the infantry was directed to operate close to the major waterways to make feasible a rapid move by water craft toward the population centers. While two companies of the 3d Battalion, 47th Infantry, remained to provide security for the fire support base, the rest of the 3d Battalion and the 3d Battalion, 60th Infantry, on 30 January conducted saturation patrolling to the east toward Cai Be. At the end of the day, forty suspects had been detained, forty bunkers destroyed, and several small caches found.

On 31 January My Tho, Ben Tre, Cai Lay, Cai Be, and Vinh Long were attacked by the Viet Cong. The Viet Cong units involved included at My Tho, the 261st, 263d, and 514th Battalions; at Vinh Long, the 308th, 306th, 857th Battalions and local forces; at Cai Be and Cai Lay, the 261st Main Force Battalion, supported by the 540th and 530th District Companies and local forces. The Mobile Riverine Force however had not found any Viet Cong in the populated area along the Rach Ruong. When the force moved from this area local guerrillas made an effort to delay its return movement to the My Tho River. Although planning and coordination were evident in the enemy attempt to prevent movement of the Mobile Riverine Force to the scene of the major battles, the boats and their embarked battalions were able to reach the

Mobile Riverine Base by 0220 on 1 February. After less than four hours rest, the Army and Navy elements were ordered to conduct operations in the vicinity of Cai Be. Before the landing could be executed, however, orders were changed by the senior adviser of IV Corps Tactical Zone and the river assault divisions were diverted to My Tho where allied forces were still heavily engaged. Elements of the two battalions were landed by water craft in My Tho at 1520, 1 February. The five major ships of the Mobile Riverine Base moved from Vinh Long to Dong Tam in order to be in a better position to support operations.

After landing unopposed in My Tho, Company B, 3d Battalion, 47th Infantry, secured landing sites on the southwestern edge of the city for its parent battalion and was joined by Companies A and C. As the troops began to move north into the city, they met heavy fire. With the 3d Battalion, 60th Infantry, moving on the west, both battalions advanced north through the western portion of the city, receiving small arms, automatic rifle, and rocket fire. Fighting was intense and continuous and of a kind new to the riverine battalions. The city had to be cleared slowly and systematically; pockets of enemy resistance had to be wiped out to prevent the Viet Cong from closing in behind allied troops.

While advancing through the city, Company A, 3d Battalion, 47th Infantry, met heavy fire at 1615 and Company E was ordered to reinforce. On the way, the lead elements of Company E also met intense fire and were eventually pinned down at the western edge of My Tho. At this time both maneuver battalions were involved in pitched battles and were taking casualties. The 3d Battalion, 60th Infantry, continued its movement north, advancing under heavy enemy fire, and air strikes were requested at 1740 to assist the battalion's forward elements. Troops moved in and out of doorways, from house to house, and from street to street. Artillery was employed against enemy troops who were fleeing the city. At 1955 a group of Viet Cong who had been in a previous engagement with troops of the Vietnam Army 7th Division attempted to enter the streets where Company B, 3d Battalion, was fighting, but by 2100 Company B had eliminated them.

Company A, 3d Battalion, 47th Infantry, requested a light fire team to support its point element, which had met intense resistance from small arms and rocket fire and had suffered several casualties. At 1825 the light fire team arrived and was used to relieve the pressure on Companies A and E by firing on Viet Cong positions. The team made runs directly over Company A and placed fire

within twenty-five meters of the company's lead elements. To evacuate the wounded, a platoon leader sprinted through enemy fire, jumped into a Vietnam Army jeep parked in the street, and drove to the wounded. Ignoring the fire directed against him, he helped the wounded men aboard the jeep and drove back to his company lines. By 2100 most of the firing had ceased and the enemy began to withdraw under cover of darkness. Throughout the night sporadic sniper fire and occasional grenade attacks were directed against the Americans and Vietnamese but no major engagements developed.

On 2 February the 2d Brigade continued to attack strongholds within the city, which was encircled by U.S. and Vietnamese units. At 0630 both U.S. battalions continued a sweep to the north in the western portion of the city, encountering only light resistance. At 0915 a Viet Cong force was engaged on the northern edge of the city. Tactical air strikes with napalm were called in and dislodged Viet Cong troops holding a guard tower near a highway bridge. The resulting damage to the bridge was repaired under fire by the Mobile Riverine Force's supporting engineer platoon. Upon completion of the sweep, the city was cleared of enemy units and the Mobile Riverine Force battalions loaded onto armored troop carriers at 1201 for redeployment to Cai Lay District in Dinh Tuong Province.

By this time, the Viet Cong offensive had lost much of its original intensity in Dinh Tuong Province and the enemy appeared to be withdrawing to the north and west. Base Area 470, where the Mobile Riverine Force had taken up positions before *Tet*, was thought to be a likely location for the Viet Cong to regroup. The Mobile Riverine Force therefore moved the twenty-five kilometers from My Tho to the Cai Lay area to cut off enemy escape routes from eastern Dinh Tuong Province. There were no significant engagements during the following two days as the Viet Cong apparently remained close to My Tho and did not immediately withdraw to their normal base areas.

On 4 February at the direction of the senior adviser of IV Corps Tactical Zone, the 3d Battalion, 60th Infantry, and 3d Battalion, 47th Infantry, moved to Vinh Long to relieve continued Viet Cong pressure on Vietnamese units. The two battalions were transported by helicopter and boat to positions south of Vinh Long, and together with allied elements they established a cordon around the city. The Mobile Riverine Base relocated during the day to an anchorage on the Mekong River north of Vinh Long. The 3d Bat-

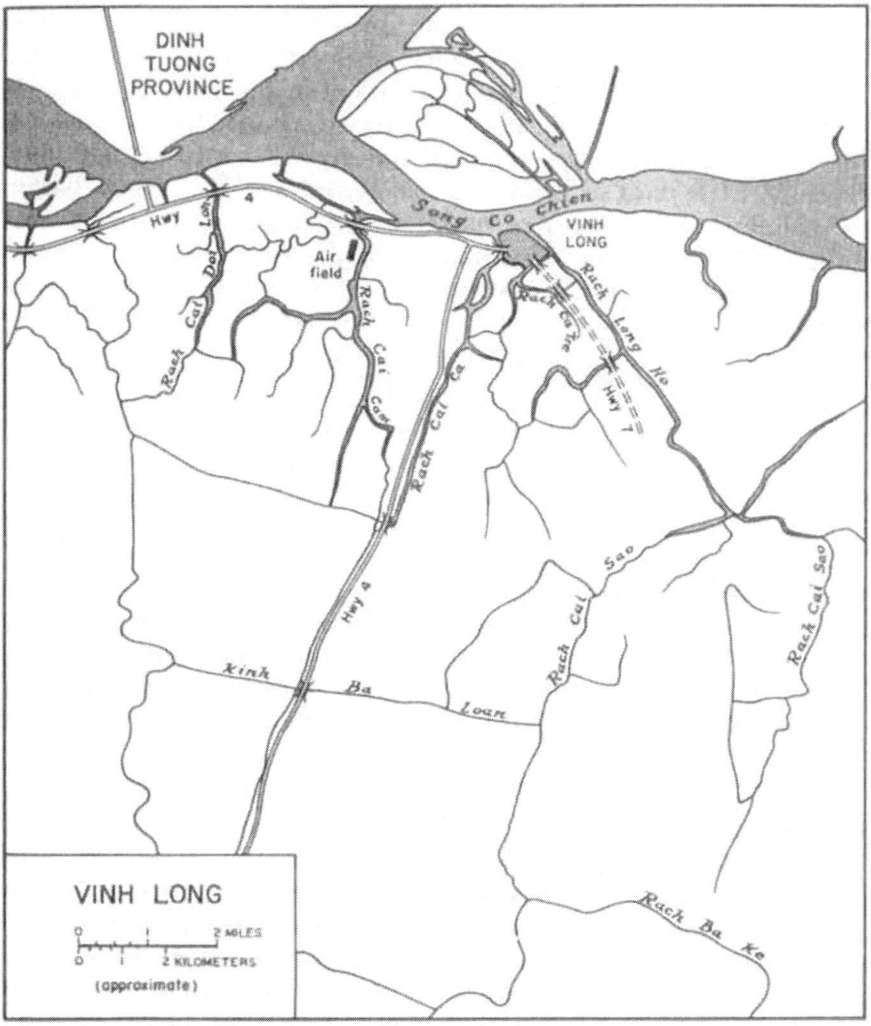

MAP 14

talion, 60th Infantry, completed the first airlift at 1655 with no opposition, but at 1733, on the east bank of the Rach Ca Tre Company A began to receive sniper fire. At the same time Company E was attacked with small arms and automatic weapons fire. The company returned fire and commenced movement against the Viet Cong, while artillery was used against the enemy's position. Company A meanwhile came under intense fire as it moved to its assigned sector. At 2030 Company E met a heavily armed Viet Cong company, and the fight was still in progress at 2145. As a result of this engagement, the 3d Battalion, 47th Infantry, sent

Company C to reinforce the 3d Battalion, 60th Infantry. (*Map 14*)

Thus far the battalions were successfully blocking enemy movement from Vinh Long at several points. Boats of the river assault divisions conducted patrols on the Rach Long Ho and Song Co Chiem throughout the night, reporting only a few incidents. Division boats received small arms and automatic weapons fire from the west and east banks of the Rach Long Ho which was suppressed with no casualties. During the remaining hours of darkness on 4 February the 3d Battalion, 60th Infantry, continued to engage the enemy in a fight that subsided gradually. Companies A and C were low on ammunition and required resupply. At 0200 a resupply helicopter was taken under fire and downed while attempting to land, but an infantry force rescued the craft and its crew. Throughout the night the enemy probed both battalion positions, but suffered a considerable number of casualties. At dawn both battalions conducted search operations within their immediate areas and confirmed Viet Cong losses.

Operations south of Vinh Long continued on 5 February with American and Vietnamese forces attempting to clear the area by pushing south against blocks of infantrymen inserted by helicopter. Intelligence indicated there was a Viet Cong battalion in the area. Company E, 3d Battalion, 60th Infantry, conducted airmobile search operations and Company A cleared the area near the Vinh Long airfield. The 3d Battalion, 47th Infantry, operated east of the Rach Cai Cam River, encountering several small enemy forces during the afternoon. Upon completion of the sweep, the infantry was loaded aboard ATC's and by 1725 had returned to the Mobile Riverine Base, after eight days and nights of continuous movement and combat. Company A, 3d Battalion, provided security for Vinh Long airfield and Company C, 3d Battalion, 60th Infantry, provided security for the Mobile Riverine Base and for the 3d Battalion, 34th Artillery.

On 6 February, in response to intelligence reports that the Viet Cong had succeeded in moving large forces southwest through the cordon around Vinh Long, Company B, 3d Battalion, 60th Infantry, departed the Mobile Riverine Base by boat in the early morning in an attempt to overtake the Viet Cong. The area to be searched was west of the Vinh Long airfield and south of National Highway 4. Company B made its first landing just south of the Highway 4 bridge over the Rach Cai Cam at 0810. Conversation with Popular Forces troops at a nearby outpost revealed that a large Viet Cong unit was in the woods and along the stream approximately 500 meters south of the bridge. Re-embarking, Com-

pany B commenced reconnaissance along the banks of the Rach Cai Cam to the south of the bridge. At 1100 the lead element of monitors and assault support patrol boats received fire from automatic weapons, recoilless rifles, and RPG's (Russian-built antitank rockets) on both banks of the river. Company B, supported by artillery, gunships, and fire from the river craft, assaulted the west bank of the Rach Cai Cam just north of the area of the ambush and swept south. No Viet Cong were found and the company re-embarked on the ATC's and continued along the stream, landing patrols at irregular intervals.

When the Viet Cong opened fire at 1100, Company E, 3d Battalion, 60th Infantry, was dispatched by boat from the Mobile Riverine Base and arrived in the area at 1430. Company B was again landed on the west bank of the Rach Cai Cam near the 1100 firefight area, and Company E was beached at that time just opposite Company B on the east bank of the Rach Cai Cam. Almost immediately both companies came under heavy RPG, grenade, and 60-mm. mortar fire. Company B charged the Viet Cong positions in its zone, killed the enemy troops, and took four weapons, suffering two soldiers wounded. Company E had twelve wounded in the first clash and was held up while the wounded were evacuated. By 1510 the enemy had withdrawn to the southwest.

Company A, 3d Battalion, 60th Infantry, beached just south of the Regional Forces and was able to move rapidly south through the rice paddies and act as a blocking force to the west of Company B. Men of Company A observed twenty Viet Cong in the river to their west, apparently trying to escape, and called in artillery fire that caused several secondary explosions. Shortly thereafter, at 1700, all three companies received heavy automatic weapons and RPG fire from the wooded stream line in front of them and became heavily engaged.

The battalion commander, having committed all of his available infantry, requested that the brigade reserve force be committed in an attempt to encircle the Viet Cong positions. By nightfall, however, enemy forces had broken contact and the 3d Battalion, 60th Infantry, returned to the Mobile Riverine Base early on 7 February.

During the fighting from 30 January to 6 February the Mobile Riverine Force made several rapid moves over a wide area. The Mobile Riverine Force moved from the Rach Ruong Canal area of western Dinh Tuong Province to My Tho; then to Cai Lay; and, finally, to Vinh Long. The effectiveness of the Mobile Riverine Force at each city resulted in a reduction of the Viet Cong

attacks to harrassing actions and elimination of the threat to the city. By using in combination the Mobile Riverine Base, which moved large numbers of troops and support elements between the areas, assault water craft, and supporting helicopters the Mobile Riverine Force not only moved quickly to each new area, but also arrived in strength and was able to sustain operations as needed.

From 8 through 12 February the Mobile Riverine Force deployed to the northwest of My Tho and to Gia Dinh Province near Saigon. The entire CORONADO X operation, which ended on 12 February, proved the ability of the Mobile Riverine Force to move rapidly over a large geographic area. The counteroffensive to relieve My Tho and Vinh Long highlighted the importance of having adequate air resources to use in conjunction with the ships and small assault craft of the Navy Task Force 117.

The back of the Viet Cong *Tet* offensive had been broken in the My Tho and Vinh Long area by the middle of February through the combined efforts of the Mobile Riverine Force and South Vietnamese armed forces. However, Viet Cong units continued to maintain pressure on Can Tho, the capital of Phong Dinh Province. The Senior Advisor, IV Corps Tactical Zone, having current operational control of the Mobile Riverine Force, decided to employ it to relieve this pressure on Can Tho. Thus CORONADO XI, the last of the CORONADO series of Mobile Riverine Force operations, was initiated. The use of the Mobile Riverine Force in the central delta was made possible by the movement of the 1st Brigade, 9th U.S. Infantry Division, from the III Corps Tactical Zone to Dinh Tuong Province. All three brigades of the 9th Division were thus placed in the delta for the first time, signaling the opportunity for the Mobile Riverine Force to move south into the Bassac River area to carry out further the original Mobile Afloat Force concept.

Before CORONADO XI, intelligence indicated that the Viet Cong, instead of withdrawing to base areas after being routed from the cities, had remained near the heavily populated areas. Later it was apparent that the enemy had left only enough men to maintain pressure on the cities so that American and Vietnamese forces would also remain nearby. Numerous intelligence reports indicated that the enemy had improved his strength throughout Phong Dinh Province, in some cases by assigning young recruits to units and sending them to battle with minimum training. This enabled the Viet Cong to display considerable force on the battlefield. The Tay Do Battalion was believed to have recruited substantial num-

bers on the strength of its good reputation as a fighting unit. Enemy forces within Phong Dinh Province were identified as the 306th Main Force Battalion, 303d Main Force Battalion, Tay Do I and II Battalions, U Minh 10th Battalion, and Military Region III headquarters, accompanied by a security element of 200 men. The over-all mission for CORONADO XI was based on the suspected location of Viet Cong units and the Military Region III headquarters.

The mission of the Mobile Riverine Force in CORONADO XI, as developed in co-ordination with the Senior Adviser, IV Corps Tactical Zone, was to conduct riverine, air, and ground search operations in Cai Rang and Phung Hiep Districts of Phong Dinh Province, to locate and destroy Military Region III headquarters, and to conduct waterborne cordon and infantry search operations on the island of Cu Lao May in the Bassac River.

Operation CORONADO XI commenced on 12 February 1968, with the movement of the Mobile Riverine Force from Dong Tam to an anchorage in the vicinity of Can Tho. The force proceeded up the Mekong River through the waterway which connects the My Tho and Bassac Rivers, and down the Bassac to Can Tho. The force arrived shortly after 1300 on 13 February, having completed a journey of 109 miles from Dong Tam. *(Map 15)*

On 14 February the two infantry battalions landed from boats and conducted sweeps north along the canals immediately south of Can Tho. Army of Vietnam troops of the 9th and 21st Divisions conducted operations north of the city, sweeping to the southwest. Company A, 3d Battalion, 47th Infantry, found a cache which contained 460 B-40 rounds, 200 pounds of explosives, 89 120-mm. mortar rounds, and 500 pounds of medical supplies.

During the period 15-19 February, the Mobile Riverine Force conducted river and air operations west of Can Tho without encountering significant enemy forces. When the Can Tho airfield suffered a heavy rocket and mortar attack on 16 February, the main task of the Mobile Riverine Force was to sweep the area in the vicinity of the airfield.

On 19 February, at the request of the senior adviser of the IV Corps Tactical Zone, General Eckhardt, Task Force 117 units initiated joint waterborne patrols with units of the Vietnam Navy River Assault Group along a 13-mile stretch of the Song Can Tho to halt all sampan traffic. At 2217 a monitor was struck by B-40 rocket rounds that penetrated the 40-mm. turret. Although seven people, including a Vietnam Navy interpreter on board, were wounded, the monitor was able to suppress the enemy automatic

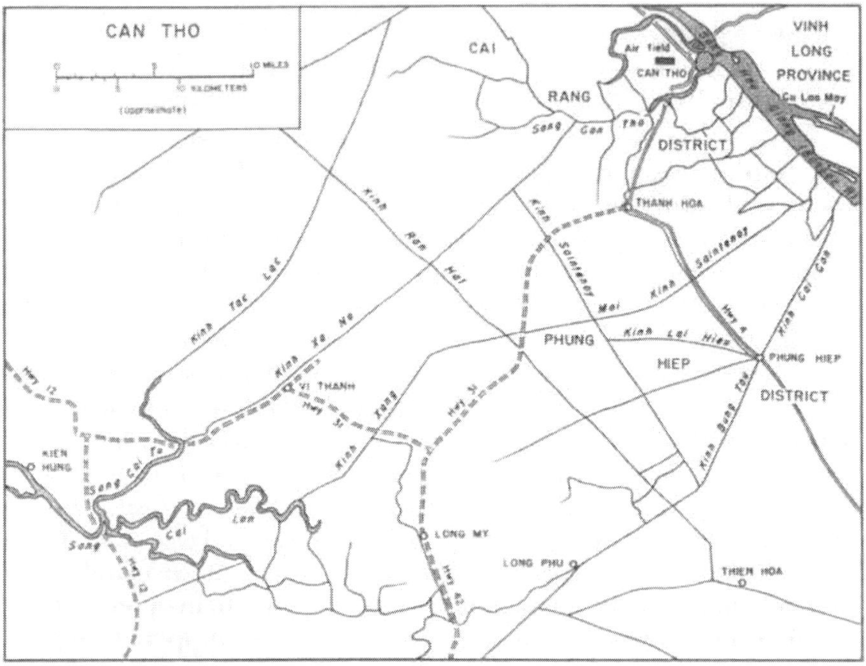

MAP 15

weapons and small arms fire. Four of the wounded were evacuated after a rendezvous with a medical aid boat. The patrols continued throughout the night in an effort to prevent escape of Viet Cong units from the area.

On 22 February the Mobile Riverine Force and government of Vietnam forces initiated a riverine and air operation in Phung Hiep District to locate the Military Region III headquarters. The 3d Battalion, 60th Infantry, and the 3d Battalion, 47th Infantry, moved to the area of operations by means of the Kinh Saintenoy Canal. Fire support was provided by Battery C, 2d Battalion, 35th Artillery, and 3d Battalion, 34th Artillery.

To obtain the overhead clearance that the boats needed to proceed beyond the Phung Hiep Bridge, the Vietnam Army engineers raised the center span of the bridge with jacks. As the assault boats proceeded west from Phung Hiep, sporadic sniper fire from the north bank of the canal wounded four U.S. Navy men. The riverine movement beyond the bridge apparently caught the Viet Cong by surprise. When Companies B and C, 3d Battalion, 47th Infantry, beached along the Kinh Lai Hieu Canal, just northeast of Hiep Hung, the enemy was in bunkers whose fighting ports faced the rice paddies in anticipation of airmobile attacks. The

Viet Cong ran into the open fields away from the canal where they were taken under fire by artillery and gunships and suffered heavy casualties. No additional U.S. casualties occurred.

On 23 February, the Mobile Riverine Force concluded the unsuccessful attempt to locate Military Region III headquarters and was withdrawn from the area of operations. Throughout the day there was no sight of the enemy or significant finding. The operation marked the deepest penetration yet of the Mekong Delta by U.S. forces.

On 24 February the Mobile Riverine Force conducted a waterborne cordon and search operation on the island of Cu Lao May during which both infantry battalions conducted medical civil action and dental civil action programs at enclosures where Vietnamese were detained for questioning. There was no contact with the enemy in this one-day operation. Northwest of Can Tho on 26 February, the Mobile Riverine Force met a large Viet Cong force. Company B, 3d Battalion, 60th Infantry, landed under fire that wounded many men and damaged twelve helicopters. Other elements of the Mobile Riverine Force moved to support Company B; Company E was airlifted at 1130 and landed approximately 800 meters east of Company B and Company A followed at 1405. At 1545, more than five hours after the first fire, Companies B and E were still heavily engaged and had not linked up. Linkup was effected at 1705, and because of the number of casualties sustained by Company B the company was placed under operational control of Company E. Companies A, B, and E established a night defensive position and were resupplied by 2200. The 3d Battalion, 47th Infantry, to the northwest encountered little fire throughout the day, although its scheme of maneuver was modified to allow it to provide additional support to the 3d Battalion, 60th Infantry.

Task Force 117 meanwhile had established patrols on the Song Can Tho along the southern boundary of operations to prevent escape of the enemy. About 2030 an armored troop carrier received a hit in the port bow causing an explosion. An assault support patrol boat went to the aid of the ATC and was taken under fire by automatic weapons. The enemy fire was returned and suppressed. Ten minutes later a second ATC was fired upon by a recoilless rifle. The round passed over the ATC, which immediately returned and suppressed the fire. There were no casualties among those on board in either attack. The units continued their patrols without incident until 0215 when a major action occurred as a large force of Viet Cong attempted to escape. An assault support

patrol boat came under heavy fire from both banks and sustained several B-40 rocket hits causing major flooding and wounding two crewmen. Other river assault craft moved into the area and suppressed the hostile fire, but one assault support patrol boat sank as it was being towed by a monitor. All crew members were rescued, but the monitor received several hits while moving to assist the disabled assault support patrol boat. During the same period, another assault support patrol boat, operating four kilometers east of the first attack, was hit by heavy rocket and automatic weapons fire. The boat captain and radio men were killed instantly and the three other crewmen were wounded but were able to beach the boat. The enemy attack lasted an hour and a half, with four assault craft involved in the battle. A light helicopter fire team helped to cut down the enemy fire.

The Viet Cong were unable to escape south across the Song Can Tho and continued to suffer casualties as the blockade was maintained successfully throughout the action. U.S. losses were moderate. During the night the Viet Cong continued harassing the Mobile Riverine Force battalions with probing attacks and sniper fire. On 27 February the area was swept again, confirming additional enemy losses and discovering caches, including five crew-served weapons, sixteen small arms, and assorted ammunition.

On 1 March, also, other elements of the Mobile Riverine Force and three Vietnam Army Ranger battalions operated approximately five kilometers southwest of Can Tho. Heavy fighting developed in the afternoon and continued into the night. The enemy slipped away before daylight and only scattered groups were found the next day.

CORONADO XI required some unusual supply operations because of the geographic location, tempo of operations, and change in control headquarters from the 9th Infantry Division to the senior adviser of IV Corps. Operations took the Mobile Riverine Force away from the normal lines of communication thus straining resupply. The severe and continuous fighting caused a heavy expenditure of ammunition and Class I and Class III supplies. The control exercised by IV Corps created a new set of logistics channels to support operations. The potential logistics problems were solved by personal liaison between the Mobile Riverine Force and Army logistics representatives located at Can Tho. Items such as C rations, fuel, and construction materials were provided by 1st Logistics Command. The IV Corps provided for evacuation of dead and wounded and responded to all Army and Navy logistics needs.

Within the Can Tho area, most movements by waterway and canal were accomplished without significant opposition because the enemy had no prepared fighting positions near the canals. The enemy, however, had taken excellent defensive measures against airmobile operations conducted near open rice paddies. Bunkers were well constructed, with good overhead protection against artillery and air strikes, and commanded adequate fields of fire. In one instance the Viet Cong could not engage the Mobile Riverine Force; their bunkers faced away from the waterway. The enemy attempted to escape, making themselves vulnerable to the fire of the advancing U.S. troops and supporting gunships.

The Viet Cong demonstrated their ability to employ automatic weapons fire in all battles. They characteristically used an initial heavy volume of automatic weapons fire, followed by sporadic sniper fire. Sniper fire was directed particularly at unit leaders and radio operators; snipers were well trained and extremely effective in the Can Tho area. The Viet Cong continued to fight until the late hours of the night, 2200 to 2400. They then withdrew but continued to harass U.S. positions with mortar and rocket fire.

All operations conducted in the Can Tho area made maximum use of naval and air assets, and demonstrated that logistic support planning and timely intelligence were essential. The use of airmobility vastly influenced the course of tactical situations, especially when reserves were employed to reinforce a unit under fire or to exploit the enemy situation.

Combat operations in CORONADO XI ended on 3 March when the Mobile Riverine Force left the Can Tho area for Dong Tam. From 4 to 6 March the force assumed the division reserve role while the 3d Battalion, 47th Infantry, was replaced by the 4th Battalion, 47th Infantry, in the rotation of battalions within the brigade.

Operation TROUNG CONG DINH was initiated by the 9th Infantry Division on 7 March 1968. Its purpose was to destroy enemy forces in Dinh Tuong Province and to reduce the threat west of My Tho to Highway 4, which was being constantly menaced by the Viet Cong. The operation involved the 1st and 2d Brigades, 9th U.S. Infantry Division, the 7th Vietnam Army Division, and provincial forces. The commanding general of the 9th U.S. Infantry Division was able to employ the 2d Brigade for the first time in a continuing co-ordinated role with another division brigade. The Mobile Riverine Force conducted a series of combined riverine and airmobile operations, beginning east of My Tho. Operations

were supported by the 3d Battalion, 34th Artillery, which established a fire support base on the north bank of the My Tho River, ten miles east of My Tho.

During initial airmobile assaults on 7 March enemy fire was light and no major firefight resulted. During the afternoon, the two battalions shifted their area of operations west-northwest of My Tho in response to intelligence reports. At 1920 the 4th Battalion, 47th Infantry, came under heavy small arms and automatic weapons fire just west of My Tho. Heavy fire continued until 2225. The following day troops met no Viet Cong in the area; at midday in response to intelligence the two battalions moved again into the area of operations of early 7 March. There were several firefights during the rest of the day. The most significant occurred at 1715 when Company B, 4th Battalion, 47th Infantry, made an assault landing by helicopter and was fired upon by a Viet Cong force from well dug-in positions along the southern edge of the landing zone. Five helicopters were downed during the first landing. Throughout the evening and into the night, as other elements of the two battalions maneuvered in support of Company B, there was sporadic but heavy fighting in the area. At 0300 the fire support base was attacked by mortars that sank two artillery barges carrying four 105-mm. howitzers. The infantry continued their sweep of the previous day's battlefield without finding any Viet Cong and returned by ATC to the Mobile Riverine Base in the late afternoon.

On 10 March the Mobile Riverine Force shifted operations near Cai Lay and Long Dinh, towns of Dinh Tuong Province. Troops of the 1st and 2d Brigades, 9th Infantry Division, conducted day and night patrolling and reconnaissance in force against suspected Viet Cong locations along Highway 4 until 16 March. On 16 March the 1st Brigade terminated its participation in the TROUNG CONG DINH operation. On 18 March, boats of River Division 92 received heavy automatic weapons and rocket fire while patrolling west of Dong Tam. The rockets damaged several assault support patrol boats and one monitor. The operation continued without opposition on 19 March, and the infantry battalions returned by ATC to the Mobile Riverine Base in the vicinity of Dong Tam.

On 22 March the Mobile Riverine Base, still located in the My Tho River south of Dong Tam, was attacked at 0320 by enemy using mortars and recoilless rifles. The *Benewah* received two 75-mm. recoilless rifle hits that caused minor damage, and near

misses were registered by Viet Cong mortars on an LST, the *Washtenaw County*.

Through the remainder of March the Mobile Riverine Force continued operations in Dinh Tuong Province with occasional light to moderate firefights. On 1 April the 3d Battalion, 60th Infantry, debarked and assumed the Dong Tam security mission and the 3d Battalion, 47th Infantry, joined the Mobile Riverine Force. Operation TROUNG CONG DINH was terminated on 2 April, after a one-battalion airmobile operation in the Ham Long District in Kien Hoa Province, just south of Dong Tam.

During TROUNG CONG DINH the Mobile Riverine Force used extensively a riverine assault reconnaissance element, a small unit first employed in December 1967 that consisted of three or four monitors and several assault support patrol boats. The riverine assault reconnaissance element led ATC convoys and employed reconnaissance by fire against likely Viet Cong ambush positions. The technique reduced casualties because of the firepower and mobility of the craft in the riverine assault reconnaissance element and the placement of ATC's carrying infantry well back in the column. Airmobility was used during TROUNG CONG DINH to increase the flexibility of the Mobile Riverine Force by providing increased intelligence, firepower, and escort coverage for convoys during troop movement.

On 4 April a significant battle was fought near the intersection of the Song Ba Lai and canals, called the Crossroads, in the Truc Giang District of Kien Hoa Province. As the boats of River Division 92 carrying the 2d Battalion, 47th Infantry, entered the Song Ba Lai they were ambushed with rockets, small arms, and automatic weapons. The battalion put into effect its counterambush plan. Company B passed through the ambush and landed unopposed just beyond the Viet Cong positions. Company E was to land just short of the ambush and the two companies were to pinch the Viet Cong between them. Instead, Company E was landed directly into the ambush, suffered heavy casualties, and was forced to withdraw. Heavy fighting with the well-protected Viet Cong continued throughout the day. The 4th Battalion, 47th Infantry, crossed from the north to assist the 3d Battalion on the south bank. Casualties among the infantry were heavy as they encountered an extensive bunker complex containing .50-caliber and other automatic weapons. The action ceased during the evening as the Viet Cong slipped away in the darkness. The two battalions continued to sweep the marshes south of the Song Ba Lai through 7 April.

From 10 through 12 April the Mobile Riverine Force conducted successive operations near the towns of Vinh Kim and Long Dinh, and east of My Tho without major encounters. An operation in the Giong Trom and Ben Tre Districts of Kien Hoa Province on 14–15 April failed to find the enemy. At 1400 on 15 April the Mobile Riverine Base was attacked by 57- and 75-mm. recoilless rifles while at anchor. The USS *Benewah* received three hits; the supply LST received eight hits—all above the water line. One LCM-6 refueler was hit and later burned and sank. There were no casualties and only minor damages to the *Benewah* and the LST.

During the remainder of April, the Mobile Riverine Force conducted four separate operations in the Cai Lay area of Dinh Tuong Province, the western part of Kien Hoa and Dinh Tuong Provinces, and the Cam Son Base area of Dinh Tuong Province. Significant engagements were fought in three of the four operations, with the Mobile Riverine Force inflicting losses on the enemy in each encounter. Two of the operations were conducted in co-ordination with the 2d Vietnamese Marine Corps Battalion.

On 2 and 3 May, the Mobile Riverine Force again returned to Kien Hoa Province in response to enemy attacks on government outposts. On 5 May a survey was made of Regional Forces and Popular Forces outposts in Cai Be District to determine the extent of damage. Operations were directed from 7 through 11 May against Military Region II Viet Cong headquarters in the Giao Duc District of Dinh Tuong Province. The Mobile Riverine Force and the 2d Battalion, Vietnam Marine Corps, conducted a reconnaissance in force but encountered no large groups of enemy troops. From 14 through 16 May the Mobile Riverine Force was employed in the Mo Cay District of Kien Hoa Province and engaged the Viet Cong 516th Battalion in several firefights. The Mobile Riverine Force and elements of the 3d Brigade, 9th Infantry Division, carried out operations from 16 through 18 May, in the Than Duc District of Long An Province in co-operation with III Corps Tactical Zone Vietnam Army forces in order to reduce Viet Cong and North Vietnam Army infiltration of the Saigon area. The brigade continued operations in Long An and Dinh Tuong Province until 26 May. At that time a two-day operation was launched to reconnoiter and exploit the area of a B–52 strike in the Giong Trom District of Kien Hoa Province.

During the month of June the enemy generally avoided contact with the maneuver battalions and intensified his attacks against naval assault support craft of Task Force 117. On 5 and 6 June

the Mobile Riverine Force conducted reconnaissance in force in the Sa Dec area. Effective use was made of the E8—a portable expendable launcher that fired riot-control cannisters—and flame-boats. The period 4 through 8 June saw the Mobile Riverine Force conducting operations in the Sa Dec area and eastern Kien Phong Province. Returning to Kien Hoa Province, the force initiated operations on 10 June in the Giong Trom and Truc Giang Districts. On 16 June it moved to Can Tho, and began operations again directed by the senior adviser of IV Corps Tactical Zone to prevent a suspected Viet Cong offensive against the city. A four-day operation commenced on 17 June with a major firefight in which the majority of the enemy's losses resulted from fire support by Company A, 7th Battalion, 1st Air Cavalry, in support of the maneuver battalions. Completing operations in Phong Dinh Province on 19 June, the Mobile Riverine Force moved to Dong Tam. After four days of operations near Dong Tam, it proceeded on 26 June to Long An Province where it was again employed to prevent Viet Cong and North Vietnam Army forces from moving in and out of the vicinity of Saigon. Unlike the brigade's experience in early 1968 operations in Long An Province, the enemy avoided contact with the maneuver battalions throughout the operation.

On 2 July the 2d Brigade was directed to place the 3d Battalion, 60th Infantry, under operational control of the 3d Brigade, 9th Infantry Division, which was in contact with the enemy in Dinh Tuong Province. The battalion remained under operational control of the 3d Brigade through 5 July. During that time the Mobile Riverine Force continued its operations in Long An Province, moving to Dong Tam on 5 July. On 6 July the 3d Battalion, 47th Infantry, relieved the 4th Battalion, 47th Infantry, of its security mission at Dong Tam Base. The 2d Brigade continued combat operations on 7 July in Kien Hoa Province near the canal Crossroads on the Song Ba Lai, where the 2d Brigade had previously fought one of its heaviest engagements. The 4th Battalion, 47th Infantry, entered its objective area and beached without incident. The 3d Battalion, 60th Infantry, air-landed assault troops at several spots south of the 4th Battalion beach sites. The only incidents of the day occurred when Company E of the 4th Battalion, 47th Infantry, received sporadic small arms fire and D Troop, 3d Battalion, 5th Air Cavalry, was subjected to sporadic automatic weapons fire. The limited enemy activity in the area was unlike prior experience of the Mobile Riverine Force along the Song Ba Lai.

On 10 July the Mobile Riverine Force began operations to

locate and destroy enemy forces and equipment within the Huong My and Mo Cay Districts of Kien Hoa Province. Operations continued, through 13 July, with both battalions using boats and helicopters to search for the enemy. During the period 15 through 17 July the Mobile Riverine Force conducted riverine operations in Vinh Binh Province in co-operation with the 9th Vietnam Army Division. Completing operations in Vinh Binh Province the Mobile Riverine Force moved to Dong Tam Base on 18 July. While anchored off Dong Tam, the 2d Brigade employed the 4th Battalion, 47th Infantry, on 20 July to conduct saturation patrolling and to establish night ambushes on the south bank of the My Tho River in the Truc Giang and Ham Long Districts of Kien Hoa Province. On 21 July the 3d Battalion, 60th Infantry, was employed in a cordon and search operation in co-operation with Regional Forces and Popular Forces elements on Thoi Son Island. Operations continued on 23 July in the Giong Trom and Truc Giang Districts of Kien Hoa Province in co-operation with the Vietnam Army. On 25 July the 2d Brigade terminated its operations in Kien Hoa Province; the next day the Mobile Riverine Force moved to Vinh Long Province.

After completing final plans and preparations for its journey the Mobile Riverine Force left its Vinh Long anchorage on 28 July and arrived in Phong Dinh Province the same day. On 30 July the force began offensive operations in co-ordination with the 5th Battalion, Vietnam Marine Corps, in the Vi Thanh area of Chuong Thien Province. To accomplish its mission, the 2d Brigade debarked from the Mobile Riverine Base and established its headquarters and tactical operations center inland in the vicinity of Vi Thanh. The Mobile Riverine Force moved into the U Minh Forest, forty-eight miles southwest of Can Tho in early August in a co-ordinated combined operation that included U.S. Army, Navy, and Air Force units and Vietnam Marine Corps and Army forces. This was the first major Allied ground operation in that area in more than a decade and another amplification of the original Mobile Riverine Force operational concept that had envisioned operations no more than twenty to thirty miles from the Mobile Riverine Base. The ten-day operation was also significant in that it substantially exceeded in scope the two- to three-day operations of the force contemplated in original plans. The operation was the first penetration by an American ground force into the U Minh Forest region and it was the deepest penetration by the Mobile Riverine Force into the Mekong Delta. The 2d Brigade established

a land base and forward logistic support area with lines of communication extending fifty miles to the riverine base anchored near Can Tho. The operation was also destined to be the last major strike operation conducted by the Mobile Riverine Force.

By early August the Mobile Riverine Force had conducted many tactical operations in co-ordination with the other brigades of the 9th U.S. Division and with Vietnamese Army, Marine Corps, and paramilitary units. The tempo of strike operations had been maintained at a high level as the force moved from one province to another, across corps boundaries, and in response to missions originating with the commanding general of the 9th U.S. Infantry Division and with II Field Forces Vietnam through the senior adviser of the IV Corps Tactical Zone. On several occasions operations were carried on to block infiltration routes in order to cut down the threat of enemy forces present in the Capital Military District around Saigon, and near Ben Tre, Sa Dec, and Can Tho. The Mobile Riverine Force had operated in eight provinces of the Mekong Delta since early February 1968. In all operations its efforts were complemented by the co-ordinated operations of other U.S. or Vietnamese units.

CHAPTER IX

Pacification and Kien Hoa Province

The Third Battalion

As the Mobile Riverine Force began its second year of operations in 1968, the 2d Brigade still lacked a third maneuver battalion and full use of the 3d Battalion, 34th Artillery. Successive commanders of the brigade had reiterated the need for a third battalion to increase the flexibility of the force and to enable it to operate with greater independence, but the brigade's overall strength had remained basically unchanged. Two of its three battalions remained afloat with the Mobile Riverine Force, while the third infantry battalion was employed in defense of Dong Tam Base. In the absence of a third maneuver battalion the Mobile Riverine Force operated extensively in conjunction with Vietnam Army units.

During the first half of 1968 the Navy component of the Mobile Riverine Force continued to grow. In June a third river assault squadron was formed and in July the Navy component was reorganized into two groups, Mobile Riverine Group Alpha and Group Bravo. By the fall of 1968, the force was scheduled to reach its full strength of 4 river assault squadrons, 184 river assault craft, 4 barracks ships, 3 repair ships, 2 barracks barges, 2 support ships, 2 resupply ships, and various other support craft.

At a riverine concept briefing on 5 January 1968, General Abrams, deputy commander of the U.S. Military Assistance Command, Vietnam, had directed the U.S. Army, Vietnam, to reexamine the organization for the Army riverine battalions and the composition of the land-based brigade. The 9th Infantry Division, Headquarters, II Field Force, Vietnam, and Headquarters, U.S. Army, Vietnam, concluded that the mission of the Mobile Riverine Force could best be accomplished by a reorganization of the seven infantry battalions of the division, deleting equipment and units not essential in riverine operations. General Westmoreland approved the reorganization on 21 February 1968.

This new organization of the battalions lent itself readily to operations other than riverine, particularly airmobile operations,

without the investment in equipment and men required for continuing land-based operations. In firepower the riverine battalion, less four 81-mm. mortars, was equivalent to the standard infantry battalion even without the considerable firepower support of the Navy assault craft. By borrowing equipment, the battalion was capable of sustained operations on dry land at a saving of some 600 spaces which would be required by a standard battalion and its support elements.

The value of and continuing need for one Vietnam Marine Corps battalion to operate with the Mobile Riverine Force was also recognized. The battalion was not, however, permanently assigned to the force because Vietnam Marine Corps units were part of the Army of Vietnam general reserve and therefore subject to recall at any time. Vietnamese marines had been operating with the Mobile Riverine Force on this basis since 15 November 1967 and had contributed to successes of the force against the Viet Cong.

The CORONADO series of operations conducted from June 1967 to July 1968 had demonstrated the effectiveness of the Mobile Riverine Force in the use of waterways to deny the enemy areas where he had previously operated with relative freedom. This success led to an examination of how the riverine effort could be expanded cheaply in order to step up the tempo of operations in the Mekong Delta. A study conducted by U.S. Army, Vietnam, and Commander, Naval Forces, Vietnam, concluded that additional units of the 9th Infantry Division could be accommodated at Dong Tam without dredging. Further, with a modest addition of naval craft to the already scheduled river assault squadrons, an expanded force could support two brigades afloat and one at Dong Tam. Through judicious use of naval river craft and Army planes and helicopters the entire 9th Division could be effectively employed in the delta; the Vietnam Marine Corps and, eventually, Vietnamese river assault group units could also be drawn into operations. An organization of two Mobile Riverine Force brigades, each with three infantry battalions and supporting artillery, was considered the most desirable. The third brigade was to operate out of Dong Tam, either independently or in conjunction with the Mobile Riverine Force. This arrangement would provide flexibility, reduce reliance on land bases, demand fewer aircraft, and help sustain pressure on the Viet Cong. It would also eliminate regular river assault squadron operations out of Dong Tam, which could develop a pattern easily detected by the enemy. The 3d Brigade with two maneuver battalions was to be sent from the Tan An

area to Dong Tam Base to facilitate its transition to the second Mobile Riverine Force. The division headquarters was transferred from Camp Martin Cox at Bearcat to its new base camp at Dong Tam on 25 July.

As of July 1968 no definite date had been established for embarking the long-awaited third infantry battalion of the 2d Brigade aboard ships of the Mobile Riverine Force. The availability of the third infantry battalion was contingent upon readjustments in 9th Infantry Division base security missions.

While general organization changes to the Mobile Riverine Force were being formulated, a major change was made in its mission. On 16 July the 2d Brigade was assigned the province of Kien Hoa as a primary area of operations, with the added mission of conducting pacification activities of a more permanent nature than those previously undertaken by the Mobile Riverine Force. The Accelerated Pacification Program guided from Headquarters, U.S. Military Assistance Command, Vietnam, prompted this assignment. After six months of operations that concentrated on the area north of the My Tho River, including Long An Province and the Rung Sat Special Zone, and after an additional six months of operations ranging more widely afield and deeper into the delta, the Mobile Riverine Force returned to concentrate on Kien Hoa. At this time Major General Harris W. Hollis, commanding the 9th Division, decided to provide the force with its third maneuver battalion and to require all three battalions to conduct airmobile and riverine operations in Kien Hoa Province. One of the battalions was based ashore in the vicinity of Ben Tre and two battalions remained afloat. The concentration on Kien Hoa Province was directed shortly after formal approval of the riverine modified tables of organization and equipment for the 2d Brigade and shortly before the arrival of the U.S. Navy craft required to lift the maneuver elements of an additional brigade.

The operational rotation of one infantry battalion from the Mobile Riverine Force to the Ben Tre base was quite different from the rotation of a battalion to Dong Tam. The May 1967 preparations of the 2d Brigade for Mobile Riverine Force operations had provided for each infantry battalion to leave at Dong Tam equipment not required when it was based afloat. Semipermanent facilities at Dong Tam reduced the need of each battalion for organic equipment to operate a nearly independent battalion base. The first battalion sent to the Ben Tre base, therefore, required equipment that Mobile Riverine Force battalions

had not needed since before June 1967. The addition of the third infantry battalion as a land-based unit consequently brought a new organization and a logistics problem that the Mobile Riverine Force had not anticipated in its plan for employing three infantry battalions in continuing strike operations from a base entirely afloat.

Concentration by the Mobile Riverine force on Kien Hoa Province presented special considerations for operational flexibility. Previous operations in Kien Hoa had revealed the limited waterway network off the major rivers. Helicopters had proved necessary both to deploy blocking forces inland in those areas lacking navigable waterways and to assist in waterway reconnaissance. Previous operations had also demonstrated that effective anchorages for the Mobile Riverine Base to support operations in Kien Hoa were limited to four: the vicinity of Dong Tam; east of My Tho; east of Vinh Long; and east of Mocay District on the Co Chien River. Although use of each of these anchorages could facilitate operations into different districts of Kien Hoa Province, changing from one anchorage to another would not be likely to deceive the enemy.

In the past bold changes in the location of the Mobile Riverine Base had achieved surprise against Viet Cong base areas that had not been recently or frequently attacked. This major asset of the Mobile Riverine Force—the mobility of the entire force, including its logistics base—could not be fully used in the concentration of Kien Hoa Province.

Beginning in early September in Kien Hoa Province under operation HOMESTEAD, the 2d Brigade conducted operations throughout the month to find and destroy the enemy in his home territory. The enemy attacked the river craft from ambush with rockets, recoilless rifles, and small arms. According to intelligence reports, the enemy had formed special five-man teams in Kien Hoa to ambush the boats of the Mobile Riverine Force. The main ambush weapon of these teams was usually the RPG2 or RPG7 rocket launcher.

In September an additional river assault squadron and two barracks ships arrived in Vietnam to support the 3d Brigade, 9th Infantry Division. At the same time the 9th Infantry Division completed reorganization of the headquarters of 3d Brigade and one battalion into the riverine modified tables of organization and equipment in anticipation of the establishment of the second Mobile Riverine Force. By the end of 1968 three additional bat-

talions were reorganized, making a total of two brigade headquarters and seven infantry battalions configured for riverine operations.

During October Mobile Riverine Group Alpha supported pacification operations in Kien Hoa Province with the 3d and 4th Battalions, 47th Infantry, the 3d Battalion, 60th Infantry, elements of the 3d Battalion, 34th Artillery, and the 3d Battalion, Vietnam Marine Corps. Mobile Riverine Group Bravo operated with other 9th Infantry and Vietnam ground forces in Vinh Binh, Vinh Long, Long An, Dinh Tuong, and Phong Dinh Provinces in carrying troops and blocking Viet Cong use of waterways.

In order to make the greatest possible use of its resources during the extended period of pacification in Kien Hoa, Dinh Tuong, and Long An Provinces, the Mobile Riverine Force was reorganized on 15 October. Mobile Riverine Group Alpha, supporting the 9th Infantry Division pacification operations, was assigned USS *Benewah*, USS *Askari*, USS *Sphinx*, USS *Westchester County*, *APL-26*, and *YLLC-4*. Mobile Riverine Group Bravo, carrying out more mobile operations ranging throughout the western delta region, was assigned USS *Mercer*, USS *Nueces*, USS *Vernon County*, and USS *Caroline County*. Five river assault divisions were assigned to Mobile Riverine Group Alpha, three to Mobile Riverine Group Bravo.

By the end of October Task Force 117 had conducted liaison and training with the 21st Vietnam Army Division and had joined the 4th Vietnam Marine Corps Battalion in Navy operations known as SEA LORDS (Southeast Asia Lake Ocean River Delta Strategy). An extensive survey of the western delta canals was also made in anticipation of future operations in that area.

In November the Mobile Riverine Force continued the pattern of operations which had begun after the reorganization in October, with Mobile Riverine Group Alpha operating in the eastern delta and Mobile Riverine Group Bravo operating to the west. Of the five river assault divisions assigned to Mobile Riverine Group Alpha, River Assault Division 91 supported the 3d Vietnam Marine Corps Battalion in operations in Kien Hoa Province; River Assault Division 92 was assigned base defense duties; River Assault Division 111 supported the 3d Battalion, 34th Artillery; River Assault Division 112 operated with the 3d Battalion, 60th Infantry, in Kien Hoa Province; and River Assault Division 151 worked with the 3d Battalion, 39th Infantry, in eastern Long An Province. Of the three river assault divisions assigned to Mobile Riverine Group Bravo, River Assault Division 121 provided for Dong Tam base

defense, River Assault Division 132 supported the SEA LORDS interdiction operations, and River Assault Division 152 supported the 4th Vietnam Marine Corps Battalion. River assault craft also supported other troop units in particular operations during the month. (*Table 2*)

TABLE 2–SEA LORDS OPERATIONS, NOVEMBER 1968–JANUARY 1969

Operations	Fire-fights	Munitions Caches Uncovered	Other Caches Uncovered	Confirmed Enemy Killed	Enemy Captured
SEARCH TURN (began 2 November 1968)	200	14 (11.0 tons)	1 (1.0 tons)	219	27
TRAN HUNG DAO (began 16 November 1968)	276	3 (11.4 tons)	0	470	26
GIANT SLINGSHOT (began 6 December 1968)	1,044	244 (137.0 tons)	22 (384.9 tons)	1,910	232
BARRIER REEF (began 2 January 1969)	77	1 (0.4 tons)	0	189	46

A significant incident occurred in November when USS *Westchester County* was mined as she lay at anchor with the other ships of Mobile Riverine Group Alpha on the My Tho River. At 0323 on 1 November, two explosions ripped separate holes on the starboard side of the LST. Three of the assault craft that were tied up alongside and two helicopters that were on board were damaged. Although the LST was not in danger of sinking, several compartments were flooded and internal blast damage was extensive. There was no damage to the ship's main engines or other machinery and she sailed to Dong Tam for emergency repairs before proceeding to a repair facility. The mining was apparently the work of enemy swimmers or sappers who penetrated, undetected, the base security forces. American casualties were 25 killed, 27 wounded, and 4 missing.

Enemy swimmers or sappers again struck the Mobile Riverine Force on the night of 15 November when the *YLLC-4* (salvage lift craft, light) was mined and sunk while at anchor on the Ham Luong River. Casualties in this instance were two killed and thirteen wounded. Because damage was extensive and the location hazardous, it was determined that salvage would be uneconomical; and the craft was destroyed to eliminate it as a navigational hazard.

On 13 November 1968 the 1st Brigade replaced the 3d Brigade at Dong Tam and the 3d Brigade returned to operate in Long An Province. This deployment moved the riverine-organized 3d Brigade away from the area that the Mobile Afloat Force concept had designated for follow-up riverine operations—the area south of the My Tho River. Part of Mobile Riverine Group Alpha continued to support the 3d Brigade in Long An Province and other U.S. forces in III Corps Tactical Zone.

Riverine warfare in December again involved interdiction, escort, patrol, base area search, and pacification. Small unit Navy actions were conducted over wide areas in Kien Hoa, Kien Giang, Chuong Thien, and An Xuyen Provinces. Mobile Riverine Group Alpha operations were confined for the most part to activity in Kien Hoa Province and support of the 2d Brigade, 9th Infantry Division. Mobile Riverine Group Bravo continued to carry out a variety of special operations in the southern delta in co-ordination with units of the 2d, 3d, and 4th Battalions of the Vietnam Marine Corps.

The 2d Brigade, supported by Mobile Riverine Group Alpha, engaged in a series of operations in December designed to keep constant pressure on the Viet Cong. With frequent insertions of troops in the Ham Long, Mo Cay, and Truc Giang Districts of Kien Hoa Province, 2d Brigade ground forces continued to seek out the enemy, especially members of the Viet Cong political organization. Of the many people detained for questioning, about 20 percent were subsequently classified as Viet Cong and the rest as innocent civilians.

Throughout January of 1969 the Mobile Riverine Force conducted operations in the provinces of Dinh Tuong, Vinh Long, Long An, and Kien Hoa as the Mobile Riverine Force again moved farther afield. On 7 January two battalions were landed in the Don Nhan district of western Kien Hoa Province, an area not previously penetrated by U.S. forces. Two platoons of enemy guerrillas were engaged by Army troops while assault craft established several water blocks to prevent the escape of the guerrillas. This strategy was intended to break up the enemy underground organization, disrupt enemy plans, demoralize enemy forces, and aid pacification. The Mobile Riverine Force used helicopters extensively in the operation, especially in January and February when the greatest number became available.

The activities of Mobile Riverine Group Alpha became more and more routine as troops were transported within Kien Hoa

Province in support of the 2d Brigade. Constant pressure was applied to enemy forces there in support of the allied program to commit more troops and time to the pacification effort. Mobile Riverine Group Bravo, operating with the 3d and 4th Battalions, Vietnam Marine Corps, advanced into Chuong Thien, Kien Giang, Phong Dinh, and northern Ba Xuyen Provinces.

In early January 1969 Mobile Riverine Group Alpha initiated a cordon and search operation on Thoi Son Island, in the My Tho River south of Dong Tam. Twenty-four river assault craft of River Assault Divisions 92 and 111, two infantry battalions, eight river patrol boats, and Vietnam Navy units participated in this operation which was directed against Viet Cong guerrillas and swimmer and sapper units based on the eastern end of the island. The island people were moved temporarily to three collection points, screened by the National Police and issued new identification cards. River craft assumed blocking stations and the island was swept by infantry units seeking Viet Cong. A total of 1,353 people voluntarily assembled at the three collection points or were apprehended in the sweep. Seventy Viet Cong were apprehended by the end of the operation on 7 January 1969.

On 12 January boats of River Assault Division 91 were withdrawn from operations preliminary to the planned turnover of 25 river assault craft to the Vietnam Navy on 1 February. Vietnamese naval crews had undergone on-the-job training with River Assault Division 91 units since early December 1968. River Assault Squadron 13 joined Mobile Riverine Group Alpha and River Assault Division 132 was relieved by River Assault Division 92 in Operation SEA LORDS. River Assault Division 132 returned to the Mobile Riverine Base in Kien Hoa and supported the 4th Battalion, 47th Infantry. River Assault Division 131 supported the 3d Battalion, 34th Artillery; River Assault Division 112 assumed support duties for the 3d Battalion, 60th Infantry, and River Assault Division 111 assumed base defense duties.

River Assault Squadron 15 had begun reconnaissance operations in an enemy base area in southern Kien Giang Province in late December 1968 and had continued these operations into January. Eighteen river assault craft, working in co-ordination with the 2d and 3d Battalions, Vietnam Marine Corps, conducted patrols and landed troops along the Can Cao Canal. Ground fighting with enemy forces was sporadic throughout the campaign, which lasted until 7 January. Despite intelligence reports of significant enemy forces in the area, U.S. and Vietnamese troops encountered only

small groups of the enemy that employed hit-and-run tactics.

This same river assault squadron, with the 2d Vietnam Marine Corps Battalion, began operations along the Song Cai Tu and Song Cai Lon Rivers in Kien Giong Province on 11 January to round up enemy forces suspected of being in the region. The enemy had used the territory before for storage and for offensive operations in Chuong Tien, Kien Giang, and Phong Dinh Provinces. U.S. and Vietnamese troops were landed at a number of beaches but found only enemy ground forces of squad size or smaller. After troop landings, river assault craft established blocking stations and checked a total of 189 sampans throughout the operation, which lasted until 18 January. The enemy attacked assault craft of the riverine force eight times. Of the total of twenty-six U.S. sailors wounded, five were injured when an ATC was sunk by a water mine on the Song Cai Tu. Two sailors and one Vietnam marine died when the mined craft sank almost immediately.

On 1 February the twenty-five river assault craft of Division 91 were transferred by the U.S. Navy to the Vietnam Navy. Combining these craft with eight assault support patrol boats received directly under the military assistance program, the Vietnam Navy formed two river assault and interdiction divisions, which conducted operations in conjunction with the U.S. Navy. Coincident with the turnover, River Assault Division 91 was dissolved.

Standard infantry operations by water and air continued in February as the Mobile Riverine Force operated in several districts of Kien Hoa Province in support of the 2d Brigade. On two occasions river assault craft came under particularly heavy and well-aimed enemy fire from the river banks. On 24 February, units of River Assault Division 131 received combined rocket and automatic weapons fire from both banks of the Song Ba Lai at a position three miles northeast of Ben Tre. Six craft were hit and slightly damaged; eleven U.S. sailors were wounded. Again, on 27 February, units of River Assault Division 112 came under heavy rocket, recoilless rifle, and automatic weapons fire from both banks of the Song Ba Lai at a position four miles northeast of Ben Tre. In this attack there were twelve U.S. Navy sailors wounded and five of the river assault craft received hits causing minor damage.

Enemy swimmers with SCUBA gear were sighted in close to the USS *Vernon County* (LST-1161) at Mobile Riverine Base Alpha at Dong Tam on two successive days—25 and 26 February. In both cases concussion grenades were dropped into the water, and subsequent hull inspections turned up no damage.

During the same period Mobile Riverine Force elements ambushed a Viet Cong water convoy. Elements of the 3d Battalion, 60th Infantry, in night positions along the Rach An Binh, three miles southeast of Mo Cay, detected three large motorized sampans moving northward along the waterway, firing into the banks. The soldiers quickly set up an ambush and took the sampans under fire. Twenty-one of the enemy were killed and the three sampans were captured with no American casualties.

Water blockades along the Song Can Tho and Kinh Xa Mo in Phong Dinh Province, begun on 30 January by Mobile Riverine Group Bravo, continued into February. The 21st Vietnam Army Infantry Division had been conducting reconnaissance in force to prevent massing of Viet Cong in the area. The river assault craft, working in conjunction with elements of the 21st Vietnam Army Division, established a water blockade along the Song Can Tho from the Cai Range Bridge in Phong Dinh Province to a point eight miles west of Can Tho, and along the Kinh Xa Mo from Thuan Mon to the junction of the Kinh Xa Mo and Song Can Tho. One company each of the 2d Vietnam Marine Corps Battalion and the 295th Phong Dinh Regional Force were embarked as reserves. During the operation, which lasted until 3 February, over 7,000 sampans were inspected. Ground forces reported only light and sporadic fire. Enemy swimmers were seen by Mobile Riverine Group Bravo on 5, 7, and 9 February, but in each case no swimmers were captured and there was no hull damage to boats.

Operating with 160 river assault craft in March because of the transfer of the twenty-five boats to the Vietnam Navy in February, the Navy elements of the Mobile Riverine Force again ranged the delta in river assault operations in co-ordination with the U.S. Army and Vietnam armed forces. Group Alpha continued operations in Kien Hoa Province with the 2d Brigade. Attacks against river assault craft continued into March and ground units engaged in heavy firefights on several occasions. Group Bravo and the Vietnam marines conducted a successful strike into Chuong Thien Province.

A plan for again combining both task groups of the Mobile Riverine Force took effect on 4 March as the units of Task Group Alpha put in at the My Tho anchorage. The single mobile riverine base allowed for a more efficient utilization of resources and reduced the number of troops needed for base defense from two river assault divisions to one, thereby allowing the extra division to be employed in offensive operations. Tactical flexibility was also maintained should the need arise for two task groups.

During April naval units of the Mobile Riverine Force participated in diversified operations with different missions and in various areas. Reconnaissance in force, patrolling, ambushes, troop transport, inspection, fire support, blocking, escort duties, and psychological warfare were carried out by river assault craft, often in conjunction with U.S. Army or Vietnam armed forces, as they operated from northern Long An Province to the southernmost reaches of the Ca Mau peninsula. The long-term riverine assault operations in Kien Hoa Province continued as the combined Army and Navy units continued to inflict losses on the enemy.

On 22 April River Assault Division 132 was on the Cai Tu River eight and one-half miles southwest of Vi Thanh. During the firefight, a water mine exploded sixty feet off the starboard bow of a command and control boat and caused minor flooding that was brought under control. Potential losses were averted on 24 April when a Viet Cong command-detonated mine was discovered attached to the antiswimmer net of an APL between the bow and ponton. The homemade mine, which was estimated to weight from 150 to 175 pounds, was discovered when an antiswimmer net was being raised as the ship prepared to get under way; inspection by mine disposal experts revealed the mine with its detonator lead severed. The detonator wire was probably cut during a minesweep patrol by a base defense boat. On 29 April an assault support patrol boat was sunk on the Cai Tu River as an enemy mine was detonated beneath its stern during an operation in Chuong Thien Province.

The need for further base and boat defensive measures was apparent in the number of thwarted and successful enemy swimmer attacks against the Mobile Riverine Force. Increased emphasis was placed on use of nets, concussion grenades, and hull inspections by underwater demolition teams.

The main thrust of Mobile Riverine Force operations in May continued against the enemy in Kien Hoa Province. However, three excursions lasting up to a week each were made by detachments of six to eight river assault craft into the Can Giuoc District of Long An Province in support of the 5th Battalion, 60th Infantry, of the 3d Brigade.

During June the U.S. Navy turned over to the Vietnam Navy sixty-four additional river assault craft valued at $18.5 million. (*Table 3*) Sailors of the Vietnam Navy had already trained on the job with U.S. Navy river assault craft crews with whom they had been integrated during the preceding months. River Assault

Squadrons 9 and 11 were decommissioned on 10 and 27 June, respectively.

TABLE 3—U.S. RIVER CRAFT TRANSFERRED TO VIETNAM NAVY IN JUNE 1969

Craft	10 June	15 June	21 June	Total
ATC	12	12	11	35
ASPB	8	8		16
Monitor	2	3	3	8
CCB	2	1	1	4
Refueler	1			1
Total	25	24	15	64

Mobile Riverine Force activities were concentrated in Kien Hoa Province during June and consisted primarily of assault landings, troop sweeps and ambushes, and blocking, escorting, and defoliation missions, although some psychological warfare and medical civic action programs were conducted as well. Operations of limited duration also took the force into Go Cong, Long An, and Vinh Binh Provinces. The total of 554 of the enemy killed by the Mobile Riverine Force in June, although substantial, was the lowest figure since January 1969, indicating a reduced tempo of operations.

Mobile Riverine Force operations were even more limited in July. On 12 June 1969 the 9th Infantry Division had been notified of its impending departure from Vietnam. The 2d Brigade was the first to be inactivated, probably because the 1st and 3d Brigades had had more experience in the area just south of Saigon where there was still a need for at least one brigade. The division was inactivated in phases. The 3d Battalion, 60th Infantry, left the country in early July. The river force divisions supporting the 9th Division were used chiefly in defense of Dong Tam Base, but they also supported Vietnam Army, Regional Forces, and Popular Forces units. River Assault Flotilla One at the end of July consisted of 113 craft in five divisions, two divisions assigned to River Assault Squadron 13 and three to Squadron 15. Sixty-three of the enemy were killed in July operations.

When the departure of the 9th Infantry Division was announced, there appeared in the press considerable conjecture on the fate of the Mekong Delta. Most articles were openly pessimistic about the ability of Republic of Vietnam forces to hold onto the gains made in the Mekong Delta after two and a half years of U.S.

operations there. A feature article of 28 July 1969, "Where the Reds Are Stopped in Vietnam" in *U.S. News & World Report,* was guardedly optimistic, and summed up the results of riverine warfare in the delta. Viet Cong defections, it declared, had increased to 300 per month from a pre- 1969 rate of 50, and the enemy was no longer able to mass forces as he had in the past. Roads and canals that had been closed for years were now open to public use. The Vietnam government appeared to be ruling more effectively in the rural areas and refugees were moving back to their land in larger numbers.

When asked "What happens once the 9th Division leaves?" Major General Harris W. Hollis, the division commander, was quoted in a *New York Times* article as saying: "The mission for which this division was sent to the Mekong Delta has been largely discharged—that of dealing with the Viet Cong and the North Vietnamese until such time as Republic of Vietnam Armed Forces can take over this function. These forces, whom we have assisted, are now ready to do this."

The Navy meanwhile was well on its way toward creating a U.S. and Vietnamese force to keep the Viet Cong from using the vital rivers and canals. With the realization that no additional U.S. Army forces would be placed afloat the Navy had begun in late 1968 to search for other ways to use its great number of river craft.

Operation SEA LORDS

As early as December 1968 the U.S. Navy, in co-ordination with the Vietnam Army and province forces, had begun a limited program to keep the enemy away from the rivers and canals in western Long An and Kien Tuong Provinces. In July and August 1969 as more river craft became available, the Navy initiated Operation SEA LORDS under the control of the newly formed Task Force 194 located at Can Tho. With the formal disestablishment of River Assault Flotilla One on 25 August 1969 all but one of the barracks ships were returned to the continental United States and decommissioned. The repair ships were taken over by the Naval Support Activity and dispersed throughout the delta to support SEA LORDS. The boats of two river assault squadrons were turned over to the Vietnam Navy and the remainder were absorbed by SEA LORDS.

A study conducted in preparation for SEA LORDS determined that the principal water route to exploit for the purpose of stopping

the Communist flow of supplies entering the IV Corps Tactical Area from Cambodia was the canal route linking Ha Tien on the Gulf of Thailand with Chau Doc on the upper Mekong River. This route, however, closely parallels the border and the risk of real or contrived border incidents dictated against placing a SEA LORDS barrier there until the plan had been tested in a less sensitive area. It was decided, therefore, that two parallel canals some thirty-five and forty miles southeast of the border would be used to form a double barrier and to inaugurate the SEA LORDS campaign. "Interdiction in depth had its attactiveness from an operations analysis point of view, and, at the same time, two waterways would be opened for friendly traffic. In conjunction with this barrier, river patrols would be strengthened from Long Xuyen through the Vam Nao Crossover to the Mekong."

The second objective of SEA LORDS—control of vital trans-delta inland waterways—would be accomplished by the removal of obstructions to navigation in the Cho Gao Canal linking the Vam Co and My Tho Rivers; by strike operations along the Mang Thit-Nicolai Canal, which joins the Co Chien and Bassac Rivers; and by reopening of the Bassac-Bac Lieu rice route in the lower delta. Penetration of rivers in the Cau Mau peninsula, the third objective, actually began before formal proposals were made to the senior adviser of IV Corps Tactical Zone, Major General George S. Eckhardt. A patrol boat incursion of the Cua Long River on 18 October 1968 is usually considered the first of the SEA LORDS operations, although a few earlier such operations were conducted by MARKET TIME.

General Eckhardt endorsed the Navy's proposals, and organization for SEA LORDS proceeded rapidly. Captain Robert S. Salzer, U.S. Navy, previously commander of the Navy component of the Mobile Riverine Force, was designated First Sea Lord and assigned a staff of nine officers and six enlisted men at General Eckhardt's headquarters in Can Tho. The Commander of Naval Forces, Vietnam, was given the task force number 194. Captain Salzer became commander of Task Group 194.0 and exercised operational control commencing 15 October 1968 of three other task groups: 194.5, Coastal Raiding and Blocking Group; 194.6, Riverine Raiding and Blocking Group; and 194.7, Riverine Strike Group. He was directed to designate one of the three task force commanders to command each specific SEA LORDS operation, with the commander of Task Force 115 usually expected to direct SEA LORDS invasions from the sea and the commander of Task

Force 116 the riverine strikes involving large commitments of ground forces.

On 2 November the first of the barrier campaigns, later given the name SEARCH TURN, was launched. Heavy craft of the Mobile Riverine Force and supporting ground troops succeeded in securing the waterways in a five-day operation that resulted in twenty-one Viet Cong killed and the capture of sizable quantities of arms and ammunition. A permanent naval patrol was then established, its main purpose to keep secure the western ends of the barrier and the network of perpendicular canals running north from Rach Gia to Ha Tien. According to intelligence reports, the principal Communist messengers passed through this territory.

While the operation was in progress other SEA LORDS forces were employed in clearing the Cho Gao Canal, which was open to navigation on 6 November. Later in the month Tan Dinh and Dung Island in the Bassac River were sealed off by a naval blockade while ground forces conducted sweep operations. MARKET TIME boats moved into areas in the Ca Mau peninsula that had long been considered exclusively the domain of the Viet Cong. The second of the interdiction barriers was established on 16 November 1968, and a third on 6 December 1968. Given the code name GIANT SLINGSHOT, this last campaign achieved the most dramatic and telling effects on enemy infiltration of all the interdiction barriers.

Thirty miles west of Saigon a peculiarly drawn border thrusts the Cambodian "Parrot's Beak" deep into Vietnam's III Corps area. The Parrot's Beak had long been known as a Communist base from which supplies moved across the border, and well-documented infiltration routes had been traced. These entered South Vietnam between the Vam Co Tay and Vam Co Dong Rivers; one turned south into the delta, the other east to supply the Viet Cong in the countryside surrounding the capital. The two rivers flow on either side of the Parrot's Beak on converging courses to the southeast. They appear to form a slingshot into which the Parrot's Beak neatly fits, hence the code name GIANT SLINGSHOT. For ground support in the GIANT SLINGSHOT operation, Rear Admiral William H. House, who had assumed the post of First Sea Lord upon Captain Salzer's departure, called on the commanding general of II Field Force, Vietnam, Lieutenant General Frederick C. Weyand, and got his endorsement and promise of help.

Because of the distances involved and because it was impossible to bring large ships up river beyond the low bridges at Tan An and Ben Luc, a new basing and support scheme was devised for GIANT

SLINGSHOT. An advance tactical support base was designed, consisting of three or more 30x90-foot Ammi barges on which berthing and messing facilities, storerooms, and a tactical operations center could be constructed and which would contain water purification equipment, generators, and other assorted machinery.

Advance tactical support bases were established at Tuyen Nhon and Moc Hoa on the Vam Co Tay and at Tra Cu and Hiep Hoa on the Vam Co Dong. The site at Hiep Hoa was later abandoned in favor of one at Go Dau Ha and in July 1969 an additional base was placed at Ben Keo near the important city of Tay Ninh. The USS *Askari* and the berthing and messing barge for underwater repair crews were at Tan An, awaiting the arrival of Mobile Base II, a sophisticated four-Ammi complex, specially constructed in the United States with improved berthing and messing facilities and equipped to make extensive boat repairs. The USS *Harnett County* provided support at Ben Luc for the patrol boats until the completion of a shore support base there.

From its very inception, GIANT SLINGSHOT was characterized by frequent, heavy clashes between patrol boats and enemy forces intent on maintaining the lines of communication to their Cambodian storehouse. Extremely large quantities of arms, munitions, and supplies were uncovered in caches buried along the river banks, proving beyond doubt that vital enemy infiltration lines had been cut.

The fourth and last of the interdiction barriers in the Mekong Delta was established on 2 January 1969 when naval patrols began operations on the La Grange–Ong Lon Canal from Tuyen Nhon on the Vam Co Tay River to An Long on the Mekong. Called BARRIER REEF, this operation joined GIANT SLINGSHOT in the east with the two interdiction barriers in the west, SEARCH TURN and TRAN HUNG DAO. The northern ring was then complete; the enemy was no longer free to move men and supplies with impunity from Cambodia into III and IV Corps areas. A naval "tariff" had been imposed on those shipments, measurable in terms of men and supplies captured or destroyed, but the supplies and men that were prevented from getting through cannot be calculated.

Once the barriers were in place, the boats were used to counter enemy pressures and probes. With relatively little disruption to existing organization and logistics, boats were shifted rapidly from one area of operations to another. For example in Operation DOUBLE SHIFT in July 1969, 105 U.S. Navy and Vietnamese Navy boats were quickly concentrated on the waters of the Vam Co Dong north of Go Dau Ha in response to serious enemy threats

to the city of Tay Ninh. This sudden and impressive display of naval power, directed by the commander on the scene, Lieutenant Commander Thomas K. Anderson, U.S. Navy, was credited by the commander of II Field Force, Lieutenant General Julian J. Ewell, with having prevented the enemy attack on Tay Ninh from the southwest.

On two separate occasions the mobility of the boats was increased when giant Army skycrane helicopters were used to lift some of them into new areas of operations. In May 1969 six river patrol boats were skyhooked to the upper Saigon River, and in June six more were lifted to the otherwise inaccessible Cai Cai Canal. Both operations achieved tactical surprise.

As each operation progressed, concerted efforts were made to integrate units of the growing Vietnamese Navy. The military desirability and political necessity to "Vietnamize" the naval war were evident long in advance of SEA LORDS planning. It seemed obvious that the Vietnamese Navy's hope of relieving the U.S. Navy of its operational responsibilities in the war as soon as possible would be considerably enhanced by success of SEA LORDS. SEA LORDS was a remarkable organization which came to mean unity of command and rapid response to changing tactical situations. Relatively junior officers and men were often placed in positions of extraordinary responsibility. Tactics and techniques were developed and tested in the heat of combat, and at times even borrowed from the enemy, improved upon, and used to defeat him. The waterborne guardpost, for instance, was a refinement of the favorite enemy tactic of ambush. Using silent boats and night observation devices, the U.S. and Vietnamese boats stalked the would-be Viet Cong ambusher and trapped him. Imagination and leadership provided the plan and a dedicated collection of sailors, soldiers, and airmen made it work.

In addition to use in SEA LORDS, the United States rivercraft turned over to the Vietnam Navy at Can Tho were used to support Vietnam Marine Corps and Vietnam Army units in the central delta. Operations were conducted in the style of the Mobile Riverine Force, but greater emphasis was placed on leaving security forces in areas that had been cleared so that the Vietnam government could continue to control them. The Army-Navy Mobile Riverine Force, which had operated in the Mekong Delta since February 1967, was disestablished on 25 August 1969 and River Flotilla One and 2d Brigade, 9th Infantry Division, were inactivated. The experience of these units was inherited by SEA LORDS and ultimately by the Vietnam Army and Navy.

Armored Troop Carriers Move Up the My Tho River in the Early Morning

CHAPTER X

Conclusions and Summary

The joint river operations conducted by the U.S. Army and Navy in South Vietnam contributed to the success of the military campaign in the Mekong Delta and added substantially to U.S. knowledge of riverine operations.

The strategic concept embodied in the plan for the Mekong Delta Mobile Afloat Force and approved by General Westmoreland in 1966 proved sound and provided a workable blueprint for a variety of projects carried out by the Mobile Riverine Force in the two years it operated. The concept recognized the importance of the Mekong Delta and its resources to the whole conflict. Although the operations planned for the river force ranged from the vicinity of Saigon south through the delta to Mui Bai Bung (Pointe de Ca Mau or Camau Point) at the southern tip of Vietnam, early priority was given to areas in southern III Corps and northern IV Corps. In executing the plan, the Mobile Riverine Force was faithful to the priority assigned to the northern delta, but never carried out fully the intent of the plan approved by General Westmoreland to extend and sustain river operations south of the Bassac River. The operations remained primarily north of the Bassac River in keeping with the area of responsibility assigned to the 9th Infantry Division in that area; the increasing role of the 2d Brigade in the division's plans; and the decrease in operational control of the force by the senior adviser of the IV Corps Tactical Zone.

Perhaps the most significant organizational aspect of the Mobile Afloat Force concept was the integration of Army and Navy units to provide a force uniquely tailored to the nature of the area of operations. Specifically, the capabilities of the two services were used to the fullest by combining tactical movement of maneuver and fire support units by land, air, and water. To this combination was added the close support of the U.S. Air Force. The Mobile Afloat Force plan stipulated that these mobile resources were not to be dissipated on independent U.S. operations but were to be used in close co-ordination with other U.S. and Free World Military Assistance Forces.

The Mobile Riverine Force, which was the organizational implementation of the Mobile Afloat Force concept, was a fortuitous union of the Navy's River Assault Flotilla One and the Army's 2d Brigade, 9th Infantry Division. Somewhat controlled by circumstances, events,. and time, the alliance depended upon a spirit of co-operation and the initiative of the two component commanders in the Mobile Riverine Force. Guidance for the commanders was largely contained in the Mobile Afloat Force plan itself, and from the outset it was up to the commanders to make the plan work. Although many innovations were made to improve equipment and procedures, they caused little deviation from the basic Mobile Afloat Force plan. The most important innovation was the mounting of artillery on barges to provide the force with the direct support of an artillery battalion that was so urgently needed. Another innovation was the building of helicopter landing platforms on armored troop carriers and the use of a helicopter landing barge as an integral part of the forward brigade tactical command post. The adaptation by the naval commander of placing the Ammi pontons alongside the Mobile Riverine Base ships eliminated the need for cargo nets and enabled an entire company to transfer from a barracks ship to assault craft safely in less than twenty minutes.

The decision of Secretary of Defense McNamara in late 1966 to cut the requested number of self-propelled barracks ships by three eliminated berthing space for two of three infantry battalions. The Navy, however, resourcefully provided space for one of the two battalions by giving the force an APL—a barracks barge—and a larger LST of the 1152 class. The Secretary of Defense's decision could have been disastrous to the execution of the Mobile Afloat Force plan had these innovations by the Navy not been made. Even so, the brigade was forced to operate without the third maneuver battalion; the Army commander was obliged to double his efforts to secure the co-operation of Vietnam Army and other U.S. units in order to make the tactical operations of the Mobile Riverine Force successful. It can be postulated that this effort by the brigade commander to assemble a sufficient force consumed much attention and energy that could have been applied to other problems. It can also be argued that the shortage of men and the necessity to operate in conjunction with the Vietnam Army generated successes that might not otherwise have been achieved if the full maneuver force had been provided and if the force had operated less frequently in co-ordination with Vietnam Army units. Although the original Mobile Afloat Force concept provided for co-operative

and co-ordinated efforts with the Vietnam Army, the shortage of troops increased the need for provincial and divisional units of the Vietnam Army throughout the various areas in which the Mobile Riverine Force operated.

It should be noted that the mobile riverine base provided for in the original plan made the force unique. When one considers that this Army and Navy force of approximately 5,000 men, capable of combat and containing within itself combat service support, could be moved from 100 to 200 kilometers in a 24-hour period and could then launch a day or night operation within 30 minutes after anchoring, its true potential is apparent. With such capabilities the force was able to carry out wide-ranging operations into previously inaccessible or remote Viet Cong territory.

The original Mobile Afloat Force concept had drawn heavily on the successful features of French riverine operations and the subsequent experience of the Vietnamese river assault groups. The Mobile Riverine Force was able also to capitalize on the knowledge of shortcomings of these earlier experiences. Both the French and the Vietnamese had used a fixed operational base on land. French units were small, rarely consisting of more than a company and five or six combat assault craft. Vietnam Army ground commanders merely used the river assault groups to transport their forces. Neither the French nor the Vietnamese had joint ground and naval forces. Both lacked the helicopter for command and control, logistic resupply, medical evacuation, and, most important, reconnaissance and troop movement. The river bases of French and Vietnam armed forces proved more vulnerable, especially during darkness, than did the afloat base of the Mobile Riverine Force. Finally, the Mobile Riverine Force, because of its mobility, strength of numbers, and Army-Navy co-operation, was capable of sustained operations along a water line of communications that permitted a concentration of force against widely separated enemy base areas. This was not true of the French or Vietnam Army riverine operations because of the small size of the forces and their dependence on a fixed base.

While the problem of command relationships did not inhibit the operations of the Mobile Riverine Force, it was a tender point in the conduct of all activities. Considering that the Army effort to develop riverine doctrine was not accepted by the Navy component commander at the outset, the Mobile Riverine Force might have been faced with insurmountable co-ordination problems, but such was not the case. Relying on the Mobile Afloat Force plan,

each component commander had a guide to follow as to common objectives and procedures until doctrine was refined by combat experience.

No joint training for riverine operations was given in the United States because of lack of time and the wide geographic separation of Mobile Riverine Force units, but early joint training in Vietnam was planned. Fortuitously, this training commenced at the river assault squadron–battalion level under the supervision of the advance staff of River Assault Flotilla One at Vung Tau. Although the training in the Rung Sat Special Zone was under combat conditions, tactics and techniques were developed at the boat and platoon, company and river division, battalion and river assault squadron levels. During the period February–May 1967 the flotilla and brigade staffs were able to arrive at a common understanding as to organizational procedures and operational concepts. By starting at the boat and platoon level, the Mobile Riverine Force procedures were built on the needs of the lowest level units, and brigade and flotilla command and staff procedures were developed to meet these needs. With the 2d Brigade conducting tactical operations in the delta during the dry season of early 1967 and the flotilla staff co-ordinating the Rung Sat operations involving 2d Brigade battalions, both staffs gained valuable riverine experience independently. Later, as part of the transition into the Mobile Riverine Force, the advance staff of the flotilla joined the brigade staff in April at Dong Tam. This also provided a necessary step from the single battalion and river assault squadron level of operations to multibattalion, brigade and flotilla operations in late April and May of 1967.

The Mobile Riverine Force command relationships as published in a planning directive from Headquarters, Military Assistance Command, Vietnam, were a compromise to obtain Navy participation in the Mobile Riverine Force. Recognizing an ambiguous, undefined division of command responsibilities between the Army and Navy commanders, the directive compensated for this in part by instructing the chain of command of both the Army and the Navy at each level within the theater to insure co-operative effort. It also instructed the commanders at the lowest level, preferably at the Army and Navy component level of the Mobile Riverine Force, to resolve any problems that might arise. This planning directive, which incorporated the doctrine of close support and mutual co-operation and co-ordination, successfully established the tenor of what was to follow. The Navy was placed in a close support

role that adhered closely to the Army principle of direct support. The Navy's decision in mid-1966 to make command and security of the Mobile Riverine Base an Army responsibility resolved a problem that could have become very difficult. Actually the Navy commander assumed tactical control only while the Mobile Riverine Base was being relocated. In all other instances he was cast in the role of supporting commander as long as Navy doctrine was not violated. The Army commander, acting in response to directives, operational plans, and orders from his higher headquarters. determined the plan of operations, which required the naval commander to support his objective. The major factor influencing the relationship between the Army and Navy commanders was the Navy commander's wish to operate independently of the 9th U.S. Division. This insistence provided the one element of disharmony in planning and conducting operations as an integral part of the land campaign for the delta. While it was never stated, it can be assumed that the Navy commander's position reflected the wishes of his Navy chain of command to have separate operations. In defense of the Navy's total effort, however, it can be said that the Navy commander was quick to accede to the basic concept of employing the Mobile Riverine Force in co-ordination with land operations. The activities of the Mobile Riverine Force, in fact, were directly related to the total land campaign being conducted in both III and IV Corps Tactical Zones. Joint operational orders were published and the relationship of the Army and Navy staffs in all planning and operational functions, including the establishment and manning of a joint operations center, was very close.

Perhaps the only area where a clear delineation of responsibility was not possible was at the boat and river division, river assault squadron, and battalion levels. The Navy river assault squadron organization did not parallel that of the Army battalion. Both the platoon leader and the company commander were dealing with Navy enlisted men, while the battalion commander was dealing with a lieutenant senior grade or a lieutenant commander without a comparable supporting staff equivalent to the staff of the battalion he was supporting. The Navy was thus placed in an awkward position, particularly in the event of enemy attack against an assault craft convoy. During the first six months to a year of operations, and prior to rotation of the initial Navy and Army commanders, it was normal precedure for the Navy element to answer directly to the Army officer in command during assault craft movements. The Navy craft did not use direct fire unless explicitly authorized by

the brigade commander. As time passed and commanders rotated, however, the precedures established in the Rung Sat and in the first six months of Mobile Riverine Force operations became vague and ambiguous since they were not committed to written joint standing operating procedures. The difference in staffs below the brigade and river flotilla level fostered a centralization at the flotilla and brigade level in planning, control, and execution. While the Mobile Riverine Force operations gradually adhered more to the principle of decentralized planning and execution, future Mobile Riverine Force organizations should correct the disparity in the command and control organizations between echelons of the two services.

The first year of operations of the Mobile Riverine Force was highly successful because the original Mobile Afloat Force concept was carried out faithfully. The operations of the Mobile Riverine Force were wide-ranging—the force was in combat in nine provinces and the Rung Sat Special Zone. These were essentially strike operations against remote enemy base areas that in some instances had not been penetrated in force for two or three years. From these base areas the main force Viet Cong units and the political underground had influenced the local population whose support was vital to the strategy of the Viet Cong. Because of the bold and frequent movement of the large Mobile Riverine Base from which strike operations could be launched with ease, the element of surprise so important to combat success was achieved. In most cases enemy defenses and tactics were directed toward evasion or resistance to air and land assaults. Early riverine operations often capitalized on these enemy dispositions. Later, when the enemy learned to orient his defenses toward the waterways, the 9th Division commander provided the helicopter support necessary to enable troops to maneuver rapidly from the land side against the enemy. As the first U.S. maneuver unit to conduct sustained operations in the IV Corps Tactical Zone, the Mobile Riverine Force developed good relationships with the commander of the 7th Vietnam Army Division and elements of the 25th Vietnam Army Division. Co-ordinated large-scale operations were conducted in a number of remote areas, contributing to the erosion of Viet Cong strength, which before the advent of the U.S. forces in the area had been equal to that of the Vietnam Army. While the efforts of the Mobile Riverine Force were primarily concentrated against Long An and Dinh Tuong Provinces, key economic provinces for control of the delta, the force also was able to strike in Go Cong and Kien Hoa Provinces. Although only indirectly related to pacifi-

cation, the limitations imposed on Viet Cong movement and the losses inflicted on Viet Cong units resulted in a reduction in the influence of the Viet Cong on the people in the area.

During the *Tet* offensive of January–February 1968, the Mobile Riverine Force was used in succession against Viet Cong forces in the populous cities of My Tho, Vinh Long, and Can Tho, which were seriously threatened. After the battles for these cities, the Mobile Riverine Force was credited with having "saved the delta" by its direct action against the enemy in these important centers before the Vietnam Army was able to rally its forces. Here again the fact that this large, concentrated force with its own base could be moved so rapidly over such great distances was the key to the Mobile Riverine Force's success against the Viet Cong in the IV Corps Tactical Zone.

During the spring of 1968 when the Mobile Riverine Force was placed under the operational control of the senior adviser of IV Corps, it again successfully penetrated remote areas. The IV Corps, however, had not the aircraft or supplies to sustain Mobile Riverine Force operations, and the force was therefore available only intermittently to the senior adviser of the IV Corps Tactical Zone.

When 9th Division headquarters moved from Bearcat to Dong Tam in August 1968, its mission was concentrated in Long An, Dinh Tuong, and Kien Hoa Provinces. (*Map 16*) With this focusing of the area of responsibility on the Mekong Delta, it can be assumed that the division commander strongly wished to integrate the Mobile Riverine Force into the divisional effort. Further, a renewed emphasis on pacification shifted the strategy away from strike operations, and as a consequence the Mobile Riverine Force largely concentrated on Kien Hoa Province. During the late summer of 1968 helicopters for troop lift were almost eliminated from the support of the force. The 9th U.S. Division decided to provide airlift chiefly for the other two operating brigades and to place almost total reliance on water movement for the 2d Brigade. This decision was a deviation from the initial operational plan of employing the Mobile Riverine Force for strikes utilizing boats and helicopters, a plan that had proved successful in the previous year's operations. Not until October of 1968 was the Mobile Riverine Force again provided with helicopters in keeping with the initial concept.

Restriction to one geographical area had limited the force in mid-1968, especially in respect to attempting surprise, and it was

MAP 16

obliged to resort to other means of deception. The Viet Cong were able, however, to analyze and anticipate movements on the waterways, reportedly by using a warning system established along the banks. Because of limited and predictable water routes and a growing enemy knowledge of the Mobile Riverine Force, river ambushes became more common. When aircraft were lacking, the force was unable to retaliate except from the water. Nor was it any longer permitted to make the long moves to new areas as set forth in the Mobile Afloat Force concept. With the full use of helicopters beginning in October, the force produced results comparable with

or superior to those of other 9th Division brigades and with the results obtained by the Mobile Riverine Force during its first year in South Vietnam.

The Mobile Riverine Force made significant contributions to the war in Vietnam. Its presence in 1967 and 1968 tipped the balance of power in the northern portion of the Mekong Delta in favor of the U.S. and South Vietnam forces. Dong Tam was developed as a division base without reducing the firm ground available to Vietnamese units in the delta and activities of the base increased the security of the important nearby city of My Tho. The Dong Tam area at one time had been under strong Viet Cong influence, and main force and local enemy units moved virtually at will until U.S. occupation of Dong Tam began in January 1967. As operations by battalions slated to join the Mobile Riverine Force continued, both the 514th and the 263d Viet Cong Battalions were brushed back from the populated area into the more remote Plain of Reeds. Even though the Viet Cong 261st Main Force Battalion was brought as a reinforcement from Kien Hoa into Dinh Tuong Province in June of 1967, the combined operations of the Mobile Riverine Force and 7th Vietnam Army Division kept the Viet Cong from moving freely around Dong Tam. Riverine operations inflicted significant casualties on Viet Cong units and made them less effective. Highway 4, the main ground artery of the delta, which was often closed to traffic in the period 1965 through early 1967, was opened and the farm produce of the delta, both for domestic and export purposes, could reach the markets. With the completion of Dong Tam Base, the 9th Division headquarters and three brigades were finally able to move into Dinh Tuong Province and the security of the northern portion of the delta was vastly improved. When the Navy extended its efforts to the Plain of Reeds and far to the west toward the Cambodian border in late 1968 and 1969, its operations were made easier by the earlier operations of the Mobile Riverine Force during 1967 and 1968. The Navy SEA LORDS operations evolved from the concept that fielded the Mobile Riverine Force and GAME WARDEN operations.

The Mobile Riverine Force wrote a distinct chapter in U.S. military history. The joint contributions made by the Army and Navy resulted in the accumulation of a body of knowledge that has been translated into service publications setting forth joint doctrine on riverine operations. In the event of future riverine operations, the service doctrine recognizes the need for a joint task force com-

mander to provide unity of command. Those involved in the early operations of the Mobile Riverine Force possessed no prior riverine experience and were forced to rely on historic examples, their own judgment, and related Army and Navy doctrine to build a new American force. While basic service differences did arise from time to time, those immediately responsible at all echelons from the soldier and sailor on up found reasonable solutions and carried them out effectively and harmoniously to the credit of both services.

Glossary

AK47	Chinese-made rifle
APA	Attack transport
APB	Self-propelled barracks ship
APC	Armored personnel carrier
APL	Barracks ship, non-self-propelled
ARL	Landing craft repair ship
ASPB	Assault support patrol boat
ATC	Armored troop carrier
B–40	Chinese-made rocket
Barrier Reef	The fourth and last of the SEA LORDS interdiction barrier operations in the western delta, commencing 2 January 1969
CCB	Command and communications boat
CIDG	Civilian Irregular Defense Group
Coronado I	A Mobile Riverine Force operation designed to destroy Cho Gao District Company and enhance the security of Cho Gao Canal in Dinh Tuong Province, 1 June–26 July 1967
Coronado III	A Mobile Riverine Force operation in the Rung Sat Special Zone designed to disrupt possible Viet Cong attacks on shipping in the Long Tau channel, 5–17 August 1967
Coronado V	A combined airmobile, mechanized, and riverine operation in Cam Son Secret Zone, Dinh Tuong Province, and Ham Long District, Kien Hoa Province, 12 September–8 October 1967
Coronado VI	An operation in the lower Rung Sat Special Zone, 11–18 October 1967
Coronado VIII	A Mobile Riverine Force operation in Nhon Trach and the Rung Sat Special Zone, 27–29 October 1967
Coronado IX	A Mobile Riverine Force operation involving deep riverine penetrations into enemy re-

	doubts and sanctuaries to achieve decisive action in Dinh Tuong Province, 1 November 1967–21 January 1968
CORONADO X	An operation in which the Mobile Riverine Force responded to Viet Cong attacks on My Tho and Vinh Long, drove the enemy from those provincial capitals, then operated near Cai Lay on Highway 4, 18 January–13 February 1968
CORONADO XI	A combined riverine and airmobile operation to relieve pressure on Can Tho and locate and destroy Viet Cong units in Phong Dinh and Ba Xuyen provinces, 13 February–6 March 1968
Dinassauts	*Divisions navales d'assaut* (French naval assault divisions)
Duong Cua Dan (People's Road)	An operation by the 1st and later 3d Brigade, 9th Infantry Division, to protect and upgrade Highway 4 northwest of My Tho and to conduct reconnaissance in force operations on both sides of Highway 4 in Dinh Tuong Province, 16 March–30 July 1968
E8	Expendable launcher. Lightweight and portable, it fires 16 riot-control canisters in each of four volleys, producing a rapid buildup of a large chemical cloud.
G–2	Assistant chief of staff for military intelligence at an army, corps (field force in Vietnam), or division headquarters
G–3	Assistant chief of staff for operations at an army, corps (field force in Vietnam), or division headquarters
G–4	Assistant chief of staff for supply at an army, corps (field force in Vietnam), or division headquarters
GAME WARDEN	U.S. Navy (Task Force 116) operation designated to thwart Viet Cong use of waterways as supply and infiltration routes
GIANT SLINGSHOT	The third SEA LORDS operation, commencing 6 December 1968, in the Parrot's Beak

GREAT BEND	A Mobile Riverine Force operation in the western Rung Sat Special Zone, 13 June–26 July 1967
HOMESTEAD	A Mobile Riverine Force operation in Kien Hoa Province, September 1968
IPW	Prisoner of war interrogation
J–3	Assistant chief of staff for military operations, MACV
JACKSTAY	A U.S. Marine Corps operation in the Rung Sat Special Zone in late 1966
JP–4	A kerosene-based fuel for turbine engines
LCA	Landing craft, assault
LCI	Landing craft, infantry
LCM	Landing craft, mechanized
LCT	Landing craft, tank
LCU	Landing craft, utility, a light lift craft
LCVP	Landing craft, vehicle or personnel
LSIL	Landing ship, infantry, large
LSM	Landing ship, medium
LSSL	Landing ship, support, large
LST	Landing ship, tank
LZ	Landing zone
M16	Current standard American rifle
M43	Soviet-made rifle firing intermediate 7.62–mm. cartridges
M102	American-made 105-mm. howitzer
MACV	U.S. Military Assistant Command, Vietnam
MARKET TIME	U.S. Navy (Task Force 115) operation designed to seal the coast of South Vietnam against infiltration of enemy troops and supplies
MCS	Mine countermeasures support ship
MDMAF	Mekong Delta Mobile Afloat Force
Monitor	An armored gunboat armed with 20- and 40-mm. guns and 81-mm. direct fire mortars
MRB	Mobile Riverine Base
N–1	Navy flotilla officer in charge of administration

N–2 Navy flotilla officer in charge of intelligence
N–3 Navy flotilla officer in charge of operations
N–4 Navy flotilla officer in charge of supply

POL Petroleum, oils, and lubricants

RIVER RAIDER I An operation in the upper Long Tau shipping channel and in the southwestern Rung Sat Special Zone, 16 February–20 March 1967, the first joint operation of Army and Navy forces which were to constitute the Mobile Riverine Force
ROAD Reorganization Objective Army Divisions
RPD Chinese-made light machine gun
RPG2 Soviet-made antitank rocket
RSSZ Rung Sat Special Zone

S–1 Adjutant of a brigade or smaller unit
S–2 Officer in charge of the military intelligence section of a brigade or smaller unit
S–3 Officer in charge of the operations and training section of a brigade or smaller unit
S–4 Officer in charge of the supply and evacuation section of a brigade or smaller unit
S–5 Civil affairs officer of a brigade or smaller unit
SEAL Navy sea-air-land teams
SEA LORDS The U.S. Navy's Southeast Asia Lake, Ocean, River, Delta Strategy Operations
SEARCH TURN The first of the SEA LORDS barrier operations in the western delta, commencing 2 November 1968
SKS Chinese-made carbine

TRAN HUNG DAO The second of the SEA LORDS barrier operations in the western delta, commencing 16 November 1968
TRUONG CONG DINH A riverine operation initially designed to drive enemy forces away from My Tho and eliminate enemy interdiction of Highway 4, 7 March–2 April 1968

WATER TRAP A major cordon and search operation of Thoi Son Island in the My Tho River south of

Dong Tam during the first week of January 1969

YFNB Large covered lighter
YLLC Salvage lift craft, light
YTB Large harbor tug

Index

☆ U.S. GOVERNMENT PRINTING OFFICE : 1985 O - 470-397

Made in United States
Cleveland, OH
09 May 2025

16802026R00122

Printing: Krips bv, Meppel
Binding: Stürtz, Würzburg

VIETNAM STUDIES

RIVERINE OPERATIONS 1966-1969

by

Major General William B. Fulton

DEPARTMENT OF THE ARMY

WASHINGTON, D.C., 1985

Series Note

This book will introduce Chinese Legal Research to lawyers, scholars, businesspersons, and others who have practical needs and interests in learning about law and lawmaking in the People's Republic of China. This book is the second to appear in the new *International Legal Research Series*, started by Carolina Academic Press to provide clear, practical advice and direction on research topics not normally taught in any law school. This book is particularly valuable because it also focuses on the most accessible, trustworthy, and cost-effective research sources for researchers who cannot read Chinese. This book is meant to assist researchers who need to find answers and direction for their work.

This book is also a companion to Paul Kossof's book on the new Chinese Trademark Law, a book published by Carolina Academic in its *International Legal Practice Series.* The two volumes together on Chinese Trademark Law and Chinese Legal Research are a valuable contribution to understanding law and lawmaking in China.

Prof. Mark E. Wojcik
The John Marshall Law School
Chicago, Illinois

Preface and Acknowledgements

Having conducted extensive research on Chinese law over the past several years, I came to the realization that foreign (non-Chinese) researchers would greatly benefit from a condensed guide to performing research on Chinese law.

When I first began researching Chinese laws and regulations, I was at a distinct disadvantage because I did not have a background in Chinese legal studies. This forced me to rely heavily on Chinese colleagues. Although I had access to English translations of Chinese regulations, it was difficult to grasp the larger picture of how to conduct research on Chinese law.

During my first "tour" in China, I was only proficient in reading Chinese legal text. Fellow Chinese researchers encountered mixed levels of success in explaining the background of legal issues because I had difficulties reading secondary sources.

I have written *Chinese Legal Research* to be a thorough tool for foreigners researching Chinese law. As foreign researchers possess varying levels of Mandarin fluency, this book is written to be fully accessible to researchers that do not read Mandarin. *Chinese Legal Research* is an introduction to performing legal research on Chinese law. Although it guides the reader through the essentials of Chinese legal research, it is my firm belief that a highly effective legal researcher must be able to read Chinese regulations and secondary sources.

I thank Laney Zhang, Senior Foreign Law Specialist at the United States Law Library of Congress, for her invaluable review of my manuscript and Professor Mark E. Wojcik from The John Marshall Law School for his advice on how to create this book on Chinese legal research.

I dedicate this book to my wife, Xiaoyun Cao.

Paul Kossof
April 2014

Administrative Map of China

Source: The World Factbook, Central Intelligence Agency, https://www.cia.gov/library/publications/the-world-factbook/geos/ch.html

Chinese Legal Research

中国法律研究

Chapter One

Introduction to Legal Research in China

China's staggering development over the past two decades has attracted an immense volume of foreign investment. Initially, only large transnational companies diverted funds to establishing a presence in China. However, as China became an increasingly large presence in the world economy, many other foreigners began to enter the Chinese market. Foreign law firms have followed their clients, generally opening offices in China's most developed cities: Beijing, Guangzhou, and Shanghai.

China lends an extraordinary amount of money to many other countries and produces the majority of many of the world's products. Its laws and regulations are often in the international spotlight not only due to economic reasons but also because of societal and human rights issues.

I. Difficulties Facing Foreign Researchers

Despite the high demand for foreign legal services, primarily originating from businesses headquartered abroad, foreign legal counsel are generally very hesitant to conduct research on Chinese legal issues.

First and foremost, most foreign attorneys simply cannot read Mandarin. They rely on translations, advice from Chinese researchers, and foreign legal blogs. Second, foreign attorneys received their legal training abroad. Most of them do not have any formal Chinese legal education.

Additionally, some foreign attorneys such as those from Britain, Canada, and the United States come from common law jurisdictions. This creates additional confusion because China employs a civil law system.

Chinese laws and regulations are constantly changing. Not only do local governments and administrations create regulations, but the national government frequently amends national legislation. In fact, many of the national laws under review in 2013 were implemented only a few years ago. This uncertain legal atmosphere can be overwhelming to foreign attorneys.

II. Welcome to *Chinese Legal Research*

The overall goal of this book is to provide foreign legal researchers with all of the background, tools, resources, and tips to effectively research Chinese law without learning written Mandarin. This book aspires to provide foreign legal researchers with a broad understanding of China's legal system, and specifically tailors itself to foreign researchers with little to no experience in China by minimizing their reliance on Mandarin text.

Chinese Legal Research begins by exploring China's legal system, its constitution, government, and legislation. It then provides a detailed examination of China's administrations and their relationship with national and local governments.

This book continues by explaining Chinese treaties. Finally, it discusses secondary sources, internet sources, and citations in China. Each chapter also includes a section of additional resources.

The appendix section includes some of China's most important laws, lists of treaties ratified by China, and useful contact information.

Chapter Two

The Chinese Legal System

Legal researchers often overlook the differences between their own legal systems and the systems that they are researching. Foreign legal researchers should be cognizant of the influence of both the civil law system and China's legal history on its contemporary legal structure.

I. History of Chinese Law

Chinese law is a combination of Chinese traditional legal values, Marxist ideals, foreign influence, and contemporary Chinese legal thought. Foreign researchers may be confused when reading Chinese laws because they often combine traditional values with foreign legal theories.

For example, the first clause of every national law lists the traditional and Communist values that inspired the law. Conversely, the rest of the law may originate mostly from contemporary foreign and Chinese legal theories. Understanding these influences will assist researchers to recognize the origins of Chinese laws and regulations.

A. Ancient China

Chinese traditional philosophy, legal theory, and even theology still influence Chinese legal procedures. Primarily, Confucian philosophy has an effect on Chinese conception of morals and ethics. One of the clearest remnants of Confucian ideology is the belief that people and their behaviors cannot be fully controlled through strict regulation. According to Confucianism, laws are most effective against criminal behavior, while other aspects of life should be balanced through the application of moral philosophy.

Ancient China generally kept social order through the *baojia* (保甲) system. This created a hierarchy wherein each family's head of the household was responsible for that family's actions. Each group of one hundred families, the *jia*, elected a leader. Ten *jia* (one thousand families) composed a *bao*, which also had one leader.

Chinese legal philosophy changed drastically during the Han dynasty. Before the Han dynasty, Chinese legal code was generally limited to criminal laws and land regulations. The Han dynasty saw a substantial increase in the amount of regulations on other social issues. "Han legalism" remained relatively constant during China's various dynasties, and ended with the Qing dynasty and the Kuomintang government.

B. 1912–1949

The Kuomintang government established the Republic of China ("ROC") and what could be considered as the beginning of China's contemporary legal system. Before the Kuomintang government, Chinese legal scholars strongly resisted foreign influence. They believed that Chinese law should be exclusively a result of Chinese legal philosophy.

Starting in 1912, Chinese legal scholars began to translate and study Western legal systems, primarily Germany's civil code. The Kuomintang government created various national legislations. Much of the legislation created by the Kuomintang even became the foundation for China's current national laws.

Although the Kuomintang opened the Chinese legal system to foreign influence, and created the first thorough system of national laws, the ROC did not implement many of its national regulations because of its conflicts with the Communist Party of China ("CPC") and the Japanese. In 1949, the CPC overthrew the Kuomintang government and created the People's Republic of China ("PRC").

Left: ROC Flag
Source: Maps of World, http://www.mapsofworld.com/flags/taiwan-flag.html
Right: PRC Flag
Source: Maps of World, http://www.mapsofworld.com/flags/china-flag.html

C. 1949–Present

Although the PRC determined to continue employing a civil law system, it nullified the national legislation under the ROC. Additionally, the CPC removed many of the elements of traditional Chinese legal philosophy from China's legal system and replaced them with legal philosophy from the Soviet Union.

Over time, China's laws and regulations evolved through a combination of foreign influence and China's own legal ideals. China is still determining how to balance foreign legal theories and its conception of what Chinese law should be.

II. China's "Chinese Style" Civil Law System

It is especially important for legal counsel from common law jurisdictions to know the differences between common law and China's civil law system.

The most significant difference between the two systems is that court decisions do not have precedential value in civil law. This signifies that unlike in the U.S., which follows a common law system, Chinese courts do not have to follow the decisions of other courts. China's civil law system often confuses foreign lawyers that do not have experience in civil law countries because they are unable to use previous cases in order to assess how a court would decide on an issue.

The next most important difference is that the national laws of civil law jurisdictions are often more vague. Chinese national legislation focuses on goals and policies, and relies on additional regulations to create the system for implementing national laws. As a result, researchers must never limit the scope of their research to national legislation. For example, Chinese legislation does not include definitions of the terms that it uses. The only way for a researcher to find definitions is to look at administrative and local regulations. Even then, the only way to find the most commonly used definition of a legal term may be to research definitions in secondary sources.

Judges in civil law systems pay less attention to both writing and reading court opinions. Instead, many civil law judges prefer to study secondary sources such as scholarly articles and legislative opinions. Chinese judges follow this trend. In comparison to common law jurisdictions, it is more likely that a Chinese judge will have read scholarly articles on a legal issue being discussed in court. Although foreign counsel typically cannot represent their clients in court,[1] they often work with their Chinese colleagues in preparing for litiga-

1. China only allows Chinese nationals to hold Chinese law licenses. Nonetheless, foreign counsel may be able to represent a client in court by assuming the position of the client

tion. Therefore, foreign researchers should focus more on secondary sources when researching Chinese law.

Another distinction of civil law, perhaps especially Chinese civil law, is that it places more restrictions on legal agreements. Common law jurisdictions generally only set minimum restrictions on the right to contract such as that an enforceable agreement cannot be unconscionable. In comparison, Chinese national and local laws and regulations often provide extensive restrictions that affect agreements.

It is important for foreigner researchers to understand that China is not strictly a civil law jurisdiction. In addition to applying the German civil legal system, the founders of the PRC also looked to the Soviet Union's legal philosophy. The combination of German civil law, Soviet Union and Communist philosophy, and Chinese influences created what Chinese legal philosophers refer to as a civil legal system with Chinese characteristics.

China's civil law system also exhibits other differences both to common law systems and other civil law systems. Many of these are relatively subtle such as how court opinions are written and how the China's national legislature, the National People's Congress ("NPC"), considers amendments to legislation. Ultimately, foreign researchers will learn many of the subtle nuances of China's civil law by comparing their own legal experience with the laws and regulations that they are researching.

III. Additional Resources

Burton, Charles, *China* in *Legal Systems of the World*, ABC-CLIO, Inc. (Herbert M. Kritzer, ed. 2002).

Chinese Law and Legal Theory, Ashgate Publishing Company (Perry Keller, ed. 2001).

Essays on China's Legal Tradition, Princeton University Press (Jerome Cohen, Randle Edwards & Chen Fu-mei, eds. 1982).

Lubman, Stanley, *Bird in a Cage: Legal Reform in China after Mao*, Stanford University Press (1999).

Pitman, Potter, *The Chinese Legal System: Globalization and Local Legal Culture*, Routledge (2001).

Understanding China's Legal System, New York University Press (Stephen Hsu, ed. 2003).

---

and then waiving its right to counsel. The court may then allow the foreign counsel to attend as the client *pro se*.

Zimmerman, James M., *China Law Deskbook: A Legal Guide for Foreign-Invested Enterprises*, AMERICAN BAR ASSOCIATION SECTION OF INTERNATIONAL LAW (2010).

Zimmerman, James M., *China*, in *A Legal Guide to Doing Business in the Asia Pacific*, AMERICAN BAR ASSOCIATION SECTION OF INTERNATIONAL LAW (Albert Vincent Y. Yu Chang and Andrew H. Thorson eds., 2010).

Chapter Three

Overview of China's Constitution

The next step to conducting legal research in China is to comprehend the basics of China's constitution. It is likely that the Chinese constitution may not directly interact with research on a particular point of law. However, it is nonetheless the backbone of China's legal system. China's constitution also provides invaluable insight into its history because the constitution was replaced and amended several times according to changing political views.

China's constitutional law changed drastically and frequently from the 1940s to the 1980s. The ROC had its own constitution, and the CPC created an interim constitution when it founded the PRC. Additionally, China has had three different constitutions between 1954 and 1978. The current constitution was adopted in 1982 ("1982 Constitution"). Since then, the NPC has added four amendments to the 1982 Constitution.

I. Previous Constitutions

As noted in the last chapter, the Kuomintang government established much of what would become China's contemporary legal system. In 1946, the ROC implemented China's first formal constitution. The CPC created the PRC only three years later. In 1949, the PRC nullified the ROC constitution and instituted the Common Program of the Chinese People's Political Consultative Conference ("Common Program").

A. The Common Program

When the CPC took over Mainland China in 1949, it organized a conference referred to as the Chinese People's Political Consultative Conference ("CPPCC"). The CPPCC had a very important political function during the establishment of the PRC. It chose the name "People's Republic of China," determined that Beijing would be China's capital, decided its national flag and anthem, and even elected the PRC's first government. The CPPCC still exists as a legislative advisory conference.

The CPPCC also created and implemented the Common Program which functioned as China's interim constitution from 1949 to 1954. The Common Program was written to provide a temporary constitutional framework. The Common Program also had an additional function. It was the CPC's first opportunity to write about the creation of China's new government.

A large portion of the Common Program contained language about the overthrow of the ROC government and the future of the PRC. For example, the first sentence of the preamble reads: "[t]he great victories of the Chinese people's war of liberation and of the people's revolution have put an end to the era of the rule of imperialism, feudalism and bureaucratic capitalism in China."

The Common Program also contains Soviet Communist language such as referring to China as the "father-land." It even states that China "shall unite with all peace-loving and freedom-loving countries..., first of all, with the U.S.S.R." Similar language is found in many of the Common Program's sixty articles.

The Common Program provided many of the constitutional rights found in China's current constitution. It grants the right to elections, freedom of speech, and equality of the sexes. It also states various governmental, administrative, and judicial responsibilities.

The Common Program was replaced by China's first constitution in 1954.

B. 1954 Constitution

The 1954 Constitution had almost twice as many articles as the Common Program. It established the NPC as China's highest legislative body and clarified the various specific responsibilities of governmental bodies. It also created the Supreme People's Court ("SPC") and Supreme People's Procuratorate ("SPP"). This constitution also first addressed how China would treat ethnic minorities. It stated that they would be able to create their own self-autonomous regions, which would have more local power than other provinces.

The NPC did not intend to use the 1954 Constitution for more than a few years. China was undergoing massive economic changes under socialism. The NPC believed that the constitution would have to be redrafted once China settled into a new socialist system. Political revolution impeded China from creating a more permanent constitution. It finally implemented the second constitution two decades later.

National Celebration of the 1954 Constitution
Source: Show China, http://www.showchina.org/zt/015/05/201106/t939703.htm

C. 1975 Constitution

China's transition into socialism was not as smooth as the NPC predicted when it implemented the 1954 Constitution. In 1957, Chinese politicians and other political activists began to call for campaigns against rightists. They targeted people who they believed were against China's socialist movement. They also targeted anyone who seemed to promote capitalism. This movement is referred to as the Anti-Rightist Movement.

Socialist activism was a strong force in Chinese politics even after the Anti-Rightist Movement formally ended around 1960. In 1966, the Cultural Revolution began. The chaos and political turmoil of the Cultural Revolution caused most government operations to cease. In 1975, one year before the end of the Cultural Revolution, Mao Zedong decided to create a new constitution that would solidify his party's power.

The 1975 Constitution substantially reduced the fundamental rights of Chinese citizens. The rights that Mao Zedong removed included freedom of speech and freedom from discrimination. This constitution also abolished the office of the Chairman. Instead, CPC Chairman (Mao Zedong) would have the highest office in China.

The Cultural Revolution ended in 1976 with the death of Mao Zedong. China repealed Mao Zedong's constitution two years later.

D. 1978 Constitution

The 1978 Constitution only lasted for four years. The NPC wrote this constitution to be a balance between Mao Zedong's political theories and the general desire to eliminate many of the radical leftist ideologies in the 1975 Constitution.

The 1978 Constitution essentially attempted to revert back to the 1954 Constitution while maintaining the general tone in Mao Zedong's constitution. It granted back many of the rights that were not present in the 1975 Constitution. On the other hand, it included additional language about promoting Marxism and defending China against capitalists.

II. The Current Constitution

After Mao Zedong's death in 1976, Deng Xiaoping became China's *de facto* leader. Deng Xiaoping never held any of China's highest offices. Despite his lack of a formal title, Deng Xiaoping had a massive influence on the removal and replacement of party leaders. He also played an important role in determining many of China's most important political decisions during that time period. One of these decisions was to create of China's current constitution.

A. 1982 Constitution

China implemented its current constitution in 1982. Its current constitution is much longer and detailed than its predecessors. The 1982 Constitution borrows heavily from the 1954 Constitution. It also incorporates several aspects of the 1978 Constitution. Additionally, many parts of the 1982 Constitution are original such as establishing new political theories and changing China's governmental structure.

The 1982 Constitution removed most of the language associated with Mao Zedong and the Cultural Revolution. In fact, it doesn't even mention China's Communist Party. The 1975 Constitution focused on socialism and class struggle. Many of these elements were also in the 1978 Constitution. In comparison, the 1982 Constitution prioritizes national development and promotes contributions to development.

China's current constitution takes a step further in promoting national development by creating the legal backbone for its economic goals. It states that although the national government has the primary role in creating and developing China's economy, Chinese citizens may also engage in private business activities. For example, the 1982 Constitution was the first to allow farmers to cultivate crops and livestock for private use and sale.

In regards to China's governmental structure, the 1982 Constitution provides that the NPC is China's primary legislative body. Unlike the "Western" system of government, the NPC has a substantial effect on China's other gov-

Revision Committee for the 1982 Constitution
Source: Renmin University, http://ruc.zuolang.com/Photo/ShowPhoto.asp?PhotoID=78

ernmental branches. However, the 1982 Constitution also states that the government should not grant its power to only a few individuals.

Another importance difference between the current constitution and its predecessors is how it addresses foreign influences. Previous constitutions either did not mention foreign influences or opposed them. In comparison, the 1982 Constitution promotes the use of foreign assistance in determining how to modernize China.

Finally, the 1982 Constitution creates more rights than any of the previous constitutions. For example, it is the first constitution to guarantee freedom of religion.

The 1982 Constitution has been amended four times.

B. Amendments

1. 1988 Amendments

In 1988, the NPC approved two amendments to the 1982 Constitution. The first change was the addition of an article stating that although China's socialist public economy is its main economy, the national government also permits a private economy. It also provides that the national government will protect its citizens' rights and interests in the private economy as well as that it will supervise and control the private economy. The second amendment allows the transfer of the right to use land.

2. 1993 Amendments

In 1993, China made nine amendments to its constitution:

1. Slightly revising the language of its preamble;

2. Adding a statement that China's political parties will exist and develop for "a long time to come";
3. Modifying the term "State economy" to "State-owned economy";
4. Modifying the terms "communes" and "cooperatives" to "households" and "forms of cooperative economy";
5. Modifying the term "socialist public ownership" to "socialist market economy";
6. Changing requirement concerning state-owned enterprises ("SOEs") from having to fulfill obligations under the "State plan" to practicing "democratic management through congresses of workers ...";
7. Changing the requirement that collective economic organizations abide by the "State plan" to that they will abide by relevant laws;
8. Modifying "State enterprise" to "State-owned enterprise"; and
9. Shortening language on municipal and county governments such as "cities divided into districts" to "municipal districts."

3. 1999 Amendments

In 1999, China made the following additional amendments to its constitution:

1. Changing the language in the preamble from "China is currently in the primary stage of socialism" to "China will stay in the primary stage of socialism for a long period of time";
2. Adding that the PRC "practices ruling the country in accordance with the law and building a socialist country of law";
3. Adding a sentence that "the State adheres to the basic economic system with the public ownership remaining dominant and diverse sectors of the economy developing side by side, and to the distribution system with the distribution according to work remaining dominant and the coexistence of a variety of modes of distribution";
4. Clarifying the double-tier management system for rural collective economic organizations as both unified and separate operations;
5. Changing references to "urban and rural working people" to "individual, private and other non-public economies"; and
6. Changing "counter-revolutionary activities" to "criminal activities that endanger State security."

4. 2004 Amendments

The last amendments made to China's constitution were implemented in 2004. They are as follows:

1. Changing the language in the preamble from "socialism with Chinese characteristics" to "Chinese-style socialism" and adding that "Marxism-Leninism, Mao Zedong Thought, [and] Deng Xiaoping Theory" are "important thought";
2. Adding "promote the co-ordinated development of the material, political and spiritual civilizations" after "to modernize the industry, agriculture, national defence and science and technology step by step";
3. Adding "all builders of socialism" to the preamble;
4. Changing the article on China's land requisition from "[t]he State may, in the public interest, requisition land for its use in accordance with the law" to "[t]he State may, in the public interest and in accordance with the provisions of law, expropriate or requisition land for its use *and shall make compensation for the land expropriated or requisitioned*" (emphasis added);
5. Modifying the term "private" to "non-public";
6. Adding that private property is "inviolable," but that the state may expropriate or requisition it as long as it compensates the owner;
7. Adding "[t]he State establishes a sound social security system compatible with the level of economic development";
8. Adding "[t]he State respects and preserves human rights";
9. Adding the term "special administrative regions";
10. Changing "martial law" to "state of emergency";
11. Shortening language specifying the levels of local governments to "local people's congresses"; and
12. Explicitly stating China's national anthem.

A careful reading of the various amendments to China's constitution shows that the Chinese government has paid very close attention to modifying terms to reflect current trends and policies. The majority of the amendments either clarify an article, remove wordy language, or change a term.

The fourth amendment in the 2004 Amendments is in bold because it is likely the most important amendment made to China's constitution. Since 2004, the Chinese government has been required by the constitution to compensate citizens for land acquisitions. This does not mean that the Chinese government reasonably compensates its citizens every time that it acquires their private property.

C. Enforcement Issues

Foreign legal researchers may be surprised to learn that China's constitution provides more rights to its citizens than most of the world's constitutions.

Experience shows that some portions of the constitution, akin to many of China's laws and regulations, are part of the "word game" (文字游戏) in Chinese politics. Simply put, China's constitution looks very good on its face. However, the implementation and enforcement of its constitution may be a different story. Nonetheless, the listing of these rights in the constitution may be used by litigants and researchers, and provides an invaluable model for China to strive towards.

1. Lack of Judicial Review

In the United States and other common law jurisdictions, a party with standing may challenge the constitutionality of a national or local law. This has given common law courts the ability to strike down laws deemed to violate a constitution. The value of striking down an unconstitutional law has been repeatedly demonstrated in U.S. history.

China, unlike common law jurisdictions, does not have judicial review. This means that Chinese courts do not have the power to invalidate laws even if they clearly violate China's constitution. Thus, Chinese courts do not allow Chinese citizens to challenge laws that would otherwise likely be deemed unconstitutional.

2. Limited Government Enforcement

The Chinese constitution states that the NPC may determine whether specific laws and activities are unconstitutional. In 2002, the NPC established a committee to determine whether laws are or are not constitutional. This committee has not made a single determination.

There are only a few instances where the NPC has repealed a law because it was unconstitutional. Furthermore, the NPC will typically only consider repealing a law after widespread outcry. Hopefully the NPC will begin to review controversial laws to determine whether they are unconstitutional. Nevertheless, the current trends indicate continued resistance to legislative review.

3. Property Acquisition Gone Wrong: The Nail House

Perhaps the best example of problems associated with the enforcement of the constitution is the Nail House (钉子户). The Chinese government often demolishes the homes of private individuals in order to make way for public and governmental projects. The local government often offers money, sometimes in exorbitant amounts, to incentivize the evacuation of these areas.

In some situations, the residents refuse to leave the property and the government encounters problems in forcing the residents to leave. In other cases, hundreds of homes will be evacuated while only one resident or family remains. All of these homes are demolished, leaving just one house standing: the Nail House.

The Nail House has become a symbol in China for the fundamental right to own private property. In some cases, the owners of a Nail House may simply be haggling over the price of the property. In other situations, the government has not offered a fair price. Nail Houses can be found throughout China, especially in downtown areas and in economic development zones.

A Nail House Obstructs Construction in Chongqing
Source: Soufun, http://news.cq.soufun.com/gqzt/2013-09-10/34511.htm#p=1

A Nail House in the Middle of a Public Street in Zhejiang Province
Source: WCC Daily, http://www.wccdaily.com.cn/shtml/hxdsb/20121123/44671.shtml

III. Additional Resources

Building Constitutionalism in China, Palgrave Macmillan (Stéphanie Balme & Michael Dowdle, eds. 2009).

Changing Meanings of Citizenship in Modern China, Harvard University Press (Merle Goldman & Elizabeth Perry, eds. 2002).

China's Search for Good Governance, Palgrave Macmillan (Guo Sujian & Deng Zhenglai eds. 2011).

Lampton, David M., *Following the Leader: Ruling China from Deng Xiaoping to Xi Jinping*, University of California Press (2014).

Saich, Tony, *Governance and Politics of China*, Palgrave Macmillan (2011).

Sommers, Amy L. & Kara L. Phillips, *Real Property in China: A Guide to Foreign Investment*, American Bar Association Section of International Law (2011).

Zhang Qianfan, *The Constitution of China: A Contextual Analysis*, Hart Publishing (2012).

Chapter Four

The Chinese Government

China's government can be roughly categorized into national and local governments. Chapter Five will discuss the various issues associated with legislation from both the national and local levels. As administrations and their regulations play a crucial role in China's legal system, Chapter Six will provide all of the information that a foreign legal research will need to understand the basics of administrative regulations. However, before delving into legislation and administrative regulations, legal researchers should first comprehend the structure of China's government.

I. National Government

The overall structure of China's national government can somewhat be compared to the U.S.'s executive, legislative, and judicial branches. However, unlike in the United States, China's legislative body (the NPC) has a substantial influence over China's executive and legislative branches. Furthermore, the CPC's power is intertwined in all of China's lawmaking bodies.

A. Presidency and Premiership

The 1982 Constitution established the position of president to be China's head of state.[1] The NPC elects China's presidents to serve five year terms. Contrary to popular belief in foreign countries, China's presidency does not confer the most political power. The highest political position in China is premier.

---

1. The Chinese term for "president" (主席) can be translated to president, chairman or premier. As the main Chinese word for "premier" is (总理), and is used to refer to the highest administrative position in China, legal researchers should use "president" for the head of state and "premier" for the Premier of the People's Republic of China.

Left: Xi Jinping, President
Source: People.com.cn, http://politics.people.com.cn/n/2012/1115/c351134-19594581.html
Right: Li Keqiang, Premier
Source: Xinhua, http://news.xinhuanet.com/english/china/2013-03/15/c_132235590.htm

The premiership is the highest administrative position and grants the most political power. In comparison, the most important aspect of the presidency is that the president may nominate the premier. The NPC has never refused a nomination from a president.

B. National People's Congress

The NPC is the world's largest parliament. As China's primary legislative body, its main role is to ratify bills and elect the heads of executive and judicial institutions. The NPC is composed of the CPC and the United Front.[2] The NPC has two meetings a year that are conducted during the same two weeks.

Although the NPC is formally responsible for passing China's primary national laws, a large portion of the NPC's power rests with a small number of NPC members known as the Standing Committee.

2. The United Front is a representative group chosen by the Communist Party of China. It generally has no power apart from the Communist Party.

C. Standing Committee

The Standing Committee is China's permanent legislative body. The NPC only assembles once a year. In comparison, the members of the Standing Committee meet often and determine most of China's national laws. The members of the Standing Committee are generally considered to be the center of China's political power.

D. State Council

The State Council, also referred to as the Central People's Government, is China's head administrative authority. The purpose of the State Council is to ensure that China's administrations and government departments are acting in accordance with the NPC's goals and legislation. The State Council meets twice a year and is composed of China's premier, vice-premier, ministers, and administrative heads.

Although the NPC is tasked with producing and revising China's national laws, the State Council creates all of China's national administrative regulations. Foreign researchers will be surprised to learn that in addition to administrations and departments, the State Council also controls some of China's academies and research centers. It even controls Xinhua, which is the government's official press.

E. Central Military Commission

China also has a very confusing structure for the administration of its military. The Central Military Commission ("CMC") is responsible for all of China's military matters. The CMC is composed of China's president, vice-president, and top generals. It directly controls important military functions such as spending, deployment, and protection of top government officials.

China also has a Ministry of National Defense that acts as its foreign representative in military matters. This signifies that although the CMC is officially in charge of military matters, it is not China's representative to the rest of the world. Also, the Standing Committee makes most of the CMC's important decisions.

F. Other Committees

This section has only discussed the largest committees in the Chinese government. Legal researchers should be aware of the government committees that may affect a single legal issue. Chinese politicians, administrators, generals, etc. are typically on many committees and commissions.

Although the number of committees in China may be daunting, legal researchers should also remember that most of these committees' decisions are based on what the CPC and the Standing Committee tell them to do.

Structure of China's Government

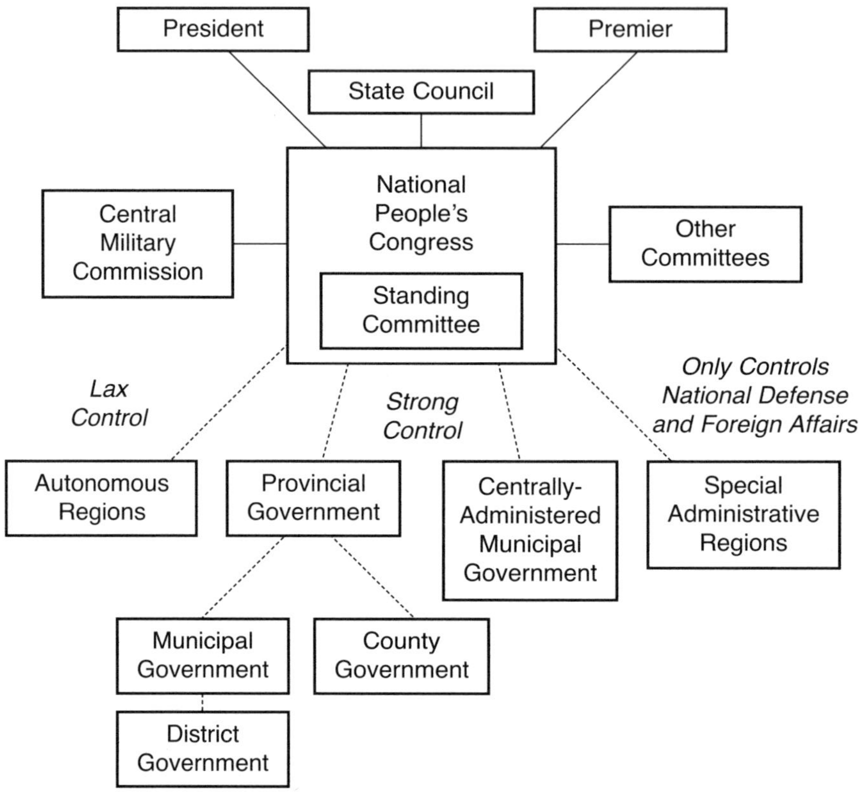

II. Local Governments

Most foreign legal consultants working in China often neglect to research local government regulations. In fact, many foreigners performing legal research are unaware of China's local governmental system. This section provides an explanation of the general hierarchy of China's local government.

Unlike some other aspects of the Chinese legal system, China's local governmental system is relatively simple. Foreign researchers can even compare it to their own jurisdictions.

China is composed of 33 administrative regions:

1. 22 provinces;[3]
2. Four centrally-administered municipalities;[4]
3. Five autonomous regions; and
4. Two special administrative regions ("SARs").

Administrative Divisions and Regions

Provinces	Anhui, Fujian, Gansu, Guangdong, Guizhou, Hainan, Hebei, Heilongjiang, Henan, Hubei, Hunan, Jiangsu, Jiangxi, Jilin, Liaoning, Qinghai, Shaanxi, Shandong, Shanxi, Sichuan, Yunnan, Zhejiang. **The PRC considers Taiwan as a province.*
Autonomous Regions	Guangxi, Inner Mongolia, Ningxia, Tibet, Xinjiang
Centrally-Administered Municipalities	Beijing, Chongqing, Shanghai, Tianjin

A. Provincial Governments

Most of China is governed by provincial governments. China's provinces can be compared to the U.S.'s states, Canada's provinces, and Italy's regions. They are essentially the highest level of local government in China.

China has 22 provinces and four centrally-administered municipalities. As its centrally-administered municipalities have municipal and provincial powers, they only create one level of legislation which could be categorized as provincial.

3. Please note that whether Taiwan is a province is a highly controversial topic. If Taiwan were considered to be a province, then China would have 34 administrative regions and 23 provinces. The author states that China has 22 provinces in this section only to facilitate the reading of the book, and does not intend for this book to convey his personal views regarding the relationship between China and Taiwan. As this book does not discuss Taiwan's legal system, please refer to the "additional sources" section at the end of this chapter for more information.

4. The term "centrally-administered municipality" (直辖市) refers to cities that have both municipal and provincial powers. China has four centrally-administered municipalities which are Beijing, Chongqing, Shanghai, and Tianjin.

Each of China's provinces has its own people's congress and committees. Each provincial congress delegates its own NPC members. In addition to representing the interests of their province, provincial governments are primarily responsible for implementing national laws. Provincial governments also determine local goals and policies.

B. Municipal Governments

China's cities and counties also have their own governments. Cities typically have both a municipal government and county governments. The county-level government is the lowest level of government in China.

Municipal governments are responsible for both implementing national and provincial legislation. Additionally, they also create their own regulations that only affect areas within the city limits. China's municipal governments have the same general structure as other civil law countries. Each municipal government has its own congress, various committees, and departments. Chinese cities generally have relatively up-to-date websites that provide information on the municipal government including its leaders, hierarchy, reports, policies, events, etc.

C. County Governments

County governments are either directly below municipal or provincial governments. If a county government is under a municipal government, then it is referred to as a district government. District governments are generally very limited in size. Their primary functions are to enforce municipal policies and regulations. They also have duties such as maintaining streets and ensuring that construction sites do not endanger the public.

In comparison, county governments are typically larger and have more responsibilities. As they are directly controlled by the provincial government, they function as a small-scale municipal government. The main difference between a county and municipal government is that counties governments generally supervise areas with lower populations and less industry. As a result, they do not have to implement as many regulations or maintain as many departments.

D. Autonomous Regions

Foreign researchers may also be surprised to learn that China's autonomous regions take up almost half of its surface area. China's ethnic minorities generally live in one specific area. Some of these areas are only inhabited by one ethnic minority, while other areas have up to a dozen different minorities.

In consideration of the sometimes substantial differences in the ways of life of these minority groups compared to mainstream Chinese culture, China has established five autonomous regions which grant more freedom than its provinces. These autonomous regions are allowed to create local regulations that may otherwise conflict with national legislation.

The autonomous regions are:

1. Guangxi Zhuang (or simply "Guangxi") (广西);
2. Inner Mongolia (内蒙古);
3. Ningxia Hui (or simply "Ningxia") (宁夏).
4. Tibet (西藏); and
5. Xinjiang Uyghur (or simply "Xinjiang") (新疆);

E. Special Administrative Regions

China has two special administrative regions: Hong Kong and Macao. Macao, which consists of three small islands and is located about 38 miles from Hong Kong, was a Portuguese colony. In 1987, the Portuguese government transferred Macao to China. The agreement between Portugal and China stated that Macao would be an SAR and enjoy complete political and economic autonomy for at least 50 years.

Similarly, in 1997, the British government gave Hong Kong to China. In the agreement between Britain and China, Hong Kong would remain autonomous and keep its capitalist system until 2047. In accordance with this agreement, China established Hong Kong as an SAR.

These two territories are almost entirely autonomous from China's central government. They have their own laws, economic policies, police force, etc. In fact, the Chinese government only has power over defense and foreign affairs issues. Specifically, China makes national defense decisions that involve its SARs and may enter into international agreements that affect its SARs.

III. Additional Resources

China-Taiwan Relations in a Global Context: Taiwan's Foreign Policy and Relations, Routledge (George Wei, ed. 2012).

Fewsmith, Joseph, *The Logic and Limits of Political Reform in China*, Cambridge University Press (2013).

Kissinger, Henry, *On China*, Penguin Books (2012).

Lampton, David M., *Following the Leader: Ruling China from Deng Xiaoping to Xi Jinping*, University of California Press (2014).

Lo, Chang-Fo, *The Legal Culture and System of Taiwan*, Kluwer Law International (2006).

Saich, Tony, *Governance and Politics of China*, Palgrave Macmillan (2011).

The Politics of China: Sixty Years of the People's Republic of China, Cambridge University Press (Roderick MacFarquhar, ed. 2011).

Zimmerman, James M., *China Law Deskbook: A Legal Guide for Foreign-Invested Enterprises*, American Bar Association Section of International Law (2010).

Chapter Five

Legislation

Legislation is almost always the starting point of the actual legal research conducted on an area of Chinese law. As court judgments do not have precedential value in China, the majority of legal research will always be based on legislative and administrative regulations. Research should be conducted on legislation first because it creates the structure for further administrative guidelines.

In-depth research of national legislation is perhaps the most important aspect of conducting Chinese legal research.

I. National Legislation

China's national laws differ substantially from legislation found in common law countries. For example, U.S. lawmakers generally seek to use language that is precise as possible when drafting statutes, anticipating any number of possible scenarios that might arise under the statute. In contrast, Chinese laws and regulations use language that is perhaps more vague or ambiguous. Chinese lawyers and judges resolve the ambiguities in the civil code through inferences of general legal principles. This practice especially confuses attorneys from common law jurisdictions.

Chinese legislation typically provides the overall foundation for the Chinese legal system. The NPC passes national laws that govern entire types of law or industries. For example, the NPC created China's civil, criminal, and contract laws, as well as its main regulations on foreign investment. The NPC sometimes also promulgates implementation regulations that explain the general provisions in its main regulations.

A. Primary Regulations

All of China's primary regulations originate from the NPC. Generally, the NPC will solicit administrations and other governmental and private institutions to submit drafts and opinions for potential laws or amendments. The Standing Committee will then discuss the potential law and either revise and promulgate it or create a different draft and solicit further opinions.

The NPC almost always asks the administrations associated with the implementation of the law to provide detailed opinions on what they would like the law to look like. The reasoning behind this practice is that the administrations:

1. know the most about the issue;
2. will be in charge of implementing the law; and
3. will most likely create their own regulations that fill in the general provisions of the law.

China's primary regulations are intentionally vague, and function as a guide for the implementation of the NPC's policies and goals. In fact, the Chinese legislature purposely avoids including specific guidelines in its primary regulations. Nonetheless, although the Chinese government invests a large amount of lawmaking power in its administrations, it sometimes prefers to create its own regulations that provide broad guidelines for implementing a specific law.

B. Implementation Regulations

Many of China's most important laws are supplemented by an additional implementation regulation. For example, China has a trademark law and an implementation regulation for the trademark law. The primary law explains the goals and policies associated with trademarks. In comparison, the implementation regulation provides more detail concerning the responsibilities of the Chinese Trademark Office ("CTMO"), how local Administrations for Industry and Commerce ("AICs") should interact with trademark issues, etc.

The NPC does not amend an implementation regulation every time that it issues a revision of the primary regulation. Instead, the NPC will either (1) promulgate an additional amendment which only amends a few specific clauses in the implementation regulation or (2) not change the implementation regulations. Generally, the NPC will not amend the implementation regulation if it does not conflict with the revision to the primary regulation.

II. Local Legislation

Effective Chinese legal research also encompasses potential local government legislation. Local governments implement their own regulations because either (1) national legislation has purposely left room for local government control, or (2) a particular local government determines that its area would benefit from additional regulations.

A. Relationship with National Legislation

In the event of a contradiction of laws, national legislation supersedes local legislation. However, foreign legal counsel should be aware that local governments may enforce detailed regulations more stringently than the national government.

Business franchising regulations are a great example of how local governments may impose additional regulations that do not apply to other areas. China has two national measures and an administration regulation that govern business franchising. Tianjin and Shenzhen determined that their provinces[1] required additional regulations. As a result, their local legislatures created their own regulations that provide definitions, limit business franchising for public utilities, and set detailed registration requirements.

B. Structure of Local Governments

Higher-level governments outline the particular duties for lower-level governments. Sometimes, lower-level governments will merely adhere to the regulations imposed upon them. They also often create their own regulations that provide detailed instructions to the departments that implement the responsibilities mentioned in higher-level governmental legislation.

For example, provincial governments have regulations on environmental protection. These regulations typically include guidelines for the protection of forests, wetland, and wildlife. They will state which local government departments are responsible for specific duties. County governments may then elect to create regulations that further instruct local departments about environmental protections. Municipal governments will also create their own regulations that address protecting the environment in urban settings. These regulations address issues such as energy-saving buildings, sanitation, and urban parks.

---

1. Although Beijing, Chongqing, Shanghai, and Tianjin are centrally-administered municipalities, this book refers to them as provinces and to their governments as provincial governments (see note 5 for a definition of "centrally-administered municipality").

III. Researching Legislation

Attorneys from common law jurisdictions often look to court cases for references to legislation. As explained in Chapter Seven, Chinese court opinions are generally very short and do not provide many citations. The majority of citations to legislative documents focuses on the basis of making a judgment. Therefore, Chinese court opinions are not a good starting point for finding controlling legislation.

The best way to find relevant legislation is to first identify which legal issues are associated with the research being conducted. The next step is to look for related legislation through the resources provided in Chapter Nine.

IV. Additional Resources

Brown, Ronald, *Understanding Labor and Employment Law in China*, CAMBRIDGE UNIVERSITY PRESS (2011).

China's Legal System: New Developments, CAMBRIDGE UNIVERSITY PRESS (Donald Clarke, ed. 2008).

Roos, Maarten, *Chinese Commercial Law: A Practical Guide*, KLUWER LAW INTERNATIONAL (2010).

The Limits of the Rule of Law in China, UNIVERSITY OF WASHINGTON PRESS (Karen Turner-Gottschang, James Feinerman, & Kent Guy, eds. 2000).

Wang, Yan, *Chinese Legal Reform*, ROUTLEDGE (2001).

Zimmerman, James M., *China Law Deskbook: A Legal Guide for Foreign-Invested Enterprises*, AMERICAN BAR ASSOCIATION SECTION OF INTERNATIONAL LAW (2010).

Chapter Six

Administrative Law and Regulations

China allocates legislative power to various administrations, and these administrations promulgate regulations according to the scope determined by the NPC. Perhaps the largest obstacle to conducting thorough legal research of administrative regulations is that a legal issue often involves several administrations. Conducting legal research of administrative regulations requires an initial understanding of the potential legal issues.

A simple example would be importing a foreign food product into China. This will require researching the General Administration of Customs' ("GAC") regulations such as importation documentation and labeling guidelines. Legal research will also extend to the China Food and Drug Administration ("CFDA") regulations because imported foodstuffs should meet the CFDA's guidelines and may need to be registered with the CFDA. The SAIC also has its own regulations that affect foreign foodstuffs. Additionally, foreign counsel must also consider the intellectual property implications of importing a foreign good into China.

The legal research for the importation example would not end with researching national administrations. In addition to national regulations, effective legal counsel should also research the customs practices for the specific port that the client will use. Legal counsel should also consider where the foodstuffs will be sold because they will be affected by local government and AIC regulations.

I. National Administrations

After researching the national laws that affect a legal issue, the next step is to determine which administrations to research. National administrations implement regulations that broadly govern how they interact with national legislation. They also have regulations for specific issues that fall under their power.

Chinese trademark law provides a good example of how a legal issue will interact with both broad and narrow administrative regulations. The CTMO has several regulations that govern specific trademark issues and CTMO departments. So, for example, if a trademark applicant under the Madrid System applies for an administrative appeal with the Trademark Review and Adjudication Board ("TRAB"), then the CTMO's regulations on both the implementation of the Madrid System and TRAB procedures will apply.

II. Local Administrative Branches

Legal researchers should also be aware of the procedures for local administrative branches. Many of China's larger administrations, such as the CFDA, have branches in every province and centrally-administered municipality. Although China's administrations are promoting conformity between local procedures, it is still likely that procedures such as those for registration documentation will differ slightly between provincial offices.

An effective legal researcher should determine local procedures in order to avoid complications such as delays in receiving registration certificates. Most websites for Chinese administrations are well-organized and contain updated contact information for their branch offices. Some information about local procedures may be available online. However, legal counsel should also be prepared to call the local branch to inquire about a specific procedure.

III. Administrations for Industry and Commerce

Foreign legal researchers may overlook local AIC regulations because their countries do not have equivalents to AICs. In addition to determining whether any SAIC regulations may apply, a legal researcher should also research local AIC regulations.

It is also important not to limit the search scope to regulations that are directly related to the SAIC regulations, because local AICs often create regulations that do not have an SAIC counterpart. Finally, some local regulations issue may originate from the local government instead of the AIC or vice versa. Thus, legal researchers should search both local government and AIC regulations.

IV. Additional Resources

Ching Mia Kuang, *CFO Guide to Doing Business in China*, WILEY (2009).

PwC Worldtrade Management Services (Shanghai), *Quick Reference to the Trade and Customs Law of China*, KLUWER LAW INTERNATIONAL (2012).

Yong Kwek Ping, *Due Diligence in China: Beyond the Checklists*, WILEY (2013).

Yu Jianxing, *Civil Society and Governance in China*, PALGRAVE MACMILLAN (2012).

Zimmerman, James M., *China Law Deskbook: A Legal Guide for Foreign-Invested Enterprises*, AMERICAN BAR ASSOCIATION SECTION OF INTERNATIONAL LAW (2010).

Chapter Seven

The Chinese Judiciary

China's judicial system is relatively similar to those in other civil law countries. Common law researchers should be aware of the minimal role that court opinions and judgments have on judicial proceedings.

The Chinese court system is composed of the SPC, people's courts, and people's procuratorates. It is important for foreign researchers to note that people's procuratorates are not merely courts, and that SPC opinions have a very strong influence on people's courts.

I. People's Courts

A. Hierarchy of the Chinese Court System

China's courts are organized into a clear hierarchy beginning from grassroot level courts and ending at the SPC. China's lowest level courts are referred to as basic people's courts. The next level is the intermediate people's court, followed by the higher people's court.

Each province and centrally-administered municipality has many basic people's courts. Provinces often have several intermediate people's courts that are based in their main cities. Centrally-administered municipalities only have one or two intermediate people's courts. There is only one higher people's court per province or centrally-administered municipality.

In addition to the ordinary people's courts, China also has maritime, military, and railroad courts. These courts are referred to as "courts of special jurisdiction."

The SPC is China's supreme court. Unlike many other supreme courts, which have less than a dozen presiding judges, the SPC is composed of different courts such as civil, criminal, administrative, and economic. It consists of over 340 judges and employs more than a thousand people. The SPC's chief judge has a term limited to two consecutive five-year terms.

Hierarchy of the Chinese Court System

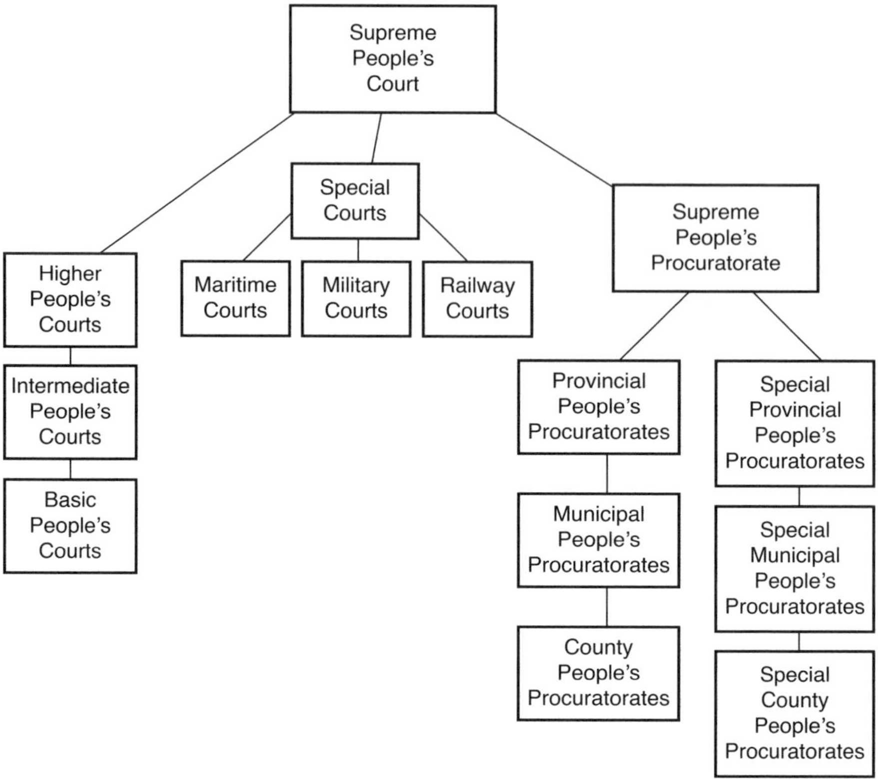

B. Relationship with AICs

In addition to looking at the Chinese court system itself, foreign researchers should also consider the relationship between courts and AICs.

AICs play an important role in supervising local economies. Their position as enforcers of national and local laws often results in potential conflicts with the courts. For example, a trademark owner may pursue relief against trademark infringement with both the local AIC and local court.

The general practice is for an AIC to wait for the court to issue a judgment. The AIC will then usually assume the court's position and sometimes help enforce the judgment. However, depending on the issue of law, AICs are not required to wait for the court to issue the judgment. In fact, the law generally does not state that they have to abide by the court decision.

This creates the possibility that AICs will act before a judgment is entered and that their decisions may even be adverse to a court decision. Fortunately, there are only a few instances where AICs have acted in a way that substantially affected the actual outcome of a court proceeding. Nonetheless, it may be beneficial to identify whether the controlling law allows the possibility of a conflict.

C. Court Opinions

Chinese court opinions often frustrate common law attorneys. Researchers from common law jurisdictions are accustomed to in-depth court opinions that explicitly provide the controlling law and cases. In comparison, Chinese court opinions can be half a page in length. They almost never cite a single case. They sometimes even omit to mention the most important laws and regulations. Of course, they are also only in Mandarin.

Lower-level courts usually do not provide any explanation regarding their application of the law. They list the arguments from each side, their own factual findings, and the judgment.

However, Chinese higher-level courts are showing substantial progress. They are beginning to explain their application of the law. Perhaps even more important for foreign researchers, higher-level courts are beginning to provide how they reached damages calculations. Courts are also now required to publish their opinions.

Foreign researchers that cannot read Mandarin will be unable to interact with court opinions unless they can find unofficial translations or summaries. There are also only a minimal amount of studies and surveys conducted on Chinese court judgments. Depending on the subject of law being researched, foreign researchers may be able to use statistical databases provided by foreign firms. For example, CIELA, maintained by Rouse & Co., provides statistical data on intellectual property judgments.[1]

II. People's Procuratorates

It is likely that foreign researchers have not heard of the term "people's procuratorate" before conducting legal research in China. People's procura-

1. *Available at* http://www.ciela.cn/ (CIELA provides comprehensive up-to-date Chinese IP litigation statistics, and also allows the user to search IP court judgments by venue comparison, ranking by city, and trend by year).

torates are state supervisory authorities that investigate and prosecute potential cases that endanger national or public security.

A. Function

People's procuratorates have the authority to look over the cases scheduled for investigation by administrations and security agencies. They can also decide whether to make an arrest and institute prosecution proceedings. They also supervise other procedures such as imprisoning people convicted under their authority.

Foreign researchers may mistakenly associate people's procuratorates with prosecutors. In comparison to prosecutors, people's procuratorates act independently from administrative or civic interference. Not only do they prosecute potential criminals, but they also control what happens before, during, and after prosecution.

In reality, people's procuratorates are often used by the government and the CPC to handle cases that may otherwise be dealt with through the police or judiciary. The courts, public security agencies, and people's procuratorates were designed to function independently and somewhat equally. However, given the especially independent status of the people's procuratorates, many contemporary Chinese legal scholars believe that people's procuratorates have in effect superseded judicial and police powers.

B. Structure

China's procuratorates are structured almost identically to people's courts. The SPP is the highest authority. Below the SPP are provincial people's procuratorates and special provincial people's procuratorates. There are also municipal and county procuratorates and special procuratorates.

The difference between "procuratorate" and "special procuratorate" is determined by the type of cases that they supervise. People's procuratorates investigate all ordinary cases of potential harm to national or public security. Special people's procuratorates are convened to handle cases of "special interest." What determines is a special interest depends on the parties involved, the stakes, the type of crime, etc. For example, special people's procuratorates handle China's most famous cases of corruption.

III. Supreme People's Court Opinions

The SPC will also release opinions to lower courts in order to clarify how courts should interpret the law. It is important not to confuse SPC opinions with general court opinions. Every court case, including cases from the SPC, has a court opinion. Additionally, the SPC will sometimes issue an opinion or interpretation on a specific judicial issue.

Many other supreme courts can only provide their opinions attached to specific cases. One of the advantages of China's judicial system is that the SPC can write guiding opinions without waiting for a case. Additionally, this signifies that courts cannot find that an opinion does not apply because the corresponding case is different.

Legal researchers should always research SPC opinions regardless of whether the legal issue will involve Chinese courts. SPC opinions provide invaluable hints regarding the state of a legal issue and the disputed areas of law.

IV. Additional Resources

Bin Liang, *The Changing Chinese Legal System, 1978—Present: Centralization of Power and Rationalization of the Legal System*, ROUTLEDGE (2007).

Brown, Ronald, *Understanding Chinese Courts and Legal Process: Law with Chinese Characteristics*, KLUWER LAW INTERNATIONAL (1997).

Keith et al., *China's Supreme Court*, ROUTLEDGE (2013).

Ming Fei, *Chinese Court Judgments: Intellectual Property*, LEGEND BUSINESS (2011).

Zhou Lin, *China Court Cases on Intellectual Property Rights*, KLUWER LAW INTERNATIONAL (2011).

Chapter Eight

Treaties and Other International Agreements

The final aspect of conducting legal research in China is interacting with Chinese treaties. China's initial encounters with foreign trade, similar to Japan and Korea, has significantly influenced how it views treaty making.

I. Unequal Treaties (1842–1949)

China's first experiences with foreign treaties had a substantial and lasting effect on how China enters into treaties with other countries. Although the actual treaties and most of their effects no longer exist, Chinese politicians are well-versed in the unfair treaties that foreigners imposed on China starting with the aftermath of the First Opium War.

A. Foreign Relations before the Unequal Treaties

Beginning in 1757, China employed the Canton System—a trade structure that controlled how foreigners conducted trade in China. Under the Canton System, foreigners could only live and trade from restricted areas in Guangdong Province. Additionally, they could only export goods through Chinese intermediaries.

Britain's high consumption of Chinese tea and lowered profitability from its cotton exports from India, its colony at the time, led to the importation of Indian opium into China. The opium trade was a great success for British traders because it increased exports from India and imports into China. Chinese dependence upon opium also enhanced Britain's geopolitical influence.

The Chinese government eventually banned opium imports and destroyed several shipments from Britain. Various events such as the Kowloon Incident, where British and American sailors vandalized a Chinese temple and killed a man, fueled hostility between China and Britain.

British warships moved to Chinese ports to prevent the Chinese from blocking their opium shipments. Many of China's ships were sunk. British soldiers also captured several forts. The British surrounded Nanking (now referred to as "Nanjing"), which was China's capital at the time, and forced China to enter into its first unequal treaty.

B. Unequal Treaties and Treaty Ports

In 1842, China and Britain signed the Treaty of Nanking. This treaty opened up five Chinese ports to foreign trade and removed various restrictions on foreign traders. It allowed foreigners to send Christian missionaries to China and impose European tariffs. Perhaps most importantly, it allowed Britain to establish Hong Kong as a colony.

Over the next century, foreign powers forced China to enter into more than a dozen unequal treaties with countries including France, Germany, Japan, Russia, and the United States. China's concessions allowed foreigners to carve out areas in Chinese port cities where they would live and trade. These treaties focused on specific ports that became referred to as "treaty ports."

These treaty ports established which foreign powers controlled what parts of China. For example, Russia mainly controlled Harbin, the capital of Heilongjiang Province, and Germany controlled Qingdao, a prosperous city in Shandong Province.

The foreign presence in many of these cities still exists today. Harbin's skyline is dominated by Russian architecture, and Tsingtao (an English transliteration of "Qingdao") beer's success is directly due to German beer-making techniques introduced during Germany's control of Qingdao's port.

C. Lasting Resentment

Many of the foreign countries ended their unequal treaties with China before the Second World War. However, the remaining unequal treaties played an important role in how the Kuomintang and CPC contended for public support. Both groups asserted that they could best eliminate the unequal treaties and their effect on China.

The rest of the unequal treaties were abrogated during the first few decades of CPC rule. Nonetheless, Chinese resentment of the unequal treaties still has a minor role in the creation of contemporary treaties.

"The Signing and Sealing of the Treaty of Nanking in the State Cabin of H.M.S. Cornwallis, 29th August, 1842"
Source: John Platt (painter) and John Burnet (engraver), Massachusetts Institute of Technology, http://ocw.mit.edu/ans7870/21f/21f.027/opium_wars_01/ow1_essay04.html

II. Contemporary Treaties

China is increasingly using treaties to promote its international economic presence. Over the past two decades, China has entered into dozens of bilateral communiqués for the establishment of diplomatic relations.[1] These short agreements function as stepping stones to creating political and economic relationships.

Additionally, China has also entered into friendship agreements with Japan, North Korea, and Russia. Its agreement with Japan is essentially a peace treaty. In comparison, its friendship agreements with North Korea and Russia seek to establish joint goals and promises to aid the other country.

China is also party to dozens of multilateral international economic agreements under the United Nations and other international associations.[2]

Appendix IV provides a table of selected multinational treaties that China is a party to, and Appendix V lists selected bilateral treaties between the U.S. and China.

1. For a complete list of China's bilateral treaties and agreements, *see* Bilateral Treaties and Agreements, Ministry of Foreign Affairs of the People's Republic of China, *available at* http://www.fmprc.gov.cn/eng/wjb/zzjg/tyfls/tyfl/2631/.

2. *See* United Nations Treaty Collection, United Nations, *available at* https://treaties.un.org/Home.aspx.

III. Applicability of Treaties to Hong Kong and Macao

A treaty entered into by the PRC will only apply to Hong Kong or Macao if they enter into a separate agreement that provides that said treaty extends them. However, any treaties surrounding foreign affairs and defense matters will immediately extend to both Hong Kong and Macao.

Nonetheless, the PRC allows Hong Kong to conduct its own relevant external affairs. For example, Hong Kong may develop its own relationships in areas such as culture, finance, trade, and tourism. Hong Kong may also determine its own representatives in international conferences and organizations.

The PRC and Hong Kong have also entered into their own bilateral agreements. Although these agreements are not necessarily treaties, as Hong Kong is not an independent country, they are worded similarly to treaties. These agreements include free-trade agreements and double taxation avoidance agreements.

IV. Taiwan

The complicated relationship between the PRC and ROC has created many difficulties in how Taiwan interacts with the PRC and other countries. The Taiwanese stance on PRC treaties is that they do not apply to Taiwan unless Taiwan determines that they do. The PRC, on the other hand, asserts that its treaties apply to Taiwan just as if it were a province.

Another complicated issue surrounding Taiwan is whether it can enter into its own "treaties." The general contemporary worldview of Taiwan is that it is not an independent country. However, countries with particularly good relationships with Taiwan sometimes want to enter into agreements with Taiwan that do not apply to Mainland China. Likewise, Taiwan may want to enter into agreements that do not include Mainland China. As a result, countries including the United States have made bilateral agreements with Taiwan.

Researchers may also find that international agreements apply to Taiwan as a separate customs territory. For example, in 2001 negotiations concluded on the admission of the Separate Customs Territory of Taiwan, Penghu, Kinmen, and Matsu (Chinese Taipei) to the World Trade Organization.[3] Chinese

3. WTO successfully concludes negotiations on entry of the Separate Customs Territory of Taiwan, Penghu, Kinmen and Matsu, World Trade Organization *available at* http://www.wto.org/english/news_e/pres01_e/pr244_e.htm.

Taipei has been a member of the World Trade Organization since January 2002.[4]

V. Additional Resources

Burton, Charles, *China* in *Legal Systems of the World*, ABC-CLIO, INC. (Herbert M. Kritzer, ed. 2002).

Bush, Richard C., *At Cross Purposes: U.S.-Taiwan Relations Since 1942*, M E SHARPE INC. (2004).

China in the WTO, PALGRAVE MACMILLAN (Carlos Magarinos, Yongtu Long, & Francisco Sercovich, eds. 2003).

Cho Hui-Wan, *Taiwan's Application to GATT/WTO: Significance of Multilateralism for an Unrecognized State*, PRAEGER (2001).

Chung Chien-Peng, *China's Multilateral Co-operation in Asia and the Pacific: Institutionalizing Beijing's 'Good Neighbour Policy,'* ROUTLEDGE (2010).

Lardy, Nicholas, *Integrating China into the Global Economy*, BROOKINGS INSTITUTION PRESS (2001).

Wang, Dong, *China's Unequal Treaties: Narrating National History*, LEXINGTON BOOKS (2008).

---

4. Separate Customs Territory of Taiwan, Penghu, Kinmen and Matsu (Chinese Taipei) and the WTO, WORLD TRADE ORGANIZATION, *available at* http://www.wto.org/english/thewto_e/countries_e/chinese_taipei_e.htm.

Chapter Nine

Secondary and Internet Sources

Interacting with China's secondary and Internet sources for legal research can be a daunting endeavour. Not only are many resources either only in Mandarin or provide various different translations, but China has a booming legal research industry. The same information is provided in multiple different official, subscription-based, and free sources.

It is crucial for legal researchers to use the best possible sources. Unofficial translations of national legislations often include various serious mistakes such as incorrectly translating a legal term, providing the wrong number of articles in a regulation or misstating a damages amount. Legal researchers should also verify summaries of court opinions, statistics, and translations of scholarly articles. Researchers must also verify that sources relied upon are current.

I. Official Sources

It is always best practice to determine whether an official source provides the regulation on its website. The first step to finding an official regulation is to determine its Chinese and English names. The official name can be found by using an English search on Baidu or one of the subscription-based legal search sites discussed in the next section.[1] The next step is to identify which government body or administration promulgated the regulation. The last step is to visit the corresponding website and look for the official version of the regulation.

However, researchers that either (1) have not determined all of the regulations that need to be researched or (2) are unable to find English names of regulations should first visit the government or administrative website. The

1. Full access to these websites requires a paid subscription. Nonetheless, they allow visitors to conduct searches and yield results limited to the name, information related to the regulation, and the first few clauses of the regulation.

following subsections provide source information for China's national and local governments, administrations, and people's courts.

A. National Government

The national government and its administrations generally provide all of their laws and regulations on their websites. They have also made great efforts to include English portals that can be found on the top-right corner of their homepages. The English version of their websites usually has much less information than the Chinese version. However, they typically provide background information and a list of laws and regulations.

Many of these websites only list the regulations with official English translations. If the English translation is on the website, then it is the official translation that should be used when conducting research. However, if the website does not have an English translation, then the next step would be to look for unofficial translations elsewhere.

Foreign researchers looking for the official Chinese translation may have to use the Chinese version of the website. On the main Chinese homepage, look for a link to "法律法规" ("legal legislation"), "法规" ("legislation") or "规定" ("regulations"). That page will have a list of regulations. The next step will be to search the page with the Chinese name of the regulation found through Baidu or a subscription-based services.

1. Central People's Government

The website maintained by the Central People's Government is an essential tool for legal researchers. Although this site is not dedicated solely to legal issues, it assists legal researches by providing:

- official reports;
- explanations on the Chinese legislature, judiciary, and local governments;
- articles on contemporary issues; and
- a thorough database of government websites through the two drop-down boxes found on the left-hand side of the homepage.

Source: Central People's Government of the People's Republic of China, http://english.gov.cn/

2. *Legislative Affairs Office of the State Council*

The State Council's Legislative Affairs Office also maintains a website designed to provide information concerning China's national legislation. In reality, the website does much more because it also lists administrative regulations and judicial opinions.

The Mandarin version of the website can be very helpful because it essentially contains all of the PRC's laws and regulations as well as commentary, summaries, and explanations. In contrast, the English version is mostly limited to summaries.

B. Local Governments

Foreign researchers conducting research on local government regulations can also visit their respective websites. Unlike national government websites, local governments may not have English portals. Generally speaking, all provinces and first-tier cities[2] have English websites. The English portals for

2. Although Chinese municipalities are often compared in terms of first, second, and third tier, it is both difficult to determine a definition for "first-tier city" as well as which cities are first-tier cities. Some factors include, but are not limited to, (1) when the city began modern development, (2) economy, (3) population, and (4) how "cosmopolitan"

Source: Legislative Affairs Office of the State Council of the People's Republic of China, http://www.chinalaw.gov.cn/

Source: Legislative Affairs Office of the State Council of the People's Republic of China, http://www.chinalaw.gov.cn/article/english/Regulations/Laws/

local governments websites usually only provide basic information about the area. It is highly unlikely that the local government will have official English translations of local regulations. They probably will have lists of local regulations in Chinese only.

---

the city is. Beijing, Shanghai, and Guangzhou are considered to be first-tier cities. However, other cities such as Chongqing, Tianjin, and Shenzhen may also be on this list.

Researching local government regulations generally requires a high level of written Mandarin skills. Most of China's local regulations have not been translated either officially or unofficially. Researching secondary sources such as law review articles may yield partial translations from researchers that have already written on local regulations.[3]

C. Administrations

Administrative sources often provide English information including official English translations. Also, administrations will sometimes release English explanations about new regulations that affect foreign businesses. Although these articles may be relatively short or omit information found in other sources, they (or their Mandarin version) are often written by people involved in the enforcement of the regulation.

As administrations only provide information concerning their prospective duties, this subsection does not provide introductions to specific administrations. See Appendix VI for a list of Chinese agencies that are especially important to foreign legal researchers.

D. Courts

Finally, Chinese courts also provide court decisions and opinions to the public. However, unlike some of China's laws and regulations, the PRC does not have official translations of Chinese court decisions (except for selected SPC interpretations). On the same token, China's official judicial website, chinacourt.org, has a limited English website.

Source: Supreme People's Court of the People's Republic of China, http://en.chinacourt.org/

3. *See, e.g.*, Paul Kossof, *Business Franchise Compliance in Mainland China*, 2 J. Transnat'l Legal Issues 131, 140 n. 66 (2013) (translating and analyzing every Chinese local regulation that affects business franchising).

Nonetheless, non-Mandarin readers may find chinacourt.org useful for its judicial news articles, which are both relatively timely and detailed. As for judicial decisions and opinions, I highly recommend using a subscription-based legal search website.

II. Subscription-Based Legal Search Sites

There are currently over a dozen widely-used bilingual Chinese legal research services. These websites market themselves towards foreign legal offices and foreign consultants. These services generally provide the same information from the same sources. However, some of them vary substantially in the quality and quantity of resources available. Some of them are more up-to-date and have more employees searching, translating, and uploading documents.

The services discussed in this section are not an exhaustive list. To facilitate readers' understanding of Chinese legal research, this section provides the three services that the author prefers to use based upon personal experience. Additionally, some of the sources listed in other publications include websites that are either very limited, under construction or only provide information in Mandarin. Some also may not be accessible outside of Mainland China. The list here is meant to provide the reader with the most suitable sources for conducting Chinese research in English.

A. Chinalawinfo

Chinalawinfo and its sister site Pkulaw.cn are likely the best bilingual Chinese legal research websites. Maintained by Peking University, these websites provide up-to-date bilingual resources including national and local regulations and scholarly articles. For Mandarin-readers, these websites also have the latest court cases and scholarly articles.

The English websites for Chinalawinfo and Pkulaw.cn allow researchers to input Chinese or English keywords and select detailed search parameters. The value of searching for Chinese characters is that there are often several unofficial translations for a law. Thus, a search for an English translation may not yield the desired search result. These websites will then find the English documents that correspond to the Chinese characters.

Another aspect of Chinalawinfo and Pkulaw.cn that facilitates research is that they provide links to regulations whenever they are mentioned. For example, China's implementing regulations for its trademark law provide hyperlinks to all of the laws that it refers to. For researchers that do not know a topic very well, these websites can also be very useful because they display revoked laws in red font.

Chinalawinfo and Pkulaw.cn are paid services. Nonetheless, they provide invaluable free services as well. For example, they usually have the best unofficial English translations for the names of laws and regulations. Even though these websites only provide the first few articles of a regulation to guests, a visiting researcher can find out the name of an English translation and then look for a free version on Baidu or Google.

English Homepage for Pkulaw.cn,
Source: Pkulaw.cn, http://en.pkulaw.cn/

B. LegalStudio

In addition to using Chinalawinfo and Pkulaw.cn to conduct legal research, foreign researchers may also want to be informed about the latest trends occurring in a specific area of Chinese law. LegalStudio is a paid service maintained by LexisNexis. Unlike legal research databases, LegalStudio provides in-depth daily bilingual updates on China's legal developments. In addition to Mandarin text, almost all of its articles and regulations also have high quality English translations. Foreign consultants also submit English-only articles.

LegalStudio's website allows subscribers to search for the latest articles, regulations, etc. on areas of the law and industries. It also includes forthcoming legislation, policies issues, and lists of most viewed regulations and articles.

Perhaps the most useful aspect of LegalStudio is that users can sign up for daily email updates. LegalStudio will send a tailored email each business day that provides the latest articles and regulations for the topics selected by the user.

Homepage for LegalStudio
Source: LegalStudio, http://www.legalstudio.com/web

C. Westlaw China and Lexis China

For U.S. researchers accustomed to using Westlaw or LexisNexis, both companies have China legal research products as well. These services are completely separate from those provided in other jurisdictions such as the U.S. This means that information on U.S. Westlaw or Lexis will not be on their Chinese products, and vice versa.

Both of these services provide useful databases that foreign researchers may prefer because they are similar to what foreign researchers may have used in their own jurisdictions. To that effect, many foreign firms subscribe to Westlaw China or Lexis China. However, they may not be as up-to-date as Chinalawinfo or Pkulaw.cn. Additionally, they have less scholarly articles and secondary sources. Nonetheless, they are very user-friendly and foreign researchers may prefer to begin conducting legal research with a familiar format.

Many researchers prefer Westlaw China over Lexis China because it acquired Isinolaw. Isinolaw, which was widely used by legal researchers especially foreigner researchers in Hong Kong, maintains a large database of Mainland China laws and regulations. As such, a comparison of the same search in Westlaw China and in Lexis China will often show more results in Westlaw China.

D. SSRN

Other electronic research databases may have articles on China that cannot be found elsewhere. The Social Science Research Network (SSRN) not only includes published articles but also has articles still in draft form. Using "China" or "Chinese" as a search term in the database will turn up many sources that researchers may not find elsewhere. SSRN can be found at www.ssrn.com. The website is free and can be used anonymously or by registering a user ID and password.

III. Blogs

Foreign researchers interested in keeping up with current Chinese legal issues should also consider visiting English blogs on Chinese law. Some of the most popular Chinese legal blogs are:

Blog Name	Website
China Law Blog	chinalawblog.com
China Law Insight	chinalawinsight.com
China Law Prof Blog	lawprofessors.typepad.com/china_law_prof_blog/
China Legal Blog	www.chinalegalblog.com/en/

Although these blogs are generally very well maintained, legal researchers should verify information posted on these blogs before using it for academic or business purposes. Additionally, bilingual blog posts sometimes include incorrect translations of Chinese legal terms. An easy way to verify whether a legal term is probably correct is to search for the English term on Baidu and Google and see whether regulations and scholarly articles use that term.

IV. Additional Resources

Hu, Robert, *Research Guide to Chinese Trademark Law and Practice*, Carolina Academic Press (2009).

Lee, Tahirih, *Basic Concepts of Chinese Law*, Garland Publishing (1997).

Luo, Wei, *Chinese Law and Legal Research*, Fred B. Rothman & Co. (2006).

Chapter Ten

Research Strategies

This book has provided the fundamental background of the PRC's legal system and its history. It has also discussed all of the sources of Chinese law as well as how to access them. At this point, the reader should have a relatively complete understanding of where to look for legal sources. But, where to begin? How do you know that you have found the correct law? Is the source accurate? And, perhaps most importantly, when do you need to hire local counsel?

I. Establish Research Parameters

The first step to conducting legal research is to establish research parameters. Depending upon the situation, the research parameters will be defined by a law-based or issue-based approach.

A. Law-Based Approach

A law-based approach begins, as the term implies, with a specific law or regulation. Typically, the law-based approach research is reactionary.

1. When to Use a Law-Based Approach

There are many situations that may prompt a law-based approach to legal research. You or your client may receive correspondence that mentions a law or regulation. Here are some examples of correspondence that you will receive while doing business in China:

- administrative letters concerning employees, licenses, and taxes;
- cease and desist letters;
- investment opportunities emerging from new regulations; and
- notices of inspections.

Lawsuits are another situation where a researcher should use a law-based approach unless the same plaintiff has also filed an administrative action under a different issue. As the lawsuit will reference the law allegedly violated, and you will need to address the law in question, it's best to start your research with that law.

2. How to Conduct Research under the Law-Based Approach

Having decided to apply the law-based approach, the next step is to determine where to look for relevant laws and regulations. The general order for researching Chinese sources under the law-based approach is as follows:

1. Obtain the law(s) and/or regulation(s) listed in the correspondence;
    - For national laws, a search on Google or Baidu will most likely be adequate. Make sure that the document is from an official government website.
2. Check whether the specific articles in the correspondence cite other laws or regulations;
3. See whether the laws and regulations in Point One mention any law, regulation or administration (typically provided in the beginning of the document), and then determine whether it applies;
4. If laws, regulations, and administrations in Point Three do apply, then follow those sources and repeat Point Three until this method is exhausted;
5. Search for applicable local government regulations; and
    - The vast majority of local government regulations do not have official English translations. Additionally, depending on the reader's Mandarin written proficiency, it may be easier to use a subscription-based website to find the official document as opposed to a local government website. Regardless of whether you can obtain the official regulation, most legal research on local governments will require the assistance of local counsel (see Section III).
6. Determine whether there are any SPC opinions that discuss how courts should interpret the laws and regulations.

Finally, if you are conducting research due to an administrative complaint or lawsuit, you must also determine the proper procedure for responding. Agencies provide their guidelines on their websites. However, depending upon the agency, the regulation may only be provided in Mandarin. National legislation provides the guidelines for court procedures (civil lawsuits are governed by the Civil Procedure Law).

3. *Cease and Desist Letter Example*

The list provided in the following subsection sounds easy enough. But, what does it look like in practice? The following example is what a legal researcher is likely to find by using this list to conduct research after receiving a cease and desist letter alleging that your client is infringing upon a trademark registered in the PRC:

1. Obtain the laws and regulations listed in the correspondence;
    - Trademark Law (2013 Amendment)
        - Alleges violation of Article 57
    - Anti-Unfair Competition Law
        - Alleges violation of Article 5
2. Check whether the specific articles in the correspondence cite other laws or regulations;
    - None
3. See whether the laws and regulations in Point One mention any law, regulation, or administration (typically provided in the beginning of the document), and then determine whether it applies;
    - Civil Procedure Law
    - CTMO
    - Local AIC
    - SAIC
    - TRAB
4. If laws, regulations, or an administration in Point Three do apply, then follow those sources and repeat Point Three until this method is exhausted;
    - Trademark Law Implementation Regulations (2014 Amendment)
5. Search for applicable local government regulations; and
    - Measures of Beijing Municipality on Famous Trademark Recognition and Protection
        - This regulation does not have an official English translation, and cannot be found in official sources without conducting a search in Mandarin.
6. Determine whether there are any SPC opinions that discuss how courts should interpret the laws and regulations.
    - Interpretations of the Supreme People's Court Concerning the Application of Laws in the Trial of Cases of Civil Disputes Arising from Trademarks

As shown through this example, following this six point list will allow you to determine which laws, regulations, administrations, and SPC opinions are associated with the specific law or regulation that the third party mentioned. However, what do you do if you don't have a starting point?

B. Issue-Based Approach

1. When and How to Use the Issue-Based Approach

Legal researchers are not always so "fortunate" to receive information about specific laws or regulations from another party. In most situations, research on Chinese law will be issue-based. After all, how many clients know the laws that apply to their legal issue?

Issue-based legal research is typically more difficult than law-based research because it requires (1) an understanding of related issues and (2) knowledge of which institutions will have related laws and regulations.

Unlike law-based research, this approach requires that the researcher conduct thorough discovery of the issue at hand. If the issue has been raised by a client, then it is imperative to have a detailed conversation with that client to determine whether there is anything else that may affect the issue. This is especially important for foreign clients doing business in China because they may not be aware of what is and is not significant.

After conducting a complete discovery of the issue, the next step is to take a step back and brainstorm. The following is an effective method to brainstorm which laws and regulations may affect any given issue:

1. Create a comprehensive list of words that may be in the titles of applicable laws and regulations;
2. Search for each word on the national and local level either through their websites or a subscription-based legal search engine;
3. Conduct administrative searches using a subscription-based legal search engine;
4. Use Google, Baidu, or a subscription-based search engine to search for SPC opinions;
5. Brainstorm which administrations may affect this issue, and visit their websites or use a filtered search on a subscription-based legal search engine; and
6. Look for articles written by academics or law firms on the specific issue, and see whether they either mention any information that you have not uncovered so far.

As with the law-based approach, you should also determine which laws and regulations determine the procedures for administrative complaints or lawsuits. However, as you should not file a complaint or lawsuit without consulting with your client, you may decide not to conduct procedural research until you know whether you will be pursuing one or both of these.

2. Domain Infringement Example

Your client calls you one day and tells you that she found a Chinese website with a registered domain name that is exactly the same as her company name. Additionally, the website also looks like her website and even uses her company logo. She asks you to see whether the site could be taken down, and wants to know whether she can receive money from the company for the negative effect that it has had on her reputation, sales, etc.

After getting off the phone, you pull out your copy of *Chinese Legal Research* and apply the method for conducting issue-based legal research:

1. Create a comprehensive list of words that may be in the titles of applicable laws and regulations;

Trademark	Domain	Copyright	Likeness
Trade Mark	Website	Logo	Misrepresentation
Service Mark	Internet Service Provider	Identifier	Appropriation

2. Search for each word on the national and local level either through their websites or a subscription-based legal search engine;
   - Anti-Unfair Competition Law
   - Copyright Law
   - Internet Domain Name Regulations
   - Trademark Law
3. Conduct administrative searches using a subscription-based legal search engine;
   - China Internet Network Information Center Domain Name Resolution Policy
4. Use Google, Baidu, or a subscription-based search engine to search for SPC opinions;

- Interpretation of the Supreme People's Court on the Application of Laws in the Trial of Civil Disputes Over Domain Names of Computer Networks

5. Brainstorm which administrations may affect this issue, and visit their websites or use a filtered search on a subscription-based legal search engine; and
   - China Internet Network Information Center
   - Local AIC
   - SAIC
6. Look for articles written by academics or law firms on the specific issue, and see whether they either mention any information that you have not uncovered so far.

II. Verifying English-Based Sources

Conducting research in English will always create the possibility of error either due to translation problems, misunderstanding or incorrect information. As such, the overall quality of Chinese legal research conducted only in English will benefit from observing the practices presented in this section.

1. Translations

Translations of national legislation and national administrative regulations are generally very good. Not only do these official translations provide the correct law, but they also tend to use the same vocabulary choices (which facilitates legal research).

On the other hand, the quality of official local governmental and administrative translations may not be as good as their national counterparts. Although they can be helpful, especially since most local government regulations are not translated, it is recommended to compare the translation with the Mandarin version before acting on it.

There are also countless sources of unofficial translations including foreign governments, foreign bar associations, academics and law firms. These translations range in quality. For example, translations made in association with suggestions to national amendments are usually done very well. On the other hand, local law firms' translations may be very low quality (as their primary purpose can be to promote hits on the firms' sites). Where possible, researchers should try to locate and compare more than one translation.

2. Summaries

Summaries often function as windows into areas of Chinese law that researchers are not familiar with. Summaries from local and foreign firms alike can provide introductions and insights that help establish a starting point for research into any given subject. Thus, you should spend some time reading summaries in order to acquaint yourself with a research topic.

However, summaries can also mislead the reader or even misrepresent information. Firstly, they are sometimes written to attract business—as such, they may present an issue as simpler or complicated than it really is. Secondly, their authors may either be inexperienced employees, foreign interns, or others. Third, dated summaries may discuss expired laws or regulations. Finally, some summaries purposely withhold valuable information because their authors would prefer that you pay for their advice!

3. Articles

Law firms, particularly large international firms, will also compile high-quality articles on updates to Chinese law. Unlike summaries, these articles often provide specific information on laws and regulations as well as practical advice and how the authors expect the change will affect legal practice. As firms use these articles to represent themselves to other firms, in-house counsel, and potential clients, their authors are usually very knowledgeable and the articles have been looked over by other attorneys.

III. When to Ask for Help

The overarching theme of this book is that foreign legal researchers can often perform effective Chinese legal research without assistance from local counsel. Nonetheless, there are some instances where local counsel should or even must participate in legal research.

A. Local Regulations

The first and most obvious reason to hire local counsel is the general unavailability of English information on local regulations. Not only is it often impossible to find any information on a local regulation, but you may also need to either call or visit a local administration in order to procure information such as procedural guidelines. Therefore, although it may be possible to conduct most or all of the research on national issues without local counsel, you may have to consult with a Chinese attorney if you encounter local regulations.

B. Language and Cultural Barriers

There are also other language and cultural barriers that may require assistance from local counsel. Depending on the province that you are researching, some local officials, judges, and others may actually prefer to speak in their own dialect. Additionally, some areas such as Inner Mongolia may even have documents written in their own language. Local counsel is also necessary when you have to consider local culture and customs.

C. Guanxi

The final reason why a foreign legal consultant should consider utilizing local counsel is *guanxi* (关系)—relationships. Although *guanxi* is never an issue for conducting legal research, it is very important for administrative and judicial actions.

The personal relationships between an attorney and an examiner, official, judge, enforcement personnel, etc. are very important for a variety of legitimate reasons. For example, local counsel may be able to call an examiner or enforcement official and ask about the status of an application or raid. An official may share something about how his or her government will change a law. As Chinese cases do not have precedential value, an attorney that knows a judge's stance on the issue in question may have a good understanding of how that judge will rule.

Chapter Eleven

Citations

The final essential element to conducting Chinese legal research is knowing how to cite Chinese regulations, cases, and common secondary sources. Most foreign academics and practitioners apply the *Bluebook* rules for citing Chinese sources in academic articles and other publications.[1] Although it is important to apply the correct *Bluebook* citation guidelines, the general practice in China is relatively less stringent than in the United States.

Legal researchers that are familiar with using the *Bluebook* will not encounter too much difficulty creating Chinese legal citations. The most difficult aspect of citing Chinese sources, especially if the researcher does not read Mandarin, is knowing when and how to romanize Chinese characters.

I. Romanization

Romanization of Chinese characters refers to transliterating Chinese into text that uses the Roman alphabet. Citation is an inevitable aspect of legal research, and romanization is an inevitable part of citing Chinese sources.

A. When to Romanize Chinese Characters

1. The Rule v. Actual Practice

The *Bluebook* states that Chinese sources should always be romanized.[2] Researchers that are concerned with carefully abiding by the *Bluebook* should consider always romanizing their Chinese sources.

However, researchers can also make their own determinations of when not to romanize Chinese characters. In fact, intentionally choosing not to romanize

1. *See The Bluebook: A Uniform System of Citation*, The Harvard Law Review Association (19th ed., 2010).

2. *See id.* at 180, Rule 20.2.4.

Chinese sources may actually benefit the researcher because it facilitates an easy reading of the research.

If a researcher decides that his or her work would benefit from not romanizing every source, then this is the golden rule:

> *The more important the source, the less likely that you need to romanize it.*
> *Always romanize local regulations.*
> *Do not romanize case names if you can provide the court name and case number.*

2. Trademark Law Example

National regulations that are very important to foreigners almost never need to be romanized. China's trademark law is closely followed by foreign researchers. So, it is sufficient to refer to it as the "Trademark Law of the People's Republic of China."

On the other hand, foreigners seldom interact with local regulations. Therefore, a citation to local regulations must always include (1) romanization, (2) the original Chinese name of the regulation, and (3) the official or unofficial translation of the regulation.

Following the trademark law example, China has various local regulations on trademarks. Each province has a regulation on local famous trademarks.[3] None of these regulations have official English names or translations. The Chongqing regulation should therefore be cited as follows:

> Chongqingshi Zhuming Shangbiao Rending He Baohu Tiaoli (重庆市著名商标认定和保护条例) [Measures of Chongqing Municipality on Famous Trademark Recognition and Protection]

This citation example will be explained throughout the rest of this chapter. The first element is also the most difficult: finding the correct romanization for a Chinese name.

B. How to Romanize Chinese

Romanizing Chinese characters sounds like an impossible task to researchers that cannot read Mandarin text. Although it is difficult, and requires the use

3. For more information on Chinese local famous trademarks, see Paul Kossof, *Chinese National Well-known Trademarks and Local Famous Trademarks in Light of the 2013 Trademark Law: Status, Effect, and Adequacy*, 13 J. Marshall Rev. Intell. Prop. L. 225 (2013).

of translation software, foreign researchers should be able to create almost-perfect romanizations without knowing a single word of Chinese.

1. Basics of Romanization

Romanization requires only a basic understanding of how Chinese characters sound.

a. Chinese Characters and Corresponding Sounds

This subsection provide a crash course to Mandarin characters and sounds in just a few sentences. As this is a legal research book, it is only fitting to use the Chinese word for "law" as a teaching tool.

The Chinese word for "law" is "法律." This word is composed of two Characters: "法," which means "law" or "method," and "律" which also means "law." Each character is monosyllabic. So, as the Chinese word for "law" contains two characters, it has two syllables.

The Chinese language uses multiple tones. To facilitate reading tones, Chinese can be transliterated into Roman characters that use tone indications. The word for law would be "Fălǜ." If the tones were removed, then it would simply be "Falv." Note that the "ǜ" becomes a "v" when the tones are removed.

Chinese legal researchers can either use tonal or non-tonal romanization. The author recommends using non-tonal romanization because (1) the researcher will not have to interact with tone marks and (2) Chinese characters are typed using non-tonal key strokes.

b. Using Translation Software to Determine Rough Romanization

At this point, the reader should understand how a single word is romanized and that he or she can use the non-tonal Roman alphabet. But how about entire Chinese phrases?

The answer is much simpler than would be expected. Most of the most popular translation software and websites provide excellent romanization for Chinese phrases. Here are two of the tried and true (and free) tricks for romanizing Chinese characters:

1. Google Translate: First select "English" for the input and "Chinese (Simplified)" for the output. Place the English word in the translation box and select "Translate."

Source: Google, http://translate.google.com/

Then place the Chinese text in the translation box and select "Chinese (Simplified)" for *both* the input and output languages. The Chinese text will then show up on the output box as well. Click on the "Ä" in the output box. The tonal Chinese romanization will display.

Source: Google, http://translate.google.com/

- Example: The Chinese characters for the Chongqing regulation are "重庆市著名商标认定和保护条例." Using the steps outlined above, the tonal romanization displayed by Google Translate is "Chóngqìng shì zhùmíng shāngbiāo rèndìng hé bǎohù tiáolì."

Source: Google, http://translate.google.com/

- The second step is to remove the tones. The only irregularity to pay attention to is whether there is a "ǜ," which will always turn into "v." This example does not have a "ǜ."
  - Google Translate: Chóngqìng shì zhùmíng shāngbiāo rèndìng hé bǎohù tiáolì
  - Without tones: Chongqing shi zhuming shangbiao rending he baohu tiaoli
- The third step is to modify the spaces between the words. This is discussed in Subsection 2.
- The last step is to capitalize the romanization. This is discussed in Subsection 3.

2. Mandarin Popup for Firefox: For Mozilla Firefox users, Mandarin Popup is an invaluable tool for conducting Chinese legal research. It translates and provides the romanization for any Chinese character or phrase that the cursor hovers over. After installing and activating Mandarin Popup, place your cursor over the beginning of the Chinese phrase. A box will then instantly pop up. As the Mandarin Popup typically only translates one word at a time, the researcher will have to record each word separately. However, Mandarin Popup works wonderfully for spacing issues because it identifies each individual word. It is also better at identifying three-character words than Google Translate.

   - Using the same example as above, place "重庆市著名商标认定和保护条例" anywhere in Firefox. Then hover the cursor over the first character ("重"). Unlike Google Translate, which uses the romanization "Chongqing shi" for "Chongqing City," Mandarin Popup states "Chong qing shi." The general rule will be to remove the spaces, and the resulting romanization "Chongqingshi" will be the correct term. Continue for each subsequent word, adding one space between each word.

Source: Mandarin Popup

You can also search Mandarin Popup's database by right-clicking anywhere in your browser and selecting "Browse Mandarin Popup Dictionary."

The two tricks discussed above allow legal researchers with no Mandarin skills to create relatively correct romanizations. However, there are two additional steps to creating a good romanization: determining spaces and capitalizing.

2. Determining Spaces

Chinese sentences do not have any spaces. When they are romanized, spaces are added between each word. Since words usually vary between one and three characters, spaces must be added to the appropriate breaks.

Researchers creating romanization will have to pay close attention to where each word breaks. Although the romanization may not always be perfect, paying close attention to the breaks will result in a nearly correct romanization.

The best way for a non-Mandarin speaker to find breaks is to use translation software such as Mandarin Popup or Nciku.com in conjunction with Google Translate. The Google Translate result (without tones) for the above example was "Chongqing shi zhuming shangbiao rending he baohu tiaoli." The only change that has to be made to this romanization is to delete the space between "Chongqing" and "shi." However, even if the researcher did not delete the space, the romanization would be acceptable.

3. Capitalization

The last step to romanizing Chinese characters is to capitalize the first letter for each word. For example, "Chongqingshi zhuming shangbiao rending he baohu tiaoli" would become "Chongqingshi Zhuming Shangbiao Rending

He Baohu Tiaoli." This is the final and correct romanization for the Chongqing local famous trademark regulation.

Romanization can be intimidating. Primarily, it requires dealing with a very unfamiliar language. Nonetheless, carefully following the steps outlined in this subsection will result in a perfect or nearly perfect romanization of the Chinese source you wish to cite.

C. Format of Citations with Romanization

The specific citation format for a Chinese source depends on the type of source. However, regardless of the type of source, a source with romanized text will always start with the (1) romanization, followed by (2) the original Chinese name of the regulation, and then (3) the official or unofficial translation of the regulation.

This is the full citation for the Chongqing regulation:

> Chongqingshi Zhuming Shangbiao Rending He Baohu Tiaoli (重庆市著名商标认定和保护条例) [Measures of Chongqing Municipality on Famous Trademark Recognition and Protection] (promulgated by the Standing Comm. People's Cong. Chongqing Municipality, Sept. 22, 2011, effective June 1, 2012) (China).

And this is a case citation:

> Shenqing Zai Shenren Fuyuping, Lixiurong Yu Bei Shenqingren Xiejinlian, Caohuozhu Ji Mingzui Guoji Canyin Guanli (Beijing) Youxian Gongsi Texu Jingying Hetong Jiufenan (申请再审人付玉平、李秀荣与被申请人谢金莲、曹火珠及名嘴国际餐饮管理（北京）有限公司特许经营合同纠纷案) [Appeal of the Franchising Agreement Case of Petitioners Fuyuping and Lixiurong Against Plaintiffs Caohuozhu and Mingzui International Food and Beverage Administration (Beijing) Co., Ltd.], 2011 MINSHENZI 1262 (Sup. People's Ct. Dec. 12, 2011) (Chinalawinfo).

It does not matter whether the Chinese source is a regulation or a case. It will always begin with the romanization, followed by the original Chinese in parentheses, and the English translation in square brackets.

Finally, it is important to note that many regulations and all case names do not have English translations. Therefore, it is the researcher's responsibility to create an unofficial translation. This can typically be achieved by translating and editing the source name. Of course, the researcher must also compare the translation with his or her understanding of the contents of the regulation or name of the case.

The rest of this chapter provides the citation format and abbreviations for common Chinese sources.

II. Chinese Constitution

The citation to the Chinese constitution is very simple. The *Bluebook* format for footnotes is this:

XIANFA art. <article>, § <section> (<year>) (China).

Example:

XIANFA art. 1, § 1 (1982) (China).

However, the format for briefs and memoranda would generally be in normal type.

Example:

Xianfa art. 1, § 1 (1982) (China).

"Xianfa" is the romanization for "宪法" which means "constitution." The year will be 1982 unless the cited constitution is one of China's previous constitutions.

III. Regulations

Citing Chinese regulations is very simple once a researcher knows how to romanize Chinese characters. It also requires information about the promulgation of the regulation and abbreviations for the name of the promulgating authority.[4]

A. General Rule

Both Chinese laws and regulations follow the same general citation format. This guideline includes the romanization for instances that it is necessary or the reader wishes to adhere to *Bluebook* rule 20.2.4:

---

4. Promulgation information can be found in the regulation itself (directly below its title and above the main content). Abbreviations are discussed in Section 5 of this chapter.

Format:

<romanization> (<original Chinese name>) [<English translation>] (promulgated by <authority, abbreviated whenever possible>, <promulgation date>, effective <effective date>) (China) (<source (ex: Chinalawinfo)>).

Example:

Chongqingshi Zhuming Shangbiao Rending He Baohu Tiaoli (重庆市著名商标认定和保护条例) [Measures of Chongqing Municipality on Famous Trademark Recognition and Protection] (promulgated by the Standing Comm. People's Cong. Chongqing Municipality, Sept. 22, 2011, effective June 1, 2012) (China) (Chinalawinfo).

B. Well-Known Regulations

Researchers may not need to provide as much information for laws and regulations that are widely discussed. In these instances, the following format may be followed:

Format:

<Official English translation> (promulgated by <authority, abbreviated whenever possible>, <promulgation date>, effective <effective date>).

Example:

Trademark Law of the People's Republic of China (2013 Amendment) (promulgated by the Standing Comm. Twelfth Nat'l People's Cong., Aug. 30, 2013).

Note that this citation does not include "(China)" or "(Chinalawinfo)." The researcher may include "(China)" or both. However, they are not necessary because it is clear that the law belongs to China and because it is provided by various sources.

IV. Cases

The two main differences between citing regulations and cases is that (1) it may be more difficult to translate court case names and (2) the case citation requires some different information.

In regards to the translation, it is important for the translation to properly identify the parties and the type of lawsuit. The parties to the left of the character "与" will be the plaintiffs or appellants, and the parties to the right of the character will be the defendants or respondents. The type of case comes immediately after the party names.

The researcher will also require the name of the court, date of the decision, and the decision number.

A. Full Citation

The general format for full case citations is as follows:

Format:

<romanization> (<original Chinese name>) [<English translation>], <promulgation year> <source indicator> <decision number> (<court abbreviation> <date of decision>) (<source (ex: Chinalawinfo)>).

Example:

Shenqing Zai Shenren Fuyuping, Lixiurong Yu Bei Shenqingren Xiejinlian, Caohuozhu Ji Mingzui Guoji Canyin Guanli (Beijing) Youxian Gongsi Texu Jingying Hetong Jiufenan (申请再审人付玉平、李秀荣与被申请人谢金莲、曹火珠及名嘴国际餐饮管理（北京）有限公司特许经营合同纠纷案) [Appeal of the Franchising Agreement Case of Petitioners Fuyuping and Lixiurong Against Plaintiffs Caohuozhu and Mingzui International Food and Beverage Administration (Beijing) Co., Ltd.], 2011 Minshenzi 1262 (Sup. People's Ct. Dec. 12, 2011) (Chinalawinfo).

The "source indicator" refers to the case number and corresponding characters which identify the court and type of case. For example, the source indicator "兴民一初字第(number)号" would be translated as follows:

- Original: 兴民一初字第(number)号
  - The "第" indicates that a number will follow, and "号" means "number."
  - The romanization of "兴民一初" is "Xingmin Yichu"
    - The first character refers to the court, and the second refers to the type of case.
    - The third and fourth characters mean "early."
- Portion of citation: Xingmin Yichu (number)
  - Remember to use small caps.

B. Shortened Citation

Depending on the level of adherence to the *Bluebook*, a researcher can also use a shortened citation. These are generally adequate for academic and business articles published in China.

Format:

<English translation>, <promulgation year> <source indicator> <decision number> (<court abbreviation> <date of decision>).

Example:

Appeal of the Franchising Agreement Case of Petitioners Fuyuping and Lixiurong Against Plaintiffs Caohuozhu and Mingzui International Food and Beverage Administration (Beijing) Co., Ltd., 2011 MINSHENZI 1262 (Sup. People's Ct. Dec. 12, 2011).

Note that it is especially important to provide the proper source indicator because the reader may not be able to find the case through the unofficial English translation.

V. Abbreviations

Table T2.9 of the *Bluebook* contains the various abbreviations that a foreign researcher may require while creating citations to Chinese sources. The following is a list of some of the more widely used abbreviations for Chinese sources:

A. Institutions

- Guowuyuan (国务院)—St. Council
- Quanguo Renmin Daibiao Dahui (全国人民代表大会)—Nat'l People's Cong.
- Zhongji Renmin Fayuan (中级人民法院)—Interm. People's Ct.
- Zuigao Renmin Fayuan (最高人民法院)—Sup. People's Ct.

B. Laws

- China Law and Practice—China L. & Prac.
- Code of the People's Republic of China—Fadian
- Guowuyuan Gongbao (国务院公报)—St. Council Gaz.

- Laws and Regulations of the People's Republic of China—P.R.C. Laws & Regs
- Xin Fagui Huibian (新法规汇编)—Xin Fagui Huibian

C. Cases

- Renmin Fayuan Anli Xuan (人民法院案例选)—Renmin Fayuan Anli Xuan
- Zuigao Renmin Fayuan Gongbao (最高人民法院公报)—Sup. People's Ct. Gaz.
- Zuigao Renmin Jianchayuan Gongbao (最高人民检察院公报)—Sup. People's Proc. Communique

VI. Additional Resources

Luo Wei, *Chinese Law and Legal Research*, Fred B. Rothman & Co. (2006).

The Bluebook: A Uniform System of Citation, The Harvard Law Review Association (19th ed., 2010).

Zimmerman, James M., *China Law Deskbook: A Legal Guide for Foreign-Invested Enterprises*, American Bar Association Section of International Law (2010).

Conclusion

Conducting legal research on any foreign jurisdiction can be overwhelming. Chinese legal research is especially intimidating to foreign researchers because of the substantial differences between their home country and China's government and legal system. Written Chinese can also be daunting, especially to researchers that simply do not have the time to learn it.

This book's message is that effective Chinese legal research can be conducted by foreigners as long as they have a thorough understanding of China's legal system. Going forward, it is essential for foreign researchers to remember that conducting legal research in China will require different and additional steps than what they are used to in their own jurisdictions.

Chinese court opinions do not provide all of the related regulations, and researchers will sometimes need to think outside of the box to determine which regulations may apply to a legal issue. Most subjects will require research of national and local government laws as well as administrative regulations. Researchers must always investigate each level to obtain a complete understanding of the applicable law.

In regards to Chinese skills, this book mitigates most of the difficulties that foreign researchers may encounter. It explains how to find Chinese regulations without an understanding of Chinese, interacting with case summaries and unofficial translations, and romanizing sources. Nonetheless, there is always a chance of error or failure. Therefore, foreign researchers may in some instances still need to ask a Chinese researcher for advice.

Conducting legal research in China is an exciting and even scary process. This book provides all of the background knowledge, resources, and tricks necessary for you to succeed. The rest can only be learned by experience.

Appendices

1. Constitution of the People's Republic of China
2. Law on Legislation of the People's Republic of China
3. General Principles of the Civil Law of the People's Republic of China
4. Selected Multinational Treaties
5. Selected U.S.-China Bilateral Treaties
6. Useful Contact Information

Appendix I

Constitution of the People's Republic of China

Amended in accordance with the Amendments to the Constitution of the People's Republic of China adopted respectively at the First Session of the Seventh National People's Congress on April 12, 1988, the First Session of the Eighth National People's Congress on March 29, 1993, the Second Session of the Ninth National People's Congress on March 15, 1999 and the Second Session of the Tenth National People's Congress on March 14, 2004.

Contents

Preamble

China is a country with one of the longest histories in the world. The people of all of China's nationalities have jointly created a culture of grandeur and have a glorious revolutionary tradition.

After 1840, feudal China was gradually turned into a semi-colonial and semi-feudal country. The Chinese people waged many successive heroic struggles for national independence and liberation and for democracy and freedom.

Great and earthshaking historical changes have taken place in China in the 20th century.

The Revolution of 1911, led by Dr. Sun Yat-sen, abolished the feudal monarchy and gave birth to the Republic of China. But the historic mission of the Chinese people to overthrow imperialism and feudalism remained unaccomplished.

After waging protracted and arduous struggles, armed and otherwise, along a zigzag course, the Chinese people of all nationalities led by the Communist Party of China with Chairman Mao Zedong as its leader ultimately, in 1949, overthrew the rule of imperialism, feudalism and bureaucrat-capitalism, won a great victory in the New-Democratic Revolution and founded the People's Republic of China. Since then the Chinese people have taken control of state power and become masters of the country.

After the founding of the People's Republic, China gradually achieved its transition from a New-Democratic to a socialist society. The socialist transformation of the private ownership of the means of production has been completed, the system of exploitation of man by man abolished and the socialist system established. The people's democratic dictatorship led by the working class and based on the alliance of workers and peasants, which is in essence the dictatorship of the proletariat, has been consolidated and developed. The Chinese people and the Chinese People's Liberation Army have defeated imperialist and hegemonist aggression, sabotage and armed provocations and have thereby safeguarded China's national independence and security and strengthened its national defence. Major successes have been achieved in economic development. An independent and relatively comprehensive socialist system of industry has basically been established. There has been a marked increase in agricultural production. Significant advances have been made in educational, scientific and cultural undertakings, while education in socialist ideology has produced noteworthy results. The life of the people has improved considerably.

The victory in China's New-Democratic Revolution and the successes in its socialist cause have been achieved by the Chinese people of all nationalities, under the leadership of the Communist Party of China and the guidance of Marxism-Leninism and Mao Zedong Thought, by upholding truth, correcting errors and surmounting numerous difficulties and hardships. China will be in the primary stage of socialism for a long time to come. The basic task of the nation is to concentrate its effort on socialist modernization along the road of Chinese-style socialism. Under the leadership of the Communist Party of China and the guidance of Marxism-Leninism, Mao Zedong Thought, Deng Xiaoping Theory and the important thought of Three Represents, the Chinese people of all nationalities will continue to adhere to the people's democratic dictatorship and the socialist road, persevere in reform and opening to the outside world, steadily improve socialist institutions, develop the socialist market economy, develop socialist democracy, improve the socialist legal system and work hard and self-reliantly to modernize the country's industry, agriculture, national defence and science and technology step by step and promote the coordinated development of the material, political and spiritual civilizations, to turn China into a socialist country that is prosperous, powerful, democratic and culturally advanced.The exploiting classes as such have been abolished in our country. However, class struggle will continue to exist within certain bounds for a long time to come. The Chinese people must fight against those forces and elements, both at home and abroad, that are hostile to China's socialist system and try to undermine it.

Taiwan is part of the sacred territory of the People's Republic of China. It is the inviolable duty of all Chinese people, including our compatriots in Taiwan, to accomplish the great task of reunifying the motherland.

In building socialism it is essential to rely on workers, peasants and intellectuals and to unite all forces that can be united. In the long years of revolution and construction, there has been formed under the leadership of the Communist Party of China a broad patriotic united front which is composed of the democratic parties and people's organizations and which embraces all socialist working people, all builders of socialism, all patriots who support socialism, and all patriots who stand for the reunification of the motherland. This united front will continue to be consolidated and developed. The Chinese People's Political Consultative Conference, a broadly based representative organization of the united front which has played a significant historical role, will play a still more important role in the country's political and social life, in promoting friendship with other countries and in the struggle for socialist modernization and for the reunification and unity of the country. The system of the multi-party cooperation and political consultation led by the Communist Party of China will exist and develop for a long time to come.

The People's Republic of China is a unitary multi-national State created jointly by the people of all its nationalities. Socialist relations of equality, unity and mutual assistance have been established among the nationalities and will continue to be strengthened. In the struggle to safeguard the unity of the nationalities, it is necessary to combat big-nation chauvinism, mainly Han chauvinism, and to combat local national chauvinism. The State will do its utmost to promote the common prosperity of all the nationalities.

China's achievements in revolution and construction are inseparable from the support of the people of the world. The future of China is closely linked to the future of the world. China consistently carries out an independent foreign policy and adheres to the five principles of mutual respect for sovereignty and territorial integrity, mutual non-aggression, non-interference in each other's internal affairs, equality and mutual benefit, and peaceful coexistence in developing diplomatic relations and economic and cultural exchanges with other countries. China consistently opposes imperialism, hegemonism and colonialism, works to strengthen unity with the people of other countries, supports the oppressed nations and the developing countries in their just struggle to win and preserve national independence and develop their national economies, and strives to safeguard world peace and promote the cause of human progress.

This Constitution, in legal form, affirms the achievements of the struggles of the Chinese people of all nationalities and defines the basic system and basic tasks of the State; it is the fundamental law of the State and has supreme legal authority. The people of all nationalities, all State organs, the armed forces, all political parties and public organizations and all enterprises and institutions in the country must take the Constitution as the basic standard of conduct, and they have the duty to uphold the dignity of the Constitution and ensure its implementation.

Chapter I
General Principles

Article 1 The People's Republic of China is a socialist state under the people's democratic dictatorship led by the working class and based on the alliance of workers and peasants.

The socialist system is the basic system of the People's Republic of China. Disruption of the socialist system by any organization or individual is prohibited.

Article 2 All power in the People's Republic of China belongs to the people. The National People's Congress and the local people's congresses at various levels are the organs through which the people exercise state power.

The people administer State affairs and manage economic and cultural undertakings and social affairs through various channels and in various ways in accordance with the provisions of law.

Article 3 The State organs of the People's Republic of China apply the principle of democratic centralism.

The National People's Congress and the local people's congresses at various levels are constituted through democratic elections. They are responsible to the people and subject to their supervision.

All administrative, judicial and procuratorial organs of the State are created by the people's congresses to which they are responsible and by which they are supervised. The division of functions and powers between the central and local State organs is guided by the principle of giving full scope to the initiative and enthusiasm of the local authorities under the unified leadership of the central authorities.

Article 4 All nationalities in the People's Republic of China are equal. The State protects the lawful rights and interests of the minority nationalities and upholds and develops a relationship of equality, unity and mutual assistance among all of China's nationalities. Discrimination against and oppression of any nationality are prohibited; any act which undermines the unity of the nationalities or instigates division is prohibited.

The State assists areas inhabited by minority nationalities in accelerating their economic and cultural development according to the characteristics and needs of the various minority nationalities.

Regional autonomy is practised in areas where people of minority nationalities live in concentrated communities; in these areas organs of self-government are established to exercise the power of autonomy. All national autonomous areas are integral parts of the People's Republic of China.

All nationalities have the freedom to use and develop their own spoken and written languages and to preserve or reform their own folkways and customs.

Article 5 The People's Republic of China governs the country according to law and makes it a socialist country under rule of law.

The State upholds the uniformity and dignity of the socialist legal system.

No laws or administrative or local regulations may contravene the Constitution.

All State organs, the armed forces, all political parties and public organizations and all enterprises and institutions must abide by the Constitution and other laws. All acts in violation of the Constitution or other laws must be investigated.

No organization or individual is privileged to be beyond the Constitution or other laws.

Article 6 The basis of the socialist economic system of the People's Republic of China is socialist public ownership of the means of production, namely, ownership by the whole people and collective ownership by the working people. The system of socialist public ownership supersedes the system of exploitation of man by man; it applies the principle of "from each according to his ability, to each according to his work".

In the primary stage of socialism, the State upholds the basic economic system in which the public ownership is dominant and diverse forms of ownership develop side by side and keeps to the distribution system in which distribution according to work is dominant and diverse modes of distribution coexist.

Article 7 The State-owned economy, namely, the socialist economy under ownership by the whole people, is the leading force in the national economy. The State ensures the consolidation and growth of the State-owned economy.

Article 8 The rural collective economic organizations apply the dual operation system characterized by the combination of centralized operation with decentralized operation on the basis of operation by households under a contract. In rural areas, all forms of cooperative economy, such as producers', supply and marketing, credit and consumers' cooperatives, belong to the sector of socialist economy under collective ownership by the working people. Working people who are members of rural economic collectives have the right, within the limits prescribed by law, to farm plots of cropland and hilly land allotted for their private use, engage in household sideline production and raise privately owned livestock.

The various forms of cooperative economy in cities and towns, such as those in the handicraft, industrial, building, transport, commercial and service trades, all belong to the sector of socialist economy under collective ownership by the working people.

The State protects the lawful rights and interests of the urban and rural economic collectives and encourages, guides and helps the growth of the collective economy.

Article 9 All mineral resources, waters, forests, mountains, grasslands, unreclaimed land, beaches and other natural resources are owned by the State, that is, by the whole people, with the exception of the forests, mountains, grasslands, unreclaimed land and beaches that are owned by collectives as prescribed by law.

The State ensures the rational use of natural resources and protects rare animals and plants. Appropriation or damaging of natural resources by any organization or individual by whatever means is prohibited.

Article 10 Land in the cities is owned by the State.

Land in the rural and suburban areas is owned by collectives except for those portions which belong to the State as prescribed by law; house sites and privately farmed plots of cropland and hilly land are also owned by collectives.

The State may, in the public interest and in accordance with law, expropriate or requisition land for its use and make compensation for the land expropriated or requisitioned.

No organization or individual may appropriate, buy, sell or otherwise engage in the transfer of land by unlawful means. The right to the use of land may be transferred according to law.

All organizations and individuals using land must ensure its rational use.

Article 11 The non-public sectors of the economy such as the individual and private sectors of the economy, operating within the limits prescribed by law, constitute an important component of the socialist market economy.

The State protects the lawful rights and interests of the non-public sectors of the economy such as the individual and private sectors of the economy. The State encourages, supports and guides the development of the non-public sectors of the economy and, in accordance with law, exercises supervision and control over the non-public sectors of the economy.

Article 12 Socialist public property is inviolable.

The State protects socialist public property. Appropriation or damaging of State or collective property by any organization or individual by whatever means is prohibited.

Article 13 Citizens' lawful private property is inviolable.

The State, in accordance with law, protects the rights of citizens to private property and to its inheritance.

The State may, in the public interest and in accordance with law, expropriate or requisition private property for its use and make compensation for the private property expropriated or requisitioned.

Article 14 The State continuously raises labour productivity, improves economic results and develops the productive forces by enhancing the enthusiasm of the working people, raising the level of their technical skill, disseminating advanced science and technology, improving the systems of economic administration and enterprise operation and management, instituting the socialist system of responsibility in various forms and improving the organization of work.

The State practises strict economy and combats waste.

The State properly apportions accumulation and consumption, concerns itself with the interests of the collective and the individual as well as of the State and, on the basis of expanded production, gradually improves the material and cultural life of the people.

The State establishes a sound social security system compatible with the level of economic development.

Article 15 The State practises socialist market economy.

The State strengthens economic legislation, improves macro-regulation and control.

The State prohibits in accordance with law any organization or individual from disturbing the socio-economic order.

Article 16 State-owned enterprises have decision-making power with regard to their operation within the limits prescribed by law.

State-owned enterprises practise democratic management through congresses of workers and staff and in other ways in accordance with law.

Article 17 Collective economic organizations have decision-making power in conducting independent economic activities, on condition that they abide by the relevant laws.

Collective economic organizations practise democratic management and, in accordance with law, elect or remove their managerial personnel and decide on major issues concerning operation and management.

Article 18 The People's Republic of China permits foreign enterprises, other foreign economic organizations and individual foreigners to invest in China and to enter into various forms of economic cooperation with Chinese enterprises and other Chinese economic organizations in accordance with the provisions of the laws of the People's Republic of China.

All foreign enterprises, other foreign economic organizations as well as Chinese-foreign joint ventures within Chinese territory shall abide by the laws of the People's Republic of China. Their lawful rights and interests are protected by the laws of the People's Republic of China.

Article 19 The State undertakes the development of socialist education and works to raise the scientific and cultural level of the whole nation.

The State establishes and administers schools of various types, universalizes compulsory primary education and promotes secondary, vocational and higher education as well as pre-school education.

The State develops educational facilities in order to eliminate illiteracy and provide political, scientific, technical and professional education for workers, peasants, State functionaries and other working people. It encourages people to become educated through independent study.

The State encourages the collective economic organizations, State enterprises and institutions and other sectors of society to establish educational institutions of various types in accordance with law.

The State promotes the nationwide use of Putonghua [common speech based on Beijing pronunciation—Tr.].

Article 20 The State promotes the development of the natural and social sciences, disseminates knowledge of science and technology, and commends and rewards achievements in scientific research as well as technological innovations and inventions.

Article 21 The State develops medical and health services, promotes modern medicine and traditional Chinese medicine, encourages and supports the setting up of various medical and health facilities by the rural economic collectives, State enterprises and institutions and neighbourhood organizations, and promotes health and sanitation activities of a mass character, all for the protection of the people's health.

The State develops physical culture and promotes mass sports activities to improve the people's physical fitness.

Article 22 The State promotes the development of art and literature, the press, radio and television broadcasting, publishing and distribution services, libraries, museums, cultural centres and other cultural undertakings that serve the people and socialism, and it sponsors mass cultural activities.

The State protects sites of scenic and historical interest, valuable cultural monuments and relics and other significant items of China's historical and cultural heritage.

Article 23 The State trains specialized personnel in all fields who serve socialism, expands the ranks of intellectuals and creates conditions to give full scope to their role in socialist modernization.

Article 24 The State strengthens the building of a socialist society with an advanced culture and ideology by promoting education in high ideals, ethics, general knowledge, discipline and the legal system, and by promoting the formulation and observance of rules of conduct and common pledges by various sections of the people in urban and rural areas.

The State advocates the civic virtues of love of the motherland, of the people, of labour, of science and of socialism. It conducts education among the people in patriotism and collectivism, in internationalism and communism and in dialectical and historical materialism, to combat capitalist, feudal and other decadent ideas.

Article 25 The State promotes family planning so that population growth may fit the plans for economic and social development.

Article 26 The State protects and improves the environment in which people live and the ecological environment. It prevents and controls pollution and other public hazards.

The State organizes and encourages afforestation and the protection of forests.

Article 27 All State organs carry out the principle of simple and efficient administration, the system of responsibility for work and the system of training functionaries and appraising their performance in order constantly to improve the quality of work and efficiency and combat bureaucratism.

All State organs and functionaries must rely on the support of the people, keep in close touch with them, heed their opinions and suggestions, accept their supervision and do their best to serve them.

Article 28 The State maintains public order and suppresses treasonable and other criminal activities that endanger State security; it penalizes criminal activities that endanger public security and disrupt the socialist economy as well as other criminal activities; and it punishes and reforms criminals.

Article 29 The armed forces of the People's Republic of China belong to the people. Their tasks are to strengthen national defence, resist aggression, defend the motherland, safeguard the people's peaceful labour, participate in national reconstruction and do their best to serve the people.

The State strengthens the revolutionization, modernization and regularization of the armed forces in order to increase national defence capability.

Article 30 The administrative division of the People's Republic of China is as follows:

1. The country is divided into provinces, autonomous regions, and municipalities directly under the Central Government;
2. Provinces and autonomous regions are divided into autonomous prefectures, counties, autonomous counties, and cities; and
3. Counties and autonomous counties are divided into townships, nationality townships, and towns.

Municipalities directly under the Central Government and other large cities are divided into districts and counties. Autonomous prefectures are divided into counties, autonomous counties, and cities.

All autonomous regions, autonomous prefectures and autonomous counties are national autonomous areas.

Article 31 The State may establish special administrative regions when necessary. The systems to be instituted in special administrative regions shall be prescribed by law enacted by the National People's Congress in the light of specific conditions.

Article 32 The People's Republic of China protects the lawful rights and interests of foreigners within Chinese territory; foreigners on Chinese territory must abide by the laws of the People's Republic of China.

The People's Republic of China may grant asylum to foreigners who request it for political reasons.

Chapter II
The Fundamental Rights and Duties of Citizens

Article 33 All persons holding the nationality of the People's Republic of China are citizens of the People's Republic of China.

All citizens of the People's Republic of China are equal before the law.

The State respects and preserves human rights.

Every citizen is entitled to the rights and at the same time must perform the duties prescribed by the Constitution and other laws.

Article 34 All citizens of the People's Republic of China who have reached the age of 18 have the right to vote and stand for election, regardless of ethnic status, race, sex, occupation, family background, religious belief, education, property status or length of residence, except persons deprived of political rights according to law.

Article 35 Citizens of the People's Republic of China enjoy freedom of speech, of the press, of assembly, of association, of procession and of demonstration.

Article 36 Citizens of the People's Republic of China enjoy freedom of religious belief. No State organ, public organization or individual may compel citizens to believe in, or not to believe in, any religion; nor may they discriminate against citizens who believe in, or do not believe in, any religion.

The State protects normal religious activities. No one may make use of religion to engage in activities that disrupt public order, impair the health of citizens or interfere with the educational system of the State.

Religious bodies and religious affairs are not subject to any foreign domination.

Article 37 Freedom of the person of citizens of the People's Republic of China is inviolable.

No citizen may be arrested except with the approval or by decision of a people's procuratorate or by decision of a people's court, and arrests must be made by a public security organ.

Unlawful detention or deprivation or restriction of citizens' freedom of the person by other means is prohibited, and unlawful search of the person of citizens is prohibited.

Article 38 The personal dignity of citizens of the People's Republic of China is inviolable. Insult, libel, false accusation or false incrimination directed against citizens by any means is prohibited.

Article 39 The residences of citizens of the People's Republic of China are inviolable. Unlawful search of, or intrusion into, a citizen's residence is prohibited.

Article 40 Freedom and privacy of correspondence of citizens of the People's Republic of China are protected by law. No organization or individual may, on any ground, infringe upon citizens' freedom and privacy of correspondence, except in cases where, to meet the needs of State security or of criminal investigation, public security or procuratorial organs are permitted to censor correspondence in accordance with the procedures prescribed by law.

Article 41 Citizens of the People's Republic of China have the right to criticize and make suggestions regarding any State organ or functionary. Citizens have the right to make to relevant State organs complaints or charges against, or exposures of, any State organ or functionary for violation of law or dereliction of duty; but fabrication or distortion of facts for purposes of libel or false incrimination is prohibited.

The State organ concerned must, in a responsible manner and by ascertaining the facts, deal with the complaints, charges or exposures made by citizens. No one may suppress such complaints, charges and exposures or retaliate against the citizens making them.

Citizens who have suffered losses as a result of infringement of their civic rights by any State organ or functionary have the right to compensation in accordance with the provisions of law.

Article 42 Citizens of the People's Republic of China have the right as well as the duty to work. Through various channels, the State creates conditions for employment, enhances occupational safety and health, improves working conditions and, on the basis of expanded production, increases remuneration for work and welfare benefits.

Work is a matter of honour for every citizen who is able to work. All working people in State-owned enterprises and in urban and rural economic collectives should approach their work as the masters of the country that they are. The State promotes socialist labour emulation, and commends and rewards model and advanced workers. The State encourages citizens to take part in voluntary labour.

The State provides necessary vocational training for citizens before they are employed.

Article 43 Working people in the People's Republic of China have the right to rest.

The State expands facilities for the rest and recuperation of the working people and prescribes working hours and vacations for workers and staff.

Article 44 The State applies the system of retirement for workers and staff members of enterprises and institutions and for functionaries of organs of State according to law. The livelihood of retired persons is ensured by the State and society.

Article 45 Citizens of the People's Republic of China have the right to material assistance from the State and society when they are old, ill or disabled. The State develops social insurance, social relief and medical and health services that are required for citizens to enjoy this right. The State and society ensure the livelihood of disabled members of the armed forces, provide pensions to the families of martyrs and give preferential treatment to the families of military personnel.

The State and society help make arrangements for the work, livelihood and education of the blind, deaf-mutes and other handicapped citizens.

Article 46 Citizens of the People's Republic of China have the duty as well as the right to receive education.

The State promotes the all-round development of children and young people, morally, intellectually and physically.

Article 47 Citizens of the People's Republic of China have the freedom to engage in scientific research, literary and artistic creation and other cultural pursuits. The State encourages and assists creative endeavours conducive to the interests of the people that are made by citizens engaged in education, science, technology, literature, art and other cultural work.

Article 48 Women in the People's Republic of China enjoy equal rights with men in all spheres of life, in political, economic, cultural, social and family life.

The State protects the rights and interests of women, applies the principle of equal pay for equal work to men and women alike and trains and selects cadres from among women.

Article 49 Marriage, the family and mother and child are protected by the State.

Both husband and wife have the duty to practise family planning.

Parents have the duty to rear and educate their children who are minors, and children who have come of age have the duty to support and assist their parents.

Violation of the freedom of marriage is prohibited. Maltreatment of old people, women and children is prohibited.

Article 50 The People's Republic of China protects the legitimate rights and interests of Chinese nationals residing abroad and protects the lawful rights and interests of returned overseas Chinese and of the family members of Chinese nationals residing abroad.

Article 51 Citizens of the People's Republic of China, in exercising their freedoms and rights, may not infringe upon the interests of the State, of society or of the collective, or upon the lawful freedoms and rights of other citizens.

Article 52 It is the duty of citizens of the People's Republic of China to safeguard the unification of the country and the unity of all its nationalities.

Article 53 Citizens of the People's Republic of China must abide by the Constitution and other laws, keep State secrets, protect public property, observe labour discipline and public order and respect social ethics.

Article 54 It is the duty of citizens of the People's Republic of China to safeguard the security, honour and interests of the motherland; they must not commit acts detrimental to the security, honour and interests of the motherland.

Article 55 It is the sacred duty of every citizen of the People's Republic of China to defend the motherland and resist aggression.

It is the honourable duty of citizens of the People's Republic of China to perform military service and join the militia in accordance with law.

Article 56 It is the duty of citizens of the People's Republic of China to pay taxes in accordance with law.

Chapter III
The Structure of the State

Section 1
The National People's Congress

Article 57 The National People's Congress of the People's Republic of China is the highest organ of state power. Its permanent body is the Standing Committee of the National People's Congress.

Article 58 The National People's Congress and its Standing Committee exercise the legislative power of the State.

Article 59 The National People's Congress is composed of deputies elected from the provinces, autonomous regions, municipalities directly under the Central Government, and special administrative regions, and of deputies elected from the armed forces. All the minority nationalities are entitled to appropriate representation.

Election of deputies to the National People's Congress is conducted by the Standing Committee of the National People's Congress.

The number of deputies to the National People's Congress and the procedure of their election are prescribed by law.

Article 60 The National People's Congress is elected for a term of five years.

The Standing Committee of the National People's Congress must ensure the completion of election of deputies to the succeeding National People's Congress two months prior to the expiration of the term of office of the current National People's Congress. Should extraordinary circumstances prevent such an election, it may be postponed and the term of office of the current National People's Congress extended by the decision of a vote of more than two-thirds of all those on the Standing Committee of the current National People's Congress. The election of deputies to the succeeding National People's Congress must be completed within one year after the termination of such extraordinary circumstances.

Article 61 The National People's Congress meets in session once a year and is convened by its Standing Committee. A session of the National People's Congress may be convened at any time the Standing Committee deems it necessary or when more than one-fifth of the deputies to the National People's Congress so propose.

When the National People's Congress meets, it elects a Presidium to conduct its session.

Article 62 The National People's Congress exercises the following functions and powers:

1. to amend the Constitution;
2. to supervise the enforcement of the Constitution;
3. to enact and amend basic laws governing criminal offences, civil affairs, the State organs and other matters;
4. to elect the President and the Vice-President of the People's Republic of China;
5. to decide on the choice of the Premier of the State Council upon nomination by the President of the People's Republic of China, and on the choice of the Vice-Premiers, State Councillors, Ministers in charge of ministries or commissions, the Auditor-General and the Secretary-General of the State Council upon nomination by the Premier;
6. to elect the Chairman of the Central Military Commission and, upon nomination by the Chairman, to decide on the choice of all other members of the Central Military Commission;
7. to elect the President of the Supreme People's Court;
8. to elect the Procurator-General of the Supreme People's Procuratorate;
9. to examine and approve the plan for national economic and social development and the report on its implementation;
10. to examine and approve the State budget and the report on its implementation;
11. to alter or annul inappropriate decisions of the Standing Committee of the National People's Congress;
12. to approve the establishment of provinces, autonomous regions, and municipalities directly under the Central Government;
13. to decide on the establishment of special administrative regions and the systems to be instituted there;
14. to decide on questions of war and peace; and

15. to exercise such other functions and powers as the highest organ of state power should exercise.

Article 63 The National People's Congress has the power to remove from office the following persons:

1. the President and the Vice-President of the People's Republic of China;
2. the Premier, Vice-Premiers, State Councillors, Ministers in charge of ministries or commissions, the Auditor-General and the Secretary-General of the State Council;
3. the Chairman of the Central Military Commission and other members of the Commission;
4. the President of the Supreme People's Court; and
5. the Procurator-General of the Supreme People's Procuratorate.

Article 64 Amendments to the Constitution are to be proposed by the Standing Committee of the National People's Congress or by more than one-fifth of the deputies to the National People's Congress and adopted by a vote of more than two-thirds of all the deputies to the Congress.

Laws and resolutions are to be adopted by a majority vote of all the deputies to the National People's Congress.

Article 65 The Standing Committee of the National People's Congress is composed of the following:

the Chairman;

the Vice-Chairmen;

the Secretary-General; and

the members.

Minority nationalities are entitled to appropriate representation on the Standing Committee of the National People's Congress.

The National People's Congress elects, and has the power to recall, members of its Standing Committee.

No one on the Standing Committee of the National People's Congress shall hold office in any of the administrative, judicial or procuratorial organs of the State.

Article 66 The Standing Committee of the National People's Congress is elected for the same term as the National People's Congress; it shall exercise its functions and powers until a new Standing Committee is elected by the succeeding National People's Congress.

The Chairman and Vice-Chairmen of the Standing Committee shall serve no more than two consecutive terms.

Article 67 The Standing Committee of the National People's Congress exercises the following functions and powers:

1. to interpret the Constitution and supervise its enforcement;
2. to enact and amend laws, with the exception of those which should be enacted by the National People's Congress;
3. to partially supplement and amend, when the National People's Congress is not in session, laws enacted by the National People's Congress, provided that the basic principles of these laws are not contravened;
4. to interpret laws;
5. to review and approve, when the National People's Congress is not in session, partial adjustments to the plan for national economic and social development or to the State budget that prove necessary in the course of their implementation;
6. to supervise the work of the State Council, the Central Military Commission, the Supreme People's Court and the Supreme People's Procuratorate;
7. to annul those administrative regulations, decisions or orders of the State Council that contravene the Constitution or other laws;
8. to annul those local regulations or decisions of the organs of state power of provinces, autonomous regions, and municipalities directly under the Central Government that contravene the Constitution, other laws or administrative regulations;
9. to decide, when the National People's Congress is not in session, on the choice of Ministers in charge of ministries or commissions, the Auditor-General or the Secretary-General of the State Council upon nomination by the Premier of the State Council;
10. to decide, when the National People's Congress is not in session, on the choice of other members of the Central Military Commission upon nomination by the Chairman of the Commission;
11. to appoint or remove, at the recommendation of the President of the Supreme People's Court, the Vice-Presidents and Judges of the

Supreme People's Court, members of its Judicial Committee and the President of the Military Court;

12. to appoint or remove, at the recommendation of the Procurator-General of the Supreme People's Procuratorate, the Deputy Procurators-General and procurators of the Supreme People's Procuratorate, members of its Procuratorial Committee and the Chief Procurator of the Military Procuratorate, and to approve the appointment or removal of the chief procurators of the people's procuratorates of provinces, autonomous regions, and municipalities directly under the Central Government;
13. to decide on the appointment or recall of plenipotentiary representatives abroad;
14. to decide on the ratification or abrogation of treaties and important agreements concluded with foreign states;
15. to institute systems of titles and ranks for military and diplomatic personnel and of other specific titles and ranks;
16. to institute State medals and titles of honour and decide on their conferment;
17. to decide on the granting of special pardons;
18. to decide, when the National People's Congress is not in session, on the proclamation of a state of war in the event of an armed attack on the country or in fulfilment of international treaty obligations concerning common defence against aggression;
19. to decide on general or partial mobilization;
20. to decide on entering the state of emergency throughout the country or in particular provinces, autonomous regions, or municipalities directly under the Central Government; and
21. to exercise such other functions and powers as the National People's Congress may assign to it.

Article 68 The Chairman of the Standing Committee of the National People's Congress directs the work of the Standing Committee and convenes its meetings. The Vice-Chairmen and the Secretary-General assist the Chairman in his work.

The Chairman, the Vice-Chairmen and the Secretary-General constitute the Council of Chairmen which handles the important day-to-day work of the Standing Committee of the National People's Congress.

Article 69 The Standing Committee of the National People's Congress is responsible to the National People's Congress and reports on its work to the Congress.

Article 70 The National People's Congress establishes a Nationalities Committee, a Law Committee, a Finance and Economic Committee, an Education, Science, Culture and Public Health Committee, a Foreign Affairs Committee, an Overseas Chinese Committee and such other special committees as are necessary. These special committees work under the direction of the Standing Committee of the National People's Congress when the Congress is not in session. The special committees examine, discuss and draw up relevant bills and draft resolutions under the direction of the National People's Congress and its Standing Committee.

Article 71 The National People's Congress and its Standing Committee may, when they deem it necessary, appoint committees of inquiry into specific questions and adopt relevant resolutions in the light of their reports.

All organs of State, public organizations and citizens concerned are obliged to furnish the necessary information to the committees of inquiry when they conduct investigations.

Article 72 Deputies to the National People's Congress and members of its Standing Committee have the right, in accordance with procedures prescribed by law, to submit bills and proposals within the scope of the respective functions and powers of the National People's Congress and its Standing Committee.

Article 73 Deputies to the National People's Congress and members of the Standing Committee have the right, during the sessions of the Congress and the meetings of the Committee, to address questions, in accordance with procedures prescribed by law, to the State Council or the ministries and commissions under the State Council, which must answer the questions in a responsible manner.

Article 74 No deputy to the National People's Congress may be arrested or placed on criminal trial without the consent of the Presidium of the current session of the National People's Congress or, when the National People's Congress is not in session, without the consent of its Standing Committee.

Article 75 Deputies to the National People's Congress may not be held legally liable for their speeches or votes at its meetings.

Article 76 Deputies to the National People's Congress must play an exemplary role in abiding by the Constitution and other laws and keeping State secrets and, in public activities, production and other work, assist in the enforcement of the Constitution and other laws. Deputies to the National People's Congress should maintain close contact with the units which elected them and with the people, heed and convey the opinions and demands of the people and work hard to serve them.

Article 77 Deputies to the National People's Congress are subject to supervision by the units which elected them. The electoral units have the power, through procedures prescribed by law, to recall deputies they elected.

Article 78 The organization and working procedures of the National People's Congress and its Standing Committee are prescribed by law.

Section 2
The President of the People's Republic of China

Article 79 The President and Vice-President of the People's Republic of China are elected by the National People's Congress.

Citizens of the People's Republic of China who have the right to vote and to stand for election and who have reached the age of 45 are eligible for election as President or Vice-President of the People's Republic of China.

The term of office of the President and Vice-President of the People's Republic of China is the same as that of the National People's Congress, and they shall serve no more than two consecutive terms.

Article 80 The President of the People's Republic of China, in pursuance of the decisions of the National People's Congress and its Standing Committee, promulgates statutes, appoints or removes the Premier, Vice-Premiers, State Councillors, Ministers in charge of ministries or commissions, the Auditor-General and the Secretary-General of the State Council; confers State medals and titles of honour; issues orders of special pardons; proclaims entering of the state of emergency; proclaims a state of war; and issues mobilization orders.

Article 81 The President of the People's Republic of China, on behalf of the People's Republic of China, engages in activities involving State affairs and receives foreign diplomatic representatives and, in pursuance of the decisions of the Standing Committee of the National People's Congress, appoints or recalls plenipotentiary representatives abroad, and ratifies or abrogates treaties and important agreements concluded with foreign states.

Article 82 The Vice-President of the People's Republic of China assists the President in his work.

The Vice-President of the People's Republic of China may exercise such functions and powers of the President as the President may entrust to him.

Article 83 The President and Vice-President of the People's Republic of China exercise their functions and powers until the new President and Vice-President elected by the succeeding National People's Congress assume office.

Article 84 In the event that the office of the President of the People's Republic of China falls vacant, the Vice-President succeeds to the office of the President.

In the event that the office of the Vice-President of the People's Republic of China falls vacant, the National People's Congress shall elect a new Vice-President to fill the vacancy.

In the event that the offices of both the President and the Vice-President of the People's Republic of China fall vacant, the National People's Congress shall elect a new President and a new Vice-President. Prior to such election, the Chairman of the Standing Committee of the National People's Congress shall temporarily act as the President of the People's Republic of China.

Section 3
The State Council

Article 85 The State Council, that is, the Central People's Government, of the People's Republic of China is the executive body of the highest organ of state power; it is the highest organ of State administration.

Article 86 The State Council is composed of the following:

the Premier;

the Vice-Premiers;

the State Councillors;

the Ministers in charge of ministries;

the Ministers in charge of commissions;

the Auditor-General; and

the Secretary-General.

The Premier assumes overall responsibility for the work of the State Council. The ministers assume overall responsibility for the work of the ministries and commissions. The organization of the State Council is prescribed by law.

Article 87 The term of office of the State Council is the same as that of the National People's Congress.

The Premier, Vice-Premiers and State Councillors shall serve no more than two consecutive terms.

Article 88 The Premier directs the work of the State Council. The Vice-Premiers and State Councillors assist the Premier in his work.

Executive meetings of the State Council are to be attended by the Premier, the Vice-Premiers, the State Councillors and the Secretary-General of the State Council.

The Premier convenes and presides over the executive meetings and plenary meetings of the State Council.

Article 89 The State Council exercises the following functions and powers:

1. to adopt administrative measures, enact administrative regulations and issue decisions and orders in accordance with the Constitution and other laws;
2. to submit proposals to the National People's Congress or its Standing Committee;
3. to formulate the tasks and responsibilities of the ministries and commissions of the State Council, to exercise unified leadership over the work of the ministries and commissions and to direct all other administrative work of a national character that does not fall within the jurisdiction of the ministries and commissions;
4. to exercise unified leadership over the work of local organs of State administration at various levels throughout the country, and to formulate the detailed division of functions and powers between the Central Government and the organs of State administration of provinces, autonomous regions, and municipalities directly under the Central Government;
5. to draw up and implement the plan for national economic and social development and the State budget;
6. to direct and administer economic affairs and urban and rural development;

7. to direct and administer the affairs of education, science, culture, public health, physical culture and family planning;
8. to direct and administer civil affairs, public security, judicial administration, supervision and other related matters;
9. to conduct foreign affairs and conclude treaties and agreements with foreign States;
10. to direct and administer the building of national defence;
11. to direct and administer affairs concerning the nationalities and to safeguard the equal rights of minority nationalities and the right to autonomy of the national autonomous areas;
12. to protect the legitimate rights and interests of Chinese nationals residing abroad and protect the lawful rights and interests of returned overseas Chinese and of the family members of Chinese nationals residing abroad;
13. to alter or annul inappropriate orders, directives and regulations issued by the ministries or commissions;
14. to alter or annul inappropriate decisions and orders issued by local organs of State administration at various levels;
15. to approve the geographic division of provinces, autonomous regions, and municipalities directly under the Central Government, and to approve the establishment and geographic division of autonomous prefectures, counties, autonomous counties, and cities;
16. in accordance with the provisions of law, to decide on entering into the state of emergency in parts of provinces, autonomous regions, and municipalities directly under the Central Government;
17. to examine and decide on the size of administrative organs and, in accordance with the provisions of law, to appoint or remove administrative officials, train them, appraise their performance and reward or punish them; and
18. to exercise such other functions and powers as the National People's Congress or its Standing Committee may assign to it.

Article 90 Ministers in charge of the ministries or commissions of the State Council are responsible for the work of their respective departments and they convene and preside over ministerial meetings or general and executive meetings of the commissions to discuss and decide on major issues in the work of their respective departments.

The ministries and commissions issue orders, directives and regulations within the jurisdiction of their respective departments and in accordance with law and the administrative regulations, decisions and orders issued by the State Council.

Article 91 The State Council establishes an auditing body to supervise through auditing the revenue and expenditure of all departments under the State Council and of the local governments at various levels, and the revenue and expenditure of all financial and monetary organizations, enterprises and institutions of the State.

Under the direction of the Premier of the State Council and in accordance with the provisions of law, the auditing body independently exercises its power of supervision through auditing, subject to no interference by any other administrative organ or any public organization or individual.

Article 92 The State Council is responsible and reports on its work to the National People's Congress or, when the National People's Congress is not in session, to its Standing Committee.

Section 4
The Central Military Commission

Article 93 The Central Military Commission of the People's Republic of China directs the armed forces of the country.

The Central Military Commission is composed of the following:

the Chairman;

the Vice-Chairmen; and

the members.

The Chairman assumes overall responsibility for the work of the Central Military Commission. The term of office of the Central Military Commission is the same as that of the National People's Congress.

Article 94 The Chairman of the Central Military Commission is responsible to the National People's Congress and its Standing Committee.

Section 5
The Local People's Congresses and Local People's Government at Various Levels

Article 95 People's congresses and people's governments are established in provinces, municipalities directly under the Central Government, counties, cities, municipal districts, townships, nationality townships, and towns.

The organization of local people's congresses and local people's governments at various levels is prescribed by law.

Organs of self-government are established in autonomous regions, autonomous prefectures and autonomous counties. The organization and working procedures of organs of self-government are prescribed by law in accordance with the basic principles laid down in Sections 5 and 6 of Chapter III of the Constitution.

Article 96 Local people's congresses at various levels are local organs of state power. Local people's congresses at or above the county level establish standing committees.

Article 97 Deputies to the people's congresses of provinces, municipalities directly under the Central Government and cities divided into districts are elected by the people's congresses at the next lower level; deputies to the people's congresses of counties, cities not divided into districts, municipal districts, townships, nationality townships, and towns are elected directly by their constituencies.

The number of deputies to local people's congresses at various levels and the manner of their election are prescribed by law.

Article 98 The term of office of the local people's congresses at various levels is five years.

Article 99 Local people's congresses at various levels ensure the observance and implementation of the Constitution and other laws and the administrative regulations in their respective administrative areas. Within the limits of their authority as prescribed by law, they adopt and issue resolutions and examine and decide on plans for local economic and cultural development and for the development of public services.

Local people's congresses at or above the county level shall examine and approve the plans for economic and social development and the budgets of their respective administrative areas and examine and approve the reports on their implementation. They have the power to alter or annul inappropriate decisions of their own standing committees.

The people's congresses of nationality townships may, within the limits of their authority as prescribed by law, take specific measures suited to the characteristics of the nationalities concerned.

Article 100 The people's congresses of provinces, and municipalities directly under the Central Government, and their standing committees may adopt local regulations, which must not contravene the Constitution and other laws and administrative regulations, and they shall report such local regulations to the Standing Committee of the National People's Congress for the record.

Article 101 Local people's congresses at their respective levels elect and have the power to recall governors and deputy governors, or mayors and deputy mayors, or heads and deputy heads of counties, districts, townships and towns.

Local people's congresses at or above the county level elect, and have the power to recall, presidents of people's courts and chief procurators of people's procuratorates at the corresponding level. The election or recall of chief procurators of people's procuratorates shall be reported to the chief procurators of the people's procuratorates at the next higher level for submission to the standing committees of the people's congresses at the corresponding level for approval.

Article 102 Deputies to the people's congresses of provinces, municipalities directly under the Central Government and cities divided into districts are subject to supervision by the units which elected them; deputies to the people's congresses of counties, cities not divided into districts, municipal districts, townships, nationality townships, and towns are subject to supervision by their constituencies.

The units and constituencies which elect deputies to local people's congresses at various levels have the power to recall the deputies according to procedures prescribed by law.

Article 103 The standing committee of a local people's congress at or above the county level is composed of a chairman, vice-chairmen and members, and is responsible and reports on its work to the people's congress at the corresponding level.

A local people's congress at or above the county level elects, and has the power to recall, members of its standing committee.

No one on the standing committee of a local people's congress at or above the county level shall hold office in State administrative, judicial and procuratorial organs.

Article 104 The standing committee of a local people's congress at or above the county level discusses and decides on major issues in all fields of work in its administrative area; supervises the work of the people's government, people's court and people's procuratorate at the corresponding level; annuls inappropriate decisions and orders of the people's government at the corresponding level; annuls inappropriate resolutions of the people's congress at the next lower level; decides on the appointment or removal of functionaries of State organs within the limits of its authority as prescribed by law; and, when the people's congress at the corresponding level is not in session, recalls individual deputies to the people's congress at the next higher level and elects individual deputies to fill vacancies in that people's congress.

Article 105 Local people's governments at various levels are the executive bodies of local organs of state power as well as the local organs of State administration at the corresponding levels.

Governors, mayors and heads of counties, districts, townships and towns assume overall responsibility for local people's governments at various levels.

Article 106 The term of office of local people's governments at various levels is the same as that of the people's congresses at the corresponding levels.

Article 107 Local people's governments at or above the county level, within the limits of their authority as prescribed by law, conduct administrative work concerning the economy, education, science, culture, public health, physical culture, urban and rural development, finance, civil affairs, public security, nationalities affairs, judicial administration, supervision and family planning in their respective administrative areas; issue decisions and orders; appoint or remove administrative functionaries, train them, appraise their performance and reward or punish them.

People's governments of townships, nationality townships, and towns execute the resolutions of the people's congresses at the corresponding levels as well as the decisions and orders of the State administrative organs at the next higher level and conduct administrative work in their respective administrative areas.

People's governments of provinces, and of municipalities directly under the Central Government decide on the establishment and geographic division of townships, nationality townships, and towns.

Article 108 Local people's governments at or above the county level direct the work of their subordinate departments and of people's governments at lower levels, and have the power to alter or annul inappropriate decisions of their subordinate departments and of the people's governments at lower levels.

Article 109 Auditing bodies are established by local people's governments at or above the county level. Local auditing bodies at various levels, independently and in accordance with the provisions of law, exercise their power of supervision through auditing and are responsible to the people's government at the corresponding level and to the auditing body at the next higher level.

Article 110 Local people's governments at various levels are responsible and report on their work to people's congresses at the corresponding levels. Local people's governments at or above the county level are responsible and report on their work to the standing committees of the people's congresses at the corresponding levels when the congresses are not in session. Local people's governments at various levels are responsible and report on their work to the State administrative organs at the next higher level. Local people's governments at various levels throughout the country are State administrative organs under the unified leadership of the State Council and are subordinate to it.

Article 111 The residents committees and villagers committees established among urban and rural residents on the basis of their place of residence are mass organizations of self-management at the grass-roots level. The chairman, vice-chairmen and members of each residents or villagers committee are elected by the residents. The relationship between the residents and villagers committees and the grass-roots organs of state power is prescribed by law.

The residents and villagers committees establish sub-committees for people's mediation, public security, public health and other matters in order to manage public affairs and social services in their areas, mediate civil disputes, help maintain public order and convey residents' opinions and demands and make suggestions to the people's government.

Section 6
The Organs of Self-Government of National Autonomous Areas

Article 112 The organs of self-government of national autonomous areas are the people's congresses and people's governments of autonomous regions, autonomous prefectures and autonomous counties.

Article 113 In the people's congress of an autonomous region, autonomous prefecture or autonomous county, in addition to the deputies of the nationality exercising regional autonomy in the administrative area, the other nationalities inhabiting the area are also entitled to appropriate representation.

Among the chairman and vice-chairmen of the standing committee of the people's congress of an autonomous region, autonomous prefecture or autonomous county there shall be one or more citizens of the nationality or nationalities exercising regional autonomy in the area concerned.

Article 114 The chairman of an autonomous region, the prefect of an autonomous prefecture or the head of an autonomous county shall be a citizen of the nationality exercising regional autonomy in the area concerned.

Article 115 The organs of self-government of autonomous regions, autonomous prefectures and autonomous counties exercise the functions and powers of local organs of State as specified in Section 5 of Chapter III of the Constitution. At the same time, they exercise the power of autonomy within the limits of their authority as prescribed by the Constitution, the Law of the People's Republic of China on Regional National Autonomy and other laws and implement the laws and policies of the State in the light of the existing local situation.

Article 116 The people's congresses of national autonomous areas have the power to enact regulations on the exercise of autonomy and other separate regulations in the light of the political, economic and cultural characteristics of the nationality or nationalities in the areas concerned. The regulations on the exercise of autonomy and other separate regulations of autonomous regions shall be submitted to the Standing Committee of the National People's Congress for approval before they go into effect. Those of autonomous prefectures and counties shall be submitted to the standing committees of the people's congresses of provinces or autonomous regions for approval before they go into effect, and they shall be reported to the Standing Committee of the National People's Congress for the record.

Article 117 The organs of self-government of the national autonomous areas have the power of autonomy in administering the finances of their areas. All revenues accruing to the national autonomous areas under the financial system of the State shall be managed and used by the organs of self-government of those areas on their own.

Article 118 The organs of self-government of the national autonomous areas independently arrange for and administer local economic development under the guidance of State plans. In exploiting natural resources and building enterprises in the national autonomous areas, the State shall give due consideration to the interests of those areas.

Article 119 The organs of self-government of the national autonomous areas independently administer educational, scientific, cultural, public health and physical culture affairs in their respective areas, protect and sift through the cultural heritage of the nationalities and work for a vigorous development of their cultures.

Article 120 The organs of self-government of the national autonomous areas may, according to the military system of the State and practical local needs and with the approval of the State Council, organize local public security forces for the maintenance of public order.

Article 121 In performing their functions, the organs of self-government of the national autonomous areas, in accordance with the provisions of the regulations on the exercise of autonomy in those areas, employ the spoken and written language or languages in common use in the locality.

Article 122 The State provides financial, material and technical assistance to the minority nationalities to help accelerate their economic and cultural development.

The State helps the national autonomous areas train large numbers of cadres at various levels and specialized personnel and skilled workers of various professions and trades from among the nationality or nationalities in those areas.

Section 7
The People's Courts and the People's Procuratorates

Article 123 The people's courts of the People's Republic of China are the judicial organs of the State.

Article 124 The People's Republic of China establishes the Supreme People's Court and the people's courts at various local levels, military courts and other special people's courts.

The term of office of the President of the Supreme People's Court is the same as that of the National People's Congress. The President shall serve no more than two consecutive terms.

The organization of the people's courts is prescribed by law.

Article 125 Except in special circumstances as specified by law, all cases in the people's courts are heard in public. The accused has the right to defence.

Article 126 The people's courts exercise judicial power independently, in accordance with the provisions of law, and not subject to interference by any administrative organ, public organization or individual.

Article 127 The Supreme People's Court is the highest judicial organ.

The Supreme People's Court supervises the administration of justice by the people's courts at various local levels and by the special people's courts. People's courts at higher levels supervise the administration of justice by those at lower levels.

Article 128 The Supreme People's Court is responsible to the National People's Congress and its Standing Committee. Local people's courts at various levels are responsible to the organs of state power which created them.

Article 129 The people's procuratorates of the People's Republic of China are State organs for legal supervision.

Article 130 The People's Republic of China establishes the Supreme People's Procuratorate and the people's procuratorates at various local levels, military procuratorates and other special people's procuratorates.

The term of office of the Procurator-General of the Supreme People's Procuratorate is the same as that of the National People's Congress; the Procurator-General shall serve no more than two consecutive terms.

The organization of the people's procuratorates is prescribed by law.

Article 131 The people's procuratorates exercise procuratorial power independently, in accordance with the provisions of law, and not subject to interference by any administrative organ, public organization or individual.

Article 132 The Supreme People's Procuratorate is the highest procuratorial organ.

The Supreme People's Procuratorate directs the work of the people's procuratorates at various local levels and of the special people's procuratorates. People's procuratorates at higher levels direct the work of those at lower levels.

Article 133 The Supreme People's Procuratorate is responsible to the National People's Congress and its Standing Committee. People's procuratorates at various local levels are responsible to the organs of state power which created them and to the people's procuratorates at higher levels.

Article 134 Citizens of all China's nationalities have the right to use their native spoken and written languages in court proceedings. The people's courts and people's procuratorates should provide translation for any party to the court proceedings who is not familiar with the spoken or written languages commonly used in the locality.

In an area where people of a minority nationality live in a concentrated community or where a number of nationalities live together, court hearings should be conducted in the language or languages commonly used in the locality; indictments, judgments, notices and other documents should be written, according to actual needs, in the language or languages commonly used in the locality.

Article 135 The people's courts, the people's procuratorates and the public security organs shall, in handling criminal cases, divide their functions, each taking responsibility for its own work, and they shall coordinate their efforts and check each other to ensure the correct and effective enforcement of law.

Chapter IV The National Flag, the National Anthem, the National Emblem and the Capital

Article 136 The national flag of the People's Republic of China is a red flag with five stars.

The national anthem of the People's Republic of China is the March of the Volunteers.

Article 137 The national emblem of the People's Republic of China consists of an image of Tian'anmen in its centre illuminated by five stars and encircled by ears of grain and a cogwheel.

Article 138 The capital of the People's Republic of China is Beijing.

Appendix II

Law on Legislation of the People's Republic of China

Adopted at the third Session of the Ninth National People's Congress on March 15, 2000 and promulgated by Order No. 31 of the President of the People's Republic of China on March 15, 2000.

Contents

Chapter I
General Provisions

Article 1 This Law is enacted in accordance with the Constitution with a view to standardizing legislation, establishing a sound legislative system of the State, establishing and improving the socialist legal system with Chinese characteristics, safeguarding and developing socialist democracy, promoting the government of the country according to law and building a socialist country under the rule of law.

Article 2 This Law shall be applicable to the enactment, revision and nullification of laws, administrative regulations, local regulations, autonomous regulations and separate regulations. The rules of the departments under the State Council and of the local governments shall be formulated, revised and nullified in accordance with the relevant provisions of this Law.

Article 3 Laws shall be made in compliance with the basic principles laid down in the Constitution, principles of taking economic development as the central task, adhering to the socialist road and the people's democratic dictatorship, upholding leadership by the Communist Party of China, upholding Marxism-Leninism, Mao Zetong Thought and Deng Xiaoping theory and persevering in reform and in opening to the outside world.

Article 4 Laws shall be made in accordance with the statutory limits of power and procedures, on the basis of the overall interests of the State and for the purpose of safeguarding the uniformity and dignity of the socialist legal system.

Article 5 Laws shall be made in order to embody the will of the people, enhance socialist democracy and guarantee that the people participate in legislative activities through various channels.

Article 6 Law shall be made by proceeding from reality and scientifically and rationally prescribing the rights and duties of citizens, legal persons and other organizations, and the powers and responsibilities of State organs.

Chapter II
Laws

Section 1
Limits of Legislative Power

Article 7 The National People's Congress and its Standing Committee exercise the legislative power of the State.

The National People's Congress enacts and amends basic laws governing criminal offences, civil affairs, the State organs and other matters.

The Standing Committee of the National People's Congress enacts and amends laws other than the ones to be enacted by the National People's Congress, and when the National People's Congress is not in session, partially supplements and amends laws enacted by the National People's Congress, but not in contradiction to the basic principles of such laws.

Article 8 The following affairs shall only be governed by law:

1. affairs concerning State sovereignty;
2. formation, organization, and the functions and powers of the people's congresses, the people's governments, the people's courts and the people's procuratorates at all levels;
3. the system of regional national autonomy, the system of special administrative region, the system of self-government among people at the grassroots level;
4. criminal offences and their punishment;
5. mandatory measures and penalties involving deprivation of citizens of their political rights or restriction of the freedom of their person;
6. requisition of non-State-owned property;
7. basic civil system;
8. basic economic system and basic systems of finance, taxation, customs, banking and foreign trade;
9. systems of litigation and arbitration; and
10. other affairs on which laws must be made by the National People's Congress or its Standing Committee.

Article 9 If laws have not been enacted on the affairs specified in Article 8 of this Law, the National People's Congress or its Standing Committee has the power to make a decision to authorize the State Council to formulate, according to actual needs, administrative regulations first on part of those affairs, except for the affairs concerning criminal offences and their punishment, mandatory measures and penalties involving deprivation of citizens of their political rights or restriction of the freedom of their person, and the judicial system.

Article 10 In a decision on authorization, the purpose and scope of the authorization shall be clearly defined.

The authorization organ shall exercise the power strictly in compliance with the authorized purpose and scope.

The authorized organ may not impart the authorized power to any other organs.

Article 11 After the administrative regulations on an affair formulated under authorization have been tested in practice and when the conditions are ripe for making a law on the affair, the National People's Congress or its Standing Committee shall make a law on it in a timely manner. As soon as the law is made, the authorization with regard to that matter shall be terminated accordingly.

Section 2 Legislation Procedures for the National People's Congress

Article 12 The Presidium of the National People's Congress may submit to the National People's Congress legislative bills, which shall be deliberated by the session of the National People's Congress.

The Standing Committee of the National People's Congress, the State Council, the Central Military Commission, the Supreme People's Court, the Supreme People's Procuratorate and the special committees of the National People's Congress may submit to the National People's Congress legislative bills, which shall be put on the agenda of a session by decision of the Presidium.

Article 13 A delegation or a group of thirty or more deputies may submit a legislative bill to the National People's Congress. The Presidium shall decide whether or not to put it on the agenda of the session, or shall refer do so after referring the bill to a relevant special committee for deliberation and for making a proposal as to whether to put it on the agenda.

When the special committee holds a meeting to deliberate the bill, it may invite the sponsoring person to attend the meeting and express opinions.

Article 14 A legislative bill to be submitted to the National People's Congress may be submitted first to the Standing Committee when the National People's Congress is not in session, and after the Standing Committee has deliberated on it at its meetings in accordance with the procedures stipulated in Section 3 of Chapter II of this Law and decides to submit it to the National People's Congress for deliberation, the Standing Committee or the sponsor shall make explanations to a plenary meeting of the session.

Article 15 When the Standing Committee decides to submit a legislative bill to a session of the National People's Congress for deliberation, it shall distribute the draft bill to the deputies one month before the session is convoked.

Article 16 The legislative bill that has been placed on the agenda of a session of the National People's Congress shall be deliberated on by all the delegations after the explanation made by the sponsor has been heard at a plenary meeting of the session.

When the delegations are deliberating on a legislative bill, the sponsor shall send people to listen to their opinions and answer inquiries.

When the delegations are deliberating on a legislative bill, the relevant organ or organization shall, at the request of the delegations, send people to give briefings therefore.

Article 17 The legislative bill that has been placed on the agenda of a session of the National People's Congress shall be deliberated by the relevant special committee which shall submit its deliberated opinions to the Presidium, and the opinions shall also be printed and distributed at the session.

Article 18 The legislative bill that has been placed on the agenda of a session of the National People's Congress shall be subject to a unified deliberation by its Law Committee on the basis of the deliberated opinions of the various delegations and the relevant special committee. The Law Committee shall submit to the Presidium a report on the result of its deliberation and a revised draft law; major dissenting views shall be stated in the report. After examination and approval by the Presidium, the report and the draft law shall be printed and distributed at the session.

Article 19 With regard to a legislative bill that has been placed on the agenda of a session of the National People's Congress, the executive chairmen of the Presidium may, when necessary, convene a meeting of the heads of the various delegations to hear and discuss the deliberated opinions of the delegations on major questions in the legislative bill, and report the result of the discussion and the opinions expressed to the Presidium.

The executive chairmen of the Presidium may also convene a meeting of the interested deputies recommended by the delegations to discuss the major, special questions in the legislative bill, and report the result of the discussion and the opinions expressed to the Presidium.

Article 20 With regard to a legislative bill that has been placed on the agenda of a session of the National People's Congress, if the sponsor requests its withdrawal before it is put to vote, he shall state the reasons, and deliberation of the bill at the session shall terminate as soon as the Presidium has accepted the request and reported the matter to the session.

Article 21 Where important questions raised during the deliberation on a legislative bill call for further study, the Congress may, by decision of a plenary meeting according to a proposal made by the Presidium, authorize the Standing Committee to further deliberate on the bill on the basis of deputies' opinions, to make a decision and to give a report on the decision to the next session of the National People's Congress; or the Standing Committee may be authorized to further deliberate on the bill on the basis of deputies' opinions, to work out a revision proposal and to submit it to the next session of the National People's Congress for deliberation and decision.

Article 22 After a revised draft of the legislative bill has been deliberated on by the various delegations, the Law Committee shall revise revised draft according to the deliberated opinions of the delegations and prepare a draft for vote, the Presidium shall submit it for vote to a plenary meeting of the session, and the draft shall be subject to adoption by a simple majority of all the deputies.

Article 23 A law adopted by the National People's Congress shall be promulgated by Order of the President signed by the President of the People's Republic of China.

Section 3
Legislation Procedures for the Standing Committee of the National People's Congress

Article 24 The Council of Chairmen may submit legislative bills to a meeting of the Standing Committee for deliberation.

The State Council, the Central Military Commission, the Supreme People's Court, the Supreme People's Procuratorate or a special committee of the National People's Congress may submit a legislative bill to the Standing Committee, and the Council of Chairmen shall decide whether to put it on the agenda of a meeting of the Standing Committee or to refer it first to the relevant special committee for deliberation before deciding whether to put it on the agenda of a meeting of the Standing Committee in light of the report submitted by the relevant special committee. If the Council of Chairmen believes that the legislative bill contains major questions calling for further study, it may advise the sponsor of the bill to revise and improve the bill before submitting it to the Standing Committee.

Article 25 Ten or more of the members of the Standing Committee may jointly submit a legislative bill to the Standing Committee, and the Council of Chairmen shall decide whether to put it on the agenda of a meeting of the Standing Committee, or to refer it first to the relevant special committee for deliberation before deciding whether to put it on the agenda of a meeting of the Standing Committee in light of the suggestions as to whether to put it on the agenda submitted by the relevant special committee. If the Council of Chairmen decides not to put the legislative bill on the agenda of a meeting of the Standing Committee, it shall report the matter to a meeting of the Standing Committee or give an explanation to the sponsor.

When a special committee holds a meeting to deliberate on a bill, the sponsor may be invited to attend the meeting and express opinions.

Article 26 When a legislative bill is placed on the agenda of a meeting of the Standing Committee, its draft shall, except under special circumstances, be delivered to the component members of the Standing Committee seven days before the meeting.

Article 27 As a rule, a legislative bill placed on the agenda of a meeting of the Standing Committee shall be put to vote after deliberations at three meetings of the Standing Committee.

When the Standing Committee is to deliberate on a legislative bill for the first time, it shall hear the explanation made by the sponsor at a plenary meeting, and then preliminary deliberation shall be conducted at group meetings.

When the Standing Committee is to deliberate on a legislative bill for the second time, it shall hear the report made by the Law Committee on the revision of the draft and the main problems thereof at a plenary meeting, and then further deliberation shall be conducted at group meetings.

When the Standing Committee is to deliberate on a legislative bill for the third time, it shall hear the report made by the Law Committee on the result of its deliberation on the draft at a plenary meeting, and then the revised draft of the legislative bill be deliberated on at group meetings. When the Standing Committee is to deliberate on a legislative bill, it may, according to needs, convene joint group meetings or plenary meetings to discuss the main questions contained in the draft.

Article 28 If the various quarters have a consensus in the main on a legislative bill placed on the agenda of a meeting of the Standing Committee, the bill may be put to vote after deliberation at two meetings of the Standing Committee; if the various quarters have a consensus in the main on a legislative bill concerning partial amendment of a law, the bill may be put to vote after deliberated at one meeting only.

Article 29 When group meetings of the Standing Committee are held to deliberate on a legislative bill, the sponsor shall send people to listen to opinions and answer inquires. When group meetings of the Standing Committee are held to deliberate on a legislative bill, the relevant organ or organization shall, at the request of any group, send people to give briefings thereof.

Article 30 The legislative bill placed on the agenda of a meeting of the Standing Committee shall be deliberated by the relevant special committee, which shall offer its opinions after deliberation and have them printed and distributed at the Standing Committee meeting.

When a special committee holds a meeting to deliberate on a legislative bill, members of other relevant special committees may be invited to attend the meeting and express opinions.

Article 31 With regard to a legislative bill placed on the agenda of a Standing Committee meeting, the Law Committee shall conduct a unified deliberation on the basis of the opinions expressed by members of the Standing Committee and relevant special committees after deliberation as well as the opinions offered by the various quarters, work out a revision report or a report on the result of its deliberation and a revised draft of the legislative bill, and state the major dissenting view in either of the two reports. It shall give feedback to the relevant special committees if their deliberated opinions of importance are not accepted.

When the Law Committee holds a meeting to deliberate on a legislative bill, members of other relevant special committees may be invited to attend the meeting and express opinions.

Article 32 When a special committee is to deliberate on a legislative bill, it shall hold a plenary meeting and may, in light of need, request the relevant organ or organization to send the leading members concerned to make explanation.

Article 33 Where the special committees disagree on major questions contained in a draft law, the matter shall be reported to the Council of Chairmen.

Article 34 With regard to a legislative bill placed on the agenda of a Standing Committee meeting, the Law Committee, the relevant special committee and the working offices of the Standing Committee shall listen to opinions of the various quarters by holding forums, seminars, hearings, etc.

The working offices of the Standing Committee shall send copies of the draft law to the relevant organs, organizations and specialists to solicit their opinions and then sort out the opinions and submit them to the Law Committee and the relevant special committee and, where necessary, print and distribute them at a meeting of the Standing Committee.

Article 35 With regard to a legislative bill placed on the agenda of a Standing Committee meeting, it may, by decision of the Council of Chairmen, be published for soliciting opinions. Opinions gathered from the organs, organizations and citizens shall be sent to the working offices of the Standing Committee.

Article 36 With regard to a legislative bill placed on the agenda of a meeting of the Standing Committee, its working offices shall collect and sort out the deliberated opinions from group meetings and the opinions offered by the various quarters as well as other relevant information, and then send them to the Law Committee and the relevant special committee and, where necessary, print and distribute them at a meeting of the Standing Committee.

Article 37 With regard to a legislative bill placed on the agenda of a meeting of the Standing Committee, if the sponsor requests its withdrawal before it is put to vote, he shall state the reasons, and deliberation of the bill at the meeting shall terminate as soon as the Council of Chairmen has accepted the request and reported the matter to the Standing Committee.

Article 38 Where a legislative bill has been deliberated on by the Standing Committee at three meetings and there are still major questions calling for further study, the Council of Chairmen may propose, provided with the consent of a joint group meeting or a plenary meeting, not to put the bill to vote for the time being, and refer it to the Law Committee and the relevant committee for further deliberation.

Article 39 Where the deliberation of a legislative bill has been laid aside for two full years owing to significant disagreement among the various quarters on the necessity and feasibility of making the bill into a law, or where a legislative bill that has been proposed not to be put to vote for the time being has failed to be placed again on the agenda of a meeting of the Standing Committee for deliberation within two years, the Council of Chairmen shall report the matter to the Standing Committee and deliberation on the said bill shall terminate.

Article 40 After the revised draft of a law has been deliberated by the Standing Committee at its meeting, the Law Committee shall further revise it on the basis of the deliberated opinions of the members of the Standing Committee before preparing a draft for vote; then, the Council of Chairmen shall request the Standing Committee to put the draft to vote at a plenary meeting, and the draft shall be subject to adoption by a simple majority of the total membership of the Standing Committee.

Article 41 A law adopted by the Standing Committee shall be promulgated by Order of the President signed by the president of the People's Republic of China.

Section 4
Legal Interpretation

Article 42 The power of legal interpretation belongs to the Standing Committee of the National People's Congress.
A law shall be interpreted by the Standing Committee of the National People's Congress if:

1. the specific meaning of a provision needs to be further defined; or
2. after its enactment, new developments make it necessary to define the basis on which to apply the law.

Article 43 The State Council, the Central Military Commission, the Supreme People's Court, the Supreme People's Procuratorate, a special committee of the National People's Congress and the standing committee of the people's congress of a province, autonomous region or municipality directly under the Central Government may request the Standing Committee of the National People's Congress to give legal interpretation.

Article 44 The working offices of the Standing Committee shall study and work out a draft for the legal interpretation, which shall be put on the agenda of the Standing Committee meeting by decision of the Council of Chairmen.

Article 45 After a draft for legal interpretation has been deliberated by the Standing Committee at its meeting, the Law Committee shall, on the basis of the deliberated opinions of members of the Standing Committee, deliberate on the draft and revise it before working out a draft legal interpretation for vote.

Article 46 The draft legal interpretation for vote shall be subject to adoption by a simple majority of the total membership of the Standing Committee and be promulgated by the Standing Committee in an announcement.

Article 47 The legal interpretation adopted by the Standing Committee of the National People's Congress has the same effect as the laws enacted by it.

Section 5
Other Provisions

Article 48 When a legislative bill is submitted, a version of the draft law, its explanation and other necessary information shall be provided at the same time. The explanation on the draft law shall cover the necessity of its enactment and its main contents.

Article 49 With regard to a legislative bill submitted to the National People's Congress or its Standing Committee, the sponsor has the right to withdraw it before it is put on the agenda of a Congress session or Committee meeting.

Article 50 With regard to a legislative bill that has failed to pass the vote at a plenary meeting of the National People's Congress or its Standing Committee, if the sponsor still considers it necessary to enact the proposed law, he may submit the bill anew in accordance with the statutory procedures, and the Presidium or the Council of Chairmen shall decide whether to put it on the agenda of a session of the Congress or a meeting of the Standing Committee; for a bill that has failed to be adopted by the National People's Congress, the case shall be referred to the National People's Congress for deliberation and decision.

Article 51 In a law, the time for its entry into effect shall be clearly stipulated.

Article 52 In an Order of the President signed for promulgating a law, the organ that enacts the law, the date of its adoption and the time for its entry into effect shall be clearly stated.

Once a law is promulgated upon signing, it shall be published in the Bulletin of the Standing Committee of the National People's Congress and in the newspapers with a nationwide distribution.

The text of a law published in the Bulletin of the Standing Committee shall be the standard text.

Article 53 The procedures for revising or nullifying a law shall be governed by the relevant provisions in this Chapter.

Where only part of the articles of a law are revised or nullified, the new text of the law must be promulgated.

Article 54 According to the need of the contents, a law may consists of parts, chapters, sections, articles, paragraphs, subparagraphs and items.

The sequence of the different parts, chapters, sections and articles shall be marked in the order of Chinese numerals, the sequence of the paragraphs shall not be marked, that of subparagraphs shall be marked in the order of bracketed Chinese numerals and that of items marked with Arabic numerals.

In the note to the title of a law, the organ that enacts the law and the date of adoption shall be clearly stated.

Article 55 The working offices of the Standing Committee of the National People's Congress may reply, after study, to any legal inquiries regarding specific questions and shall report thereon to the Standing Committee for the record.

Chapter III Administrative Regulations

Article 56 The State Council shall, in accordance with the Constitution and laws, formulate administrative regulations.

The administrative regulations may be formulated to govern the following matters:

1. matters requiring the formulation of administrative regulations in order to implement the provisions of law; and
2. matters within the administrative functions and powers of the State Council as provided for in Article 89 of the Constitution.

When the administrative regulations governing an affair which has been formulated first by the State Council under authorization decided on by the National People's Congress or its Standing Committee, an affair on which the National People's Congress or its Standing Committee is responsible to make a law, have been tested in practice and when the conditions are ripe for making a law on the affair, the State Council shall, in a timely manner, request the National People's Congress or its Standing Committee to make the law.

Article 57 The drafting of administrative regulations shall be arranged by the State Council. Where a relevant department under the State Council considers it necessary to formulate administrative regulations to govern a matter, it shall apply to the State Council for including the matter in its legislation list.

Article 58 In drafting administrative regulations, opinions from relevant organs, organizations and citizens shall be widely listened to, and forums, seminars, hearings, etc. may be held for the purpose.

Article 59 When the drafting of the administrative regulations is completed, the drafting unit shall submit the draft, its explanation, differing opinions from the various quarters on major questions in the draft and other relevant information to the legislative affairs department under the State Council for examination.

The legislative affairs department under the State Council shall submit an examination report and a revised draft to the State Council and in its examination report explain the major questions in the draft.

Article 60 The decision-making procedures for administrative regulations shall comply with the relevant provisions in the Organic Law of the State Council of the People's Republic of China.

Article 61 Administrative regulations shall be promulgated by Order of the State Council signed by the Premier of the State Council.

Article 62 After promulgation upon signing, the administrative regulations shall immediately be published in the Bulletin of the State Council and in newspapers with a nationwide distribution. The text of the administrative regulations published in the Bulletin of the State Council shall be the standard text.

Chapter IV
Local Regulations, Autonomous Regulations, Separate Regulations, and Rules

Section 1
Local Regulations, Autonomous Regulations, and Separate Regulations

Article 63 The people's congresses or their standing committees of the provinces, autonomous regions and municipalities directly under the Central Government may, in light of the specific conditions and actual needs of their respective administrative areas, formulate local regulations, provided that such regulations do not contradict the Constitution, the laws and the administrative regulations.

The people's congresses or their standing committees of the comparatively larger cities may, in light of the specific local conditions and actual needs, formulate local regulations, provided that they do not contradict the Constitution, the laws, the administrative regulations and the local regulations of their respective provinces or autonomous regions, and they shall submit the regulations to the standing committees of the people's congresses of the provinces or autonomous regions for approval before implementation. The standing committees of the people's congresses of the provinces or autonomous regions shall examine the legality of such local regulations which are submitted for approval, and shall approve them within four months if they do not contradict the Constitution, the laws, the administrative regulations, and the local regulations of their respective provinces or autonomous regions.

When the standing committee of the people's congress of a province or autonomous region examines the local regulations of a comparatively larger city submitted for approval, it shall make a decision to deal with the matter if it finds that the said regulations contradict the rules of the people's government of the province or autonomous region.

A "comparatively larger city" used in this Law refers to a city where a provincial or autonomous regional people's government is located or where a special economic zone is located, or a city approved as such by the State Council.

Article 64 Local regulations may be formulated to govern the following matters:

1. matters requiring the formulation of specific provisions in light of the actual conditions of an respective administrative area for implementing the provisions of laws or administrative regulations; and
2. matters of local character that require the formulation of local regulations.

Except for the affairs provided for in Article 8 of this Law, the provinces, autonomous regions, municipalities directly under the Central Government and the comparatively larger cities may, in light of the specific local conditions and actual needs, first formulate local regulations on all other affairs for which the State has not yet formulated any laws or administrative regulations.

Once the laws or administrative regulations formulated on such matters by the State come into effect, the provisions in local regulations which contradict the said laws or administrative regulations shall be null or void, and the organs that have formulated such regulations shall promptly amend or annul the provisions.

Article 65 The people's congresses or their standing committees of the provinces and cities where special economic zones are located may, upon authorization by decision of the National People's Congress, formulate regulations and enforce them within the limits of the special economic zones.

Article 66 The people's congresses of the national autonomous areas have the power to formulate autonomous regulations and separate regulations on the basis of the political, economic and cultural characteristics of the local nationality (nationalities). The autonomous regulations and separate regulations of the autonomous regions shall be submitted to the Standing Committee of the National People's Congress for approval and shall go into effect upon approval. The autonomous regulations and separate regulations of the autonomous prefectures or counties shall be submitted to the standing committees of the people's congresses of the relevant provinces, autonomous regions or municipalities directly under the Central Government for approval and shall go into effect upon approval.

Where certain provisions of the laws and administrative regulations are concerned, adaptation on the basis of the characteristics of the local nationality (nationalities) may be made in autonomous regulations and separate regulations, but such adaptation may not contradict the basis principles of the laws and administrative regulations; where the provisions of the Constitution and the Law on Regional National Autonomy as well as the provisions in other laws and administrative regulations specially formulated to govern the national autonomous areas are concerned, no adaptation may be made.

Article 67 Local regulations governing especially important matters of an administrative area shall be subject to adoption by the people's congress of the area.

Article 68 The procedures for submission, deliberation and voting of bills of local regulations, autonomous regulations or separate regulations shall be enacted by the people's congress at the corresponding level in accordance with the Organic Law of the People's Republic of China on Local People's Congresses and Local People's Governments and with reference to the provisions of Sections 2, 3 and 5 in Chapter II of this Law.

The organ in charge of unified deliberation of a draft of local regulations shall prepare a report on the result of its a conclusive report on deliberation and a revised draft of the regulations.

Article 69 Local regulations formulated by the people's congress of a province, autonomous region or municipality directly under the Central Government shall be promulgated by the presidium of the congress in an announcement.

Local regulations formulated by the standing committee of the people's congress of a province, autonomous region or municipality directly under the Central Government shall be promulgated by the standing committee in an announcement.

Local regulations formulated by the people's congress or its standing committee of a comparatively larger city shall, upon approval, be promulgated by the standing committee of the people's congress of the city in an announcement.

Autonomous regulations or separate regulations shall, upon approval, be promulgated by the standing committee of the local people's congress of the autonomous region, autonomous prefecture or autonomous county in an announcement.

Article 70 Once local regulations or autonomous regulations and separate regulations of an autonomous region are promulgated, they shall be published in the gazette of the standing committee of the local people's congress and in newspapers distributed within the administrative area.

The text of local regulations or autonomous regulations and separate regulations published in the gazette of the standing committee of the people's congress shall be the standard text.

Section 2
Rules

Article 71 The ministries and commissions of the State Council, the People's Bank of China, the State Audit Administration as well as the other organs endowed with administrative functions directly under the State Council may, in accordance with the laws as well as the administrative regulations, decisions and orders of the State Council and within the limits of their power, formulate rules.

Matters governed by the rules of departments shall be those for the enforcement of the laws or the administrative regulations, decisions and orders of the State Council.

Article 72 With regard to a matter that falls within the limits of power of two or more departments under the State Council, the State Council shall be requested to formulate administrative regulations, or the departments concerned under the State Council shall jointly formulate rules.

Article 73 The people's governments of the provinces, autonomous regions, municipalities directly under the Central Government and the comparatively larger cities may, in accordance with laws and administrative regulations and the local regulations of their respective province, autonomous regions or municipalities, formulate rules.

Local governments may formulate rules to govern the following matters:

1. matters requiring the formulation of rules to implement the provisions of laws, administrative regulations and local regulations; and
2. specific administrative matters pertaining to their respective administrative areas.

Article 74 The procedures for formulating the rules of departments under the State Council and rules of local governments shall be enacted by the State Council with reference to the provisions in Chapter III of this Law.

Article 75 The Rules of departments shall be subject to decision by the executive meetings of ministries or meetings of commissions.

The rules of local governments shall be subject to decision by the executive meetings or plenary meetings of the respective governments.

Article 76 The rules of departments shall be promulgated by orders signed by the heads of the departments.

The rules of local governments shall be promulgated by orders signed by governors of provinces, chairmen of autonomous regions or mayors.

Article 77 the rules of departments shall, upon promulgation by signed orders, be promptly published in the gazette of the State Council or gazettes of the departments and in newspapers with a nationwide distribution.

The rules of local governments shall, upon promulgation by signed orders, be promptly published in gazettes of the local people's governments and in newspapers distributed within their respective administrative areas.

The text of rules published in the gazette of the State Council or of the department and in the gazette of the local people's government shall be the standard text.

Chapter V
Application and Record

Article 78 The legal effect of the Constitution is the highest, and no laws, administrative regulations, local regulations, autonomous regulations, separate regulations or rules whatever may contradict it.

Article 79 The effect of laws is higher than that of administrative regulations, local regulations, and rules.

The effect of administrative regulations is higher than that of local regulations, and rules.

Article 80 The effect of local regulations is higher than that of the rules of the local governments at or below the corresponding level.

The effect of the rules formulated by the people's governments of the provinces or autonomous regions is higher than that of the rules formulated by the people's governments of the comparatively larger cities within the administrative areas of the provinces and autonomous regions.

Article 81 Where in autonomous regulations or separate regulations provisions are formulated in accordance with law to make adaptations with regard to certain provisions of laws, administrative regulations or local regulations, the provisions of the autonomous regulations or separate regulations shall apply in the autonomous areas concerned.

Where regulations of special economic zones are formulated, upon authorization, to make adaptations with regard to certain provisions of laws, administrative regulations or local regulations, the provisions of the regulations of special economic zones shall apply in the special economic zones concerned.

Article 82 The effect of the rules of different departments is equal between the departments, and the effect of the rules of departments and of the rules of local governments is equal between the departments and local governments; their application shall be confined to their respective limits of authority.

Article 83 With regard to laws, administrative regulations, local regulations, autonomous regulations, separate regulations or rules, if they are formulated by one and same organ and if there is inconsistency between special provisions and general provisions, the special provisions shall prevail; if there is inconsistency between the new provisions and the old provisions, the new provisions shall prevail.

Article 84 Laws, administrative regulations, local regulations, autonomous regulations, separate regulations and rules shall not be retroactive, but the regulations formulated specially for the purpose of better protecting the rights and interests of citizens, legal persons and other organizations are excepted.

Article 85 Where there is inconsistency between the new general provisions and the old special provisions in different laws governing one and the same matter and it is hard to decide which provisions shall prevail, a ruling shall be made by the Standing Committee of the National People's Congress.

Where there is inconsistency between the new general provisions and the old special provisions in different administrative regulations governing one and the same matter and it is hard to decide which provisions shall prevail, a ruling shall be made by the State Council.

Article 86 Where there is inconsistency between local regulations and rules, a ruling shall be made by the organ concerned according to the limits of power prescribed below:

1. Where there is inconsistency between the new general provisions and the old special provisions formulated by one and the same organ, the said organ shall make a ruling;
2. Where there is inconsistency between the provisions of local regulations and those of the rules of departments governing one and same matter and it is hard to decide which provisions shall prevail, the State Council shall make a decision; if it considers that the provisions of the local regulations should be applied, it shall decide that the provisions of the local regulations be applied in the locality concerned; if it considers that the rules of departments should be applied, the case

shall be submitted to the Standing Committee of the National People's Congress for a ruling; and

3. Where there is inconsistency between the provisions in the rules of different departments or between the provisions in the rules of the departments and those in the rules of local governments governing one and the same matter, the State Council shall make a ruling.

Where there is inconsistency between the regulations formulated upon authorization and the provisions of laws and it is hard to decide which shall prevail, the Standing Committee of the National People's Congress shall make a ruling.

Article 87 Under any of the following circumstances, laws, administrative regulations, local regulations, autonomous regulations, separate regulations or rules shall be altered or annulled by the organ concerned in accordance with the limits of power prescribed in Article 88 of this Law:

1. where the limits of power are transcended;
2. where provisions of the legislation of lower levels contravene those of the legislation of upper levels;
3. where, because of inconsistency between the provisions of different rules governing one and the same matter, it is ruled that the provisions of one side be altered or annulled;
4. where the provisions of rules are considered inappropriate and should therefore be altered or annulled; or
5. where legal procedures are violated.

Article 88 The limits of power for altering or annulling laws, administrative regulations, local regulations, autonomous regulations, separate regulations or rules are as follows:

1. The National People's Congress has the power to alter or annul any inappropriate laws enacted by its Standing Committee, and to annul any autonomous regulations or separate regulations which have been approved by its Standing Committee but which contravene the Constitution or the provision of the second paragraph in Article 66 of this Law;
2. The Standing Committee of the National People's Congress has the power to annul any administrative regulations which contradict the Constitution and laws, to annul any local regulations which contradict the Constitution, laws or administrative regulations, and to annul any autonomous regulations or separate regulations which have been approved by the standing committees of the people's congresses of the

relevant provinces, autonomous regions or municipalities directly under the Central Government but which contravene the Constitution or the provision of the second paragraph in Article 66 of this Law;

3. The State Council has the power to alter or annul any inappropriate rules of the departments and of local governments;
4. The people's congress of a province, autonomous region or municipality directly under the Central Government has the power to alter or annul any inappropriate local regulations formulated or approved by its standing committee;
5. The standing committee of a local people's congress has the power to annul any inappropriate rules formulated by the people's government at the same level;
6. The people's government of a province or autonomous region has the power to alter or annul any inappropriate rules formulated by people's governments at the next lower level; and
7. The authorizing organ has the power to annul any of the regulations formulated by an authorized organ that transcends the authorized limits of power or contravenes the authorized purpose, and when necessary, may revoke the authorization.

Article 89 Administrative regulations, local regulations, autonomous regulations, separate regulations and rules shall, within 30 days from the date of promulgation, be reported to the organ concerned for the record in accordance with the following provisions:

1. Administrative regulations shall be reported to the Standing Committee of the National People's Congress for the record;
2. Local regulations formulated by the people's congresses or their standing committees of provinces, autonomous regions and municipalities directly under the Central Government shall be reported to the Standing Committee of the National People's Congress and the State Council for the record; local regulations formulated by the people's congresses or their standing committees of the comparatively larger cities shall be reported by the standing committees of the people's congresses of the relevant provinces and autonomous regions to the Standing Committee of the National People's Congress and the State Council for the record;
3. Autonomous regulations and separate regulations formulated by autonomous prefectures and autonomous counties shall be reported by

the standing committees of people's congresses of the relevant provinces, autonomous regions or municipalities directly under the Central Government to the Standing Committee of the National People's Congress and the State Council for the record;

4. The rules of a department and of a local government shall be reported to the State Council for the record. The rules of a local government shall at the same time be reported to the standing committee of the people's congress at the same level for the record. The rules formulated by the people's government of a comparatively larger city shall simultaneously be reported to the standing committee of the people's congress and to the people's government of the relevant province or autonomous region for the record; and
5. Regulations formulated upon authorization shall be reported to the organ specified by the authorization decision for the record.

Article 90 When the State Council, the Central Military Commission, the Supreme People's Court, the Supreme People's Procuratorate and the standing committees of the people's congresses of the provinces, autonomous regions and municipalities directly under the Central Government consider that administrative regulations, local regulations, autonomous regulations or separate regulations contradict the Constitution or laws, they may submit to the Standing Committee of the National People's Congress written requests for examination, and the working offices of the Standing Committee shall refer the requests to the relevant special committees for examination and suggestions.

When State organs other than the ones mentioned in the preceding paragraph, public organizations, enterprises and institutions or citizens consider that administrative regulations, local regulations, autonomous regulations or separate regulations contradict the Constitution or laws, they may submit to the Standing Committee of the National People's Congress written suggestions for examination, and the working offices of the Standing Committee shall study the suggestions and shall, when necessary, refer them to the relevant special committees for examination and suggestions.

Article 91 When after examination a special committee of the National People's Congress considers that administrative regulations, local regulations, autonomous regulations or separate regulations contradict the Constitution or laws, it may submit written suggestions to the organ that has formulated the regulations for examination; or the Law Committee and other relevant special committees may convene a joint meeting for examination and request the said organ to attend the meeting and give an explanation, and then after examination submit to the organ written comments and suggestions. The organ that has formulated the regulations shall, within two months, study and put forth suggestions as to whether to revise the regulations, and shall give feedback to the Law Committee and other relevant special committees of the National People's Congress.

If, after examination, the Law Committee and other relevant special committees of the National People's Congress consider that administrative regulations, local regulations, autonomous regulations or separate regulations contradict the Constitution or laws and the organ that has formulated such regulations refuses to make revision, they may submit written comments and suggestions on the basis of their examination and propose a motion for annulment of the regulations to the Council of Chairmen, which shall decide whether to submit the motion to a meeting of the Standing Committee for examination and decision.

Article 92 Other authorities which receive local regulations, autonomous regulations, separate regulations or rules submitted to them for the record shall, in accordance with the principle of safeguarding legal uniformity, formulate procedures for the examination of such regulations or rules.

Chapter VI
Supplementary Provisions

Article 93 The Central Military Commission shall, in accordance with the Constitution and laws, formulated military regulations.

The General Departments, the various services and arms and the military commands of the Central Military Commission may, in accordance with laws and the military regulations, decisions and orders of the Commission, formulate military rules within the limits of their power.

Military regulations and military rules shall be implemented within the armed forces. Measures for formulating, revising and nullifying military regulations and military rules shall be formulated by the Central Military Commission in accordance with the principles laid down in this Law.

Article 94 This Law shall go into effect as of July 1, 2000.

Appendix III

General Principles of the Civil Law of the People's Republic of China

Adopted at the Fourth Session of the Sixth National People's Congress on April 12, 1986 and promulgated by Order No. 37 of the President of the People's Republic of China on April 12, 1986.

Contents

Chapter I
Basic Principles

Article 1 This Law is formulated in accordance with the Constitution and the actual situation in our country, drawing upon our practical experience in civil activities, for the purpose of protecting the lawful civil rights and interests of citizens and legal persons and correctly adjusting civil relations, so as to meet the needs of the developing socialist modernization.

Article 2 The Civil Law of the People's Republic of China shall adjust property relationships and personal relationships between civil subjects with equal status, that is, between citizens, between legal persons and between citizens and legal persons.

Article 3 Parties to a civil activity shall have equal status.

Article 4 In civil activities, the principles of voluntariness, fairness, making compensation for equal value, honesty and credibility shall be observed.

Article 5 The lawful civil rights and interests of citizens and legal persons shall be protected by law; no organization or individual may infringe upon them.

Article 6 Civil activities must be in compliance with the law; where there are no relevant provisions in the law, they shall be in compliance with State policies.

Article 7 Civil activities shall have respect for social ethics and shall not harm the public interest, undermine State economic plans or disrupt social economic order.

Article 8 The law of the People's Republic of China shall apply to civil activities within the People's Republic of China, except as otherwise stipulated by law.

The stipulations of this Law as regards citizens shall apply to foreigners and stateless persons within the People's Republic of China, except as otherwise stipulated by law.

Chapter II
Citizen (Natural Person)

Section 1
Capacity for Civil Rights and Capacity for Civil Conduct

Article 9 A citizen shall have the capacity for civil rights from birth to death and shall enjoy civil rights and assume civil obligations in accordance with the law.

Article 10 All citizens are equal as regards their capacity for civil rights.

Article 11 A citizen aged 18 or over shall be an adult. He shall have full capacity for civil conduct, may independently engage in civil activities and shall be called a person with full capacity for civil conduct.

A citizen who has reached the age of 16 but not the age 18 and whose main source of income is his own labour shall be regarded as a person with full capacity for civil conduct.

Article 12 A minor aged 10 or over shall be a person with limited capacity for civil conduct and may engage in civil activities appropriate to his age and intellect; in other civil activities, he shall be represented by his agent ad litem or participate with the consent of his agent ad litem.

A minor under the age of 10 shall be a person having no capacity for civil conduct and shall be represented in civil activities by his agent ad litem.

Article 13 A mentally ill person who is unable to account for his own conduct shall be a person having no capacity for civil conduct and shall be represented in civil activities by his agent ad litem.

A mentally ill person who is unable to fully account for his own conduct shall be a person with limited capacity for civil conduct and may engage in civil activities appropriate to his mental health; in other civil activities, he shall be represented by his agent ad litem or participate with the consent of his agent ad litem.

Article 14 The guardian of a person without or with limited capacity for civil conduct shall be his agent ad litem.

Article 15 The domicile of a citizen shall be the place where his residence is registered; if his habitual residence is not the same as his domicile, his habitual residence shall be regarded as his domicile.

Section 2
Guardianship

Article 16 The parents of a minor shall be his guardians.
If the parents of a minor are dead or lack the competence to be his guardian, a person from the following categories who has the competence to be a guardian shall act as his guardian:

1. paternal or maternal grandparent;
2. elder brother or sister; or
3. any other closely connected relative or friend willing to bear the responsibility of guardianship and having approval from the units of the minor's parents or from the neighbourhood or village committee in the place of the minor's residence.

In case of a dispute over guardianship, the units of the minor's parents or the neighbourhood or village committee in the place of his residence shall appoint a guardian from among the minor's near relatives. If disagreement over the appointment leads to a lawsuit, the people's court shall make a ruling.
If none of the persons listed in the first two paragraphs of this Article is available to be the guardian, the units of the minor's parents, the neighbourhood

or village committee in the place of the minor's residence or the civil affairs department shall act as his guardian.

Article 17 A person from the following categories shall act as guardian for a mentally ill person without or with limited capacity for civil conduct:

1. spouse;
2. parent;
3. adult child;
4. any other near relative;
5. any other closely connected relative or friend willing to bear the responsibility of guardianship and having approval from the unit to which the mentally ill person belongs or from the neighbourhood or village committee in the place of his residence.

In case of a dispute over guardianship, the unit to which the mentally ill person belongs or the neighbourhood or village committee in the place of his residence shall appoint a guardian from among his near relatives. If disagreement over the appointment leads to a lawsuit, the people's court shall make a ruling.

If none of the persons listed in the first paragraph of this article is available to be the guardian, the unit to which the mentally ill person belongs, the neighbourhood or village committee in the place of his residence or the civil affairs department shall act as his guardian.

Article 18 A guardian shall fulfil his duty of guardianship and protect the person, property and other lawful rights and interests of his wards. A guardian shall not handle the property of his ward unless it is in the ward's interests.

A guardian's rights to fulfil his guardianship in accordance with the law shall be protected by law.

If a guardian does not fulfil his duties as guardian or infringes upon the lawful rights and interests of his ward, he shall be held responsible; if a guardian causes any property loss for his ward, he shall compensate for such loss. The people's court may disqualify a guardian based on the application of a concerned party or unit.

Article 19 A person who shares interests with mental patient may apply to a people's court for a declaration that the mental patient is a person without or with limited capacity for civil conduct.

With the recovery of the health of a person who has been declared by a people's court to be without or with limited capacity for civil conduct, and upon his own application or that of an interested person, the people's court may declare him to be a person with limited or full capacity for civil conduct.

Section 3 Declarations of Missing Persons and Death

Article 20 If a citizen's whereabouts have been unknown for two years, an interested person may apply to a people's court for a declaration of the citizen as missing.

If a person's whereabouts become unknown during a war, the calculation of the time period in which his whereabouts are unknown shall begin on the final day of the war.

Article 21 A missing person's property shall be placed in the custody of his spouse, parents, adult children or other closely connected relatives or friends. In case of a dispute over custody, if the persons stipulated above are unavailable or are incapable of taking such custody, the property shall be placed in the custody of a person appointed by the people's court.

Any taxes, debts and other unpaid expenses owed by a missing person shall defrayed by the custodian out of the missing person's property.

Article 22 In the event that a person who has been declared missing reappears or his whereabouts is ascertained, the people's court shall, upon his own application or that of an interested person, revoke the declaration of his missing-person status.

Article 23 Under either of the following circumstances, an interested person may apply to the people's court for a declaration of a citizen's death:

1. if the citizen's whereabouts have been unknown for four years; or
2. if the citizen's whereabouts have been unknown for two years after the date of an accident in which he was involved.

If a person's whereabouts become unknown during a war, the calculation of the time period in which his whereabouts are unknown shall begin on the final day of the war.

Article 24 In the event that a person who has been declared dead reappears or it is ascertained that he is alive, the people's court shall, upon his own application or that of an interested person, revoke the declaration of his death.

Any civil juristic acts performed by a person with capacity for civil conduct during the period in which he has been declared dead shall be valid.

Article 25 A person shall have the right to request the return of his property, if the declaration of his death has been revoked. Any citizen or organization that has obtained such property in accordance with the Inheritance Law shall return the original items or make appropriate compensation if the original items are no longer existent.

Section 4
Individual Businesses and Leaseholding Farm Households

Article 26 "Individual businesses" refers to businesses run by individual citizens who have been lawfully registered and approved to engage in industrial or commercial operation within the sphere permitted by law. An individual business may adopt a shop name.

Article 27 "Leaseholding farm households" refers to members of a rural collective economic organization who engage in commodity production under a contract and within the spheres permitted by law.

Article 28 The legitimate rights and interests of individual businesses and leaseholding farm households shall be protected by law.

Article 29 The debts of an individual business or a leaseholding farm household shall be secured with the individual's property if the business is operated by an individual and with the family's property if the business is operated by a family.

Section 5
Individual Partnership

Article 30 "Individual partnership" refers to two or more citizens associated in a business and working together, with each providing funds, material objects, techniques and so on according to an agreement.

Article 31 Partners shall make a written agreement covering the funds each is to provide, the distribution of profits, the responsibility for debts, the entering into and withdrawal from partnership, the ending of partnership and other such matters.

Article 32 The property provided by the partners shall be under their unified management and use.

The property accumulated in a partnership operation shall belong to all the partners.

Article 33 An individual partnership may adopt a shop name; it shall be approved and registered in accordance with law and conduct business operations within the range as approved and registered.

Article 34 The operational activities of an individual partnership shall be decided jointly by the partners, who each shall have the right to carry out and supervise those activities.

The partners may elect a responsible person. All partners shall bear civil liability for the operational activities of the responsible person and other personnel.

Article 35 A partnership's debts shall be secured with the partners' property in proportion to their respective contributions to the investment or according to the agreement made.

Partners shall undertake joint liability for their partnership's debts, except as otherwise stipulated by law. Any partner who overpays his share of the partnership's debts shall have the right to claim compensation from the other partners.

Chapter III
Legal Persons

Section 1
General Stipulations

Article 36 A legal person shall be an organization that has capacity for civil rights and capacity for civil conduct and independently enjoys civil rights and assumes civil obligations in accordance with the law.

A legal person's capacity for civil rights and capacity for civil conduct shall begin when the legal person is established and shall end when the legal person terminates.

Article 37 A legal person shall have the following qualifications:

1. establishment in accordance with the law;
2. possession of the necessary property or funds;
3. possession of its own name, organization and premises; and
4. ability to independently bear civil liability.

Article 38 In accordance with the law or the articles of association of the legal person, the responsible person who acts on behalf of the legal person in exercising its functions and powers shall be its legal representative.

Article 39 A legal person's domicile shall be the place where its main administrative office is located.

Article 40 When a legal person terminates, it shall go into liquidation in accordance with the law and discontinue all other activities.

Section 2
Enterprise as Legal Person

Article 41 An enterprise owned by the whole people or under collective ownership shall be qualified as a legal person when it has sufficient funds as stipulated by the State; has articles of association, an organization and premises; has the ability to independently bear civil liability; and has been approved and registered by the competent authority.

A Chinese-foreign equity joint venture, Chinese-foreign contractual joint venture or foreign-capital enterprise established within the People's Republic of China shall be qualified as a legal person in China, if it has the qualifications of a legal person and has been approved and registered by the administrative agency for industry and commerce in accordance with the law.

Article 42 An enterprise as legal person shall conduct operations within the range approved and registered.

Article 43 An enterprise as legal person shall bear civil liability for the operational activities of its legal representatives and other personnel.

Article 44 If an enterprise as legal person is divided or merged or undergoes any other important change, it shall register the change with the registration authority and publicly announce it.

When an enterprise as legal person is divided or merged, its rights and obligations shall be enjoyed and assumed by the new legal person that results from the change.

Article 45 An enterprise as legal person shall terminate for any of the following reasons:

1. if it is dissolved by law;
2. if it is disbanded;
3. if it is declared bankrupt in accordance with the law; or
4. for other reasons.

Article 46 When an enterprise as legal person terminates, it shall cancel its registration with the registration authority and publicly announce the termination.

Article 47 When an enterprise as legal person is disbanded, it shall establish a liquidation organization and go into liquidation. When an enterprise as legal person is dissolved or is declared bankrupt, the competent authority or a people's court shall organize the organs and personnel concerned to establish a liquidation organization to liquidate the enterprise.

Article 48 An enterprise owned by the whole people, as legal person, shall bear civil liability with the property that the State authorizes it to manage. An enterprise under collective ownership, as legal person, shall bear civil liability with the property it owns. A Chinese-foreign equity joint venture, Chinese-foreign contractual joint venture or foreign-capital enterprise as legal person shall bear civil liability with the property it owns, except as stipulated otherwise by law.

Article 49 Under any of the following circumstances, an enterprise as legal person shall bear liability, its legal representative may additionally be given administrative sanctions and fined and, if the offence constitutes a crime, criminal responsibility shall be investigated in accordance with the law:

1. conducting illegal operations beyond the range approved and registered by the registration authority;
2. concealing facts from the registration and tax authorities and practising fraud;
3. secretly withdrawing funds or hiding property to evade repayment of debts;

4. disposing of property without authorization after the enterprise is dissolved, disbanded or declared bankrupt;
5. failing to apply for registration and make a public announcement promptly when the enterprise undergoes a change or terminates, thus causing interested persons to suffer heavy losses;
6. engaging in other activities prohibited by law, damaging the interests of the State or the public interest.

Section 3
Official Organ, Institution and Social Organization as Legal Person

Article 50 An independently funded official organ shall be qualified as a legal person on the day it is established.

If according to law an institution or social organization having the qualifications of a legal person needs not go through the procedures for registering as a legal person, it shall be qualified as a legal person on the day it is established; if according to law it does need to go through the registration procedures, it shall be qualified as a legal person after being approved and registered.

Section 4
Economic Association

Article 51 If a new economic entity is formed by the enterprises or an enterprise and an institution that engage in economic association and it independently bears civil liability and has the qualifications of a legal person, the new entity shall be qualified as a legal person after being approved and registered by the competent authority.

Article 52 If the enterprises or an enterprise and an institution that engage in economic association conduct joint operation but do not have the qualifications of a legal person, each party to the association shall, in proportion to its respective contribution to the investment or according to the agreement made, bear civil liability with the property each party owns or manages. If joint liability is specified by law or by agreement, the parties shall assume joint liability.

Article 53 If the contract for economic association of enterprises or of an enterprise and an institution specifies that each party shall conduct operations independently, it shall stipulate the rights and obligations of each party, and each party shall bear civil liability separately.

Chapter IV
Civil Juristic Acts and Agency

Section 1
Civil Juristic Acts

Article 54 A civil juristic act shall be the lawful act of a citizen or legal person to establish, change or terminate civil rights and obligations.

Article 55 A civil juristic act shall meet the following requirements:

1. the actor has relevant capacity for civil conduct;
2. the intention expressed is genuine; and
3. the act does not violate the law or the public interest.

Article 56 A civil juristic act may be in written, oral or other form. If the law stipulates that a particular form be adopted, such stipulation shall be observed.

Article 57 A civil juristic act shall be legally binding once it is instituted. The actor shall not alter or rescind his act except in accordance with the law or with the other party's consent.

Article 58 Civil acts in the following categories shall be null and void:

1. those performed by a person without capacity for civil conduct;
2. those that according to law may not be independently performed by a person with limited capacity for civil conduct;
3. those performed by a person against his true intentions as a result of cheating, coercion or exploitation of his unfavourable position by the other party;
4. those that performed through malicious collusion are detrimental to the interest of the State, a collective or a third party;
5. those that violate the law or the public interest;
6. economic contracts that violate the State's mandatory plans; and
7. those that performed under the guise of legitimate acts conceal illegitimate purposes.

Civil acts that are null and void shall not be legally binding from the very beginning.

Article 59 A party shall have the right to request a people's court or an arbitration agency to alter or rescind the following civil acts:

1. those performed by an actor who seriously misunderstood the contents of the acts; and
2. those that are obviously unfair.

Rescinded civil acts shall be null and void from the very beginning.

Article 60 If a part of a civil act is null and void, it shall not affect the validity of other part.

Article 61 After a civil act has been determined to be null and void or has been rescinded, the party who acquired property as a result of the act shall return it to the party who suffered a loss.

The erring party shall compensate the other party for the losses it suffered as a result of the act; if both sides are in error, they shall each bear their proper share of the responsibility.

If the two sides have conspired maliciously and performed a civil act that is detrimental to the interests of the State, a collective or a third party, the property that they thus obtained shall be recovered and turned over to the State or the collective, or returned to the third party.

Article 62 A civil juristic act may have conditions attached to it. Conditional civil juristic acts shall take effect when the relevant conditions are met.

Section 2
Agency

Article 63 Citizens and legal persons may perform civil juristic acts through agents. An agent shall perform civil juristic acts in the principal's name within the scope of the power of agency. The principal shall bear civil liability for the agent's acts of agency.

Civil juristic acts that should be performed by the principal himself, pursuant to legal provisions or the agreement between the two parties, shall not be entrusted to an agent.

Article 64 Agency shall include entrusted agency, statutory agency and appointed agency.

An entrusted agent shall exercise the power of agency as entrusted by the principal; a statutory agent shall exercise the power of agency as prescribed by law;

and an appointed agent shall exercise the power of agency as designated by a people's court or the appointing unit.

Article 65 A civil juristic act may be entrusted to an agent in writing or orally. If legal provisions require the entrustment to be written, it shall be effected in writing.

Where the entrustment of agency is in writing, the power of attorney shall clearly state the agent's name, the entrusted tasks and the scope and duration of the power of agency, and it shall be signed or sealed by the principal.

If the power of attorney is not clear as to the authority conferred, the principal shall bear civil liability towards the third party, and the agent shall be held jointly liable.

Article 66 The principal shall bear civil liability for an act performed by an actor with no power of agency, beyond the scope of his power of agency or after his power of agency has expired, only if he recognizes the act retroactively. If the act is not so recognized, the performer shall bear civil liability for it. If a principal is aware that a civil act is being executed in his name but fails to repudiate it, his consent shall be deemed to have been given.

An agent shall bear civil liability if he fails to perform his duties and thus causes damage to the principal.

If an agent and a third party in collusion harm the principal's interests, the agent and the third party shall be held jointly liable.

If a third party is aware that an actor has no power of agency, is overstepping his power of agency, or his power of agency has expired and yet joins him in a civil act and thus brings damage to other people, the third party and the actor shall be held jointly liable.

Article 67 If an agent is aware that the matters entrusted are illegal but still carries them out, or if a principal is aware that his agent's acts are illegal but fails to object to them, the principal and the agent shall be held jointly liable.

Article 68 If in the principal's interests an entrusted agent needs to transfer the agency to another person, he shall first obtain the principal's consent. If the principal's consent is not obtained in advance, the matter shall be reported to him promptly after the transfer, and if the principal objects, the agent shall bear civil liability for the acts of the transferee; however, an entrusted agency transferred in emergency circumstances in order to safeguard the principal's interests shall be excepted.

Article 69 An entrusted agency shall end under any of the following circumstances:

1. when the period of agency expires or when the tasks entrusted are completed;
2. when the principal rescinds the entrustment or the agent declines the entrustment;
3. when the agent dies;
4. when the principal loses his capacity for civil conduct; or
5. when the principal or the agent ceases to be a legal person.

Article 70 A statutory or appointed agency shall end under any of the following circumstances:

1. when the principal gains or recovers capacity for civil conduct;
2. when the principal or the agent dies;
3. when the agent loses capacity for civil conduct;
4. when the people's court or the unit that appointed the agent rescinds the appointment; or
5. when the guardian relationship between the principal and the agent ends for other reasons.

Chapter V
Civil Rights

Section 1
Property Ownership and Related Property Rights

Article 71 "Property ownership" means the owner's rights to lawfully possess, utilize, profit from and dispose of his property.

Article 72 Property ownership shall not be obtained in violation of the law.

Unless the law stipulates otherwise or the parties concerned have agreed on other arrangements, the ownership of property obtained by contract or by other lawful means shall be transferred simultaneously with the property itself.

Article 73 State property shall be owned by the whole people.

State property is sacred and inviolable, and no organization or individual shall be allowed to seize, encroach upon, privately divide, retain or destroy it.

Article 74 Property of collective organizations of the working masses shall be owned collectively by the working masses. This shall include:

1. land, forests, mountains, grasslands, unreclaimed land, beaches and other areas that are stipulated by law to be under collective ownership;
2. property of collective economic organizations;
3. collectively owned buildings, reservoirs, farm irrigation facilities and educational, scientific, cultural, health, sports and other facilities; and
4. other property that is collectively owned.

Collectively owned land shall be owned collectively by the village peasants in accordance with the law and shall be worked and managed by village agricultural production cooperatives, other collective agricultural economic organizations or villagers' committees. Land already under the ownership of the township (town) peasants' collective economic organizations may be collectively owned by the peasants' of the township (town).

Collectively owned property shall be protected by law, and no organization or individual may seize, encroach upon, privately divide, destroy or illegally seal up, distrain, freeze or confiscate it.

Article 75 A citizen's personal property shall include his lawfully earned income, housing, savings, articles for daily use, objects of cultural relics, books, reference materials, trees, livestock, as well as means of production the law permits a citizen to possess and other lawful property.

A citizen's lawful property shall be protected by law, and no organization or individual may appropriate, encroach upon, destroy or illegally seal up, distrain, freeze or confiscate it.

Article 76 Citizens shall have the right of inheritance under the law.

Article 77 The lawful property of social organizations, including religious organizations, shall be protected by law.

Article 78 Property may be owned jointly by two or more citizens or legal persons.

There shall be two kinds of joint ownership, namely co-ownership by shares and common ownership. Each of the co-owners by shares shall enjoy the rights and assume the obligations respecting the joint property in proportion to his share. Each of the common owners shall enjoy the rights and assume the obligations respecting the joint property.

Each co-owner by shares shall have the right to withdraw his own share of the joint property or transfer its ownership. However, when he offers to sell his share, the other co-owners shall have a right of pre-option if all other conditions are equal.

Article 79 If the owner of a buried or concealed object is unknown, the object shall belong to the State. The unit that receives the object shall commend or give a material reward to the unit or individual that turns in the object.

Lost-and-found objects, flotsam and stray animals shall be returned to their rightful owners, and any costs thus incurred shall be reimbursed by the owners.

Article 80 State-owned land may be used according to law by units under ownership by the whole people; it may also be lawfully assigned for use by units under collective ownership. The State shall protect the usufruct of the land, and the usufructuary shall be obligated to manage, protect and properly use the land.

The right of citizens and collectives to contract for management of land under collective ownership or of State-owned land under collective use shall be protected by law. The rights and obligations of the two contracting parties shall be stipulated in the contract signed in accordance with the law.

Land may not be sold, leased, mortgaged or illegally transferred by any other means.

Article 81 State-owned forests, mountains, grasslands, unreclaimed land, beaches, water surfaces and other natural resources may be used according to law by units under ownership by the whole people; or they may also be lawfully assigned for use by units under collective ownership. The State shall protect the usufruct of those resources, and the usufructuary shall be obliged to manage, protect and properly use them.

State-owned mineral resources may be mined according to law by units under ownership by the whole people and units under collective ownership; citizens may also lawfully mine such resources. The State shall protect lawful mining rights.

The right of citizens and collectives to lawfully contract for the management of forests, mountains, grasslands, unreclaimed land, beaches and water surfaces that are owned by the collectives or owned by the State but used by collectives shall be protected by law. The rights and obligations of the two contracting parties shall be stipulated in the contract in accordance with the law.

State-owned mineral resources and waters as well as forest land, mountains, grasslands, unreclaimed land and beaches owned by the State and those that are lawfully owned by collectives may not be sold, leased, mortgaged or illegally transferred by any other means.

Article 82 Enterprises under ownership by the whole people shall lawfully enjoy the rights of management over property that the State has authorized them to manage and operate, and the rights shall be protected by law.

Article 83 In the spirit of helping production, making things convenient for people's life, enhancing unity and mutual assistance, and being fair and reasonable, neighbouring users of real estate shall maintain proper neighbourly relations over such matters as water supply, drainage, passageway, ventilation and lighting. Anyone who causes obstruction or damages to his neighbour, shall stop the infringement, eliminate the obstruction and compensate for the damages.

Section 2
Creditors' Rights

Article 84 A debt represents a special relationship of rights and obligations established between the parties concerned, either according to the agreed terms of a contract or legal provisions. The party entitled to the rights shall be the creditor, and the party assuming the obligations shall be the debtor.

The creditor shall have the right to demand that the debtor fulfil his obligations as specified by the contract or according to legal provisions.

Article 85 A contract shall be an agreement whereby the parties establish, change or terminate their civil relationship. Lawfully established contracts shall be protected by law.

Article 86 When there are two or more creditors to a deal, each creditor shall be entitled to rights in proportion to his proper share of the credit. When there are two or more debtors to a deal, each debtor shall assume obligations in proportion to his proper share of the debt.

Article 87 When there are two or more creditors or debtors to a deal, each of the joint creditors shall be entitled to demand that the debtor fulfil his obligations, in accordance with legal provisions or the agreement between the parties; each of the joint debtors shall be obliged to perform the entire debt, and the debtor who performs the entire debt shall be entitled to ask the other joint debtors to reimburse him for their shares of the debt.

Article 88 The parties to a contract shall fully fulfil their obligations pursuant to the terms of the contract.

If a contract contains ambiguous terms regarding quality, time limit for performance, place of performance, or price, and the intended meaning cannot be determined from the context of relevant terms in the contract, and if the parties cannot reach an agreement through consultation, the provisions below shall apply:

1. If quality requirements are unclear, State quality standards shall apply; if there are no State quality standards, generally held standards shall apply.
2. If the time limit for performance is unclear, the debtor may at his convenience fulfil his obligations towards the creditor; the creditor may also demand at any time that the debtor perform his obligations, but sufficient notice shall be given to the debtor.
3. If the place of performance is unclear, and the payment is money, the performance shall be effected at the seat or place of residence of the party receiving the payment; if the payment is other than money, the performance shall be effected at the seat or place of residence of the party fulfilling the obligations.
4. If the price agreed by the parties is unclear, the State-fixed price shall apply. If there is no State-fixed price, the price shall be based on market price or the price of a similar article or remuneration for a similar service.

If the contract does not contain an agreed term regarding rights to patent application, any party who has completed an invention-creation shall have the right to apply for a patent.

If the contract does not contain an agreed term regarding rights to the use of scientific and technological research achievements, the parties shall all have the right to use such achievements.

Article 89 In accordance with legal provisions the agreement between the parties on the performance of a debt may be guaranteed using the methods below:

1. A guarantor may guarantee to the creditor that the debtor shall perform his debt. If the debtor defaults, the guarantor shall perform the debt or bear joint liability according to agreement. After performing

the debt, the guarantor shall have the right to claim repayment from the debtor.

2. The debtor or a third party may offer a specific property as a pledge. If the debtor defaults, the creditor shall be entitled to keep the pledge to offset against the debt or have priority in satisfying his claim out of the proceeds from the sale of the pledge pursuant to relevant legal provisions.
3. Within the limits of relevant legal provisions, a party may leave a deposit with the other party. After the debtor has discharged his debt, the deposit shall either be retained as partial payment of the debt or returned. If the party who leaves the deposit defaults, he shall not be entitled to demand the return of the deposit; if the party who accepts the deposit defaults, he shall repay the deposit in double.
4. If a party has possession of the other party's property according to contract, and the other party violates the contract by failing to pay a required sum of money within the specified time limit, the possessor shall have a lien on the property and may keep the retained property to offset the debt or have priority in satisfying his claim out of the proceeds from the sale of the property pursuant to relevant legal provisions.

Article 90 Legitimate loan relationships shall be protected by law.

Article 91 If a party to a contract transfers all or part of his contractual rights or obligations to a third party, he shall obtain the other party's consent and may not seek profits therefrom.

Contracts which according to legal provisions are subject to State approval, such as transfers, must be approved by the authority that originally approved the contract, unless the law or the original contract stipulates otherwise.

Article 92 If profits are acquired improperly and without a lawful basis, resulting in another person's loss, the illegal profits shall be returned to the person who suffered the loss.

Article 93 If a person acts as manager or provides services in order to protect another person's interests when he is not legally or contractually obligated to do so, he shall be entitled to claim from the beneficiary the expenses necessary for such assistance.

Section 3
Intellectual Property Rights

Article 94 Citizens and legal persons shall enjoy rights of authorship (copyrights) and shall be entitled to sign their names as authors, issue and publish their works and obtain remuneration in accordance with the law.

Article 95 The patent rights lawfully obtained by citizens and legal persons shall be protected by law.

Article 96 The rights to exclusive use of trademarks obtained by legal persons, individual businesses and individual partnerships shall be protected by law.

Article 97 Citizens who make discoveries shall be entitled to the rights of discovery. A discoverer shall have the right to apply for and receive certificates of discovery, bonuses or other awards.

Citizens who make inventions or other achievements in scientific and technological researches shall have the right to apply for and receive certificates of honour, bonuses or other awards.

Section 4
Personal Rights

Article 98 Citizens shall enjoy the rights of life and health.

Article 99 Citizens shall enjoy the right of personal name and shall be entitled to determine, use or change their personal names in accordance with relevant provisions. Interference with, usurpation of and false representation of personal names shall be prohibited.

Legal persons, individual businesses and individual partnerships shall enjoy the right of name.

Enterprises as legal persons, individual businesses and individual partnerships shall have the right to use and lawfully assign their own names.

Article 100 Citizens shall enjoy the right of portrait.

The use of a citizen's portrait for profits without his consent shall be prohibited.

Article 101 Citizens and legal persons shall enjoy the right of reputation. The personality of citizens shall be protected by law, and the use of insults, libel or other means to damage the reputation of citizens or legal persons shall be prohibited.

Article 102 Citizens and legal persons shall enjoy the right of honour. It shall prohibited to unlawfully divest citizens and legal persons of their honorary titles.

Article 103 Citizens shall enjoy the right of marriage by choice. Mercenary marriages, marriages upon arbitrary decision by any third party and any other acts of interference in the freedom of marriage shall be prohibited.

Article 104 Marriage, the family, old people, mothers and children shall be protected by law.

The lawful rights and interests of the handicapped shall be protected by law.

Article 105 Women shall enjoy equal civil rights with men.

Chapter VI
Civil Liability

Section 1
General Stipulations

Article 106 Citizens and legal persons who breach a contract or fail to fulfil other obligations shall bear civil liability.

Citizens and legal persons who through their fault encroach upon State or collective property or the property or person of other people shall bear civil liability.

Civil liability shall still be borne even in the absence of fault, if the law so stipulates.

Article 107 Civil liability shall not be borne for failure to perform a contract or damage to a third party if it is caused by force majeure, except as otherwise provided by law.

Article 108 Debts shall be cleared. If a debtor is unable to repay his debt immediately, he may repay by installments with the consent of the creditor or a ruling by a people's court. If a debtor is capable of repaying his debt but refuses to do so, repayment shall be compelled by the decision of a people's court.

Article 109 If a person suffers damages from preventing or stopping encroachment on State or collective property, or the property or person of a third party, the infringer shall bear responsibility for compensation, and the beneficiary may also give appropriate compensation.

Article 110 Citizens or legal persons who bear civil liability shall also be held for administrative responsibility if necessary. If the acts committed by citizens and legal persons constitute crimes, criminal responsibility of their legal representatives shall be investigated in accordance with the law.

Section 2
Civil Liability for Breach of Contract

Article 111 If a party fails to fulfil its contractual obligations or violates the terms of a contract while fulfilling the obligations, the other party shall have the right to demand fulfilment or the taking of remedial measures and claim compensation for its losses.

Article 112 The party that breaches a contract shall be liable for compensation equal to the losses consequently suffered by the other party.

The parties may specify in a contract that if one party breaches the contract it shall pay the other party a certain amount of breach of contract damages; they may also specify in the contract the method of assessing the compensation for any losses resulting from a breach of contract.

Article 113 If both parties breach the contract, each party shall bear its respective civil liability.

Article 114 If one party is suffering losses owing to the other party's breach of contract, it shall take prompt measures to prevent the losses from increasing; if it does not promptly do so, it shall not have the right to claim compensation for the additional losses.

Article 115 A party's right to claim compensation for losses shall not be affected by the alteration or termination of a contract.

Article 116 If a party fails to fulfil its contractual obligations on account of a higher authority, it shall first compensate for the losses of the other party or take other remedial measures as contractually agreed and then the higher authority shall be responsible for settling the losses it sustained.

Section 3
Civil Liability for Infringement of Rights

Article 117 Anyone who encroaches on the property of the State, a collective or another person shall return the property; failing that, he shall reimburse its estimated price.

Anyone who damages the property of the State, a collective or another person shall restore the property to its original condition or reimburse its estimated price. If the victim suffers other great losses therefrom, the infringer shall compensate for those losses as well

Article 118 If the rights of authorship (copyrights), patent rights, rights to exclusive use of trademarks, rights of discovery, rights of invention or rights for scientific and technological research achievements of citizens or legal persons are infringed upon by such means as plagiarism, alteration or imitation, they shall have the right to demand that the infringement be stopped, its ill effects be eliminated and the damages be compensated for.

Article 119 Anyone who infringes upon a citizen's person and causes him physical injury shall pay his medical expenses and his loss in income due to missed working time and shall pay him living subsidies if he is disabled; if the victim dies, the infringer shall also pay the funeral expenses, the necessary living expenses of the deceased's dependents and other such expenses.

Article 120 If a citizen's right of personal name, portrait, reputation or honour is infringed upon, he shall have the right to demand that the infringement be stopped, his reputation be rehabilitated, the ill effects be eliminated and an apology be made; he may also demand compensation for losses.

The above paragraph shall also apply to infringements upon a legal person's right of name, reputation or honour.

Article 121 If a State organ or its personnel, while executing its duties, encroaches upon the lawful rights and interests of a citizen or legal person and causes damages, it shall bear civil liability.

Article 122 If a substandard product causes property damage or physical injury to others, the manufacturer or seller shall bear civil liability according to law. If the transporter or storekeeper is responsible for the matter, the manufacturer or seller shall have the right to demand compensation for its losses.

Article 123 If any person causes damage to other people by engaging in operations that are greatly hazardous to the surroundings, such as operations conducted high aboveground, or those involving high pressure, high voltage, combustibles, explosives, highly toxic or radioactive substances or high-speed means of transport, he shall bear civil liability; however, if it can be proven that the damage was deliberately caused by the victim, he shall not bear civil liability.

Article 124 Any person who pollutes the environment and causes damages to others in violation of State provisions for environmental protection and the prevention of pollution shall bear civil liability in accordance with the law.

Article 125 Any constructor who engages in excavation, repairs or installation of underground facilities in a public place, on a roadside or in a passageway without setting up clear signs and adopting safety measures and thereby causes damages to others shall bear civil liability.

Article 126 If a building or any other installation or an object placed or hung on a structure collapses, detaches or drops down and causes damages to others, its owner or manager shall bear civil liability, unless he can prove himself not a fault.

Article 127 If a domesticated animal causes harm to any person, its keeper or manager shall bear civil liability. If the harm occurs through the fault of the victim, the keeper or manager shall not bear civil liability; if the harm occurs through the fault of a third party, the third party shall bear civil liability.

Article 128 A person who causes harm in exercising justifiable defence shall not bear civil liability. If justifiable defence exceeds the limits of necessity and undue harm is caused, an appropriate amount of civil liability shall be borne.

Article 129 If harm occurs through emergency actions taken to avoid danger, the person who gave rise to the danger shall bear civil liability. If the danger arose from natural causes, the person who took the emergency actions may either be exempt from civil liability or bear civil liability to an appropriate extent. If the emergency measures taken are improper or exceed the limits of necessity and undue harm is caused, the person who took the emergency action shall bear civil liability to an appropriate extent.

Article 130 If two or more persons jointly infringe upon another person's rights and cause him damages, they shall bear joint liability.

Article 131 If a victim is also at fault for causing the damage, the civil liability of the infringer may be reduced.

Article 132 If none of the parties are at fault in causing damage, they may share civil liability according to the actual circumstances.

Article 133 If a person without or with limited capacity for civil conduct causes damages to others, his guardian shall bear civil liability. If the guardian has done his duty of guardianship, his civil liability may be appropriately reduced.

If a person who has property but is without or with limited capacity for civil conduct causes damages to others, the expenses of compensation shall be paid from his property. Shortfalls in such expenses shall be appropriately compensated for by the guardian unless the guardian is a unit.

Section 4
Methods of Bearing Civil Liability

Article 134 The main methods of bearing civil liability shall be:

1. cessation of infringements;
2. removal of obstacles;
3. elimination of dangers;
4. return of property;
5. restoration of original condition;
6. repair, reworking or replacement;
7. compensation for losses;
8. payment of breach of contract damages;
9. elimination of ill effects and rehabilitation of reputation; and
10. extension of apology.

The above methods of bearing civil liability may be applied exclusively or concurrently.

When hearing civil cases, a people's court, in addition to applying the above stipulations, may serve admonitions, order the offender to sign a pledge of repentance, and confiscate the property used in carrying out illegal activities and the illegal income obtained therefrom. It may also impose fines or detentions as stipulated by law.

Chapter VII
Limitation of Action

Article 135 Except as otherwise stipulated by law, the limitation of action regarding applications to a people's court for protection of civil rights shall be two years.

Article 136 The limitation of action shall be one year on cases concerning the following:

1. claims for compensation for bodily injuries;
2. sales of substandard goods without proper notice to that effect;
3. delays in paying rent or refusal to pay rent; or
4. loss of or damage to property left in the care of another person.

Article 137 A limitation of action shall begin when the entitled person knows or should know that his rights have been infringed upon. However, the people's court shall not protect his rights if 20 years have passed since the infringement. Under special circumstances, the people's court may extend the limitation of action.

Article 138 If a party chooses to fulfil obligations voluntarily after the limitation of action has expired, he shall not be subject to the limitation.

Article 139 A limitation of action shall be suspended during the last six months of the limitation if the plaintiff cannot exercise his right of claim because of force majeure or other obstacles. The limitation shall resume on the day when the grounds for the suspension are eliminated.

Article 140 A limitation of action shall be discontinued if suit is brought or if one party makes a claim for or agrees to fulfilment of obligations. A new limitation shall be counted from the time of the discontinuance.

Article 141 If the law has other stipulations concerning limitation of action, those stipulations shall apply.

Chapter VIII Application of Law in Civil Relations with Foreigners

Article 142 The application of law in civil relations with foreigners shall be determined by the provisions in this chapter.

If any international treaty concluded or acceded to by the People's Republic of China contains provisions differing from those in the civil laws of the People's Republic of China, the provisions of the international treaty shall apply, unless the provisions are ones on which the People's Republic of China has announced reservations.

International practice may be applied on matters for which neither the law of the People's Republic of China nor any international treaty concluded or acceded to by the People's Republic of China has any provisions.

Article 143 If a citizen of the People's Republic of China settles in a foreign country, the law of that country may be applicable as regards his capacity for civil conduct.

Article 144 The ownership of immovable property shall be bound by the law of the place where it is situated.

Article 145 The parties to a contract involving foreign interests may choose the law applicable to settlement of their contractual disputes, except as otherwise stipulated by law.

If the parties to a contract involving foreign interests have not made a choice, the law of the country to which the contract is most closely connected shall be applied.

Article 146 The law of the place where an infringing act is committed shall apply in handling compensation claims for any damage caused by the act. If both parties are citizens of the same country or have established domicile in another country, the law of their own country or the country of domicile may be applied.

An act committed outside the People's Republic of China shall not be treated as an infringing act if under the law of the People's Republic of China it is not considered an infringing act.

Article 147 The marriage of a citizen of the People's Republic of China to a foreigner shall be bound by the law of the place where they get married, while a divorce shall be bound by the law of the place where a court accepts the case.

Article 148 Maintenance of a spouse after divorce shall be bound by the law of the country to which the spouse is most closely connected.

Article 149 In the statutory succession of an estate, movable property shall be bound by the law of the decedent's last place of residence, and immovable property shall be bound by the law of the place where the property is situated.

Article 150 The application of foreign laws or international practice in accordance with the provisions of this chapter shall not violate the public interest of the People's Republic of China.

Chapter IX
Supplementary Provisions

Article 151 The people's congresses of the national autonomous areas may formulate separate adaptive or supplementary regulations or provisions in accordance with the principles of this Law and in the light of the characteristics of the local nationalities. Those formulated by the people's congresses of autonomous regions shall be submitted in accordance with the law to the Standing Committee of the National People's Congress for approval or for the record. Those formulated by the people's congresses of autonomous prefectures or autonomous counties shall be submitted to the standing committee of the people's congress in the relevant province or autonomous region for approval.

Article 152 If an enterprise owned by the whole people has been established with the approval of the competent authority of a province, autonomous region or centrally administered municipality or at a higher level and it has already been registered with the administrative agency for industry and commerce, before this Law comes into force, it shall automatically qualify as a legal person without having to re-register as such.

Article 153 For the purpose of this Law, "force majeure" means unforeseeable, unavoidable and insurmountable objective conditions.

Article 154 Time periods referred to in the Civil Law shall be calculated by the Gregorian calendar in years, months, days and hours.

When a time period is prescribed in hours, calculation of the period shall begin on the prescribed hour. When a time period is prescribed in days, months and years, the day on which the period begins shall not be counted as within the period; calculation shall begin on the next day.

If the last day of a time period falls on a Sunday or an official holiday, the day after the holiday shall be taken as the last day.

The last day shall end at 24:00 hours. If business hours are applicable, the last day shall end at closing time.

Article 155 In this Law, the terms "not less than," "not more than" "within" and "expires" shall include the given figure; the terms "under" and "beyond" shall not include the given figure.

Article 156 This Law shall come into force on January 1, 1987.

Appendix IV

Selected Multinational Treaties

Instrument	Date of Entry into Force
Agreement for the Establishment of the International Development Law	Apr. 28, 1989
Basel Convention on the Control of Transboundary Movements of Hazardous Wastes and their Disposal	May 5, 1992
Cash-on Delivery Agreement	Jan. 1, 1996
Convention against Torture and Other Cruel, Inhuman, or Degrading Treatment or Punishment	June 26, 1987
Convention against Illicit Traffic in Narcotic Drugs and Psychotropic Substances	Nov. 11, 1990
Convention for the Suppression of Unlawful Acts against the Safety of Maritime Navigation	Mar. 1, 1992
Convention on Biological Diversity	Dec. 29, 1993
Convention on the Elimination of All Forms of Discrimination against Women	Sept. 3, 1981
Convention on the Prevention of Crimes against Internationally Protected Persons	Feb. 20, 1977
Convention on the Rights of the Child	Sept. 2, 1990
Convention on the Taking of Evidence Abroad in Civil or Commercial Matters	Oct. 7, 1972
Convention Relating to the Status of Refugees	Dec. 23, 1982
International Convention against the Taking of Hostages	June 3, 1983
International Convention for the Suppression of Terrorist Bombings	May 23, 2001

International Convention on Civil Liability for Oil Pollution Damage	June 19, 1975
Money Orders Agreement	Jan. 1, 1996
Montreal Protocol on Substances that Deplete the Ozone Layer (Adjusted and Amended)	Aug. 20, 1992
Protocol on Environment Protection to the Antarctic Treaty	Jan. 14, 1998
Protocol Relating to the Madrid Agreement Concerning International Registration of Marks	Dec. 1, 1995
Vienna Convention for the Protection of the Ozone Layer	Sept. 22, 1998
United Nations Convention Against Corruption	Not in Effect
United Nations Convention against Transnational Organized Crimes	Not in Effect
United Nations Convention on Contracts for the International Sale of Goods	Jan. 1, 1988
United Nations Convention on the Law of the Sea	Nov. 16, 1994
United Nations Framework Convention on Climate Change	Mar. 21, 1994

Appendix V

Selected U.S.-China Bilateral Treaties

Cultural Property

Memorandum of understanding concerning the imposition of import restrictions on categories of archaeological material from the Paleolithic period through the Tang Dynasty and monumental sculpture and wall art at least 250 years old.

Signed at Washington January 14, 2009. Entered into force January 14, 2009.

T.I.A.S. 09-114.

Customs

Agreement regarding mutual assistance in customs matters.

Signed at Washington April 9, 1999. Entered into force April 10, 2003.

T.I.A.S. 13029.

Education

Agreement for cooperation in educational exchanges.

Signed at Beijing May 25, 2010. Entered into force May 25, 2010.

T.I.A.S. 10-525.

Environmental Cooperation

Agreement for cooperation in the Global Learning and Observations to Benefit the Environment (GLOBE) Program, with appendices.

Signed at Beijing October 18, 1995. Entered into force October 18, 1995.

Finance

Agreement relating to investment guaranties, with related notes and statement.
Exchange of notes at Beijing October 30, 1980. Entered into force October 30, 1980.
32 U.S.T. 4010; T.I.A.S. 9924; 1267 U.N.T.S. 315.

Health

Agreement on the safety of drugs and medical devices.
Signed at Beijing December 11, 2007. Entered into force December 11, 2007.
T.I.A.S. 07-1211.

Agreement of the safety of food and feed, with annex.
Signed at Beijing December 11, 2007. Entered into force December 11, 2007.
T.I.A.S. 07-1211.1.

Intellectual Property

Memorandum of understanding on the protection of intellectual property.
Signed at Washington January 17, 1992. Entered into force January 17, 1992.
T.I.A.S. 12036.

Judicial Assistance

Agreement on mutual legal assistance in criminal matters.
Signed at Beijing June 19, 2000. Entered into force March 8, 2001.
T.I.A.S. 13102.

Labor

Memorandum of understanding on prohibiting import and export trade in prison labor products.
Signed at Washington August 7, 1992. Entered into force August 7, 1992.
T.I.A.S.

Statement of cooperation on the implementation of the memorandum of understanding on prohibiting import and export trade in prison labor products.

Signed at Beijing March 14, 1994. Entered into force March 14, 1994.

T.I.A.S.

Maritime Matters

Agreement concerning maritime search and rescue cooperation.

Signed at Washington January 20, 1987. Entered into force January 20, 1987.

T.I.A.S. 12013; 2244 U.N.T.S. 71.

Agreement on establishing a consultation mechanism to strengthen military maritime safety.

Signed at Beijing January 19, 1998. Entered into force January 19, 1998.

T.I.A.S. 12924.

Privileges and Immunities

Agreement relating to privileges and immunities for the respective diplomatic missions in Washington and Beijing.

Exchange of notes at Washington January 5, 1981. Entered into force January 5, 1981.

T.I.A.S. 13079.

Satellites

Memorandum of agreement on liability for satellite launches.

Signed at Washington December 17, 1988. Entered into force March 16, 1989.

T.I.A.S.

Taxation

Agreement with respect to mutual exemption from taxation of transportation income of shipping and air transport enterprises.

Signed at Beijing March 5, 1982. Entered into force September 23, 1983;

effective January 1, 1981.

35 U.S.T. 3819; T.I.A.S. 10884; 2014 U.N.T.S. 419.

Agreement for the avoidance of double taxation and the prevention of tax evasion with respect to taxes on income, with protocol and exchange of notes.

Signed at Beijing April 30, 1984. Entered into force November 21, 1986.

T.I.A.S. 12065.

Protocol concerning the interpretation of paragraph 7 of the protocol to the agreement for the avoidance of double taxation and the prevention of tax evasion with respect to taxes on income of April 30, 1984.

Signed at Beijing May 10, 1986. Entered into force November 21, 1986.

T.I.A.S. 12065.

Technology Transfer

Memorandum of agreement on satellite technology safeguards.

Signed at Beijing February 11, 1993. Entered into force February 11, 1993.

T.I.A.S. 12486.

Telecommunication

Agreement on the establishment of the direct secure telephone link.

Signed at Beijing April 29, 1998. Entered into force April 29, 1998.

T.I.A.S.

Agreement on the establishment of a secure defense telephone link.

Signed at Shanghai February 29, 2008. Entered into force February 29, 2008.

T.I.A.S. 08-229.

Tourism

Memorandum of understanding on the development of bilateral tourism relations.

Signed at Beijing August 5, 1982. Entered into force August 5, 1982.

34 U.S.T. 1418; T.I.A.S. 10426.

Trade and Commerce

Agreement on trade exhibitions.

Signed at Beijing May 10, 1979. Entered into force May 10, 1979.

30 U.S.T. 4472; T.I.A.S. 9470; 1171 U.N.T.S. 305.

Agreement on trade relations.

Signed at Beijing July 7, 1979. Entered into force February 1, 1980.

31 U.S.T. 4651; T.I.A.S. 9630; 1202 U.N.T.S. 179.

Accord on industrial and technological cooperation.

Signed at Washington January 12, 1984. Entered into force January 12, 1984.

35 U.S.T. 4233; T.I.A.S. 10920.

Arrangement relating to a visa system for exports to the United States of cotton, wool and manmade fiber textiles and textile products.

Exchange of letters at Washington February 16, 1984. Entered into force February 16, 1984.

35 U.S.T. 4417; T.I.A.S. 10945.

Agreement concerning trade in textiles and textile products, with annexes.

Exchange of letters at Beijing February 2, 1988. Entered into force February 2, 1988; effective January 1, 1988.

NP

Amendments and extension

May 24 and 26, 1988 (NP).

August 26 and September 7, 1988 (NP).

November 7, 1988, and January 24, 1989 (NP).

April 28 and May 1, 1989 (NP).

April 23 and 24, 1991 (NP).

Memorandum of agreement regarding international trade in commercial launch services, with annexes.

Signed at Beijing March 13, 1995. Entered into force March 13, 1995; effective January 1, 1995.

T.I.A.S.

Amendment

October 27, 1997.

Agreement concerning trade in textile and apparel products, with annexes.

Exchange of notes at Beijing March 29 and June 8, 1995.

Entered into force June 8, 1995; effective January 1, 1994.

NP

Agreement on agricultural cooperation, with addenda.

Signed at Washington April 10, 1999. Entered into force April 10, 1999.

T.I.A.S.

Memorandum of understanding concerning trade in textile and apparel products, with annexes.

Signed at London November 8, 2005. Entered into force January 1, 2006.

T.I.A.S.

Memorandum of understanding regarding certain measures granting refunds, reductions or exemptions from taxes or other payments.

Signed at Geneva November 29, 2007. Entered into force November 29, 2007.

T.I.A.S. 07-1129.

Agreement on intellectual property rights cooperation.

Signed at Chengdu November 21, 2011. Entered into force November 21, 2011.

T.I.A.S.

Memorandum of understanding regarding films for theatrical release, with annex.

Signed at Geneva April 25, 2012. Entered into force April 25, 2012.

T.I.A.S. 12-425.

Appendix VI

Useful Contact Information

U.S. Embassies and Consulates in China

Mainland China

U.S. Embassy to China
No. 55 An Jia Lou Rd.
Beijing, China

Tel: 86 10 8531 3000
English Homepage: http://beijing.usembassy-china.org.cn/

U.S. Consulate General in Chengdu
No. 4 Lingshiguan Rd.
Chengdu, Sichuan Province, China

Tel: 86 28 8558 3992
English Homepage: http://chengdu.usembassy-china.org.cn/

U.S. Consulate General in Guangzhou
No. 43 Hua Jiu Rd., Zhujiang New Town
Guangzhou, Guangdong Province, China

Tel: 86 20 3814 5000
English Homepage: http://guangzhou.usembassy-china.org.cn/

U.S. Consulate General in Shanghai
No. 1469 Huai Hai Zhong Rd.
Shanghai, China

Tel: 86 21 6279 7662
English Homepage: http://shanghai.usembassy-china.org.cn/

U.S. Consulate General in Shenyang
No. 52, 14th Wei Rd.
Shenyang, Liaoning Province, China

Tel: 86 24 2322 1198
English Homepage: http://shenyang.usembassy-china.org.cn/

U.S. Consulate General in Wuhan
New World International Trade Tower I, No. 568 Jianshe Av.
Wuhan, Hubei Province, China

Tel: 86 27 8555 7791
English Homepage: http://wuhan.usembassy-china.org.cn/

Hong Kong & Macao

U.S. Consulate General Hong Kong & Macau
No. 26 Garden Road
Hong Kong, China

Tel: 854 2845 1598
English Homepage: http://hongkong.usconsulate.gov/

Taiwan

American Institute in Taiwan
No. 7 Hsin Yi Rd.
Taipei, Taiwan, China

Tel: 886 02 2162 2000
English Homepage: http://www.ait.org.tw/en/home.html

*The American Institute in Taiwan is a private nonprofit corporation that performs citizen and consular services.

U.S. Chambers of Commerce (AmCham)

Mainland China

Beijing Chapter (Main Office)
The Office Park, Tower AB, 6th Floor

No. 10 Jintongxi Rd.
Beijing, China

Tel: 86 10 8519 0800
English Homepage: http://www.amchamchina.org/

Dalian Chapter
Intercontinental Dalian, 5th Floor
No. 6 Youhao St.
Dalian, Liaoning Province, China

Tel: 86 411 8236 6666 Ext. 6490
English Homepage: http://www.amchamchina.org/dalianchapter

Shanghai Chapter
Shanghai Centre, Suite 568
No. 1376 W. Nanjing Rd.
Shanghai, China

Tel: 86 21 6279 7119
English Homepage: http://www.amcham-shanghai.org/AmchamPortal/

Tianjin Chapter
Exchange Tower 2, Room 2918
No. 189 Nanjing Rd.
Tianjin, China

Tel: 86 22 2318 5071
English Homepage: http://www.amchamchina.org/tianjinchapter

Wuhan Chapter
Wuhan Jin Jiang International Hotel, Suite 720
No. 707 Jianshe Avenue
Wuhan, Hubei Province, China

Tel: 86 27 8578 6888 Ext. 720
English Homepage: http://www.amchamchina.org/central_china_chapter

Hong Kong & Macao

AmCham Hong Kong
Bank of America Tower, Rm. 1904

No. 12 Harcourt Rd.
Hong Kong, China

Tel: 852 2530 6900
English Homepage: http://www.amcham.org.hk/

Taiwan

AmCham Taipei
No. 129 MinSheng E. Rd., Suite 706
Taipei, Taiwan, China

Tel: 886 02 2718 8226 / 886 02 2718 8182
English Homepage: http://www.amcham.com.tw/

Chinese Embassies and Consulates in the U.S.

Chinese Embassy to the U.S.
3505 International Place
N.W. Washington, D.C. 20008

Tel: 001-202-495-2266
English Homepage: http://www.china-embassy.org/eng/

Chinese Consulate General in Chicago
100 W. Erie St.
Chicago, IL 60610

Tel: 001-312-803-0095
English Homepage: http://www.chinaconsulatechicago.org/eng/

*The Education Section for the Chinese Consulate General in Chicago is located at 3322 W. Peterson Av. and can be reached at 001-773-463-9187.

Chinese Consulate General in Houston
3417 Montrose Blvd.
Houston, Texas 77006

Tel: 001-713-520-1462
English Homepage: http://houston.china-consulate.org/eng

Chinese Consulate General in Los Angeles
500 Shatto Place, 3rd Floor
Los Angeles, CA 90020

Tel: 001-213-807-8088
English Homepage: losangeles.china-consulate.org/eng

Chinese Consulate General in New York
520 12th Av.
New York, NY 10036

Tel: 001-212-244-9392 / 001-212-244-9456
English Homepage: http://www.nyconsulate.prchina.org/eng/

Chinese Consulate General in San Francisco
1450 Laguna St.
San Francisco, CA 94115

Tel: 001-415-852-5900
English Homepage: http://www.chinaconsulatesf.org/eng/

Selected Chinese Government Agencies

Administration for Industry and Commerce
No. 8 Sanlihe E. Rd.
Beijing, China

Tel: 86 10 6801 0463 / 86 10 6801 3447
English Homepage: http://www.saic.gov.cn/english../

Administration of Taxation
No. 5 W. Yangfangdian Rd.
Beijing, China

Tel: 86 10 6341 7114
English Homepage: http://www.chinatax.gov.cn/n6669073/

Food and Drug Administration
No. 26 W. Xuanwumen St.
Beijing, China

Tel: 86 10 6831 3344
English Homepage: http://eng.sfda.gov.cn/WS03/CL0755/

General Administration of Customs
No. 6 Jianguomennei Av.
Beijing, China

Tel: 86 10 6519 4114
English Homepage: http://english.customs.gov.cn/publish/portal191/

Intellectual Property Office
No. 6 Xitucheng Rd.
Beijing, China

Tel: 86 10 6208 3114
English Homepage: http://english.sipo.gov.cn/

Ministry of Commerce
No.2 E. Changan Av.
Beijing, China

Tel: 86 10 5165 1200 Ext. 612 / 613 / 623
English Homepage: http://english.mofcom.gov.cn/

Ministry of Public Security
No. 14 Dongchangan St.
Beijing, China

Tel: 86 10 6626 2550
English Homepage: http://www.mps.gov.cn/English/

Trademark Office
No. 1 S. Chama St.
Beijing, China

Tel: 86 10 6321 9000
English Homepage: http://www.saic.gov.cn/sbjEnglish/

Chinese Bar Associations

All China Lawyers Association
No. 24 Dongsishitiao St., 5th Floor
Beijing, China

Tel: 86 10 8266 8266 Ext. 808 / 809
Mandarin Homepage: http://acla.org.cn/?

Beijing Lawyers Association
Huayuan Shoufu Towers, 5th Building
No. 18 Binhe Rd.
Beijing, China

Tel: 81 10 6451 5950
English Homepage: http://www.beijinglawyers.org.cn/qtmodel/english.htm

Guangzhou Lawyers Association
Yuexiu City Plaza, 9th Floor
No. 437 Middle Dongfeng Rd.
Guangzhou, Guangdong Province, China

Tel: 62 20 8355 0600
Mandarin Homepage: http://www.gzlawyer.org/

Hong Kong Bar Association
LG2, High Court
No. 38 Queensway
Hong Kong, China

Tel: 852 2869 0210
English Homepage: http://www.hkba.org/index.html

Macau Lawyers Association
World Trade Center, 11th Floor
No. 918 Avenida da Amizade
Macao, China

Tel: 853 28 728 121
English Homepage: http://www.aam.org.mo/en/enIndex.aspx

Shanghai Bar Association
Jun Yao International Plaza, 33rd-35th Floor
No. 789 Zhaojiabang Rd.
Shanghai, China

Tel: 86 21 6403 30000
English Homepage: http://2009.lawyers.org.cn/en/

Taipei Bar Association
No. 7 Luosifu Rd., 9th Floor
Taipei, Taiwan, China

Tel: 866 02 2351 5071
English Homepage: http://www.tba.org.tw/english/index.asp

Taiwan Bar Association
No. 130 S. Chongqing Rd., Building 5
Taipei, Taiwan, China

Tel: 886 02 2331 2865
Mandarin Homepage: http://www.twba.org.tw/

The Law Society of Hong Kong
Wing On House, 3rd Floor
71 Des Voeux Rd.
Hong Kong, China

Tel: 852 2845 0387
English Homepage: http://www.hklawsoc.org.hk/pub_e/default.asp?

Foreign Bar Associations

ABA Section of International Law: China Committee
321 N. Clark St.
Chicago, IL 60654

Tel: 001-800-285-2221
Homepage: http://apps.americanbar.org/dch/committee.cfm?com=IC860000

Inter-Pacific Bar Association
Roppongi Hills North Tower, 7th Floor
6-2-31 Roppongi, Minato-ku
Tokyo, Japan

Tel: 81 3 5786 6796
Homepage: http://ipba.org/

Table of Abbreviations

AIC: Administration for Industry and Commerce
CFDA: China Food and Drug Administration
CMC: Central Military Commission
CPC: Communist Party of China
CPPCC: Chinese People's Political Consultative Conference
CSRC: China Securities Regulatory Committee
CTMO: Chinese Trademark Office
GAC: General Administration of Customs
JV: Joint Venture
MOFCOM: Ministry of Commerce
MOFTEC: Ministry of Foreign Trade and Economic Cooperation
NPC: National People's Congress
PRC: People's Republic of China
ROC: Republic of China
RMB: Currency of the People's Republic of China ("Renminbi" or "Chinese Yuan")
SAIC: State Administration for Industry and Commerce
SAR: Special Administrative Region
SIPO: State Intellectual Property Office
SOE: State-Owned Enterprise
SPC: Supreme People's Court
SPP: Supreme People's Procuratorate
TRAB: Trademark Review and Adjudication Board
WFOE: Wholly Foreign Owned Enterprise

Glossary

Administration for Industry and Commerce: Local (provincial, municipal, district, and county) level authority in charge of market supervision and regulation. (abb. AIC) (CH: 工商行政管理局, CH abb.: 工商局)

AIC: See "Administration for Industry and Commerce"

Anti-Rightest Movement: A "purge" of Chinese citizens that were believed to be rightists. Most of the political persecution under this movement occurred from 1957 to 1960. (CH: 反右派运动)

Central Military Commission: As the comprehensive military commission in China, the CMC is composed of China's top military generals. It is responsible for appointing senior officers, designating military spending, and defending government compounds. Although the CMC is technically responsible for all military decisions, most of China's real military power is held by the Standing Committee. China also has a Ministry of National Defense which acts as China's foreign military affairs representative. (abb. CMC) (CH: 中央军事委员会)

Central People's Government: See "State Council"

Centrally-Administered Municipality: A Chinese city with both municipal and provincial powers. China has four centrally administered municipalities which are Beijing, Chongqing, Shanghai, and Tianjin. (CH: 直辖市)

Chinese People's Political Consultative Conference: Originally established under a different name by the Communist Party of China to negotiate with the Kuomintang, the CPPCC established the PRC in 1949. It still functions as an advisory conference to the NPC. (abb. CPPCC) (CH: 中国人民政治协商会议)

Chinese Trademark Office: National administration for trademark registration and other procedures. Part of the SAIC, and contains the TRAB. (abb. CTMO) (CH: 国家工商行政管理总局商标局, CH abb.: 商标局)

CMC: See "Central Military Commission"

Communist Party of China: (abb. CPC) (CH: **中国共产党**, CH abb.: **共产党**)

CPC: See "Communist Party of China"

CPPCC: See "Chinese People's Political Consultative Conference"

CTMO: See "Chinese Trademark Office"

Cultural Revolution: Also known as the "Great Proletarian Cultural Revolution," the Cultural Revolution was a movement to strengthen communism in China and lasted from 1966 to 1976. The mass rallies, political purges, and widespread media propaganda fractured China's government to the point that most daily government operations ceased. (CH: **无产阶级文化大革命**, CH abb.: **文化大革命**)

GAC: See "General Administration of Customs"

General Administration of Customs: National customs administrations with responsibilities including import/export and value-added tax collection. (abb. GAC) (CH: **海关总署**)

Guomindang: See "Kuomintang"

Kuomintang: Also referred to as the "Guomindang" or "Chinese Nationalist Party," the Kuomintang was the Republic of China's ruling political party. (CH: "**国民党**")

Madrid Agreement: See "Madrid System"

Madrid Protocol: See "Madrid System"

Madrid System: Widely utilized international trademark registration system. The Madrid System is composed of the Madrid Agreement, a substantive treaty that creates trademark system guidelines, and the Madrid Protocol which allows the creation of an international community trademark which may then be utilized to apply for individual trademark rights.

National People's Congress: Highest national body and singular legislative house. The NPC is the world's largest parliament. (abb. NPC) (CH: **全国人民代表大会**, CH abb. **人大**)

NPC: See "National People's Congress"

People's Procuratorates: State organs that exercise authority over specific cases involving national and public security. (CH: **人民检察院**)

People's Republic of China: Contemporary China (1949–present). (abb. PRC) (CH: 中华人民共和国, CH abb.: 中国)

PRC: See "People's Republic of China"

Republic of China: Established in 1912 and relocated to Taiwan in 1949, the Republic of China was Mainland China's government before Mainland China became the People's Republic of China under the Communist Party of China. (abb. ROC) (CH: 中华民国)

ROC: See "Republic of China"

SAIC: See "State Administration for Industry and Commerce"

SPC: See "Supreme People's Court"

SPC Interpretation: See "Supreme People's Court Interpretation/Opinion"

SPC Opinion: See "Supreme People's Court Interpretation/Opinion"

SPP: See "Supreme People's Procuratorate"

Standing Committee: NPC committee that exercises legislative powers accorded to it by the NPC. The Standing Committee meets between NPC meetings to discuss and create national laws. (CH: 全国人民代表大会常务委员会, CH abb.: 常委会)

State Administration for Industry and Commerce: Ministerial AIC directly under the State Council. The SAIC supervises and regulates the market and serves as the head department for all AICs. It also contains the CTMO. (abb. SAIC) (CH: 国家工商行政管理总局, CH abb.: 国家工商局)

State Council: Also referred to as the "Central People's Government," the State Council is composed of China's premier and the heads of its administrations and departments. The State Council supervises all of China's administrations and departments to assure that they act in accordance with the NPC's goals and legislation. (CH: 中华人民共和国国务院, CH abb.: 国务院)

Supreme People's Court: China's Supreme Court. (abb. SPC) (CH 最高人民法院, CH abb. 最高法院)

Supreme People's Court Interpretation/Opinion: China's judicial system permits the SPC to issue interpretations/opinions on laws, regulations, and judicial decisions that are not associated with specific cases.

Supreme People's Procuratorate: China's highest prosecutorial agency. The SPP investigates and prosecutes criminal behavior. (abb. SPP) (CH: **最高人民检察院**, CH abb.: **最高检察官**)

TRAB: See "Trademark Review and Adjudication Board"

Trademark Review and Adjudication Board: CTMO appellate board that issues determinations on trademark-related procedures including reviews of refusals, reviews of oppositions, invalidation requests, and reviews of cancellations. (abb. TRAB) (CH: **商标评审委员会**)

United Front: A representative group of organizations such as minority groups and trade unions. Since the United Front is select by the Communist Party of China, it primarily of unification. (CH: **民主党派**)

About the Author

Paul Kossof first came over to Mainland China to perform research related to Chinese non-performing loans for a U.S. asset management firm. He later worked in Mainland China at Chinese, American, and British law firms. Paul focuses on Chinese law, specifically intellectual property and business law, and anti-corruption compliance. He lives with his wife in Chicago where he is completing an International Business and Trade LLM degree at The John Marshall Law School. Paul expects to relocate to China in early 2015.

Index